# MOLECULAR CYTOLOGY

## Volume 1

## The Cell Cycle

# MOLECULAR CYTOLOGY

## Volume 1
## The Cell Cycle

JEAN BRACHET
*Laboratoire de Cytologie et Embryologie Moléculaires*
*Département de Biologie Moléculaire*
*Université Libre de Bruxelles*
*Brussels, Belgium*

1985

ACADEMIC PRESS, INC.

**Harcourt Brace Jovanovich, Publishers**

Orlando San Diego New York Austin
London Montreal Sydney Tokyo Toronto

ACADEMIC PRESS, INC.
Orlando, Florida 32887

*United Kingdom Edition published by*
ACADEMIC PRESS INC. (LONDON) LTD.
24–28 Oval Road, London NW1 7DX

**Library of Congress Cataloging in Publication Data**

Brachet, J. (Jean), Date
Molecular cytology.

Includes bibliographical references and indexes.
1. Cytology. 2. Molecular biology. I. Title.
QH581.2.B73 1985 574.87 84-24418
ISBN 0–12–123370–7 (v. 1. : alk. paper)
ISBN 0–12–123372–3 (v. 1. : paperback)

PRINTED IN THE UNITED STATES OF AMERICA

85 86 87 88 9 8 7 6 5 4 3 2 1

To my dear wife, Françoise,
the faithful companion of the
good and evil days.

*I build a castle of comfort and indulgence*
*for him, and stand sentinel always to*
*keep little vulgar cares out.*
G. B. Shaw, *Candida,* Act III

# CONTENTS

# PREFACE

When I was asked to write a second edition of "Biochemical Cytology" (published by Academic Press in 1957), I wondered where my own copy of this old book might be. When it became useless for teaching purposes, it had to leave the desk for some distant bookshelf. I finally found it and put it back on the desk to be subjected to autopsy. My verdict was that only two things retained some value: the Preface, because the author's general philosophy has remained almost unchanged, and the general backbone (the book's skeleton), because cells have remained cells. However, the picture of the cells has become increasingly complex; we know much more about them and understand them better than we did 25 years ago.

Today "Biochemical Cytology" is of little more than historical interest; our ignorance of major facts a quarter of a century ago makes the reading of this old book almost painful. Reading it now does illustrate fashions in science: many facts and hypotheses which were hotly disputed in "Biochemical Cytology" do not arouse the slightest interest today. Because cell biology has undergone a complete revolution, thanks to the explosive progress of knowledge in the field of molecular biology, the book evolved into an entirely new one—hence the title change to "Molecular Cytology." Nonetheless, the general philosophies of both the author and the book have remained unchanged; the Preface of "Biochemical Cytology" remains valid for "Molecular Cytology."

We have attempted in this book to present an integrated version of what is currently known about the morphology and the biochemistry of the cell. There are many excellent cytology and biochemistry textbooks. What remains to be done is the difficult and important task of linking these two sciences more closely together, now that they have so much in common. This is what we tried to do, with the hope that the book will prove useful to advanced students and research workers.

It has been assumed that the reader already knows the fundamentals of descriptive cytology, biochemistry, embryology, genetics, and molecular biology. Our goal will have been reached if the reader enjoys the attempt we have made to show that structure and metabolism are so closely linked together that they cannot be separated.

Special emphasis has been laid on the problems which are most familiar to the author. This will perhaps excuse the apparent imbalance of the book. If too much is said on embryos and too little on cancer cells, it is because the author has spent much of his life working with embryos and has so far not touched a cancer cell.

Emphasis has been given to the more dynamic aspects of cytology, not to detailed description. More is said about nucleocytoplasmic interactions in unicellular organisms and eggs than about the pure description of cytoplasmic and nuclear constituents.

Hypotheses and personal opinions have not been forgotten, for hypotheses, provided that they can be tested experimentally, may become more important than dry facts. Ideas are as vital for scientists as engines for cars or airplanes. Nowadays, some scientists forget that thinking may sometimes be more useful than performing an experiment.

The book has been written directly in English, and the author may not have expressed the ideas and facts as precisely as he would have wished. But what has been lost in subtlety has perhaps been gained in directness and clarity.

The need for a book dealing with biochemical or molecular cytology is obviously much less acute today than it was 25 years ago. At that time "Biochemical Cytology" was not accepted easily by either biochemists or cytologists. I remember vividly a very distinguished professor of anatomy and histology accusing me publicly of having produced a dreadful bastard; my answer was that hybridization can lead to improvement of crops. Today the battle is over, and there are several excellent books dealing with the molecular aspects of cell biology. Some of them, for instance those of Dyson (1978),* DeRobertis and DeRobertis (1981),† and Alberts *et al.* (1983),‡ are textbooks for students; these texts are remarkably complete, clearly written, and illustrated. I used them frequently for the preparation of these volumes, since a scientist remains a student all his life.

This book, like "Biochemical Cytology," is intended for advanced students and for research workers, but it is not an encyclopedia: to aim at completeness is an impossible task in view of the tremendous growth of the scientific literature during the past years. Already the monumental six-volume treatise "The Cell," which was edited 20 years ago by the late Alfred Mirsky and myself, is obsolete and incomplete. The voluminous literature and the specialization of prospective authors have so far precluded a second edition. A second edition would be a giant treatise and, thanks to computer-assisted literature searches, the reader might be crushed under the weight of documentation. If we believe the Preface of Anatole France's "L'Ile des Pingouins" ("Penguin Island"), such a situation has arisen: The author, who is writing a book on the history of the penguins, seeks information from the greatest art critic in the world. His office is filled with files from top to bottom. Finally, A. France, after climbing on the top of a scale, finds the file dealing with penguin art and lets it drop. All the files escape from their

*Dyson, R. D. (1978). "Cell Biology: A Molecular Approach." Allyn & Bacon, Newton, Massachusetts.

†DeRobertis, E., and DeRobertis, E. M., Jr. (1981). "Essentials of Cell and Molecular Biology." Holt, New York.

‡Alberts, B., *et al.* (1983). "Molecular Biology of the Cell." Garland, New York.

boxes, and the critic dies under their weight. His last words are "Que d'Art!" Since many more papers have been published on cell biology than on art in Penguin Island, the utilization of a complete documentation system would have been exceedingly dangerous for the author and boring for the readers.

No computer has been used for the preparation of the present book except an old rusty one, my brain, where lipofuscins and melanins should be expected to accumulate. My major task has been selection, a process which always brings out criticism because it is arbitrary. Most of the selected references are those of recent papers (where the previous literature is summarized) and of review articles published in easily available journals. Despite their number and quality, papers published in specialized symposia have been seldom quoted.

An author should take responsibility for the choice of the topics, the way he deals with them, and the correctness of the references. He should, as much as possible, avoid factual errors. However, the great English biochemist Frederick Gowland Hopkins told me, many years ago, that all textbooks, including his own, are full of errors. Reading through "Biochemical Cytology," which was once praised as "remarkably free of factual errors," I was ashamed to find the book full of errors. They are in general due to the progress of science, which will always move ahead. Errors will accumulate in the present book and, in agreement with Orgel's "error catastrophe theory" of aging (see Chapter 3, Volume 2), they will ultimately lead to its death. An author is also expected to avoid repetition. But there is repetition (perhaps too frequently) in this book, because I believe that important facts should be told more than once and presented again, in a slightly different way, when they are examined from another angle—at least this is a lesson I learned after more than 40 years of teaching and scientific direction of a laboratory.

In short, the aim of the present book is to present a critical and synthetic view, not an encyclopedic description, of living dynamic cells to young scientists. Whether or not this aim has been reached will be decided by the readers themselves.

It is a pleasure and a duty for me to thank all those who kindly agreed to read the manuscript, in particular Professor Werner W. Franke (Heidelberg), who carefully revised the first three chapters. Thanks go also to my colleagues from the University of Brussels, Professors A. Burny, H. Chantrenne, A. Ficq, and P. Van Gansen, who corrected many errors that had escaped my notice. The heavy burden of selecting and preparing the illustrations went to Professors P. Van Gansen and H. Alexandre and Mr. D. Franckx and that of checking the correctness and completeness of the references to Mrs. A. Pays. Last, but not least, my warmest thanks are due to Mrs. J. Baltus, who had the long and unpleasant job of typing the manuscript, and to the publishers for encouragement and patience.

JEAN BRACHET

# CONTENTS OF VOLUME 2

## Cell Interactions

## CHAPTER 1

# GENERAL BACKGROUND

### I. THE CELL YESTERDAY AND TODAY: A RECENT HISTORY OF CYTOLOGY AND ITS BIOCHEMICAL APPROACH

When the author was an undergraduate medical student at the University of Brussels in 1927 (more than one-half century ago), what was he told by his teachers about the structure and chemistry of the cell?

Cells were known since the days of T. Schwann (who became a professor at the University of Liège) and M. J. Schleiden. The ''resting'' cell and mitosis were described mainly by German biologists (Virchow, Flemming, Waldeyer, Boveri, Altmann, Benda, and O. Hertwig). E. Van Beneden, the spiritual grandfather of almost all the Belgian cytologists, embryologists, and zoologists of today, also made important contributions. But in Brussels it was fashionable to assume a very skeptical view of many of the cell constituents, and to consider them artifacts. In the nucleus, only nucleoli were considered real, but their role (a store of reserve materials?) and chemical composition remained completely unknown. The chromatin ''network'' was an artifact, and it was not understood how and from where chromosomes ''individualized'' during mitotic prophase. Morgan's ideas about the chromosomal localization of the ''mysterious genes'' were not accepted without some reservation, although the reality of meiotic crossing-over, first observed by Janssens in Louvain, was not disputed. The very existence of a nuclear membrane was also the object of some controversy: some cytologists compared the cell nucleus to a lipid droplet emulsified in water, and thus to a liposome. What was the structure of cytoplasm? Was it granular, fibrillar, vacuolar? Mitochondria were probably (almost certainly) true organelles that performed some hidden function, but the Golgi ''apparatus'' was generally rejected as an artifact. In the best and most recent treatise on the cell, the great American cytologist and embryologist E. B. Wilson (1925) depicted the cell as shown in Fig. 1. He shared, however, the skepticism of his Belgian colleagues—that the cytoplasm of the cell was virtually empty. Yet light microscopes were already nearly perfect and the eyes of well-trained observers were excellent. Since J. Loeb (1899) had shown that addition of histological fixatives to proteins in solution could give rise to granular or fibrillar networks, all cytologists lived with the fear that artifacts could provide erroneous data. Colloid chemistry not only retarded the discovery of macromolecules but also slowed the development of cytology.

Even the reality of a mitotic apparatus in an object as large and transparent as a

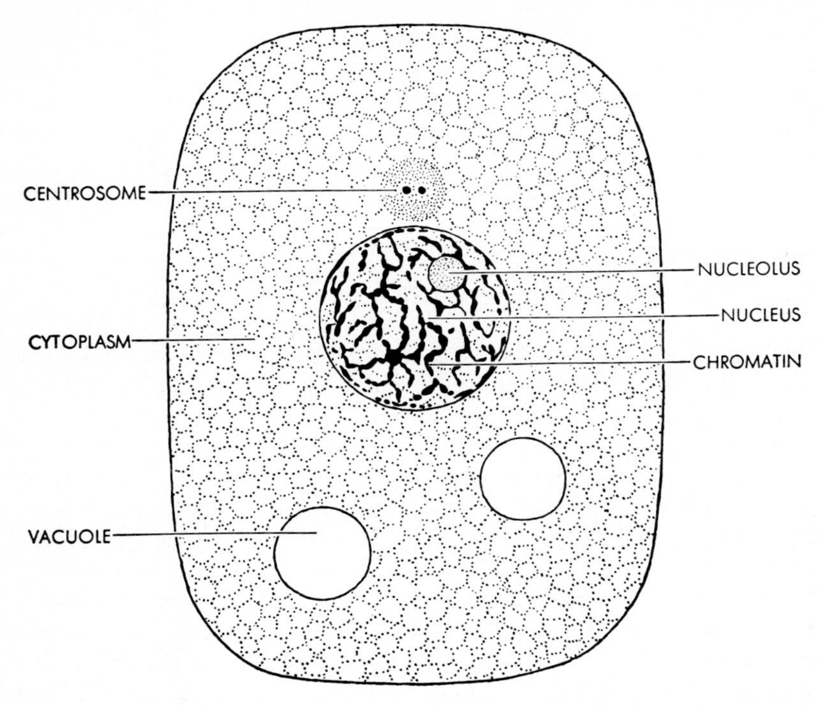


FIG. 1. Schematic representation of a typical cell according to E. B. Wilson (1925).

sea urchin egg seemed doubtful. In 1946, the numerous cytologists working at the Marine Biological Laboratory in Woods Hole, Massachusetts, were still divided into two opposite camps: "physical chemists," who did not believe in spindles and asters, and "cytologists," who took their existence for granted. I was forced to choose sides, joining those who believed in the existence of the mitotic apparatus. But one had to wait until the development of phase contrast microscopy and, more important, the isolation of the mitotic apparatus by Dan Mazia before the controversy was settled.

Modern cytology derives from the work done during World War II by Albert Claude (1943) and his co-workers (George Palade, Keith Porter, George Hogeboom, and Walter Schneider) at the Rockefeller Institute. The use of the electron microscope under ever-improving conditions of fixation, thin sectioning, staining, etc., allowed the demonstration of the morphological complexity of the cytoplasm [endoplasmic reticulum, ribosomes (Palade's granules), mitochondria, and Golgi bodies, etc.]. The same group developed another exceedingly important technique, namely, fractionation by differential centrifugation of homogenates from living tissues. Since the days of Warburg and Keilin, it had been known that broken cells released granules that might play a role in cell oxidation. The Rockefeller group combined morphological (at the ultrastructural level) and biochemical analyses to demonstrate, among other things, that the energy-producing organelles of the cell are the mitochondria and not, as had been believed since J. Loeb's days, the nucleus. Also, for the first time, Claude isolated the microsomes and showed that they are broken pieces of Porter's endoplasmic reticulum. Later, the use of similar, but improved, techniques led Christian de Duve to the discovery of lysosomes (de Duve *et al.*, 1955) and peroxisomes.

The exciting story of the discoveries that led to modern cell biology has been told in detail in a special issue of the *Journal of Cell Biology* (Gall *et al.*, 1981). The authors, which include the Nobel Prize winners C. de Duve and G. Palade, have given an excellent overview of the present status of cell biology in a historical perspective.

## II. NUCLEIC ACIDS YESTERDAY AND TODAY: THE CYTOCHEMICAL APPROACH

The work of A. Claude and his colleagues paved the way for biochemical cytology. Since the present book deals with *molecular* cytology, it is worth recalling here what we knew about nucleic acids 50 years ago and appreciating the impact of cytochemistry on molecular biology.

When the author joined the laboratory of Albert Dalcq, Professor of Anatomy and Embryology, in 1927, he was asked to choose between two research problems: a study of the production and role of mitogenetic rays in developing embryos, or an investigation of the localization of thymonucleic acid in growing oocytes, using the newly discovered Feulgen reaction. Fortunately, he preferred chemistry to physics and decided on the second problem. (Mitogenetic rays are now completely forgotten, and there would be no profit in reviving them.) At the time, it was strongly believed that there were two kinds of nucleic acids, animal and vegetal. Animal cells were thought to contain *thymonucleic acid* in their nuclei, which would be replaced in plant cells by a pentose nucleic acid. Instead of a pentose, thymonucleic acid had an unusual sugar in its molecule. It was identified, in 1929, by Levene and Mori as deoxyribose. The erroneous distinction between plant and animal nucleic acids, which was present in all textbooks, was due to hasty generalizations and to a discarding of facts that did not fit the theory. The only nucleic acids that had undergone serious investigation were thymonucleic acid (extracted from calf thymus) and zymonucleic acids (from yeast), which provided insufficient evidence from which to draw the conclusion that all animal cells contained only thymonucleic acid and all vegetal cells contained only pentose nucleic acid in their nuclei. Some observations did not fit within this scheme. R. Feulgen (Feulgen and Rossenbeck, 1924) correctly claimed that plant nuclei and even bacteria gave a positive cytochemical reaction for thymonucleic acid; for some mysterious reason, the pancreas was rich in a "plant" pentose nucleic acid. These two nucleic acids have had a very distinguished career: thymonucleic acid is just one of the deoxyribonucleic acids (DNAs) that constitute the genetic material of the cell, and zymonucleic acid was only one of the ribonucleic acids (RNAs) that play an essential role in the synthesis of specific proteins. The author's own work (1930–1945) on the localization of the two main types of nucleic acids in oocytes and eggs contributed to a better understanding of their role. His results were obtained using cytochemical methods. However, since biochemists did not believe in their validity, these methods had to be supplemented with biochemical analyses. A combination of cytochemical and biochemical (or biophysical) methods remains at the root of modern research in molecular cytology.

Brachet, in his early work (1929, 1933), showed that DNA is present in the extended lampbrush chromosomes of oocytes from many different animal species. It does not disappear, as many believed, at middle oogenesis. Furthermore, DNA is evenly distributed in the chromosomes of the daughter cells during the cleavage of fertilized eggs. These findings supported Morgan's theory of heredity and should have suggested that DNA was a good candidate for an important role in genetic material. But the main authority in the field, P. A. Levene, had come to the conclusion that DNA was a tetranucleotide ($M_r$ between 1200 and 1300 daltons); such a small molecule could obviously play no role in heredity. Therefore, Levene (1931) suggested that the role of nucleic acids might be that of a buffer that would keep the intranuclear pH at a constant value. During the post–World War II years, the molecular weight of DNA increased steadily as methods for its isolation improved. It is now known that the molecular weight of DNA is not a physical constant, since it depends on the size of the chromosomes. When the famous experiments of Avery *et al.* (1944) on bacteria and of Hershey and Chase (1952) on phages became known, widespread skepticism (which was not shared by the author) still prevailed about the possibility of a genetic role for DNA.

The story should be traced to the 1930s. During cleavage of fertilized eggs, the number of cells, and thus of nuclei, increases tremendously. The cleaving egg should thus be the site of extensive DNA synthesis. However, two opposing theoretical questions existed: is there a net DNA synthesis, as proposed by J. Loeb (1899), or a migration of cytoplasmic DNA into the nuclei, as suggested by E. Godlewski (1908)? Biochemical estimations of the deoxyribose content of cleaving sea urchin eggs, combined with the use of the Feulgen cytochemical reaction, clearly demonstrated the existence of a net DNA synthesis during early development. However, paradoxical results were obtained when the total purine or nucleic acid phosphorus was compared in unfertilized sea urchin eggs and swimming embryos (plutei): very little change took place during development. The only possible explanation was that unfertilized sea urchin eggs contained little DNA and large amounts of another nucleic acid. Estimations of the pentose content of the eggs showed that this nucleic acid is a pentose nucleic acid—thus a "plant" nucleic acid (what we know as RNA today) (Brachet, 1933).

Is RNA present in all cells, or are egg and pancreas cells exceptions? The answer to this question was again provided by work in cytochemistry. In 1941, T. Caspersson and Brachet (1941a,b), working independently and using two different methods for RNA detection [namely, ultraviolet (uv) microphotometry and staining with Methyl Green-pyronine before and after ribonuclease digestion, respectively] came to the same conclusion: all cells (animal or vegetal) contain RNA in their nucleoli and cytoplasm. Furthermore, there was a constant correlation between the RNA content of a cell and its ability to synthesize proteins. The conclusion that RNA, in some unknown and mysterious way, is involved in protein synthesis met with skepticism from the majority of biochemists, who then favored the idea that protein synthesis results from the "reverse" activity of proteolytic enzymes. In addition, work done by T. Caspersson in collaboration with the geneticist Jack Schultz on *Drosophila* provided evidence for the view that RNA (at least nucleolar RNA) is synthesized under the influence of the neighboring heterochromatin (in fact, the nucleolar organizer) (Caspersson and Schultz, 1939).

Cytochemical work done around 1940 had thus already provided all of the elements of Crick's (1963) "fundamental dogma" of molecular biology as follows:

$$\text{DNA} \xleftarrow[\text{Replication}]{} \text{DNA} \xrightarrow[\text{Transcription}]{} \text{RNA} \xrightarrow[\text{Translation}]{} \text{Protein}$$

However, the cytochemical data led only to hypotheses and provided no explanations. The use of simpler models (bacteria, viruses) and the combination of genetics, biochemistry, and x-ray crystallography has led to the astounding development of molecular biology that we have witnessed during the past 30 years. Due to the work of researchers such as Watson, Crick, Wilkins, Kornberg, Ochoa, Nirenberg, Lwoff, Monod, Jacob, Brenner, Benzer, and many others, the fundamental dogma is now fact. But the amazing story of molecular biology is not the subject of this volume; it has been told by Delbrück (1966) and Stent (1968).

In the 1940s, one of the most exciting and mysterious problems was what we now call translation, i.e., the intervention of RNA in protein synthesis. What we were observing at that time with cytochemical methods were, in fact, the ribosomal RNAs (rRNAs) and their nucleolar precursors. If, as we believed, these RNAs are involved in protein synthesis, the main site of this synthesis should be the microsomes, which had been isolated by Claude (1940) from chick embryos and found to contain RNA. During World War II, Brachet and R. Jeener (1944) showed that microsomes can be isolated from all tissues, animal or vegetal. They are thus universal components of the cell. Their abundance is correlated with the intensity of protein synthesis in the respective tissues—they always contain RNA, proteins, and lipids, as well as small amounts of the specific proteins synthesized in any given tissue. Unfortunately, we were unable to prove our hypothesis that the microsomes are the main site of cellular protein synthesis because radioactive amino acids were not available in Belgium at that time. One had to await the work of Borsook *et al.* (1950) and Hultin (1950a,b) with labeled amino acids before it was proved that microsomes are the main agents of cellular protein synthesis. It is now appropriate to look at the cell of today.

## III. UNIFORMITY AND DIVERSITY IN CELLS

Is there a "typical" cell? The answer is no. This is much clearer now than it was when the author wrote "Biochemical Cytology." At that time, very little work had been done on cell differentiation, a problem that was of interest mainly to embryologists. Although an entire chapter of this book was devoted to the role of the nucleus and the cytoplasm in embryonic differentiation, nothing was said about cell differentiation, a subject that now inspires so many cell biologists, as evidenced by the International Society of Differentiation, congresses organized by this society, and the journal *Differentiation.* At every cell biology congress, the same questions arise: is there such a thing as a completely undifferentiated cell? Can a differentiated cell dedifferentiate? Of course, there is no simple answer to these important questions.

When we observe cells *in situ,* we are struck by the diversity of their morphology and biochemical composition (due to the presence of a specific marker, such as hemoglobin or melanin). H. Holtzer *et al.* (1972), who have devoted so much time and thought to the study of the cell differentiation, call these proteins "luxury proteins." Much of the work done today in this field is on cell cultures: "undifferentiated" (?) embryonic cells can differentiate in culture, under proper conditions, into muscle, cartilage, blood, and pigment, and other cells. A cell culture is certainly more amenable to experimentation than a whole embryo or even a piece of tissue dissected out of an embryo. There are now "banks" of frozen cell cultures that can be shipped throughout the world as easily as frozen meat or fish; thousands of cell lines, with their pedigrees, are now available. But there are also some problems: (1) cells are cultured in an artificial medium, which soon alters their physiological properties and the expression of their genes (Bissell, 1981); (2) cell lines undergo mutations; and (3) the presence of variants, mutants, and revertants in the cell population must be taken into account. It is known that one of the cell lines most frequently used for research, the HeLa line, which originated from a human cancer, has diverged so much under culture conditions that the HeLa cells used in different laboratories are no longer directly comparable. B. Ephrussi, who spent the last part of his outstanding scientific life working on differentiation of cultured cells (although his career began with sea urchin eggs), noted that differentiation will ultimately have to be studied where it normally takes place—in embryos. Obviously, the two approaches have their advantages and disadvantages, and complement each other. In this work, embryonic differentiation and cell differentiation in culture will be discussed in the same chapter (Volume 2, Chapter 3).

It was easy to describe and define a cell in the days of E. B. Wilson (see Fig. 1): it is surrounded by a cell membrane and at the center is the nucleus, which contains chromatin and a nucleolus (or several nucleoli). For teaching purposes, one usually begins to describe a cell by drawing three concentric circles; this first sketch looks like a medieval castle, with its outer and inner walls surrounding the dungeon. Of course, no cell conforms to such an elementary scheme, and some cells have no resemblance at all to this simplified representation. For instance, the red blood cells of mammals are called "cells" even though they have no nucleus; striated muscles are syncytia, with many nuclei in a common cytoplasm. What should one call the unicellular alga *Acetabularia?* During the greater part of its life, it has a single nucleus and might be called a cell despite its most unusual shape (see Volume 2, Chapter 2 for details). When this primary nucleus breaks down and gives rise to daughter nuclei without cell division, the alga becomes a syncytium; when these daughter nuclei differentiate into gametes, which copulate and form new algae, the alga behaves like an organism. The cell nucleus is seldom spherical, and in certain cells (those of the silk glands of silkworms, for instance) its surface is enormously increased by fingerlike processes.

The cell that perhaps conforms best to the classical, simple scheme is the oocyte. Nevertheless, it would be a mistake to use oocytes as examples of undifferentiated, typical cells. Their large size distinguishes oocytes from the surrounding ovarian cells. More importantly, the oocyte's function is to become an egg, the only cell that, in animals, is capable of producing a complete organism of the same species. This totipotency differentiates eggs from all somatic cells, at least in animals.[1]

Since cells display such great morphological diversity, what do they have in common? The answer is that they are alive and are therefore subject to the laws of molecular biology. All living organisms must obey these laws (and Crick's central dogma). When Jacques Monod said that what is true for *Escherichia coli* must be true for the elephant, he was fundamentally right. There are many important differences between prokaryotes and eukaryotes (between bacteria and elephants). Prokaryotes have a circular DNA molecule instead of a true nucleus and lack nucleoli, a nuclear membrane, and specialized mitochondria for ATP production. Their ribosomes are smaller than those of the eukaryotes, and the

[1]In some plants—carrots or tobacco, for instance—almost any cell isolated from the adult can produce a complete organism if it is adequately cultivated. These plant cells are thus totipotent.

half-life of their messenger RNAs (mRNAs) is exceedingly short (a few minutes instead of hours or even days in *Acetabularia*). Bacteria are very small compared to cells. Yet, the two obey the same laws: the flow of genetic information in all living things runs from DNA to RNA and from RNA to protein; the source of energy is ATP. The differences between an elephant and an *E. coli,* in summary, are quantitative rather than qualitative.

## IV. A MAP OF THE CELL AND AN OUTLINE OF THESE VOLUMES

When a tourist plans to visit a foreign country or town, he should look at a map before beginning his trip. Such maps have been simplified in order to make them clear and understandable. Figure 2 is a map or diagram depicting, in a grossly oversimplified way, the labyrinth in which we shall move in a theoretical, idealized cell. A comparison of the schematic descriptions of the cell shown in Figs. 1 and 2 indicates the tremendous progress made by cell biology in a little more than 50 years; the mainspring of this progress is electron microscopy.

We shall begin (after a few technical explanations) with the *cell surface*. The plasma membrane, which plays such an important role in the interactions between the cell and the outer world, as well as in intercellular communication, can no longer be separated from the extracellular matrix that surrounds it and the underlying cytoskeleton. The latter extends inside the entire cytoplasm, and it is somewhat arbitrary to distinguish between a membrane-associated cytoskeleton and the rest of it. In both cases, the cytoskeleton is made up of microtubules (with tubulin as the main constituent); microfilaments, which owe their contractility to actin and a number of associated proteins (including myosin) and intermediate-sized filaments whose chemical nature is much more variable. For convenience, we shall consider the cell surface as composed of the plasma membrane, the extracellular matrix, and the membrane-associated cytoskeleton. Deeper in the cell, the cytoplasmic cytoskeleton will be examined. Inside the network that it forms, many organelles can be found. Some of them are bound by a single membrane and are clearly interrelated: the endoplasmic reticulum, the Golgi bodies, the lysosomes and related vesicles, the peroxisomes, and the glyoxysomes. Others display some degree of independence toward the other cell constituents and should thus be considered semiautonomous: the mitochondria, the chloroplasts of green plant cells, and the centrioles and basal bodies, which play an essential part in cell division and cell motility, respectively. Finally, the agents of this motility, cilia and flagella, will be examined.

In terms of the cell nucleus, we shall consider successively the complex structure and functions of the nuclear membrane, the properties and chemical composition of the chromatin, and the nucleoli.

Mitotic cell division is an obvious prerequisite for cell proliferation; mitosis, however, is only the final result of the complex events that constitute the cell cycle. Particularly important is the so-called S phase of the cell cycle, in which

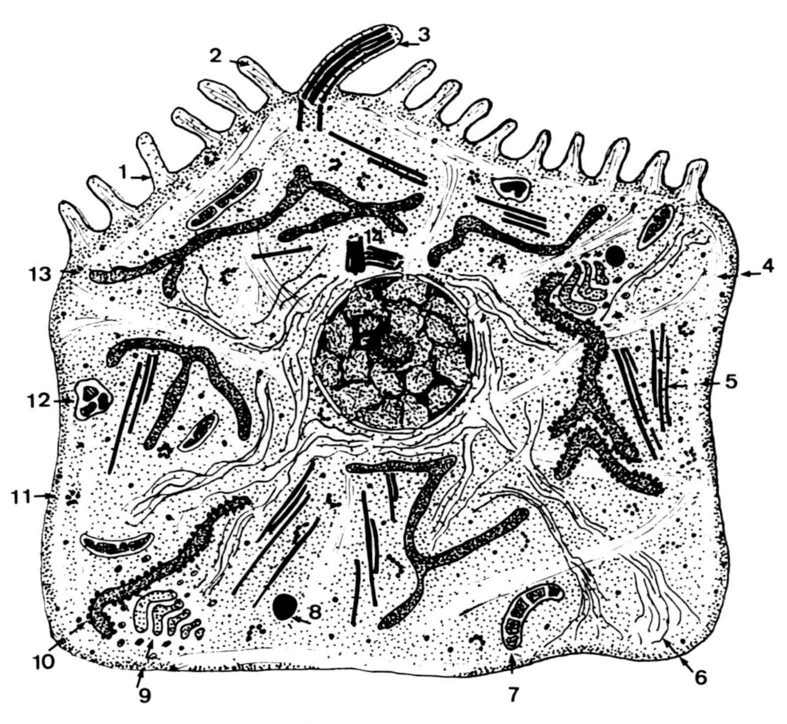


FIG. 2. Schematic representation of a typical cell as we currently know it. The complexity of the cytoplasmic organization, compared to that of Fig. 1, results mainly from advances derived from electron microscopy. (1) Plasmic membrane; (2) microvillus; (3) cilium; (4) microfilaments; (5) microtubules; (6) intermediate filaments; (7) mitochondrion; (8) primary lysosome; (9) dictyosome; (10) rough reticulum; (11) polyribosome; (12) secondary lysosome; (13) smooth reticulum; (14) centrioles. [Original drawing by P. Van Gansen.]

nuclear DNA undergoes exact replication. During mitosis itself, the structure of the chromosomes and the mitotic apparatus, as well as the mechanisms of the furrowing that separates the two daughter cells, will be examined. Important also is the regulation of cell division by inhibitory or stimulatory factors of endogenous origin and the action of specific mitotic poisons.

The description of the cell will be simplified here, as will the vocabulary. For instance, is it really necessary to speak of a zonula occludens when the expression "tight junction" is equally descriptive, namely, a junction that holds two cells together firmly? Is it necessary to use the term "nexus" for the loose gap junctions?

The last three chapters in Volume 2 are more expanded versions of the ones in "Biochemical Cytology." The chapter on nucleocytoplasmic interactions in somatic cells has been greatly expanded over its predecessor, because several new approaches (isolation of karyoplasts and cytoplasts, somatic cell hybridization and gene transfer, a field whose potentialities cannot yet be predicted) have been developed. An entire chapter is devoted to nucleocytoplasmic interactions in oocytes and eggs for two reasons: one is that the author has been working so long in this field (and likes it particularly); a better reason is that a good deal of our recent knowledge in molecular cytology has come from work done on oocytes and eggs. Finally, cell differentiation, malignant cell transformation, and cell aging have been handled in a single chapter because we believe that these three events have much in common.

## REFERENCES

Avery, O. T., MacLeod, C. M., and McCarthy, M. (1944). *J. Exp. Med.* **79,** 137.
Bissell, M. J. (1981). *Int. Rev. Cytol.* **70,** 27.
Borsook, M., Deasy, C. L., Haagen-Smith, A. J., Keighley, G., and Lowy, P. H. (1950). *J. Biol. Chem.* **184,** 529.
Brachet, J. (1929). *Arch. Biol.* **39,** 677.
Brachet, J. (1933). *Arch. Biol.* **44,** 519.
Brachet, J. (1941a). *Enzymologia* **10,** 87.
Brachet, J. (1941b). *Arch. Biol.* **53,** 207.
Brachet, J., and Jeener, R. (1944). *Enzymologia* **11,** 196.
Caspersson, T. (1941). *Naturwissenschaften* **29,** 33.
Caspersson, T., and Schultz, J. (1939). *Nature (London)* **143,** 602.
Claude, A. (1940). *Science* **91,** 77.
Claude, A. (1943). *Biol. Symp.* **10,** 111.
Crick, F. H. C. (1963). *Prog. Nucleic Acids Res.* **1,** 163 (1963).
de Duve, C., Pressman, B. C., Gianetto, R. J., Wattiaux, R., and Appelmans, F. (1955). *Biochem. J.* **60,** 604.
Delbrück, M. (1966). *In* "Phage and the Origins of Molecular Biology" (J. Cairns, G. Stent, and J. Watson, eds.). Cold Spring Harbor Lab., Cold Spring Harbor, New York.
Ephrussi, B. (1953). "Nucleo-cytoplasmic Relations in Micro-organisms." Oxford University Press, London and New York.
Feulgen, R., and Rossenbeck, H. (1924). *Hoppe-Leyler's Z. Physiol. Chem.* **135,** 203.
Gall, J. G., Porter, K. R., and Siekevitz, P. (1981). *J. Cell Biol.* **91** (3).
Godlewski, E. (1908). *Arch. Entwcklungsmech. Org.* **26,** 278.
Hershey, A. D., and Chase, M. (1952). *J. Gen. Physiol.* **36,** 39.
Holtzer, H., Mayne, R., and Mochan, B. (1972). *Curr. Top. Dev. Biol.* **7,** 229.
Hultin, T. (1950a). *Exp. Cell Res.* **1,** 376.
Hultin, T. (1950b). *Exp. Cell Res.* **1,** 599.
Levene, P. A. (1931). "Nucleic Acids." Chem. Catalog Co., New York.
Levene, P. A., and Mori, T. (1929). *J. Biol. Chem.* **83,** 793.
Loeb, J. (1899). *Arch. Entwicklungsmech. Org.* **8,** 689.
Stent, G. S. (1968). *Science* **160,** 390.
Wilson, E. B. (1925). "The Cell in Development and Heredity." Macmillan, New York.

# CHAPTER 2

# TECHNIQUES USED IN CELL BIOLOGY: A BRIEF SUMMARY

The study of cell physiology calls for a combination of many different methods. It would be fruitless to present them here in detail since new techniques are being continuously introduced, while classical methods are still undergoing improvements and modifications. All possible aspects of the methodology used in cell biology can be found in the publication, "Methods in Cell Biology." Besides the description of techniques for cell culture, cell separation, cell fusion, nuclear transplantation, dissection of cells in nucleate and anucleate halves, and others, this series contains much useful information on older methods that remain of fundamental importance: optical methods, electron microscopy, cytochemical techniques, and the isolation of cell constituents by differential centrifugation of homogenates. Their merits will be discussed only briefly here.

## I. OPTICAL METHODS

The classical technique of light microscopy remains the basis of any cell study, whether the observations are made on living cells (sometimes vitally stained with appropriate dyes) or on sections (or smears) of fixed and stained tissues. Video microscopy greatly enhances the definition of the structures seen under the light microscope.

There are several useful complements to traditional light microscopy (see Barer, 1966, for a review). Phase contrast microscopy provides improved contrast for the study of living, transparent cells; it is particularly useful when beating of cilia or flagella, cell movement, or cell division should be followed by microcinematography. Interference microscopy also shows details (especially with Nomarski's interference-contrast optics that cannot be seen under a light microscope and has the additional advantage of enabling one to measure the dry mass of the cells (Davies *et al.,* 1953). Polarization microscopy is useful for the study of birefringent objects such as mitotic or meiotic spindles. Caspersson's uv microscope allows the measurement of the complete uv spectrum of a very small part of the cell. This instrument has played an important role (see Chapter 1) in the discovery of the localization and the function of RNA in protein synthesis. Unfortunately, the high cost of commercially available uv microscopes has precluded their general use. Much more popular today are the fluorescence microscopes, which allow very sensitive cytochemical analyses of nucleic acids, specific proteins, and carbohydrate-containing components of the cell surface. For instance, acridine orange gives a green fluorescence when it binds to DNA; a red fluorescence is produced when it binds to RNA. The respective intensities of the two fluorescences can be measured by fluorocytophotometry. Some ''vital'' dyes [those that penetrate living cells and display little toxicity, for example, ethidium bromide, DAPI (4′,6-diamino-2-phenylindole), and several dyes produced by Hoechst] become highly fluorescent after binding selectively to DNA and can thus be used for its cytochemical detection. On the other hand, fluorescein derivatives can easily be linked to proteins and are now widely used, as we shall see, for the detection of the cell surface glycoproteins and of the various proteins that form the ''cytoskeleton.''

Electron microscopy has now attained a very high degree of perfection, as have the techniques, first worked out by Claude, Porter, and Palade, for the preparation of the samples. Fixation with glutaraldehyde and osmium tetroxide is a marked improvement over the previously popular fixation with osmium tetroxide alone. Embedding in appropriate resins, cutting thin sections, and examining the latter under high resolution are now common practice in biological laboratories. There are various staining techniques for increasing the contrast of the structures seen under the use of electron microscope and, as we shall see, cytochemistry is also possible at the EM level. The use of heavy metal shadowing, in which the object is covered with metals (for instance, platinum–palladium alloys), allows dimensional visualization of an object (a spread DNA molecule, a virus, etc.) [see Hayat (1974–1976) for a full treatise on electron microscopy and Plattner and Zingsheim (1983) for an exhaustive review].

In the last decade, there have been two important developments in electron microscopy: freeze-fracture and scanning electron microscopy. In freeze-fracture, the frozen sample is sectioned in a vacuum and then fractured at −100°C. This procedure allows the separation of the two leaflets that form the plasma membrane and the direct observation of the particles that lie in the space between the leaflets. It is now possible to analyze quantitatively the distribution of these ''intramembrane particles'' (de Laat *et al.*, 1981). A modification of freeze-fracture called ''dry-cleaving'' has proved useful for the study of the membrane-associated cytoskeleton (Masland *et al.*, 1981). Scanning electron microscopy allows the observation of the cell surface and thereby provides a much better resolution than the light microscope, showing cilia and irregularities of the cell surface, microvilli (see Chapter 1, Fig. 2), in particular.

Finally, high-voltage electron microscopes (1000 kV or more instead of 100 to 120 kV), in contrast to the ordinary electron microscope, can be used for the study of thick objects such as intact cells or cells extracted with detergents. This technique, in the hands of K. Porter and his colleagues, has yielded interesting results that will be discussed later, particularly the organization of the cytoskeleton.

## II. CYTOCHEMICAL METHODS

The essential principle of all cytochemical methods is to apply a chemical test to histological sections that is specific for the detection of a given chemical substance. The localization of the reaction product is then observed under the microscope. As pointed out by Lison as early as 1936, there are two absolute prerequisites if any cytochemical reaction is to be of value: (1) the reaction must be highly specific and (2) the localization of the reaction product (which is usually colored or fluorescent) must remain unchanged during all manipulations. Nonspecificity and diffusion of substrate or reaction products remain the main pitfalls of cytochemistry.

As far as we are aware, there are no very recent textbooks on cytochemistry. In Pearse's "Histochemistry" (1968), detailed "recipes" for the detection of many substances of biochemical interest were collected and their value critically evaluated. The last edition of Lison's book (1960) should also be consulted.

The problem of adequate fixation remains important. A good deal of the work now being done is on cultured cells that are well spread over their substratum and that can be fixed with simple fixation techniques (air-drying, ethanol-acetic acid, formaldehyde, glutaraldehyde, etc.). Fixation of large objects (eggs, pieces of tissues), which must be cut in sections with a microtome, is more difficult. In such cases, freeze-drying[1] or freeze-substitution (fixation at very low temperatures followed by treatment with cold methanol in order to dissolve the ice crystals) is obviously advantageous (diffusion of substances such as glycogen is avoided during fixation and hydrosoluble substances are not extracted). For a review of "ergo-fixation," see Plattner and Bachmann (1982).

The major advances in cytochemistry over the past years are the possibility of making quantitative measurements by cytophotometry of colored or fluorescent reaction products, identifying specific proteins or nucleic acid sequences, and adapting existing cytochemical methods to the EM level (ultrastructural experiments).

In principle, the products of all cytochemical methods can be analyzed cytophotometrically, following the method used by Caspersson in 1936. The cytophotometer is receiving more widespread use in the analysis of the DNA content of cells stained with the Feulgen reaction (or, as pointed out before, with fluorescent dyes that bind to DNA). Acridine orange has been widely used, especially by Ringertz and Savage (1976) and Darzynkiewicz (1973), to ascertain the RNA as well as the DNA content of individual cells. An extremely important development in this field has been the construction of flow cytophoto-

[1]The principle of freeze-drying was perhaps imagined by the French writer Edmond About (1828–1885) in his famous novel "L'homme à l'oreille cassée." A colonel of the Napoleonic Great Army was frozen in Russia and dehydrated *in vacuo* in Germany; he eventually revived after being rehydrated.

meters (cell sorters). These costly but very powerful instruments enable the sorting out of cells that differ in their ability to bind to any fluorescent probe (see Jovin and Arndt-Jovin, 1980; Lydon *et al.*, 1980; Kruth, 1982). For instance, cells stained *in vivo* with DAPI or Hoechst 33312 can be separated according to their DNA content; cells with different binding affinities for cell surface fluorescent ligands can also be separated from each other.

Detection of specific proteins by immunocytochemistry dates back to A. Coons (1956). Antibodies are raised in the rabbit against a specific antigen. These antibodies are then labeled by fluorescein isothiocyanate. Cells are stained, after proper fixation, with the fluorescent antibody and observed under a fluorescent microscope. Cytochemists usually prefer the indirect method to Coons' direct method. In the indirect method, unlabeled rabbit antiserum reacts with the cell. In a second step, labeled goat antibodies to rabbit $\gamma$-globulin are added and the fluorescence is observed. Immunocytochemistry can be extended to the ultrastructural level if the goat antibody is labeled with ferritin (an iron-containing, electron-dense protein) or peroxidase. Peroxidase, in the presence of diaminobenzidine and $H_2O_2$, gives the classical peroxidase reaction: a brown color under the light microscope and an electron-dense precipitate under the electron microscope. Labeling of the antisera with colloidal gold has been advanced (De Mey *et al.*, 1981) An example of the utilization of these techniques for the detection of an important protein, tubulin, can be found in papers by Willingham *et al.* (1981), De Mey *et al.* (1976), and De Brabander *et al.* (1977). Another example is the detection of another important protein, actin. Lazarides (1975), using immunofluorescence, has demonstrated the presence of actin in microfilaments at the light microscopic level, while Willingham *et al.* (1980) have shown this at the ultrastructural level. Since immunocytochemistry can be applied to any antigen, including small molecules such as nucleotides bound to a protein, it is obvious that this technique has immense possibilities; it is, indeed, in very wide use today. We have chosen only two examples (tubulin and actin) of its application, but hundreds can be found in the literature. The use of monoclonal antibodies (Milstein, 1981; Milstein and Cuello, 1983), which are well-defined entities, increases the potentialities and the specificity of immunocytochemistry.

The development of cytophotometrical techniques allows quantitative measurements on cells. Thus, it is now easy to measure, in adequately treated cells, the intensity of the classical reactions characteristic of protein-bound amino acids (arginine, tyrosine, cysteine, etc.); this can also be done on cells in which basic proteins have been stained with fast green according to Alfert and Geschwin (1953; see also Das and Alfert, 1977). As already mentioned, similar measurements are routinely done on cells stained with Feulgen or fluorescent dyes in order to ascertain their nucleic acid content.

The powerful technique of *in situ* hybridization is the counterpart, for nucleic acids, of what immunofluorescence is for proteins. It allows the detection of specific DNA or RNA sequences in chromosomes or interphase nuclei. The principle is to hybridize, in adequately fixed cells, DNA (which should first be denatured) or RNA with complementary specific radioactive probes (cRNA for hybridization with DNA or cDNA for hybridization with RNA). The preparation of such specific radioactive probes (tritiated or radioiodinated) is done with classic methods (synthesis of a cDNA by reverse transcription, nick translation, etc.); their description lies outside the scope of this volume. The radioactive hybrid that has formed *in situ* inside the cell can be visualized under the light or electron microscope by autoradiography (Fig. 1) or with a biotin-labeled probe (Hutchison *et al.,* 1982; Manuelidis *et al.,* 1982). Genes that exist in multiple copies and are associated in clusters (ribosomal genes, histone genes), as well as highly repetitive DNA sequences (satellite DNAs), are the easiest to detect by *in situ* hybridization. The first successful attempts to detect the localization of the ribosomal genes at the EM level were made by Jacob *et al.* (1971), while Gall and Pardue had already shown their presence in nucleoli in 1969. Gerhard *et al.* (1981) have claimed that it is possible to detect, by *in situ* hybridization, any gene that exists in a single copy in the cell, provided that this radioactive gene is available in purified form and in sufficient amounts. This has been done by molecular cloning. These authors have applied this technique with success to the detection of globin genes in metaphase chromosomes. Success in the detection of the human insulin gene on chromosome 11 by the same technique, which has been described in detail by Harper and Saunders (1981), has also been reported by Harper *et al.* (1981). The localization of many genes on metaphase chromosomes has now been ascertained by *in situ* hybridization.

*In situ* hybridization with tritiated or iodinated cDNA (review by Henderson, 1982) has allowed the easy detection of RNAs (Fig. 3). Hybridization with radioactive polyuridylic acid (poly U) has shown the localization in eggs of the polyadenylic (poly A) segments present in the majority of the mRNAs (Capco and Jeffery, 1978; Angerer and Angerer, 1981). The potential of *in situ* hybridization, as one can see, is enormous. However, it should be pointed out that its efficiency, at the present time, remains low (10–15% for rRNA, according to Coté *et al.,* 1980). Like immunocytochemistry, *in situ* hybridization of nucleic acids today remains a qualitative rather than a quantitative technique.

So many techniques are currently available for the identification and localization of many chemical cell constituents that it is impossible to mention them all. Only a few examples will be given here. For instance, it is possible to detect in the cell surface carbohydrate sites that bind, in a specific way, proteins of vegetal origin called "lectins." Such lectin binding can even be observed in intermembrane particles after freeze-fracture (Pinto da Silva *et al.,* 1981). It is possible to use fluorescent lectins (the most popular is concanavalin A, or ConA) or, at the EM level, lectins coupled to ferritin (Nicolson and Singer, 1971) or

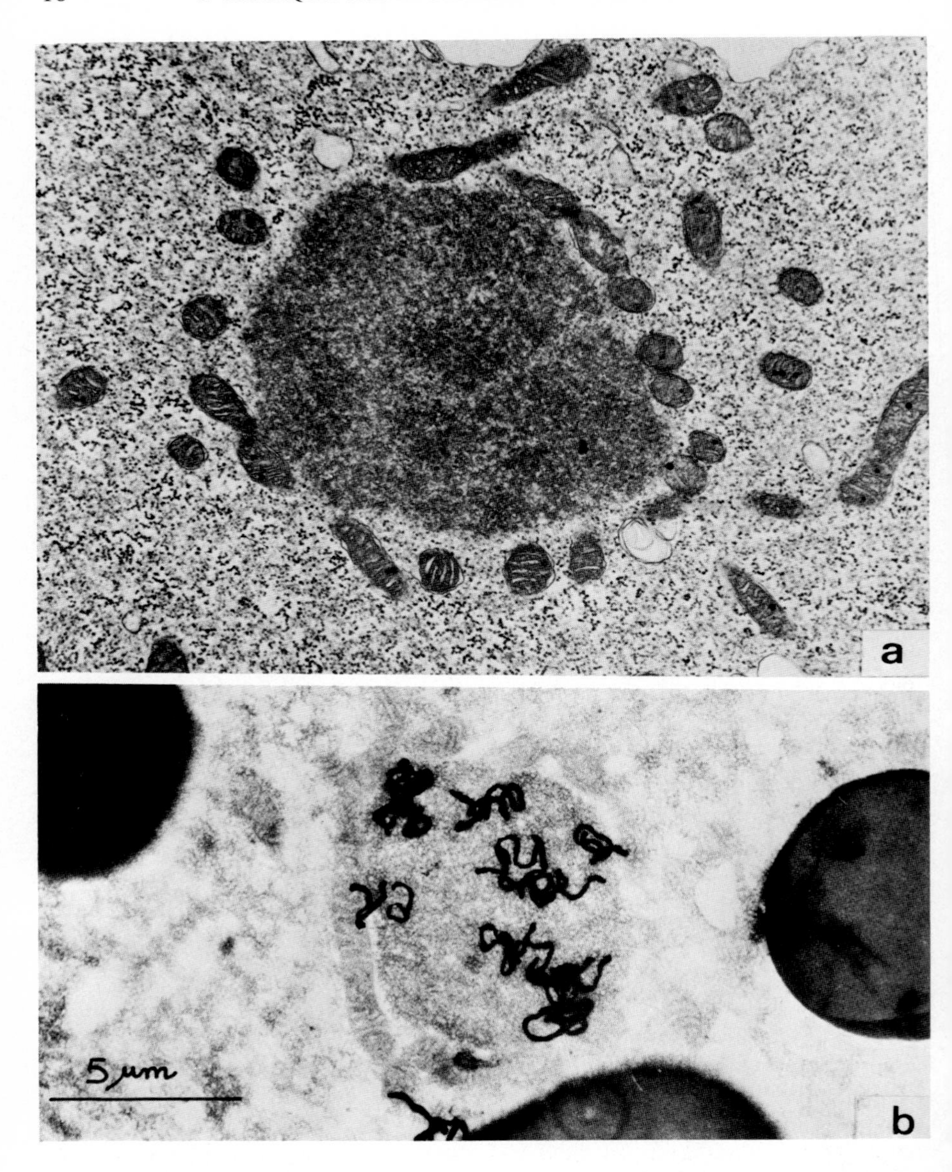


FIG. 1. *In situ* hybridization. (a) In the cortex of a progesterone-treated *Xenopus* oocyte, Feulgen-positive bodies surrounded by mitochondria can be seen under the electron microscope; (b) these bodies hybridize specifically with [$^3$H]rRNA, as shown by autoradiography at the ultrastructural level. This demonstrates that they contain ribosomal genes (rDNA). [Electron micrographs by G. Steinert.]

tritiated lectins. Anionic sites on the cell surface can be detected under the electron microscope with cationized ferritin (King and Preston, 1977). Immunofluorescence has been used by Yamada *et al.*, (1980) for the cytochemical detection of an important constituent of the surface coat of fibroblasts, fibronectin. Very detailed reviews of the methods available for the study of the asymmetric distribution of lipids and proteins in the plasma membrane have been published by Etemadi (1980a,b).

Immunofluorescence has played an important role in the analysis of the intracellular distribution not only of tubulin and actin and of the proteins that are associated with them, but also of the many proteins that constitute the intermediate filaments (see Chapter 3, Section II). For actin, techniques other than immunofluorescence can be used to detect its presence at the ultrastructural level. These methods are based on the fact that actin filaments are "decorated" (i.e., display lateral appendages) when they bind to substances such as meromyosin (Goldman, 1975), pancreatic deoxyribonuclease (DNase I) (Wang and Goldberg, 1978), or the mushroom poison phallotoxin (Wulf *et al.*, 1979). The most popular of these methods is meromyosin binding.

*In situ* hybridization is not the only technique that has contributed to our understanding of the architecture and chemical composition of the nucleus. Of particular interest is the remarkable technique of chromosome spreading, devised by Oscar Miller, which allows the visualization of gene transcription under the electron microscope (Miller and Beatty, 1969). Active genes display a "Christmas tree" configuration, whereby nascent and growing RNA chains are seen as the branches of the tree (Fig. 2). It should be pointed out that what we observe under the electron microscope are the proteins associated with nascent RNA, and not RNA itself. Similar methods have led Olins and Olins (1974) to the very important discovery that chromatin has a beaded structure that results from interactions between DNA and histones. These $\nu$ bodies or nucleosomes, as well as some of the results obtained with Miller's spreading technique, will be discussed in Chapter 4.

The technique of chromosome banding plays a very important role in cytogenetics (Fig. 3). Its development followed the discovery by Caspersson *et al.* (1968) that quinacrine mustard, an alkylating agent that binds to DNA, did not distribute evenly along the chromosomes. Thus, strongly fluorescent bands alternated with clear bands according to a pattern specific for each chromosome. There are currently many banding techniques based on the differential binding of quinacrine (Q binding) or dyes such as Giemsa (G binding). The molecular bases of the banding patterns have been widely discussed, but the reasons for these patterns remains unclear. It has been suggested that strongly and poorly stained bands result from a different base composition of DNA in various segments of the chromosomes. Others believe that the banding pattern results from the distribution of chromosomal proteins, probably nonhistone chromosomal proteins.

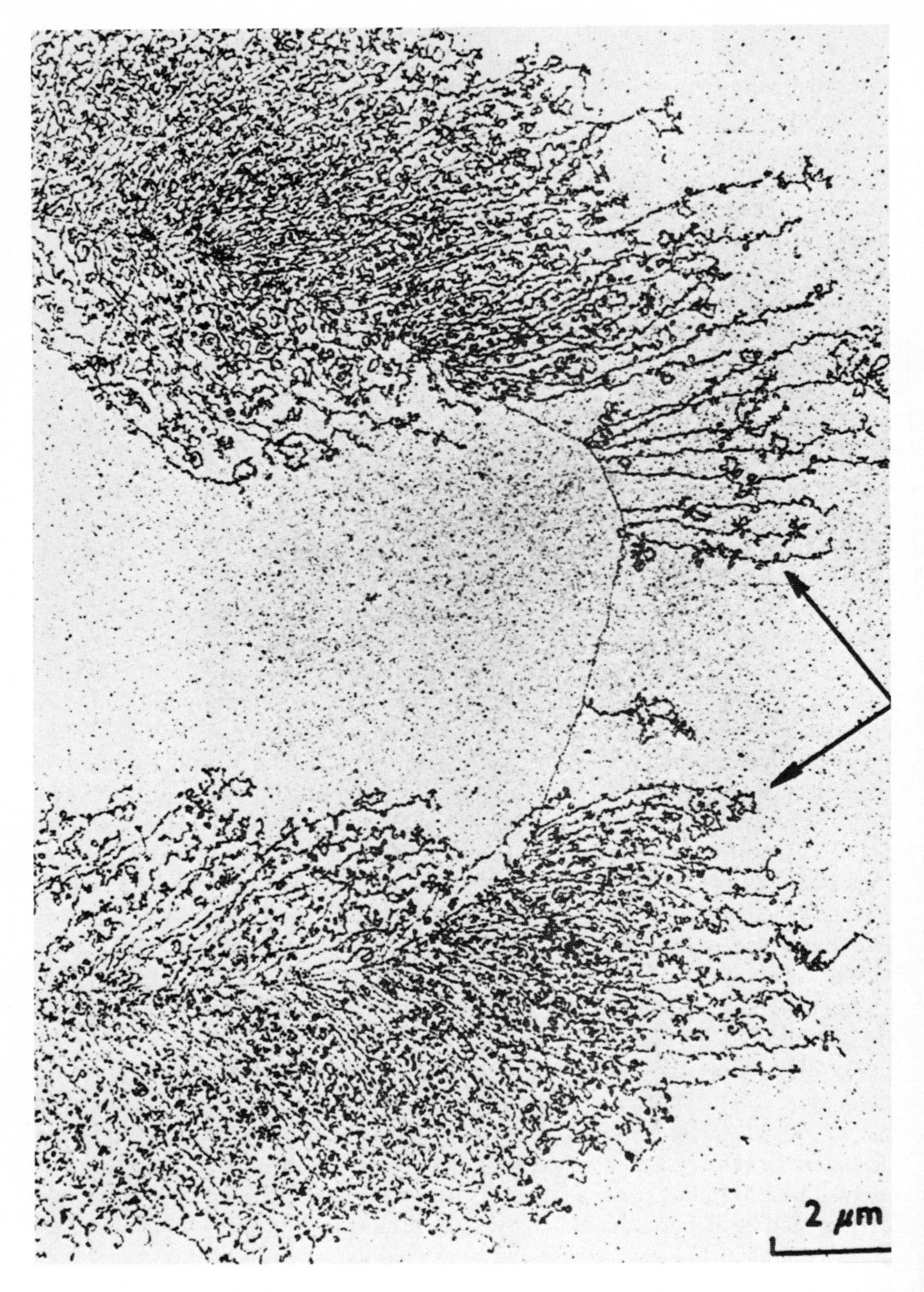


FIG. 2. Loop of a lampbrush chromosome; the "transcription units" (arrows) display the classic "Christmas tree" configuration. [Courtesy of Dr. O. L. Miller, Jr.]

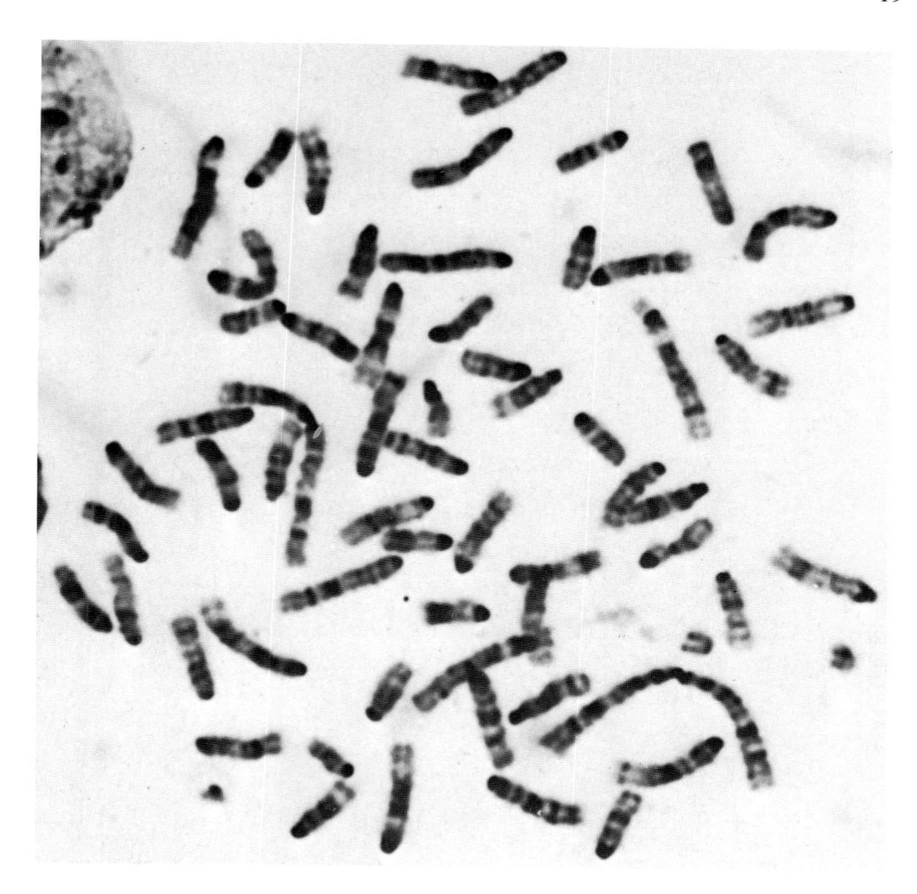

FIG. 3. Banding pattern of metaphase chromosomes in a mouse hepatoma cell. The spread chromosomes were first treated with trypsin and then stained with Giemsa. [Micrograph by C. Szpirer.]

However, Sumner *et al.* (1981) concluded that quinacrine binds uniformly to the chromosomes and that the banding pattern results from differential quenching of fluorescence. Nevertheless, chromosome banding techniques are invaluable for the fine analysis of chromosomal aberrations and are used for prenatal diagnosis of congenital diseases in humans.

A basically similar method detects sister chromatid exchanges (Fig. 4) in which DNA from one chromatid (chromosomes are made of two DNA-containing chromatids) moves to the other chromatid as a result of chromosome breakage. Sister chromatid exchange is one of the earliest signs of DNA damage after treatment with mutagenic and carcinogenic agents. It can be detected by allowing

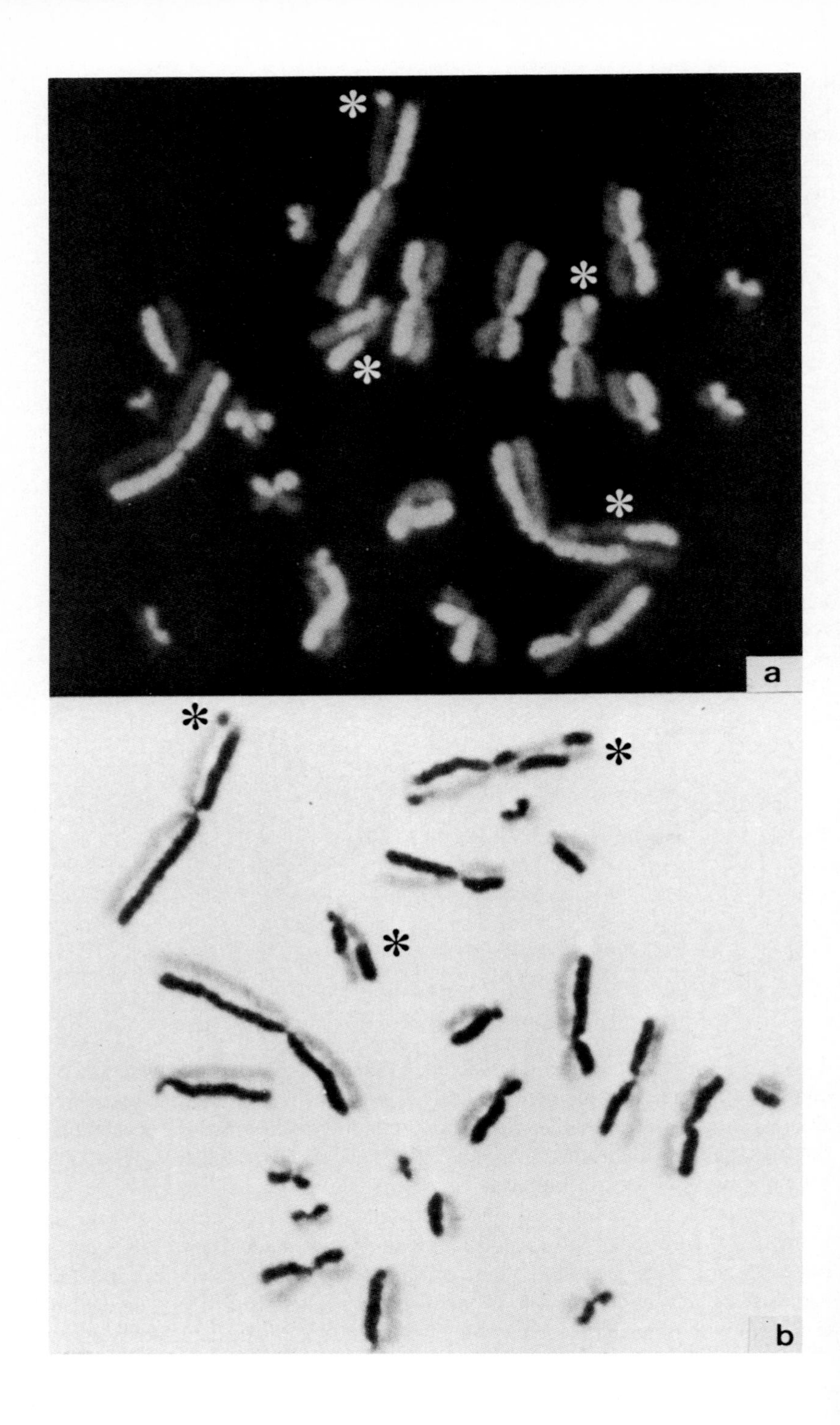
a
b

cells to replicate in the presence of bromodeoxyuridine (BrdUrd), which is incorporated into DNA instead of thymidine, during its replication and staining with the fluorescent dye Hoechst 33258. After uv irradiation, the fluorescence is selectively quenched in BrdUrd-containing DNA segments.

Within the last few years, an interesting method has been developed for the cytological detection of the nucleolar organizers (NOR), i.e., the nucleolar genes responsible for the synthesis of the rRNAs. It is essentially an ammoniacal silver impregnation method and has been called "Ag-AS NOR" by its developers, Goodpasture and Bloom (1975). Selective reduction of ammoniacal silver nitrate in NOR is due to the association of NOR with specific nonhistone nucleolar proteins (Busch *et al.*, 1979; Daskal *et al.*, 1980). It is usually believed that only active NOR, i.e., actively transcribing ribosomal genes, possess these proteins and thus give a positive silver staining. The Goodpasture–Bloom test has been used successfully at the ultrastructural level (Dresser and Moses, 1979; Bourgeois *et al.*, 1979).

Chromatin condensation can also be detected by cytochemical methods; DNA is more tightly bound to proteins in condensed than in extended chromatin. As a result, its capacity to bind basic dyes such as acridine orange (Rigler *et al.*, 1969) or tritiated actinomycin D (Ficq and Brachet, 1965; Steinert and Van Gansen, 1971) is decreased. Fluorophotometry or autoradiography (at the ultrastructural level, in the work of Steinert and Van Gansen) allows quantitative estimations of the binding of these chemicals to DNA under various experimental conditions.

Two techniques, autoradiography and cytochemical detection of enzymes, will be discussed only briefly in this volume. These techniques have now been extended to the EM level. Thus, far more enzymes can now be detected at this level. Classical autoradiography using the light microscope was discussed by A. Ficq in 1959; high-resolution autoradiography was reviewed in detail by Stevens (1966). The principles of the method have remained unchanged. However, new photographic emulsions using very small silver grains have allowed the extension of the technique to the EM level. Resolution is greatly enhanced, but very long exposure times are required.

The classical Gomori method for the detection of phosphatases can now be used at the EM level and is useful for the study of lysosomes and the plasma membrane. Among the enzymes of particular interest to molecular biologists that can now be detected cytochemically are RNA polymerase (Moore and Ringertz, 1973), DNA polymerase (Darzynkiewicz, 1973; Chevaillier and Philippe, 1976), and adenylate cyclase (Chakraborty and Nelson, 1979).

---

FIG. 4. Sister chromatid exchanges in Chinese hamster ovary (CHO) cells. (a) Staining with a fluorescent dye (after BrdUrd treatment); (b) same as (a) but cells exposed to uv light before being stained with Giemsa. Asterisks show the chromosomes where chromatid exchanges have taken place. [From Wolff, 1977. Reproduced, with permission *Annual Review of Genetics*, Volume 11. © 1977 by Annual Reviews Inc.]

## III. BIOLOGICAL, BIOCHEMICAL, AND BIOPHYSICAL TECHNIQUES

The description of cell culture methods that are now commonly used is outside the scope of this book. However, very important progress has been made in this field. Thus, it is now possible to cultivate several cell lines in a serum-free, chemically controlled medium (Barnes and Sato, 1980). This will perhaps reduce the fear of artificial changes induced in cells by the culture medium (Bissell, 1981).

It is often necessary to separate different cells from each other: we have already mentioned the cell sorter, which is based on flow cytometry and is ideal for this purpose. Laboratories that cannot afford such costly equipment often rely on the classical gradient centrifugation methods. These have been reviewed by Harwood (1974).

All the methods mentioned thus far are primarily used by morphologists or those observing intact (living or stained) cells under a microscope. However, progress in the field of molecular cytology would have languished had biochemists not broken cells into pieces in order to isolate and analyze their constituents, following the important advances made by Claude (1940, 1943). These approaches have yielded exceedingly important results.

The isolation of cell constituents (nuclei, mitochondria, lysosomes, microsomes, Golgi and plasma membranes, and soluble cytosol) by differential centrifugation of homogenates is now a very widely used technique. Its resolution has been greatly improved by de Duve. The results obtained using the centrifuge were summarized by de Duve in his Nobel Prize lecture (1975). The main problem with this technique is the clean separation of various fractions obtained by centrifugation at various speeds. It is essential to verify their purity using biochemical markers and electron microscopy. Commonly used markers are DNA for nuclei, cytochromoxidase or succinic dehydrogenase for mitochondria, NADH-cytochrome *c* reductase for the endoplasmic reticulum (microsomes), galactosetransferase for the Golgi membranes, and 5′-nucleotidase and alkaline phosphodiesterase for the plasma membrane. A paper by Wibo *et al.* (1981) on the subfractionation, by differential centrifugation, of preparations enriched in plasma membranes, outer mitochondrial membranes, and Golgi complex membranes gives an excellent overview of the potentials of the method.

The validity of current methods for isolation of nuclei has been questioned because of the inherent problems of the techniques, such as loss of soluble substances or nonspecific adsorption of cell contaminants. For this reason, biochemists interested in the quantitative estimation of the nuclear contents in soluble substances (ions or polyamines, for instance) sometimes use a method first designed by Behrens (1938). The freeze-dried tissue is finely ground and is then subjected to repeated centrifugation in a density gradient of non-water-miscible substances. This nonaqueous isolation technique is time-consuming, and the yields are very small.

For the cell biologist, nuclei are good only if they are still "living," i.e., capable of macromolecule synthesis or mitotic division when reintroduced into anucleate cells (*Xenopus laevis* oocytes, for instance). The technology for the isolation and transplantation of nuclei in oocytes has been described by Gurdon (1977) and Feldherr (1977).

Many cell biologists are interested in the overall role of the cell nucleus, a topic that will be discussed in Volume 2, Chapter 1. Merotomy, i.e., the cutting into two halves (nucleate and anucleate) of an egg or a large unicellular organism (an ameba or *Acetabularia*), requires little manual skill. However, one might well question the universality of the results obtained with such highly specialized biological materials: do they remain valid for ordinary small cells? It is now possible to separate most cells in nucleate (karyoplast) and anucleate (cytoplast) halves using the technique developed by Prescott *et al.* (1972). The principle of the method is to treat intact cells with a drug (cytochalasin B) that disrupts the constitutive microfilaments of the cytoskeleton, and then submit them to mild centrifugation. A technique for obtaining large amounts of anucleate cytoplasts has been described by Bossart *et al.* (1975). If the cells have been treated with colchicine (which prevents the formation of the mitotic spindle and scatters the metaphase chromosomes throughout the cytoplasm) prior to treatment with cytochalasin B and centrifugation, minicells containing one or a few chromosomes can be obtained (Ege *et al.,* 1977). Their advantage for genetic studies is obvious.

Cell fusion is another very important tool for the analysis of nucleocytoplasmic interactions. The exceedingly interesting results obtained with this method by B. Ephrussi and M. Weiss in the study of cell differentiation, and by H. Harris in the analysis of the cytoplasmic factors that affect nuclear activity and control malignancy will be examined later (Volume 2, Chapter 1). When a cell comes into close contact with another cell, fusion between the two is an infrequent event. The frequency of fusion between neighboring cells can be greatly increased if the integrity of the cell membrane is altered by treatment of the intact cells with agents such as uv-inactivated Sendai virus (a hemolytic virus) or polyethylene glycol. More recently, Zimmerman (1982) has shown that cell fusion can easily be obtained in an electrical field. As will be seen, fusion between cells of very different origins (insect and mammalian cells, for instance) can be induced. The result of cell fusion is the formation of somatic hybrids containing nuclei from both parental strains (heterokaryons) in a common cytoplasm. Anucleate cytoplasts can also be fused to whole cells or to karyoplasts from the same or another cell line; the products of such fusions are reconstituted cells or cybrids. For details about cell fusion, the interested reader should consult the book on cell hybrids by Ringertz and Savage (1976).

It is becoming important for cell biologists to introduce foreign substances such as proteins or nucleic acids into living cells. With the increased availability of many purified genes because of recombinant DNA technology (see, for instance, the special issue of *Science*, 1980, dealing with this subject), it is of obvious interest to introduce them into cells and to observe the outcome. We shall review this question in Volume 2, Chapter 1, since this chapter is only concerned with methodology. Many high molecular-weight substances, such as proteins or nucleic acids, can penetrate cells simply by being added to the medium under proper conditions. However, uptake of these substances by the cells is often low and occurs at random. Thus, the question remains, how can one increase the rate of uptake? The more direct method is to inject substances into cells that cannot be penetrated easily. This can be done by micromanipulation, a technique that can now be successfully applied to somatic cells (Graessmann and Graessmann, 1983); direct injection into the nucleus is possible for a skilled experimenter. Other less direct techniques are based on fusion of the cells with resealed red blood cell ghosts (obtained by hemolysis of mammalian red blood cells) or liposomes (phospholipid vesicles) in which the substance of interest has been introduced. Examples of the methodology are the introduction into cells, via red blood cell ghosts, of specific proteins (Schlegel and Rechsteiner, 1975), or, more recently, of obelin, a substance that allows the photometric measurement of the free $Ca^{2+}$ content of the cell (Campbell *et al.*, 1981). For the formation of proteolipids (association of proteins and liposomes), see the review by Hokin (1981).

Thus, it is important to isolate cell constituents in amounts sufficient for biochemical analysis. As early as 1952, D. Mazia and K. Dan isolated the mitotic apparatus of cleaving sea urchin eggs. Since that time, many modifications of the initial method, some of them applicable to somatic cells, have been proposed. They are described in reviews by Zimmerman *et al.* (1977) and Turner and McIntosh (1977).

Finally, separation and isolation of chromosomes (generally from colchicine-treated cells in which the metaphase chromosomes are scattered) are also possible. Several methods are now available—for instance, centrifugation in a metrizamide gradient (Wray, 1977) or isolation at unit gravity in a specially constructed chamber (Collard *et al.*, 1980a,b). Separation of small from large chromosomes by gradient centrifugation or by flow cytophotometry can be achieved. Large amounts of metaphase chromosomes can be obtained in purified form using current methods. The obvious genetic interest of these techniques is further increased by the fact that it is possible to introduce isolated chromosomes, bearing genetic markers, into recipient cells (chromosome-mediated gene transfer; see Volume 2, Chapter 1).

"Quantitative histochemistry," i.e., the reduction on an ultramicroscale of classical biochemical techniques, has lost much of its importance because of the great increase in the resolution and sensitivity of these techniques. Use of microvessels, microcuvettes, microrespirometers, etc. is seldom necessary in view of the high sensitivity of so many methods utilizing either radioisotopes or fluorescent reagents. An example of what can now be achieved is the work of Bravo *et al.* (1981) and Celis and Bravo (1981). They separated more than 1000 labeled polypeptides from only 100 HeLa cells; the label was $^{35}S$ and the method was two-dimensional gel electrophoresis.

Knowledge of a number of important parameters is necessary before one begins to manipulate cells. Among them are the cell internal pH ($pH_i$), the cell membrane potential ($\Delta_\psi$), and the concentration of free. calcium ions. The last, as we have seen, can be measured by introducing obelin (or aequorin) into cells and monitoring the light emission of these substances due to the presence of $Ca^{2+}$ ions. In general, the concentration of free $Ca^{2+}$ is almost negligible compared with that of total calcium; the order of magnitude is about 0.1 $\mu M$. For large cells, such as sea urchin eggs (100 μm in diameter), electrophysiological techniques can be used for the measurement of $pH_i$ and membrane potential; the small size of the electrodes currently available reduces the danger of artifacts due to local cytolysis. Other methods are needed for work on cells of smaller size. For measurement of $pH_i$, one can consider the fluorescein diacetate method of Thomas *et al.* (1979) and the technique of Johnson and Epel (1981). In this case, $pH_i$ is calculated from the distribution in the cells and in the outer medium of a radioactive weak acid, [$^{14}C$]DMO. For the measurement of the membrane potential, cyanidine derivatives whose fluorescence is dependent on $\Delta_\psi$ are widely used today. This method can be applied to cell organelles, such as isolated mitochondria, or to bacteria.

This walk through the enormous field of methodology was much longer than expected—but it was necessary. Scientific progress is very often dependent on technical improvements. After all, if Leeuwenhoek had not improved microscope lenses as far back as 1675, we would not know of the existence of protozoa and spermatozoa today.

## REFERENCES


Alfert, M., and Geschwind, I. I. (1953). *Proc. Natl. Acad. Sci. U.S.A.* **39,** 991.

Angerer, L. M., and Angerer, R. C. (1981). *Nucleic Acids Res.* **9,** 2819.

Barer, R. (1966). *In* ''Physical Techniques in Biological Research'' (A. W. Pollister, 2nd ed.), Vol. 3, Part A, p. 30. Academic Press, New York.

Barnes, D., and Sato, G. (1980). *Cell* **22,** 649.

Behrens, M. (1938). *Hoppe-Seyler's Z. Physiol. Chem.* **253,** 1851.

Bissell, M. J. (1981). *Int. Rev. Cytol.* **70,** 27.

Bossart, W., Loeffler, H., and Bienz, K. (1975). *Exp. Cell Res.* **96,** 360.

Bourgeois, C. R., Hernandez-Verdun, D., Hubert, J., and Bouteille, M. (1979). *Expl. Cell Res.* **123,** 449.

Bravo, R., Bellatin, J., and Celis, J. E. (1981). *Cell Biol. Int. Rep.* **5,** 93.

Busch, H., Daskal, Y., Gyorkey, F., and Smetana, K. (1979). *Cancer Res.* **39,** 857.

Campbell, A. K., Daw, R. A., Hallett, M. B., and Luzio, J. P. (1981). *Biochem. J.* **194,** 551.

Capco, D. G., and Jeffery, W. R. (1978). *Dev. Biol.* **67,** 137.

Caspersson, T. (1936). *Skand. Arch. Physiol.* **78,** Suppl. 8.

Caspersson, T., Farber, S., Foley, G. E., Kudynowski, J., Modest, E. J., Simmonsson, E., Wagh, U., and Zech, L. (1968). *Exp. Cell Res.* **49,** 219.
Celis, J. E., and Bravo, R. (1981). *Trends Biochem. Sci.* **6,** 197.
Chakraborty, J., and Nelson, L. (1979). *Biol. Reprod.* **20,** 131.
Chevaillier, P., and Philippe, M. (1976). *Chromosoma* **54,** 33.
Claude, A. (1940). *Science* **91,** 77.
Claude, A. (1943). *Biol. Symp.* **10,** 3.
Collard, J. G., Tulp, A., Hollander, J. H., Bauer, F. W., and Boezeman, J. (1980a). *Exp. Cell Res.* **126,** 191.
Collard, J. G., Tulp, A., Stegeman, J., Boezeman, J., Bauer, F. W., Jongkind, J. F., and Verkerk, A. (1980b). *Exp. Cell Res.* **130,** 217.
Coons, A. H. (1956). *Intern. Rev. Cytol.* **5,** 1.
Coté, B. D., Uhlenbeck, O. C., and Steffensen, D. M. (1980). *Chromesoma* **80,** 349.
Darzynkiewicz, Z. (1973). *Exp. Cell Res.* **80,** 483.
Das, N. K., and Alfert, M. (1977). *Methods Cell Biol.* **16,** 241.
Daskal, Y., Smetana, K., and Busch, H. (1980). *Exp. Cell Res.* **127,** 285.
Davies, H. G., Engström, A., and Lindström, B. (1953). *Nature (London)* **172,** 1041.
De Brabander, M., De Mey, J., Joniau, M., and Geuens, G. (1977). *J. Cell Sci.* **28,** 283.
de Duve, C. (1975). *Science* **189,** 196.
de Laat, S. W., Tertoolen, L. G. J., and Bluemink, J. G. (1981). *Eur. J. Cell Biol.* **23,** 273.
De Mey, J., Hoebeke, J., De Brabander M., Geuens, S., and Joniau, M. (1976). *Nature (London)* **264,** 273.
De Mey, J., Moeremans, M., Geuens, G., Nuydens, R., and De Brabander, M. (1981). *Cell Biol. Int. Rep.* **5,** 889.
Dresser, M. E., and Moses, M. J. (1979). *Exp. Cell Res.* **121,** 416.
Ege, T., Ringertz, N. R., Hamberg, H., and Sidebottom, E. (1977). *Methods Cell Biol.* **15,** 339.
Etemadi, A. H. (1980a). *Biochim. Biophys. Acta* **604,** 347.
Etemadi, A. H. (1980b). *Biochim. Biophys. Acta* **604,** 423.
Feldherr, C. M. (1977). *Methods Cell Biol.* **16,** 167.
Ficq, A. (1959). *In* "The Cell" (J. Brachet and A. E. Mirsky, eds.), Vol. 1, p. 67. Academic Press, New York.
Ficq, A., and Brachet, J. (1965). *Exp. Cell Res.* **38,** 153.
Gall, J. G., and Pardue, M. L. (1969). *Proc. Natl. Acad. Sci. U.S.A.* **63,** 378.
Gerhard, D. S., Kawasaki, E. S., Bancroft, F. C., and Szabo, P. (1981). *Proc. Natl. Acad. Sci. U.S.A.* **78,** 3755.
Goldman, R. D. (1975). *J. Histochem. Cytochem.* **23,** 529.
Goodpasture, C., and Bloom, S. E. (1975). *Chromosoma* **53,** 37.
Graessmann, M., and Graessmann, A. (1983). *In* "Methods in Enzymology" (R. Wu, L. Grossman, and K. Moldave, eds.), Vol. 11, p. 482. Academic Press, New York.
Gurdon, J. B. (1977). *Methods Cell Biol.* **16,** 125.
Harper, M. E., and Saunders, G. F. (1981). *Chromosoma* **83,** 431.
Harper, M. E., Ulbrich, A., and Saunders, G. F. (1981). *Proc. Natl. Acad. Sci. U.S.A.* **78,** 4458.
Harwood, R. (1974). *Int. Rev. Cytol.* **38,** 369.
Hayat, M. A. (1974–1976). "Principles and Techniques of Electron Microscopy: Biological Applications." Van Nostrand-Reinhold, Princeton, New Jersey.
Henderson, A. S. (1982). *Int. Rev. Cytol.* **76,** 1.
Hokin, L. E. (1981). *J. Membr. Biol.* **60,** 77.
Hutchison, N. J., Langer-Safer, P. R., Ward, D. C., and Hamkalo, B. A. (1982). *J. Cell Biol.* **95,** 609.
Jacob, J., Todd, K., Birnstiel, M. L., and Bird, A. (1971). *Biochim. Biophys. Acta* **228,** 761.

Johnson, C. H., and Epel, D. (1981). *J. Cell Biol.* **89,** 284.
Jovin, T. M., and Arndt-Jovin, D. J. (1980). *Trends Biochem Sci.* **5,** 214.
King, C. A., and Preston, T. M. (1977). *FEBS Lett.* **73,** 59.
Kruth, H. S. (1982). *Anal. Biochem.* **125,** 225.
Lazarides, E. (1975). *J. Histochem. Cytochem.* **23,** 507.
Lison, L. (1936). "L'histochimie animale." Gauthier-Villars, Paris.
Lison, L. (1960). "Histochimie et cytochimie animales," 3rd ed. Gauthier-Villars, Paris.
Lydon, M. J., Keeler, K. D., and Thomas, D. B. (1980). *J. Cell. Physiol.* **102,** 175.
Manuelidis, L., Langer-Safer, P. R., and Ward, D. C. (1982). *J. Cell Biol.* **95,** 619.
Mazia, D., and Dan, K. (1952). *Proc. Natl. Acad. Sci. U.S.A.* **38,** 826.
Mesland, D. A. M., Spiele, H., and Roos, E. (1981). *Exp. Cell Res.* **132,** 169.
Miller, O. L., Jr., and Beatty, B. R. (1969). *Science* **164,** 955.
Milstein, C. (1981). *Proc. R. Soc. London, Ser. B* **211,** 393.
Milstein, C., and Cuello, A. C. (1983). *Nature (London)* **305,** 537.
Moore, G. P. M., and Ringertz, N. R. (1973). *Exp. Cell Res.* **76,** 223.
Nicolson, G. L., and Singer, S. J. (1971). *Proc. Natl. Acad. Sci. U.S.A.* **68,** 942.
Olins, A. L., and Olins, D. E. (1974). *Science* **183,** 330.
Pearse, A. G, E. (1968). "Histochemistry Theoretical and Applied," 3rd ed. Churchill, London.
Pinto da Silva, P., Kachar, B., Torrisi, M. R., Brown, C., and Parkinson, C. (1981). *Science* **213,** 230.
Plattner, H., and Bachmann, L. (1982). *Int. Rev. Cytol.* **79,** 237.
Plattner, H., and Zingsheim, H. P. (1983). *Subcell. Biochem.* **9,** 1.
Prescott, D. M., Myerson, D., and Wallace, J. (1972). *Exp. Cell Res.* **71,** 480.
Rigler, R., Killander, D., Bolund, L., and Ringertz, N. R. (1969). *Exp. Cell Res.* **55,** 215.
Ringertz, N. R., and Savage, R. E. (1976). "Cell Hybrids." Academic Press, New York.
Schlegel, R. A., and Rechsteiner, M. C. (1975). *Cell* **5,** 371.
Steinert, G., and Van Gansen, P. (1971). *Exp. Cell Res.* **64,** 355.
Stevens, A. R. (1966). *In* "High Resolution Autoradiography" (D. M. Prescott, ed.), Vol. 2, p. 255. Academic Press, New York.
Sumner, A. T., Carothers, A. D., and Rutovitz, D. (1981). *Chromosoma* **82,** 717.
Thomas, J. A., Buchsbaum, R. N., Zimniak, A., and Racker, E. (1979). *Biochemistry* **18,** 2210.
Turner, J. L., and McIntosh, J. R. (1977). *Methods Cell Biol.* **16,** 373.
Wang, E., and Goldberg, A. R. (1978). *J. Histochem. Cytochem.* **26,** 745.
Wibo, M., Thinès-Sempoux, D., Amar-Costesec, A., Beaufay, H., and Godelaine, D. (1981). *J. Cell Biol.* **89,** 456.
Willingham, M. C., Yamada, S. S., and Pastan, I. (1980). *J. Histochem. Cytochem.* **28,** 453.
Willingham, M. C., Yamada, S. S., Davies, P. J. A., Rutherford, A. V., Gallo, M. G., and Pastan, I. (1981). *J. Histochem. Cytochem.* **29,** 17.
Wolff, S. (1977). *Ann. Rev. Genet.* **11,** 183.
Wray, W. (1977). *FEBS Lett.* **62,** 202.
Wulf, E., Deboden, A., Bautz, F. A., Faulstich, H., and Wieland, T. (1979). *Proc. Natl. Acad. Sci. U.S.A.* **76,** 4498.
Yamada, S. S., Yamada, K. M., and Willingham, M. C. (1980). *J. Histochem. Cytochem.* **28,** 953.
Zimmerman, A. M., Zimmerman, S., and Forer, A. (1977). *Methods Cell Biol.* **16,** 361.
Zimmerman, U. (1982). *Biochim. Biophys. Acta* **694,** 227.

## Chapter 3

# THE CYTOPLASM DURING INTERPHASE

The title of this chapter could have been "The Cytoplasm of the Resting Cell" (cf. "Biochemical Cytology"), but the adjective "resting" would have been misleading since it is during interphase (the period between two mitotic events) that the cell most actively synthesizes nucleic acids and proteins. An overview and a schematic diagram of a typical (?) interphase cell were given in Chapter 1, Fig. 2. Details concerning the organization of the cytoplasm can be found in a Cold Spring Harbor symposium (Vol. 46) published in 1981; many competent authors present (in more than 1000 pages) what was known in 1980 about this important topic. We shall begin the discussion of the cell with its outer wall, the cell surface.

## I. THE CELL SURFACE

### A. General Background

The cell surface plays an essential role in controlling the exchanges between the cell and the outer medium and the interactions between a given cell and its neighboring cells. The true membrane of the cell (plasma membrane, sometimes called "plasmalemma") is only one element of the more complex system that constitutes the cell surface. Outside the plasma membrane is the cell coat or glycocalyx, a projection of glycoproteins and glycosaminoglycans toward the outer environment. Most cells establish contact with other cells through this extracellular or pericellular matrix; glycoproteins of cellular origin (fibronectin, collagens, laminins) and sulfated mucopolysaccharides accumulate where cells (fibroblasts, for instance) come in contact with an artificial substratum. This thickened extracellular matrix is often called the "basal lamina." On the inner side of the plasma membrane lies a specialized portion of the cytoskeleton (actin microfilaments and tubulin microtubules). This membrane-associated cytoskeleton plays an important role in the continuous changes in shape that many cells undergo at the microscopic level. The cell surface displays a high degree of morphological and biochemical complexity and should be regarded as a dynamic structure. Finally, it should be pointed out that the cell surface must be considered a single unit consisting of an extracellular matrix, a plasma membrane, and a superficial cytoskeleton. It has been shown that, in fibroblasts, a 130,000-dalton protein called "vinculin" is a specific intermediate between actin in the microfilaments and fibronectin that has accumulated on the ventral side of the cell, where fibroblasts attach to the substratum through the so-called adhesion plaques (Singer and Paradiso, 1981). Fibronectin is required for spreading cells attached to a solid substratum (Rubin *et al.*, 1981), but this extracellular glycoprotein is not present locally when fibroblasts establish close contacts with the substratum through focal adhesion plaques (Avnur and Geiger, 1981).

Among the many review articles devoted to the cell surface, one can single out those of Edelman (1976); Nicolson (1976) on the membrane-associated cytoskeleton; Rebhun (1977) on the role played by cyclic adenosine monophosphate (cAMP) in cell motility and intercellular adhesion; Bosmann (1977) and Pierce *et al.* (1980) on the possible role of membrane glycosyltransferases and proteases in cell recognition; Vasiliev and Gelfand (1977) on morphology and locomotion of fibroblasts and epithelial cells; Hynes and Yamada (1982), Kleinmann *et al.* (1981), and Ruoslahti *et al.* (1982) on fibronectin, a major component of the basal lamina; Evans (1980), and Aplin and Hughes (1983) on the various types of junctions that link cells together; Hooper and Subak-Sharpe (1981) and Loewenstein (1982) on the metabolic exchanges that occur between cells through these junctions; van Deenen (1981) on the distribution of the phospholipids in biological membranes; Pearse (1980) and Besterman and Low (1983) on the mechanisms that allow the selective uptake of proteins through the cell membrane; and Plattner (1981) on the opposite process (secretion of proteins through the cell membrane). Many other reviews of these and related topics will become available in the coming months or years, since study of the cell surface is now one of the most popular fields in cell biology.

### B. Molecular Organization and Chemical Composition of the Plasma Membrane

In "Biochemical Cytology" we showed (p. 33) the model of molecular organization proposed in 1954 for the plasma membrane by J. Danielli: a double layer of lipids separates protein sheets and is crossed by pores. This model is still valid to some extent, in particular the existence of the double layer of lipids and of transmembrane channels accessible, simultaneously, from both sides of the membrane (reviewed by Läuger, 1980). However, there is now compelling evidence, based largely on morphological studies on freeze-fractured membranes, on biochemical analysis of isolated membranes, and on physical measurements of membrane viscosity for the fluid mosaic model proposed by S. J. Singer and G. L. Nicolson in 1972. As shown in Fig. 1, this model retains some features of the Danielli one, but it is now certain that proteins are inserted into the lipid bilayer. They can be seen as small granules in fractured membranes, between the outer and inner (protoplasmic) layers of the plasma membrane (Fig. 2).

These intramembrane particles, which include membrane-bound enzymes, were first observed in red blood cells but seem to be universal; they have been identified as intrinsic membrane proteins and display lateral mobility. They can,

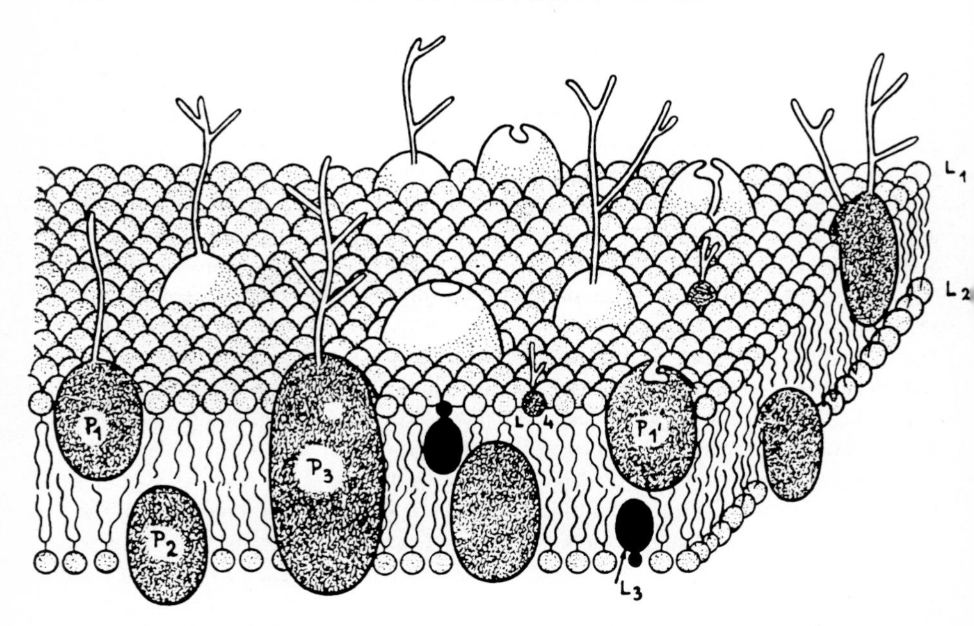


FIG. 1. Schematic representation of the plasma membrane. L, lipids ($L_1$, $L_2$, phospholipids; $L_3$, cholesterol; $L_4$, glycolipids). P, proteins ($P_1$, external proteins with carbohydrate chains; $P_2$, internal proteins; $P_3$, intrinsic membrane proteins with carbohydrate appendages). [Original drawing by P. Van Gansen.]

in a passive way, be moved around inside the mobile bilayer and thus accumulate in certain regions of the plasma membrane. This displacement is affected by the viscosity or fluidity (reviewed by Quinn, 1981) of the bilayer, which can vary in various areas of the plasma membrane. Freezing the cells will solidify the lipids and immobilize the intramembrane particles so that they can no longer flow in a "sea of lipids." There are other membrane proteins, namely, external glycoproteins, which contribute, through their carbohydrate chains, to the surface coat or extracellular matrix, where a variety of substances (proteins such as collagen, fibronectin or laminin, glycosaminoglycans) may accumulate to form a basal lamina. Facing the cytoplasm are internal proteins that are in close contact with the membrane-associated cytoskeleton, in particular the actin or actomyosin contractile microfilaments; in red blood cells, these proteins are in close contact with spectrin, a myosin-like protein that interacts with actin. Spectrin-like molecules, in particular fodrin, have recently been found in many cells. One important feature that should be mentioned is membrane asymmetry. Due to differences in lipid and protein organization, the two leaflets that form the membrane are not identical. For instance, in red blood cells, the content of the outer and inner layers in sphingomyelin, phosphatidylcholine, phosphatidylethanolamine, and phosphatidylserine are markedly different (van Deenen, 1981). The differences in the protein composition of the two layers forming the plasma membrane are even more striking; external and internal proteins differ so widely from each other and have such different functions that they have little if anything in common.

As previously mentioned, the lateral mobility of proteins and lipids can be measured by precise physical methods (Peters *et al.*, 1981; Henis and Elson, 1981; Quinn, 1981; Sowers and Hackenbrock, 1981). This is an essential feature of the fluid mosaic model. This displacement of intrinsic proteins, and thus of intramembrane particles, was discovered from studies on the binding of substances called ''lectins'' to the cell membrane. Studies on lymphocytes showed that certain glycoproteins of plant origin, called ''phytohemagglutinins'' (because they agglutinate red blood cells) are mitogenic for lymphocytes inducing enlargement of the cells (blast transformation); this is followed by cell division. Binding of these agglutinins produces a signal at the cell membrane level that is somehow transmitted to the nucleus. DNA replication is thus induced and followed by cell division. Later studies showed that many substances (including

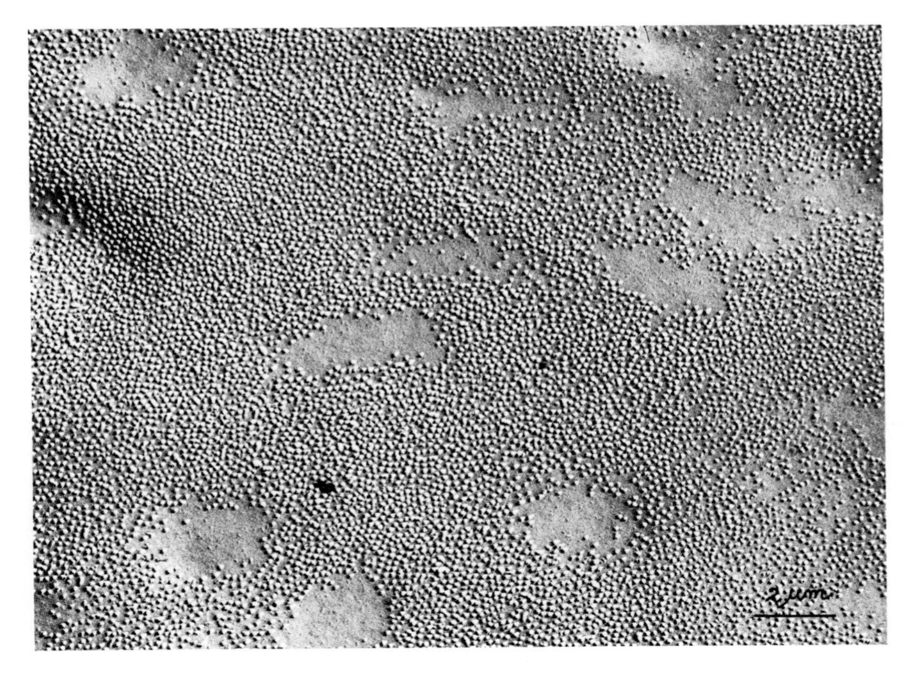


FIG. 2. Distribution of intrinsic membrane proteins, as seen by freeze-etching in the plasma membrane of a *Xenopus* oocyte. [Courtesy of Prof. J. Blueminck.]

antigens, in the case of lymphocytes), which bind specifically to glycosidic residues in the cell membrane, often have cell-agglutinating and mitogenic properties. These substances, lectins, bind to specific sugar residues (for instance, mannose acetylglucosamine or fucose) on the cell surface. The addition of the appropriate sugar suppresses this binding. ConA and wheat germ agglutinin (WGA) are the most widely known in this class of compounds. Lectins can, of course, be labeled with fluorescent dyes or radioisotopes. In 1971, Fox *et al.* detected specific membrane sites using these substances. Nicolson (1972) demonstrated that a clustering of ConA-binding sites takes place after addition of labeled ConA to isolated cells. Edelman *et al.* (1973) found that ConA affects the distribution of the microtubules and that this, in turn, affects the mobility and distribution of the lectin receptors (i.e., the glycoproteins bearing the specific glycosidic residues to which a lectin binds on the cell surface). When a fluorescent lectin is added to isolated cells, the following is observed. The lectin first binds homogeneously to the entire cell surface. It then distributes in patches (patching). Finally, the fluorescent material becomes concentrated in a cap (capping) (Fig. 3). These observations clearly demonstrate that the lectin receptors undergo a redistribution in the cell membrane. Therefore, they display lateral mobility, a phenomenon that is suppressed when the lipids of the membrane are frozen by cold treatment and when energy production is arrested. According to Phillips *et al.* (1974), the order of magnitude of the ConA-binding sites is about 1100 per cell. The molecular mechanisms of capping remain uncertain, but it is generally agreed that the membrane-associated cytoskeleton plays an essential role. This view has been strongly advocated by Edelman *et al.* (1973; see also McClain *et al.,* 1977), who noted that the surface receptors and the membrane-associated microtubules formed a "surface modulating assembly." This is largely based on experiments with inhibitors of microtubule (colchicine) and microfilament (cytochalasin B) assembly, and is strongly supported by immunofluorescence and EM observations that show that tubulin and actin "co-cap" with the ConA receptors (Toh and Hard, 1977; Albertini *et al.,* 1977). According to Bourguignon and Singer (1977), the contractile proteins actin and myosin accumulate in patches and caps after treatment of cells with ConA. This has led to the suggestion that actin is bound to an integral protein of the membrane through one of the many actin-binding proteins to be discussed later. One of them is vinculin (Geiger *et al.,* 1980; Otto, 1983; Glenney and Glenney, 1983); another is spectrin, the major membrane-associated cytoskeletal protein of red blood cells, which has been found in a large variety of cells (reviewed by Lazarides and Nelson, 1982); spectrin lies on the cytoplasmic side of the plasma membrane, where it crosslinks the cortical actin filaments; it plays a role in capping, at least in lymphocytes, according to Nelson *et al.* (1983a). Spectrin-like molecules are made of a common calmodulin-binding subunit and a variable subunit; the most widespread member of this family is fodrin, which links

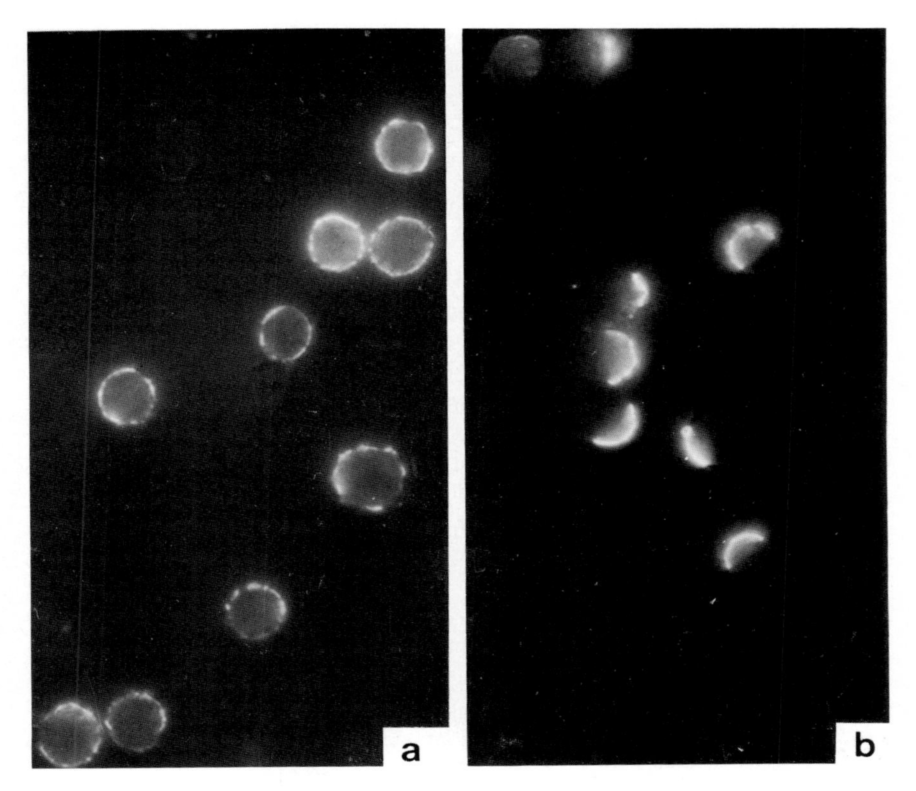


FIG. 3. "Capping" in mouse B-lymphocytes incubated in anti-immunoglobulins conjugated with fluorescein isothiocyanate (FITC). (a) Incubation at 4°C in the presence of $10^{-2}$ *M* $NaN_3$; there is no capping in the absence of energy production. (b) Incubation at 20°C for 30 min induces capping. [Photographs taken with a fluorescence microscope by G. Urbain-Van Santen.]

together the plasma membrane and the underlying microfilaments. S. S. Brown *et al.* (1983) have discovered a new cell surface protein that they have called "connectin" ($M_r$ = 70,000); it binds to both actin and laminin, a major constituent of the extracellular matrix: this transmembrane protein may mediate interactions between the cortical cytoskeleton and the extracellular matrix. While the role of the actin microfilaments in capping seems to be universally accepted, the part played by the microtubules tends to remain more controversial. It seems that capping by ConA requires disassembly rather than assembly of microtubules (Oliver *et al.*, 1980). This seems logical since the cell surface–associated microtubules tend to increase the rigidity of the cell surface and thus decrease the possibility of capping. Lectins will be reconsidered later when cell–cell communication, cell aggregation, and cell division are discussed in greater detail.

The chemical composition of plasma membranes isolated by differential centrifugation of homogenates will now be considered. As was previously mentioned, the main chemical components of plasma membranes (and of all cell membranes) are proteins, lipids, and sugars. The relative proportions of these three components may vary considerably from one cell type to another, with proteins representing 50–60% of the total. Lipids constitute about 40%, and carbohydrates are only a minor constituent. Of the lipids, phospholipids are the most abundant, followed by cholesterol (its insertion into the membrane locally can modify the molecular arrangements of the lipid bilayer) and galactolipids. The carbohydrates (oligosaccharides) form a part of glycoproteins and glycolipids (gangliosides). Their carbohydrate residues are the binding sites for lectins such as ConA. The proteins are either integral (inserted in the lipid bilayer) or peripheral (sticking inside or outside the bilayer). Integral proteins represent more than 70% of the total plasma membrane proteins, although major differences may exist between cells of varying origin. Cell surface glycoproteins can be labeled by lactoperoxidase iodination and then can be separated from each other (Hynes, 1973). Isolation of plasma membranes in reasonably purified form can be achieved by a variety of techniques, followed by differential centrifugation. Separation from other membranes (microsomal, mitochondrial, Golgi membranes, etc.) and subsequent purification can be monitored by measurement of the activity of a number of marker enzymes. For the plasma membrane, the most popular marker is 5′-nucleotidase (which hydrolyzes 5′-AMP to adenosine and phosphate), because its estimation is relatively simple. Although its physiological role is not clear, it is assumed to play a role in the intake of nucleosides and perhaps phosphates into the cells (Fleit *et al.*, 1975).

The biochemical roles of other plasma membrane enzymes are better established. One enzyme is adenylate cyclase, which catalyzes the synthesis of cAMP. This action has generated great interest because it responds directly to hormonal stimulation. The regulation and the complex chemical composition of adenylate cyclase, which is made up of three different constituents [a hormonal receptor, a guanosine triphosphate (GTP)-binding and hydrolyzing regulatory unit, and a catalytic unit] have been discussed by Levitzki and Helmreich (1979) and by Rodbell (1980). A phosphodiesterase capable of hydrolyzing cAMP (Russell and Pastan, 1973) is also localized in the plasma membrane. It is now generally known that cyclic nucleotides, cAMP in particular, control many cell activities (including cell proliferation) by activating protein kinases (enzymes capable of phosphorylating proteins), usually at the level of their serine and threonine residues. It is interesting, in this respect, that isolated plasma membranes have been reported to possess an endogenous protein kinase activity that can phosphorylate more than 25 proteins located in the plasma membrane. This kinase activity increases markedly at the time of DNA replication in the cell nucleus (Mastro and Rozengurt, 1976). Thus, the plasma membrane plays an important (but not exclusive) role in cAMP synthesis, cAMP degradation, and cAMP-controlled regulatory activities.

Another very important plasma membrane enzyme, which is often used as a marker enzyme, is the $Na^+/K^+$–ATPase or sodium pump. Its role is to exchange three sodium ions for two potassium ions. This exchange requires energy that is provided by hydrolysis of one ATP molecule. The enzyme is inhibited in a specific way by ouabain. In some cells, at least, the presence of another ATP-ase—ecto-ATPase—has been reported. It is present if intact cells are able to hydrolyze ATP added to the medium. Ecto-ATPase is insensitive to ouabain and to inhibitors of ATP production by mitochondria, and is activated by divalent ions (Carraway *et al.*, 1980).

Other marker enzymes that have been found valuable for monitoring the purification of fractions enriched in plasma membranes are alkaline phosphatase, leucine aminopeptidase (Gotlib and Searls, 1980), and the glycosyltranferases, which are believed to play a role in cell recognition and aggregation (reviewed by Pierce *et al.*, 1980).

According to Klingenberg (1981), the plasma membrane proteins are generally in the form of oligomers, which span the membrane and are disposed asymmetrically (Fig. 4). Since these oligomers cross the membrane, they should be able to transport nutrients from the outer medium into the cell. That some integral proteins are indeed transmembrane proteins that span the membrane by binding to the underlying contractile proteins actin and myosin, is shown in the work of Evans and Fink (1977) and Ash *et al.*, 1977). Two of these transmembrane proteins are the already mentioned vinculin and connectin, which link together actin and the extracellular coat proteins fibronectin and laminin, respectively (Singer and Paradiso, 1981; M. S. Brown *et al.*, 1983).

Finally, it should be mentioned that the cell membrane contains a variety of receptors that bind specifically to a host of chemicals; as we shall see, they play an important role in endocytosis (the process that allows the intake into the cell of large molecules) and in the stimulation of cell division by growth factors.

A discussion of the membrane-associated cytoskeleton and the coat or basal lamina that covers the cell surface is appropriate. Little will be said, however, about the microtubules and microfilaments that are closely associated with the plasma membrane (Volume 2, Chapter 1, Fig. 2), because they will be discussed in more detail in the next section. There is no reason to believe that cell surface–associated microtubules and microfilaments differ significantly from those that form the so-called internal cytoskeleton. In both cases, microtubules result from the linear polymerization of tubuline. Depolymerization causes these rigid structures to disappear (for instance, by cold treatment or excess of $Ca^{2+}$ ions). In contrast to microfilaments, microtubules are not contractile per se. Actin-linear polymerization and binding to myosin, leading to the formation of contractile microfilaments, is a delicate phenomenon that is controlled by many actin-

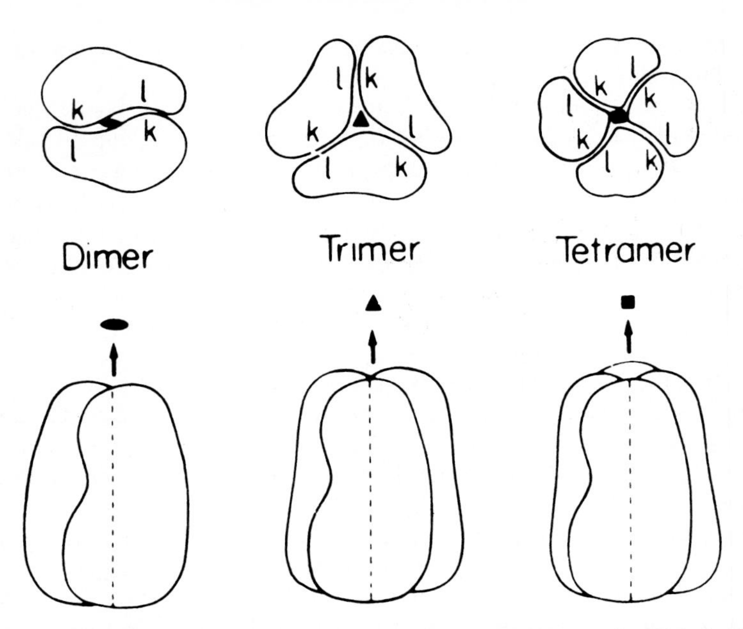


FIG. 4. Proteins that traverse membranes tend to have dimeric structures in which the dimer is arranged asymmetrically across the membrane with the axis of symmetry perpendicular to the membrane plane (left). Protein oligomers may consist of two, three, or four identical subunits. The repeating interfaces between the subunits are designated k and l. This general structure is well suited to the function of transporting nutrients across the cell membrane. [Klingenberg, 1981. Reprinted by permission from *Nature* **290**, 449. Copyright © 1981 Macmillan Journals Limited.]

associated proteins. Since polymerization and depolymerization of tubulin and actin are easily reversible, changes in the assembly of the membrane-associated cytoskeleton should play an important role in the maintenance of cell shape and in cell locomotion.

On the outer side of the plasma membrane, most, if not all, cells are surrounded by an extracellular coat (Fig. 5). Its development and chemical composition are highly variable: in aquatic animal cells (e.g., amebas) and in eggs of many animal species, this coat is often reduced to a layer of glycosaminoglycans (mucopolysaccharides) associated with proteins in the form of mucoproteins. The main glycosaminoglycans found in the cell coat are hyaluronic acid, chondroitin sulfate, dermatan sulfate, and heparin. These sulfated mucopolysaccharides can be easily detected cytochemically, based on the fact that they give metachromatic staining (i.e., they stain red instead of blue in the presence of toluidine blue) and a positive periodic acid–Schiff (PAS) reaction for sugars. This reaction is based on the Schiff aldehyde reaction, which becomes positive when the glycol groups of the hexoses are oxidized with periodic acid. At the other extreme are the plant cells, whose rigid cellulose cell wall constitutes an extracellular cell coat.

Most of the recent work on the extracellular matrix has been done with cultured mammalian cells—in particular, fibroblasts. Under culture conditions, normal (in contrast to malignant) cells must adhere to the substratum (glass or plastic) in order to move, divide, and differentiate. Cells are usually cultured in the presence of serum, which provides not only nutrients but also substances (glycoproteins, glycosaminoglycans) that help them stick to the substratum. The material that binds the cell to the substratum, the basal lamina (Fig. 6), can be isolated by treatment with detergents and hypotonic buffers (Keski-Oja *et al.*, 1981). The main constituent of the basal lamina of fibroblasts is a glycoprotein called "fibronectin." It has aroused considerable interest because its production is greatly decreased in malignantly transformed cells (reviewed by Yamada and Olden, 1978; Hynes and Yamada, 1982) and because the addition of this protein to cells facilitates their migration (Ali and Hynes, 1978). As pointed out by Vartio and Vaheri in a recent review (1983), fibronectin is involved in many important functions: cell anchorage, elaboration of the extracellular matrix, and chemotaxis. We shall see that it is involved in the extensive cell movements that occur during embryonic gastrulation. These different functions correspond to specific interactive domains in the fibronectin molecule. According to Gospodarowicz *et al.* (1980), the extracellular matrix (basal lamina) of fibroblasts is

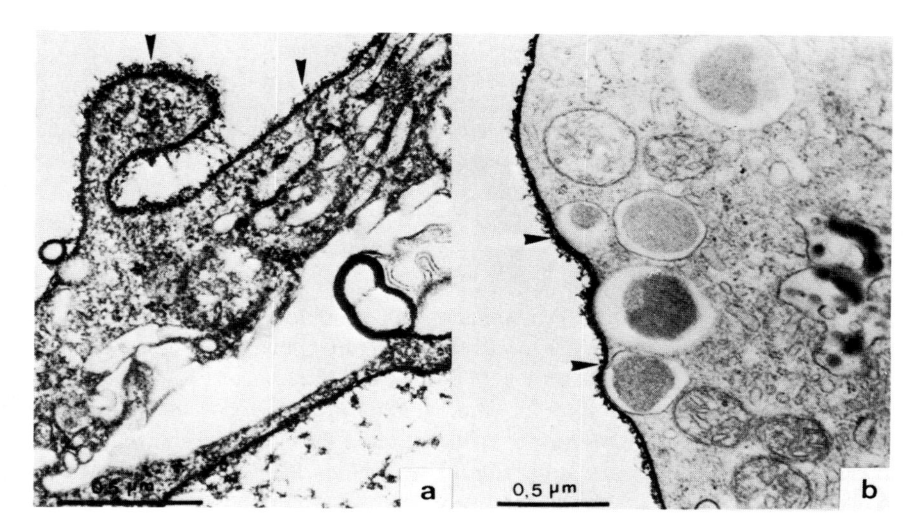


FIG. 5. Glycoealix (arrows) of sponge pinacocytes. (a) Ultrathin section stained under standard conditions, followed by ruthenium red staining. (b) Section stained with ruthenium red alone in order to demonstrate the glycocalix. [Courtesy of P. Willenz.]

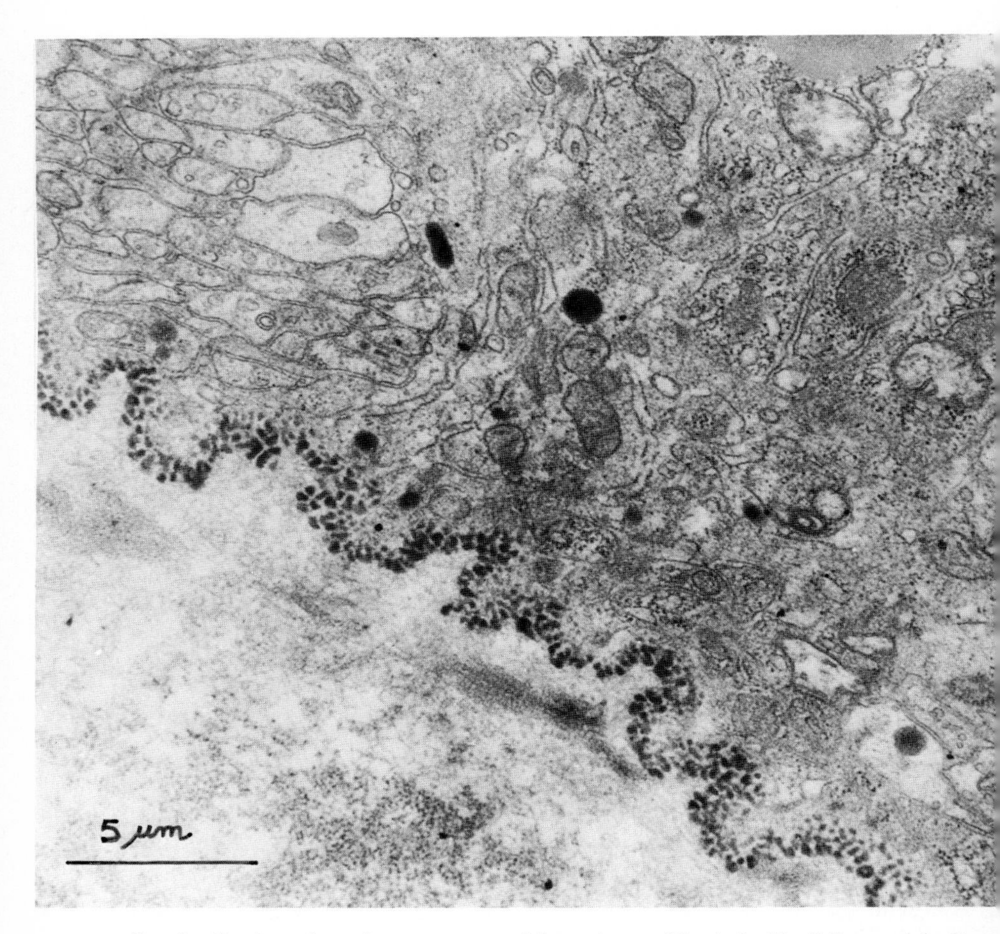


FIG. 6. Basal membrane in an enterocyte of the earthworm *Eisenia foetida*. [Micrograph by P. Van Gansen.]

made up of fibronectin, type IV collagen, glycosaminoglycans, and proteoglycans. The extracellular matrix favors proliferation by allowing the cells to adhere to the substratum. In epithelial cells, fibronectin is replaced by another high molecular-weight, glycoprotein, laminin, which is synthesized by the cells themselves and allows them to adhere to collagen (type IV, Terranova *et al.*, 1980). Endothelial cells synthesize both fibronectin and laminin, according to Gospodarowicz *et al.* (1981). The general function of these two glycoproteins seems to be the same: they establish a bridge between the cell and collagen, which is necessary for proliferation and differentiation of cells in culture (reviewed by Kleinman *et al.*, 1981). Similar but probably more complex processes take place *in vivo,* when cells of epithelial and mesenchymatous origin become separated by a basement membrane during organogenesis. The thick basement membrane results from the interaction of macromolecules (laminin, fibronectin, type IV collagen, proteoglycans, and glycosaminoglycans) and is synthesized and secreted by the two cell types (Brownell *et al.*, 1981; Hogan, 1981).

### C. Membrane Permeability to Water, Ions, and Small Organic Molecules

The necessity of the plasma membrane for the regulation of all exchanges between the cell and the surrounding world is obvious. Many of these exchanges ($O_2$, $CO_2$, $H_2O$, inorganic anions and cations, sugars, amino acids, nucleosides) are conerned with small molecules and will be discussed only briefly here. The transport of large molecules, in particular proteins, between the cell and the outer medium by endo- and exocytosis will be considered in a separate section.

Since the time of the pioneering work of Overton (1895) on plant cells, it has been known that cells swell when they are placed in a hypotonic medium and shrink (plasmolysis) when they are surrounded by a hypertonic medium. Thus, there are water exchanges between cell and medium. The same is true, of course, of animal cells. Overton also discovered that lipid solvents readily penetrate all cells, and this discovery was the first indication that lipids are major constituents of the plasma membrane.

Semipermeable membranes and liposomes (phosphatidylserine vesicles) provide good models for the plasma membrane. They can be submitted to accurate physicochemical analysis using classical mathematical equations. However, this discussion will be limited to an overview of a few aspects of membrane transport problems that are particularly important for cell biologists.

Besides the requirement of an isotonic medium, the cell needs a proper balance of various inorganic cations, namely, $Na^+$, $K^+$, $Ca^{2+}$, and $Mg^{2+}$. Cells live in media that are rich in sodium and poor in potassium. The reverse situation occurs with the cells themselves; a loss of potassium leads to arrest of cell division and a decrease in protein synthesis and energy (adenosine triphosphate, or ATP) production. Thus, it is essential for cells to retain their high potassium content. The necessity of the ouabain-sensitive sodium pump ($Na^+/K^+$–ATPase) is thus self-explanatory. In order to eliminate excess $Na^+$, which enters the cell by diffusion, and to accumulate $K^+$, the cell has to expend energy. Inhibition of ATP production by the mitochondria arrests the functioning of the pump, which leads to a swelling of the cell and, ultimately, cell death. If only passive diffusion of ions took place, as in artificial semipermeable membranes subjected to Donnan's equilibrium, life would be impossible. However, the amount of energy needed to maintain a proper ionic balance is very high. Active transport through the sodium pump ($Na^+/K^+$–ATPase) and passive diffusion due to the gradient distribution of $Na^+$ through the plasma membrane coexist.

One way to distinguish between the two processes is by the use of inhibitors: ouabain suppresses active transport, while the diuretic amiloride arrests a passive $Na^+$ transport pathway. For more details about the sodium pump, the reader is referred to reviews by Skou (1974) (who discovered the $Na^+/K^+$–ATPase) and by Glynn and Karlish (1975).

Still more important for the cell than the $Na^+/K^+$ balance is the maintenance of its free calcium ion concentration. These calcium ions play a fundamental role in the control of water permeability, the organization of the cytoskeleton, cell viscosity and motility, muscle contraction, and indeed, almost all—if not all—cell activities (reviewed by Carafoli, 1975; Sulakhe and St. Louis, 1980; Godfraind-De Becker and Godfraind, 1980). Striking examples of the importance of the $Ca^{2+}$ concentration can be found in the field of sea urchin embryology: culture of cleaving eggs in $Ca^{2+}$-free sea water leads to the dissociation of the blastomeres; treatment of unfertilized eggs with an agent that increases the free calcium content (the bivalent ion ionophore A23187) induces parthenogenetic activation (see Volume 2, Chapter 2 for details). It is generally believed that $Ca^{2+}$ influx is a passive process (involving an exchange between $Na^+$ and $Ca^{2+}$), while $Ca^{2+}$ efflux is an active process catalyzed by a $Ca^{2+}$ pump (an ATPase). However, cells have other ways of maintaining the free $Ca^{2+}$ at low and acceptable (around 0.1 $\mu M$) levels. One way is by compartmentation, the sequestration of calcium ions into intracellular organelles (mitochondria, endoplasmic reticulum, pigment granules and yolk platelets in eggs, sarcoplasmic reticulum in muscle). Thus, cells possess calcium reservoirs, and they can liberate sequestered calcium ions if needed. The quantitative significance of this compartmentation will become obvious when it is recalled that, according to Campbell *et al.* (1981), the free calcium ion concentration is only 0.3 $\mu M$ in the cell when that of the external medium is 1 m$M$. Calcium ion exchanges are further modulated by calcium-binding proteins, in particular by calmodulin (reviewed by Means *et al.*, 1982), which is involved in the majority of $Ca^{2+}$-regulated cell activities. Phosphorylation of membrane proteins by $Ca^{2+}$-calmodulin–dependent protein kinases could also affect permeability to various ions. Calmodulin ($M_r$ = 17,000) is a calcium acceptor which plays an important role in the regulation of protein secretion, endocytosis, capping of membrane receptors, locomotion, cell–cell interactions, and cell morphology and division. It is believed that these changes in cellular activities result from the phosphorylation of amino acid residues in key proteins by $Ca^{2+}$-calmodulin-dependent protein kinases. This belief is largely based on experiments where calmodulin inhibitors (trifluoperazine, calmidazolium) have been added to living cells. Interestingly, Owada *et al.* (1984) have shown that fibroblasts contain a calmodulin-binding protein which they called "caldesmon" ($M_r$ = 77,000). This endogenous calmodulin inhibitor is accumulated in regions of the cell where the actin filaments form bundles, suggesting that it controls the actin cytoskeleton.

Glenney and Glenney (1984) have recently reported that brain spectrin (fodrin) reacts with calmodulin (in contrast to the classical erythrocyte spectrin). This might provide another $Ca^{2+}$-calmodulin control of the interactions which take place between plasma membrane and underlying cytoskeleton.

Recent work has thrown an unexpected light on the molecular mechanisms of calcium release from its intracellular stores (review by Berridge and Irvine, 1984). We have seen that phospholipids are important constituents of the plasma membrane. Among them are phosphoinositides, which are hydrolyzed by the specific phospholipase C in inositol 1,4,5-triphosphate ($IP_3$) and diacylglycerol (reviews by Majerus *et al.*, 1984; Joseph, 1984). There is convincing evidence that $IP_3$ mobilizes calcium ions from their intracellular stores (mainly the endoplasmic reticulum); diacylglycerol activates the recently discovered proteinkinase C (review by Nishizuka, 1984). This enzyme, which requires $Ca^{2+}$ and phospholipids for full activity, is believed to control the $Na^+/H^+$ exchanges across the plasma membrane. If these exchanges lead to proton extrusion, the internal pH of the cell will increase. This, as we shall see in Chapter 5, is a very important signal for the initiation of DNA synthesis in the nucleus and for cell proliferation (Berridge, 1984; Berridge *et al.*, 1984).

As one can see, the three main protein kinases (dependent on c-nucleotides, $Ca^{2+}$-calmodulin, and $Ca^{2+}$-phospholipid) are all involved in the control of major cellular activities. They phosphorylate serine and threonine (and less frequently tyrosine) residues in a number of target proteins. Phosphorylation changes their conformation and ultimately the cell morphology and activities (motility, cell–cell interactions, proliferation, etc.).

Amino acid transport in mammalian cells is a very complex process (reviewed by Christensen, 1979; Kilberg, 1982; Shotwell *et al.*, 1983). Four main systems have been recognized: system A, for valine and glycine, is dependent upon the presence of sodium ions; system L, for branched amino acids, is $Na^+$ independent; system ASC is active in the transport of alanine, serine, and cysteine; finally, system N is involved in the transport of glutamine, asparagine, and histidine. In cultured liver cells (hepatocytes), no fewer than eight different systems (six of them being $Na^+$ dependent) cooperate in amino acid transport, according to Kilberg (1982). Systems A and N display an ''adaptive'' control: they increase the uptake rate when the cells have been maintained in the absence of extracellular amino acids for some time.

Glucose transport is also $Na^+$ dependent, but usually involves the intervention of a glucose ''transporter'' or ''carrier'' (functionally similar to the permeases of bacterial and yeast cells). This is shown by the fact that cytochalasin B, besides exerting its well-known inhibitory effects on actin polymerization, greatly reduces the uptake of hexoses. These two effects of cytochalasin B are completely independent of each other, and binding of the drug to a glucose transporter is now a well-established fact (Sogin and Hinkle, 1980; Mookerjee *et al.*, 1981;

Cuppoletti *et al.*, 1981). This glucose transporter is a 200,000-dalton glycoprotein containing as much as 15% sugar residues.

Besides the natural carriers, there are chemicals that are capable of transferring ions through the plasma membrane in a fairly specific way. These chemicals belong to two different categories: ionophores and channel formers. As shown by De Kruiff *et al.* (1974), the mycostatic drugs amphotericin and mystatin interact with the cholesterol present in the plasma membrane, boring holes 8 Å in diameter in the membrane and greatly increasing cell permeability to water and halides. The subject of ionophores and ion channel borers has been reviewed by Ovchinnikov (1979), Lakshminarayanaiah (1979), and Läuger (1980). Ion carriers (ionophores) transport ions back and forth from the outer medium into the cell because of their solubility in the lipid bilayer. They thus tend to equilibrate the ion concentration between medium and cell. The classical examples are the $K^+$ ionophores valinomycin and nigericin, which exchange potassium ions for protons ($H^+$). The exchange is electrogenic in the case of valinomycin (it thus affects the membrane potential) and electroneutral for nigericin. In work with isolated mitochondria, valinomycin is often used to collapse the membrane potential ($\Delta\ \psi$) and nigericin to collapse the proton gradient ($\Delta$ pH), which are the two elements of the proton motive force involved in energy production. Monensin is another monovalent ion ionophore, although it does not discriminate between $Na^+$ and $K^+$ ions. The previously mentioned A23187 is an extremely useful divalent ion ($Ca^{2+}$ and $Mg^{2+}$) ionophore. On the other hand, gramicidin, an antibiotic, opens channels in the membrane that are accessible simultaneously from both sides. The diameter of the pores created by gramicidin for potassium ions is 2.6–3.0 Å. These channels are gated, and changes in membrane voltage open or close the gates. Ionophores and channel formers are useful for the cell biologist who wishes to modify the ion distribution and the membrane potential of the living cell. It is necessary, however, to consider the side effects. For instance, A23187 increases the $Ca^{2+}$ content of the cells, but there is a simultaneous loss of $K^+$ ions (Burgess *et al.*, 1979).

### D. Endocytosis and Exocytosis

It has been known since the days of E. Metchnikoff that neutrophils and macrophages are capable of phagocytosis—ingestion of bacteria or cell debris followed by their digestion in vacuoles. The same process is used by many protozoa (amebas, for instance) for feeding. Pinocytosis, the uptake of proteins in solution by cells in culture, was discovered by Lewis in 1931. The cells "drink" the protein solution, and it is possible, in this way, to make them absorb foreign substances (for instance, horseradish peroxidase, which can be easily detected cytochemically) dissolved in this protein-containing medium. Pinocytosis was carefully studied in amebas by Holter and Marshall (1954) and by Chapman-Andresen and Holter (1955), who showed (among other things) that the uptake of glucose into these protozoa is increased considerably if it is dissolved in a 10% protein solution. At the same time, the author and his colleagues showed that pancreatic ribonuclease ($M_r$ 13,000) readily penetrates ascites tumor cells, amphibian eggs, onion root tips, amebas, and flagellates. This leads to a decrease in RNA content and in protein synthesis activity in the ribonuclease-treated living cells or organisms (reviewed in "Biochemical Cytology"). It has been found that it is possible to lyse the pinocytic vacuoles containing macromolecules by an osmotic shock without killing the cells (Okada and Rechsteiner, 1982). Phagocytosis and pinocytosis are two forms of endocytosis, the penetration into cells of solids or liquids. The general mechanism of uptake is, first, the binding of the material to be ingested by the plasma membrane. This is followed by a localized invagination of the membrane and, finally, by the formation of a vacuole (phagosome or pinosome) now generally called an "endosome." Phagosomes and pinosomes often fuse with lysosomes, forming digestive vacuoles, which will be discussed later. If digestive vacuoles are formed, their contents are degraded by the digestive enzymes present in the lysosomes. If such a fusion does not occur, the membrane surrounding the endosomes can be recycled, i.e., reintegrated into the plasma membrane.

Exocytosis is the reverse process, namely, the excretion of the contents of a vacuole to the outside of the cell. This phenomenon is particularly striking in the case of the secretory vacuoles, which, in the exocrine pancreas, contain the digestive enzymes. During exocytosis, the membrane of the secretory vacuole must first fuse with the plasma membrane; the contents of the vacuole are then extruded outside the cell. Thus, in both exocytosis and endocytosis, the plasticity of the plasma membrane and its ability to fuse with other cell membranes are of the utmost importance.

The mechanisms that allow the uptake or extrusion of large molecules in cells have been the subject of reviews by Pastan and Willingham (1983), Helenius *et al.* (1983), Steinman *et al.* (1983), and M. S. Brown *et al.* (1983).

A very favorable system for the study of endocytosis is the growing oocyte of the frog (Wallace and Jared, 1976) or hen (Pearse, 1978), since their enormous increase in weight and volume is essentially due to the uptake of a single protein (vitellogenin), which is present in the blood capillaries of the female (for details, see Volume 2, Chapter 2). Endocytosis in the hen oocyte is so intense that 1 gm of yolk proteins is taken up by one oocyte in a single day. Wallace and Jared (1976) observed that, in frog oocytes, vitellogenin is taken up and sequestered 20–50 times faster than other nonyolk proteins. They concluded that the oocyte possesses specific receptors for vitellogenin, while the uptake of the other proteins results from "simple" (fluid phase) pinocytosis (Fig. 7). This distinction between receptor-mediated and simple nonspecific pinocytosis seems to be valid for many, if not all, cells. The vast majority of the recent papers on the subject deal with receptor-mediated pinocytosis, which will be discussed later. The

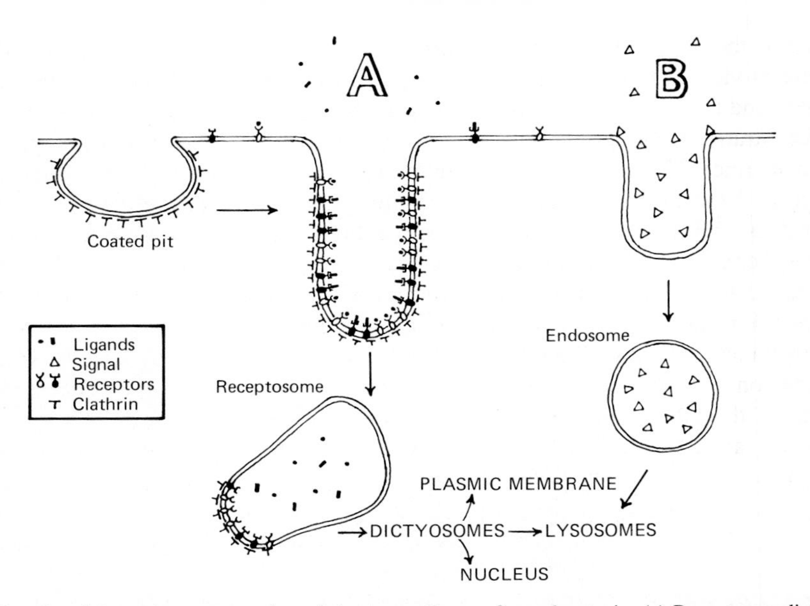


FIG. 7. Schematic representation of the two pathways for endocytosis. (a) Receptor-mediated endocytosis through clathrin-coated pits and receptosomes; (b) nonspecific pinocytosis. [Original drawing by P. Van Gansen, based on several papers.]

proteins of the pinocytic vacuoles and those of the rest of the cell membrane are identical (Mellman *et al.,* 1980). Pinocytosis can be further subdivided into micro- and macropinocytosis. In the latter, the cell surface is greatly increased by projections called "microvilli." They may be so developed in cells in which absorption of intact proteins is very intense, as in amphibian oocytes (Fig. 8), or the brush borders of intestinal or kidney cells, that they are called "macrovilli." Their extension or retraction, which can easily be observed with the scanning electron microscope (Fig. 9), greatly increases or decreases the cell surface. Large pinocytosis vacuoles are often associated with their bases, as can be seen in Fig. 10. They are surrounded by actin filaments that certainly play an important role in endocytosis. It has been shown that, in cultured hepatocytes, endocytosis is inhibited by cytochalasin B, but not by colchicine. This suggests that fluid-phase endocytosis requires the contraction of microfilaments, but not the integrity of microtubules; however a study by Starling *et al.* (1983) shows that colchicine and other microtubule-disrupting agents inhibit pinocytosis in rat visceral yolk sacks; the authors conclude that the microtubules play a significant role in pinocytosis. It is possible that the importance of this role varies from cell to cell.

Microvilli, on the other hand, are covered by extensions of the plasma membrane, but their axis is made up of "core" filament bundles. The main component of these filaments is actin, which is associated with the calcium-binding proteins villin, fimbrin, and calmodulin. Villin crosslinks actin in the absence of $Ca^{2+}$ and thus favors the formation of bundles of microfilaments. At low $Ca^{2+}$ concentrations, the actin filaments become shorter. This interesting protein has two distinct actin-binding sites (Bretscher and Weber, 1980a; Glenney and Weber, 1980; Craig and Powell, 1980; Bretscher *et al.* 1981; Glenney *et al.*, 1981a,b). Fimbrin ($M = 60{,}000$) plays a similar role in the microvilli of sea urchin eggs (De Rosier and Censullo, 1981) and other cells. This protein is believed to control actin polymerization and depolymerization in conjunction with villin (Bretscher and Weber, 1980b; Glenney *et al.*, 1981b). Finally, the involvement of calmodulin ($M_r = 16{,}000$) in $Ca^{2+}$-dependent extension and retraction is indicated by the fact that one of its inhibitors, trifluoperazine, induces the disappearance of the microvilli (Osborn and Weber, 1980). More will be said about these and other actin-binding proteins when we deal with the

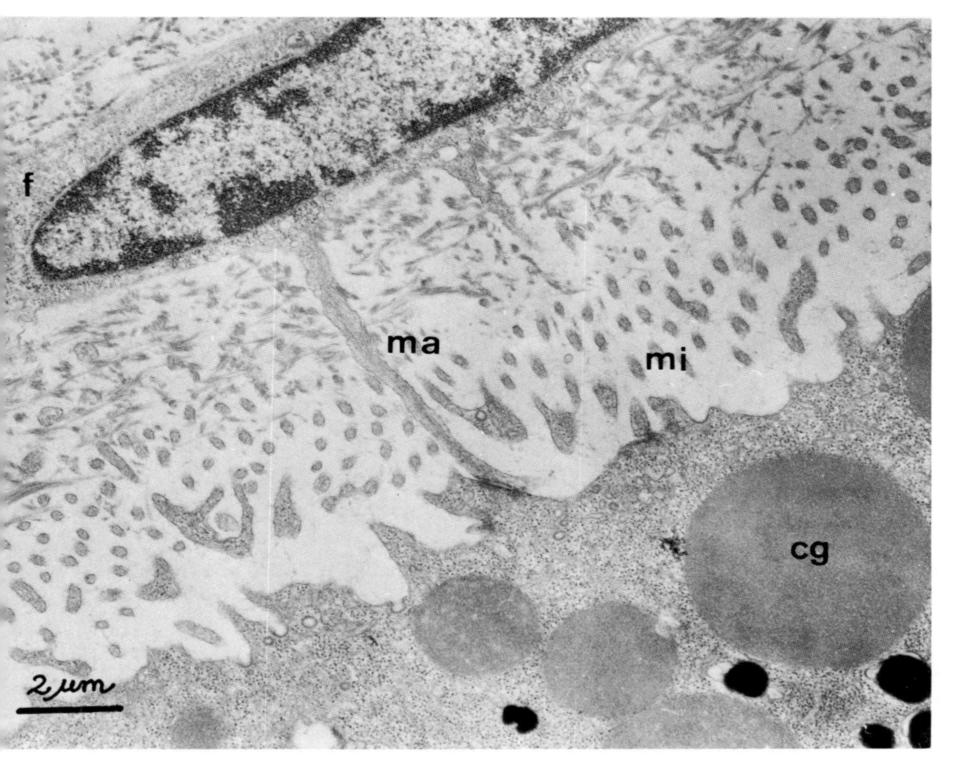


FIG. 8. Ovarian follicle at mid-oogenesis in *Xenopus laevis*. f, follicle cell; cg, cortical granule in the oocyte cortex; ma, macrovilli; mi, microvilli. [Micrograph of G. Steinert.]

FIG. 9. Surface of a *Xenopus* oocyte undergoing maturation. Only the maturation spot (M) is devoid of microvilli. [Scanning electron microscope photograph courtesy of L. De Vos.]

cytoskeleton. However, it should be emphasized here that extension and retraction of microvilli and, therefore, macro- and micropinocytosis are ultimately controlled by local changes in the free calcium ion concentration, which is modulated by calcium-binding proteins.

Microvilli are unstable structures in the sense that they can extend and retract depending on the length of their axis; it is believed that the elongation of microvilli that takes place soon after fertilization in sea urchins is due to actin polymerization. However, the mechanism of retraction in isolated intestinal brush borders is still open to discussion (Harris, 1983): it was generally believed that it results from the solvation of the microvilli cores by the actin-severing protein villin, but it has been proposed that retraction is due to actomyosin contraction; as in muscle, this contraction would be regulated by a calmodulin-dependent myosin light-chain kinase (Burgess, 1982; Keller and Mooseker, 1982). Accord-

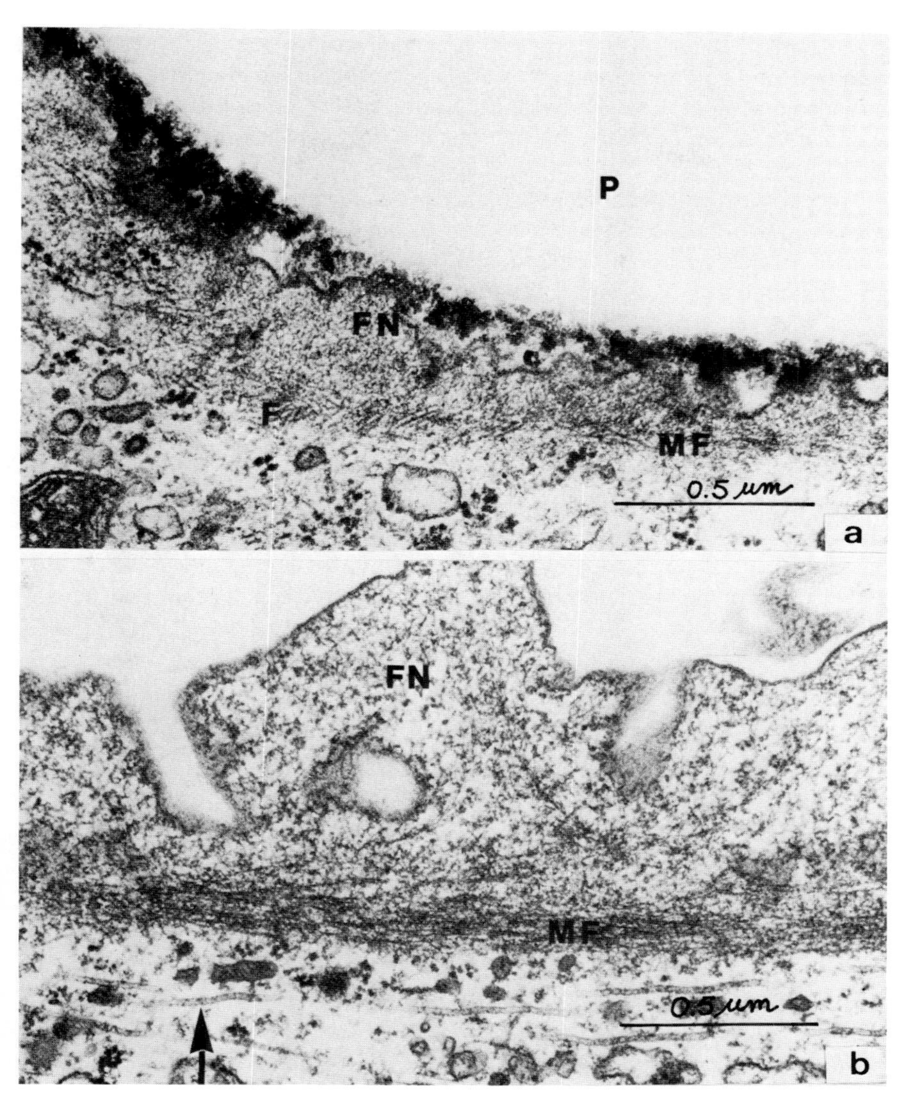


FIG. 10. Relationships between endocytosis and actin microfilaments in a macrophage. (a) Phagocytosis of a polystyrene particle P (dissolved by the techniques used for fixation and embedding). FN, filamentous network; MF, oriented microfilaments; F, 10-nm filaments. (b) Subplasmalemnal microfilaments (MF) at the substratum-attached surface of a macrophage. The arrow shows microtubules associated with organized bundles of microfilaments. [Reaven and Axline (1973). Reproduced from *The Journal of Cell Biology,* 1973, **59,** 12–24, by copyright permission of The Rockefeller University Press.]

ing to Hirokawa *et al.* (1983), brush border contraction would result mainly from the active sliding of actin and myosin filaments (again, as in muscle). It should be added that all of these experiments have been done on isolated brush borders and that we do not know the mechanisms that operate when microvilli retract *in vivo*.

The cell surface, therefore, is a highly dynamic structure. This conclusion is reinforced by results obtained on receptor-mediated endocytosis, which has been studied in cells such as fibroblasts and hepatoma cells, where the addition of growth hormones, for example, insulin and epidermal growth factor (see Chapter 5), induces cell proliferation. Many proteins bind specifically to receptors present on the cell surface of many cell types. The ligand–receptor complexes are then internalized by endocytosis according to mechanisms that are still being studied.

As shown in Fig. 11, certain regions of the cell surface are coated with small electron-dense granules. During receptor-mediated endocytosis, the coated regions invaginate to form bristle-coated pits. They are composed of 108 molecules of a 180,000-dalton protein called "clathrin" (associated with a few minor proteins) and are surrounded by a lipid vesicle (Goldstein *et al.*, 1979; Pearse, 1980). Clathrin has the remarkable capacity to undergo spontaneous polymeriza-

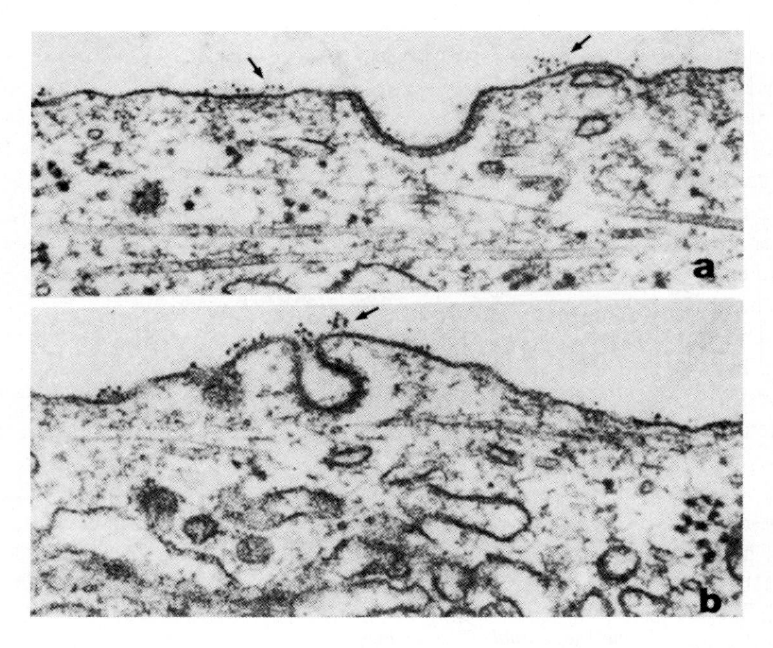


FIG. 11. Coated pit (a) and internalization (b) of a ferritan-labeled protein (arrow) in the coated pit. [Bretscher *et al.* (1981).]

tion *in vitro* at pH 6.5 with the formation of basketlike lattices or cages that are very similar to the coated vesicles seen in cells under the election microscope (Bloom *et al,* 1980; Nandi *et al.,* 1980). At the molecular level, these lattices comprise 12 pentagons and a variable number of hexagons. The 8.6 S clathrin present in the coated vesicles is composed of three heavy chains (180,000 daltons) and three light chains (33,000–36,000 daltons); one heavy chain is in contact with one light chain. Formation of cages is spontaneous (Kirchausen and Harrison, 1981). There are about 1000 coated pits per cell, covering 1–2% of the cell surface.

It is now widely accepted that the specific endocytosis of ligands bound to receptors occurs through the coated pits, which may perhaps constitute molecular filters that selectively exclude certain proteins (Bretscher *et al.,* 1980). In contrast to micropinocytic vacuoles (micropinosomes), coated pits are not surrounded by actin microfilaments (Willingham *et al.,* 1981c). There is some evidence that calmodulin plays a role in the formation of the clathrin-coated pits, and it has even been suggested that endocytosis always begins with a transitory calmodulin-mediated influx of calcium ions (Salisbury *et al.,* 1980). More recently, some evidence has been presented that the clustering of receptor-bound ligands in the clathrin-coated regions of the membrane and in the coated pits is due to transglutaminase (a protein crosslinking enzyme that binds glutamine to lysine residues). Inhibitors of this enzyme (amines, bacitracin, dansylcadaverine, etc.) prevent the clustering of ligand–receptor complexes in coated pits and their subsequent subsequent internalization (Dickson *et al.,* 1981a; Tucciarone and Lanclos, 1981; Chuang, 1981).

As can be seen, the entire subject of receptor-mediated endocytosis is still the subject of intensive studies, and it is too early to draw a simple and schematic conclusion. Even more obscure is the fate in the cytoplasm of the internalized material, because clathrin can be detected by immunocytochemistry or biochemical methods at sites other than the coated pits and coated vesicles of the plasma membrane. Clathrin-coated structures have been observed in Golgi vesicles and in the whole GERL (Golgi, endoplasmic, reticulum, lysosomes) system (Willingham *et al.,* 1981a; Keen *et al.,* 1981; Kartenbeck *et al.,* 1981). Until recently, it was thought that endocytosis results from the invagination of the coated pits into closed ''coated vesicles'' that leave the membrane and move into the cytoplasm (Roth and Berger, 1982): in this way, the ligand–receptor complex would be internalized. However, new information speaks against such a simple model: there is now good evidence for the coated pit-receptosome pathway proposed by Willingham and Pastan (1980, 1983), Willingham *et al.* (1981b), and Pastan and Willingham (1981). As shown in Fig 12, internalization would take place in an uncoated vesicle, the receptosome, where the ligand–receptor complex would be protected against immediate destruction by the lysosomal hydrolytic enzymes. Clathrin, actin, myosin, and tubulin are absent

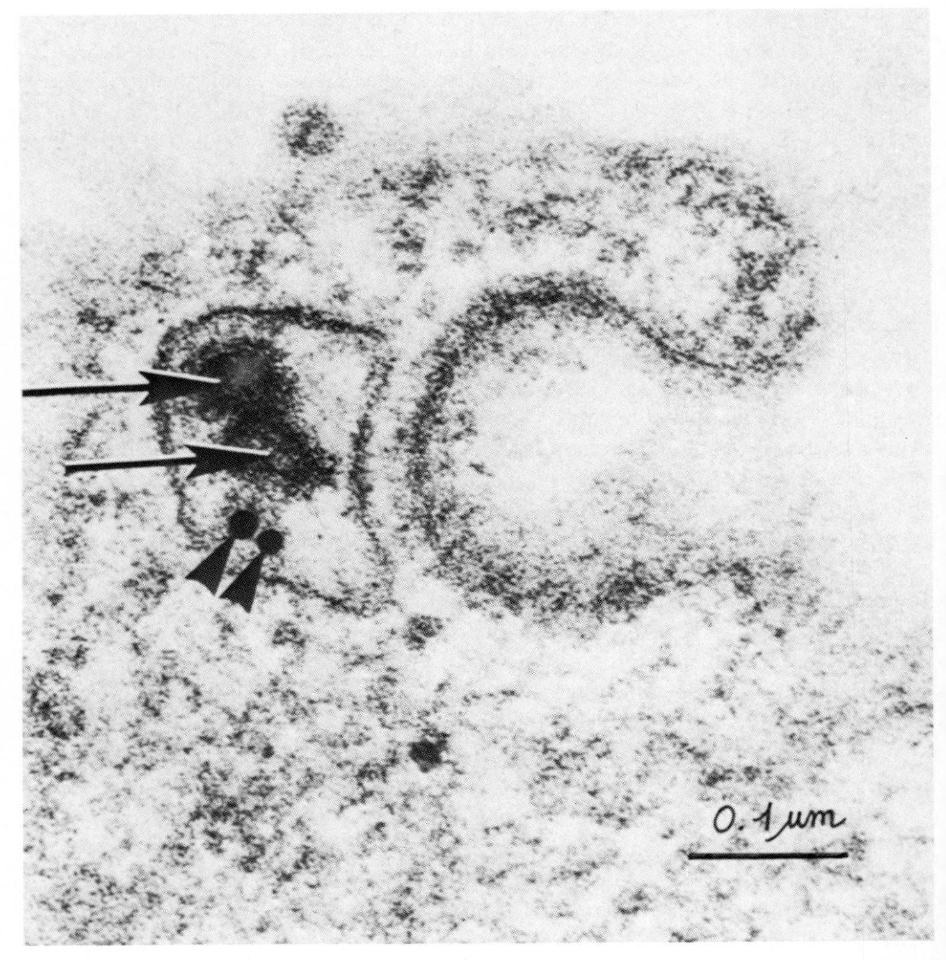


FIG. 12. Receptosome containing both internalized virus particles (arrows) and internalized protein molecules labeled with colloidal gold (arrowheads). In this fibroblast, the receptosome lies close to a coated pit. This is an example of "co-internalization." [Dickson *et al.* (1981b). Reproduced from *The Journal of Cell Biology,* 1981, **89,** p. 132, by copyright permission of The Rockefeller University Press.]

from these receptosomes, whose function is to distribute the internalized ligand toward the Golgi apparatus and the acid phosphatase–rich membrane system GERL (discussed later in this chapter). Finally, the ligand would generally be segregated into lysosomes. In summary, as proposed by Willingham *et al.* (1981d) and Dickson *et al.* (1981b), the protein that is to be internalized by receptor-mediated endocytosis is first concentrated in the coated pits by transglutaminase activity. It then moves into uncoated receptosomes and ends up in large lysosomes and small Golgi vesicles. In a number of elegant experiments by Wehland *et al.* (1981), antibodies against clathrin were injected into living fibroblasts. The authors found that receptor-mediated endocytosis was not suppressed. The coated pits are thus stable elements attached to the cell surface; there are no coated vesicles in fibroblasts.

However, this last conclusion is not accepted by all [see, e.g., the contrasting reviews by Helenius *et al.* (1983) and Pasten and Willingham (1983)]. Everyone agrees that endosomes (phagosomes, pinosomes, receptosomes) exist; their low pH allows the dissociation of ligand–receptor complexes. After entry into the endosomes, some ligands are delivered to the surrounding cytosol. Others go to the lysosomes (after undergoing modifications in the Golgi apparatus), where they are broken down. Finally, some ligands and many receptors are returned to the cell surface. Thus, there is a bidirectional flow of membranes between the plasma membrane and the intracellular organelles (lysosomes, Golgi cisternae). This allows the recycling of important plasma membrane proteins (receptors, in particular) [reviewed by Steinman *et al.* (1983) and M. S. Brown *et al.* (1983)]. Within 1 min, the ligand–receptor complexes are detectable in receptosomes deriving from coated pits; within a few minutes, all kinds of endosomes are acidified (pH between 4 and 5) by an ATP-driven proton pump (Galloway *et al.*, 1983; Yamashiro *et al.*, 1983). The existence of saltatory movements in endosomes (and lysosomes) (Herman and Albertini, 1983) that allow them to move toward the Golgi region in 10–20 min is also accepted by all. But there remains one point of disagreement: the reality (or not) of coated endocytic vacuoles (coated vesicles). Willingham and Pastan (1983) have provided some EM evidence for the view that the coated structures remain attached by a "neck" to the plasma membrane. If so, there would be no free clathrin-coated vesicles in fibroblasts during endocytosis. The authors point out that receptosomes can be identified under the electron microscope as structures (2000–4000 Å in diameter) surrounded by a continuous uncoated membrane with a frilled appearance on one edge; the center is apparently empty except for a sirgle intraluminar vesicular structure. In addition, Dickson *et al.* (1983) have isolated from cell homogenates particles that are rich in cholesterol, poor in plasma membrane enzyme markers, and devoid of both clathrin and lysosomal enzymes: these particles (or rather, vesicles) are believed to be the receptosomes (endosomes) [Dickson *et al.* (1983)]. On the other hand, Pfeffer *et al.* (1983) claim to have isolated coated vesicles made up of at least six different proteins, including clathrin, tubulins, and a phosphoprotein that may link the vesicles to the cytoskeleton. According to Stone *et al.* (1983) and Forgac *et al.* (1983), such clathrin-coated vesicles possess an ATP-driven proton pump that acidifies their content. Finally, Petersen and Van Deurs (1983) refute the receptosome concept, suggesting that the coated pits pinch off from the plasma membrane and form coated vesicles. Willingham and Pastan (1985) have recently clarified the situation by pointing out that receptosomes and endosomes are the same thing.

The present disagreement may be more apparent than real, since Dickson *et al.* and Pfeffer *et al.* worked with different materials (human carcinoma cells and bovine brain, respectively). If ultrathin sections do not cut through the necks attaching the coated structures to the cell surface, these structures will look like coated vesicles. Time will certainly settle a disagreement that, after all, is not of paramount importance compared to the great progress made, in a few years, in our understanding of the molecular mechanisms of endocytosis. A few years ago, it was not realized that endosomes (whether coated or not) are a kind of circulatory system for the cell; recycling receptors came as a surprise to many cell biologists. Interestingly, this complex endocytic system is sensitive to small ions: depletion of fibroblasts from their intracellular potassium ion store inhibits the formation of coated pits and receptor-mediated endocytosis; addition of KCl to the $K^+$-depleted cells quickly restores endocytic activity (Larkin *et al.*, 1983). We shall see that monovalent ions, as well as $Ca^{2+}$, play an important role in the intracellular traffic that takes place between the various cell organelles. Both clathrin-coated vesicles and endosomes have an acid pH (about 5.4). Acidification of their content is due to the presence in their membranes of an ATP-dependent proton pump (Van Dyke *et al.*, 1984). Addition of monensin, which induces a $Na^+/H^+$ exchange, dissipates the pH gradient within 2 min (Tycko *et al.*, 1983).

Binding of a hormone (insulin, for instance) to its receptor is a complex phenomenon, and the subsequent internalization of the hormone–receptor complex often has important consequences for the cell (stimulation of DNA synthesis in fibroblasts, for instance). The insulin receptor (reviewed by Cobb and Rosen, 1984) is made of four subunits ($\alpha_2$ and $\beta_2$). The $\alpha$ subunits ($M_r = 135{,}000$) bind the hormone; the $\beta$ subunits ($M_r = 195{,}000$) have a protein kinase activity, which is stimulated by insulin binding. This kinase activity phosphorylates tyrosine residues in the receptor itself and is thus a tyrosine-specific protein kinase (Shia and Pilch, 1983; Roth and Cassell, 1983; Kasuga *et al.*, 1983). We shall see, in subsequent chapters, that similar mechanisms operate for many, if not all, growth factors.

The cell surface is covered with a large number and variety of receptors. In order to play their role in endocytosis, they must accumulate in coated pits. This means, as pointed out by Bretscher and Pearse (1984), that the receptors must compete for sites in the coated pits and that "getting into a coated pit is a competitive business for the different receptors."

In hepatocytes (and in other cells), the insulin–receptor complexes are internalized within 30–60 min. This results necessarily in a decrease in the number of receptors on the cell surface and in a decrease in the response of the cells to the hormonal stimulus (so-called down regulation). But after 4 hr, the internalized receptors have been recycled to the cell surface and the cells again respond fully to a new insulin stimulation.

In summary, the successive steps of receptor-mediated endocytosis are the following [see the review by Willingham and Pastan (1984) for details]: clustering and internalization of the receptor-bound ligands in clathrin-coated pits, formation of endocytic vacuoles from coated pits, movement of vacuoles by saltatory motion in the Golgi system which is provided with small clathrin-coated pits where endocytosed materials are concentrated and directed towards the lysosomes.

There is also a traffic out of the Golgi system to the cell membrane of materials destined for *exocytosis,* a process about which little will be said here since it is closely linked to protein secretion, a process in which the endoplasmic reticulum and the Golgi system are closely involved. From a biochemical viewpoint, it is very likely that in endocytosis, the initial event is an increase in free $Ca^{2+}$ ions. This is shown dramatically, as we shall see in Chapter 7, by the sudden release of the contents of the cortical granules of sea urchin eggs when a surge in $Ca^{2+}$ concentration takes place as a result of fertilization or parthenogenetic activation. It has further been shown that the $Ca^{2+}$ ionophore A23187 induces exocytosis in lymphocytes. There is also ample evidence that in many cells, exocytosis is inhibited by cytochalasin B, indicating an involvement of actin microfilaments in the process (Wilkins and Lin, 1981). Thus, so far as is currently known, the basic biochemical mechanisms for exocytosis and endocytosis seem to be very similar, if not identical.

### E. Communication between Cells

Cells, like human beings, are, in general, gregarious organisms. Very few cells, such as blood cells, spermatozoa, and eggs, prefer solitude. The great majority of our cells live, like so many of us, in a crowded environment. Persons exchange words to communicate with their neighbors. Cells must also communicate with each other—sending and receiving signals that are difficult to understand. When a few cells are inoculated in a proper culture medium, they first grow with the vigor of youth. However, after a few days, the culture reaches "confluence." The overcrowded cells come into close contact with other cells, no longer move and divide, and must interact with their intimate neighbors. One way of interacting is to fuse with neighboring cells, but this is a rare occurrence unless there is human intervention. Cells also adhere to neighboring cells, but this occurs only when cells belonging to a given family make contact with others from the same or a closely related cell line. Foreign cells are rejected by a phenomenon called "sorting out"; cells of a given cell line are definitely xenophobic. Among cells belonging to the same family, junctions can be established. These junctions may be tight, allowing fast and tight adherence between neighboring cells, or they may be leaky (gap junctions); the latter allow the passage of molecules of moderate size from one cell to another.

Thus, the cell surface is involved in specific recognition processes. The cell junctions and their specific adhesion (or rejection) will be examined below.

### *1. Intercellular Junctions*

Electron microscopy indicates that four types of junctions between neighboring cells can be distinguished: desmosomes, intermediary (adherence type) junctions, tight junctions, and gap junctions. (Fig. 13).

Desmosomes link epithelial cells. There is an intercellular gap filled with mucoproteins and glycoproteins in their center, from which many tonofilaments diverge (or converge toward the gap?). The material present in the desmosomal interstitial region has been isolated by Gorbsky and Steinberg (1981), who called it "desmoglea." Its main chemical constituents are glycoproteins. The tonofilaments were isolated by Franke *et al.* (1978). They belong to the general family of intermediate-sized filaments, which will be discussed later, and are characterized by the presence of proteins that are similar to the cystine-rich keratin of the skin. There is a whole family of cytokeratins, according to papers by Franke (1981) and Franke *et al.* (1981a–d). For example, the cytokeratins that form the tonofilaments of the desmosomes in parenchymental cells of the liver and the pancreas are different from the "prekeratins" of ectodermic cells. According to Maltoltsy *et al.* (1981), filaments of prekeratin are composed of three major and seven minor proteins, which would stabilize the former. Ramaekers *et al.* (1980) indicate that the desmosomes of the lens do *not* contain cytokeratins, but another protein characteristic of intermediate-sized filaments, vimentin. More recently, Wiche *et al.* (1983) isolated a new protein (plectin, $M_r = 300{,}000$) from desmosomes; it seems to be involved in the formation of cell junctions. While tonofilaments may be very different chemically, they have the same function: to strengthen the bonds between neighboring cells.

Tight junctions (reviewed by Pinto da Silva and Kachar, 1982) also maintain adjacent cells in very close contact; there is no desmoglea to separate them. These junctions constitute a barrier to the freedom of mutual exchanges (see Farquhar and Palade, 1963, for one interpretation of the electron micrographs). Experiments showing the lack of lateral diffusion of molecules between adjacent cells of the intestinal mucosa have confirmed that the role of tight junctions is—as their name implies—to hold cells firmly together and to prevent exchanges between them.

In sharp contrast to desmosomes and tight junctions, gap junctions allow the passage of ions or molecules smaller than 1000 daltons from one cell to another. In ultrathin sections, the electron microscope shows the existence of gaps (about 2 nm) between adjacent cells (Fig. 14). These gaps are filled with particles associated with hexagonally ordered proteins present in the membranes of the two adjacent cells. Gap junctions can easily be recognized in freeze-fractured membranes. They are characterized by a clustering of intramembrane particles on both sides of the junction (Fig. 15). Zampighi *et al.* (1980) and Peracchia (1980) have reviewed the morphology and molecular structure of the gap junctions.

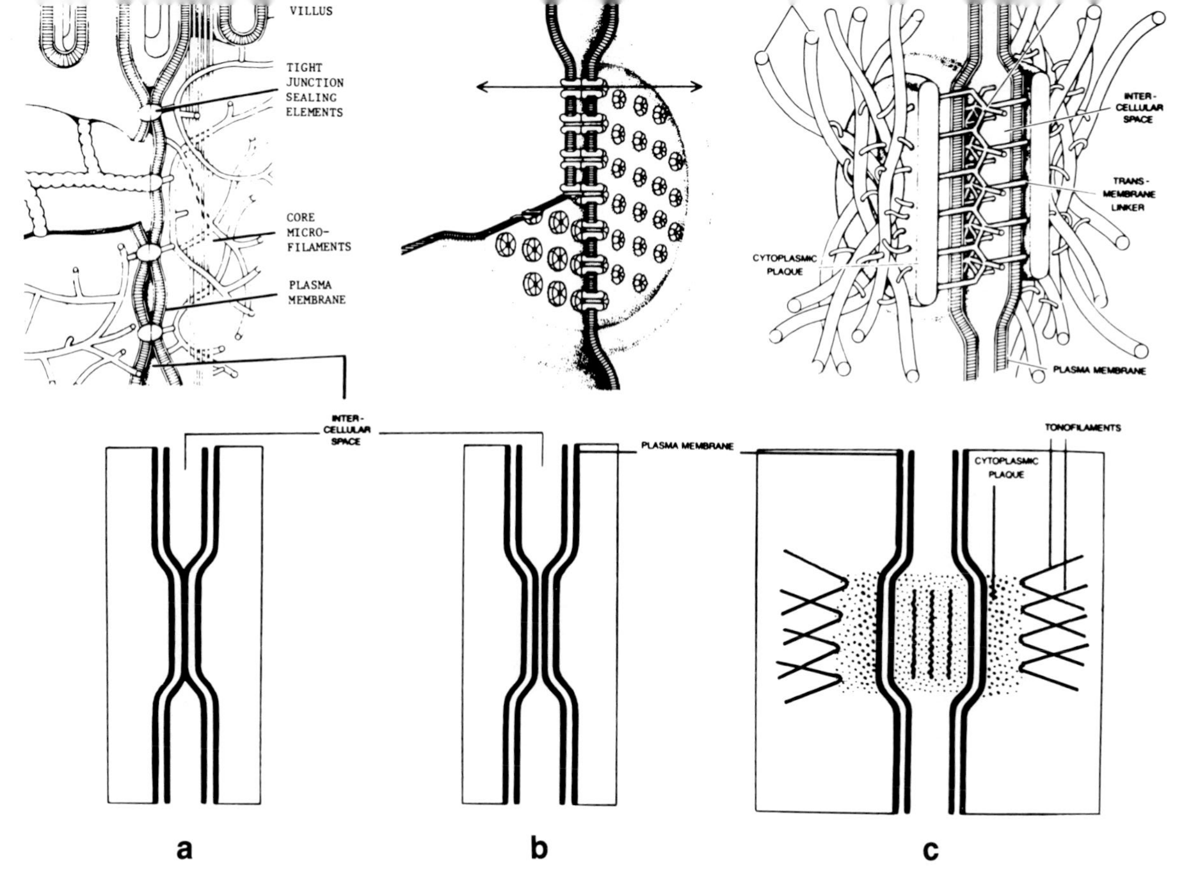


FIG. 13. Schematic representation of cell-to-cell junctions. (a) Tight junction; (b) gap junction; (c) desmosome. (*Top*) Three-dimensional models describing the three types of junctions. [Staehelin and Hull (1978).]

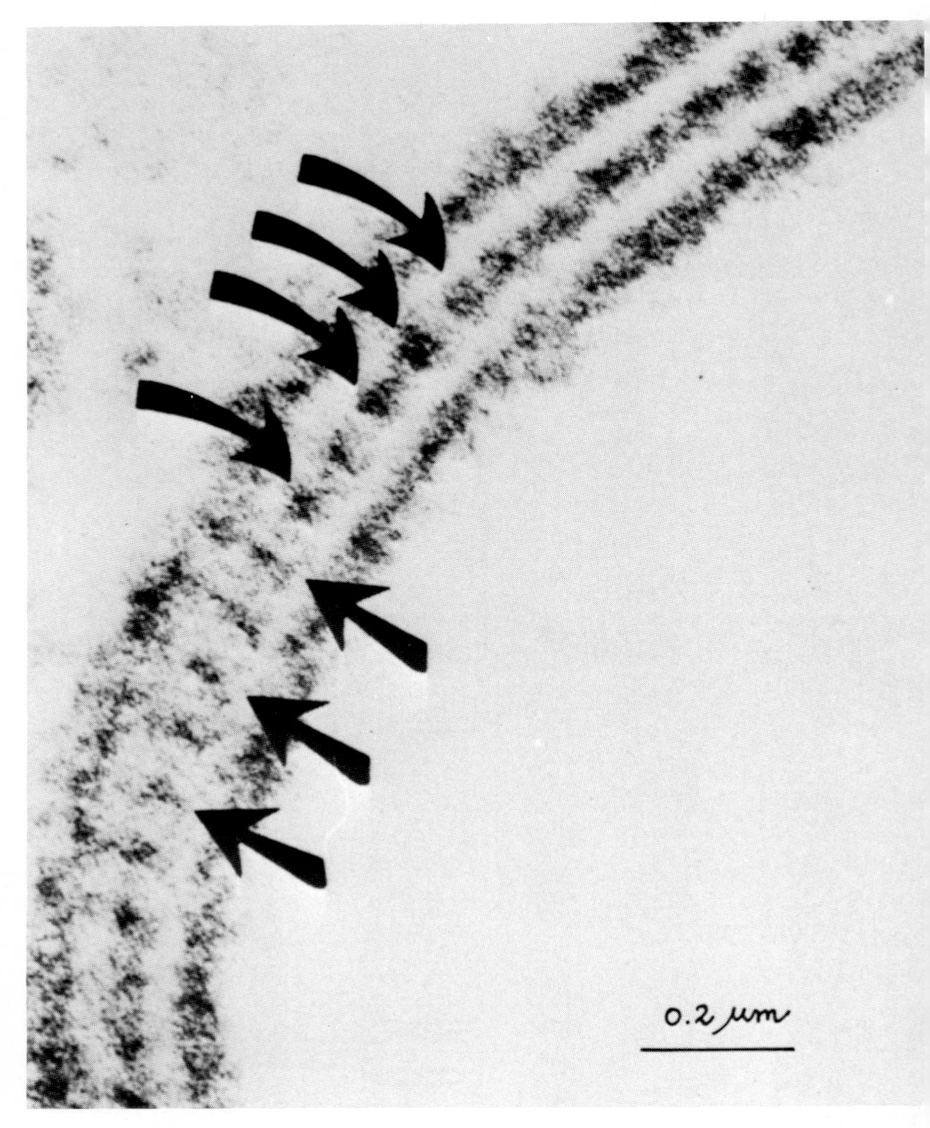


FIG. 14. EM section through gap junctions in a bile canaliculi-purified membrane fraction. Arrows, intermembrane proteins. [Zampighi *et al.* (1980). Reproduced from *The Journal of Cell Biology,* 1980, **86,** p. 193, by copyright permission of The Rockefeller University Press.]

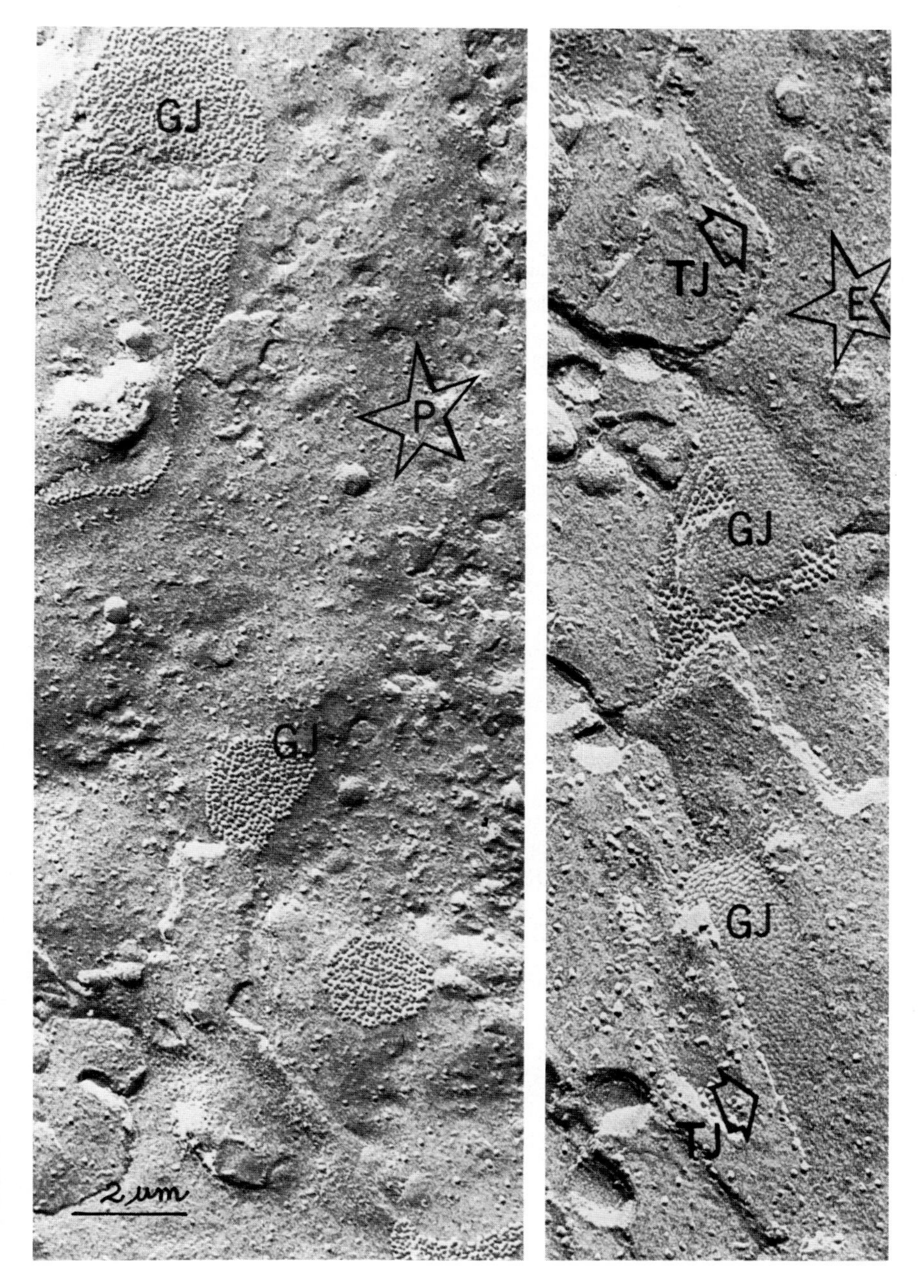


Fig. 15. Gap junctions (GJ) and tight junctions (TJ) as seen by freeze-etching electron microscopy (endothelial cells). E, external face; P, plasmic phase of the plasma membrane. [Hüttner and Peters (1978).]

Much has been learned about the molecular organization and chemical composition of the gap junctions since Goodenough (1974) succeeded in isolating them in large quantities. Work by Unwin and Zampighi (1980) has disclosed that the opening or closing of the gap junctions is due to the variable inclination of connexons, cylinders composed of six subunits (Fig. 16). When the connexons are hexagonally ordered, the gap junctions are in a highly resistant state, i.e., ions have difficulty moving from cell to cell (Kistler and Bullivant, 1980). Anoxia or a low pH produces their crystallization (Raviola *et al.*, 1980 which interrupts the exchanges of ions and small molecules between neighboring cells.

What do we know about these exchanges? The existence of permeable junctions was discovered by Loewenstein (reviewed in 1982), who found, using electrophysiological methods, that cells are often electrically coupled because of a flow of ions between them and their neighbors. A very favorable material for the study of low-resistance junctions and ionic coupling is the cleaving *Xenopus* egg. Because of its large size, it is easy to impale microelectrodes in the various blastomeres and to measure electrical coupling. Such coupling can be detected at the four-cell stage (Di Caprio *et al.*, 1974) and remains measurable until the fifth cleavage (de Laat *et al.*, 1976). If the eggs are placed in a $CO_2$-rich atmosphere, their internal pH decreases from 7.7 to 6.3 and the ionic communication between the blastomeres via the gap junctions is interrupted (Turin and Warner, 1977). This interruption is probably due to crystallization of the connexons. Another

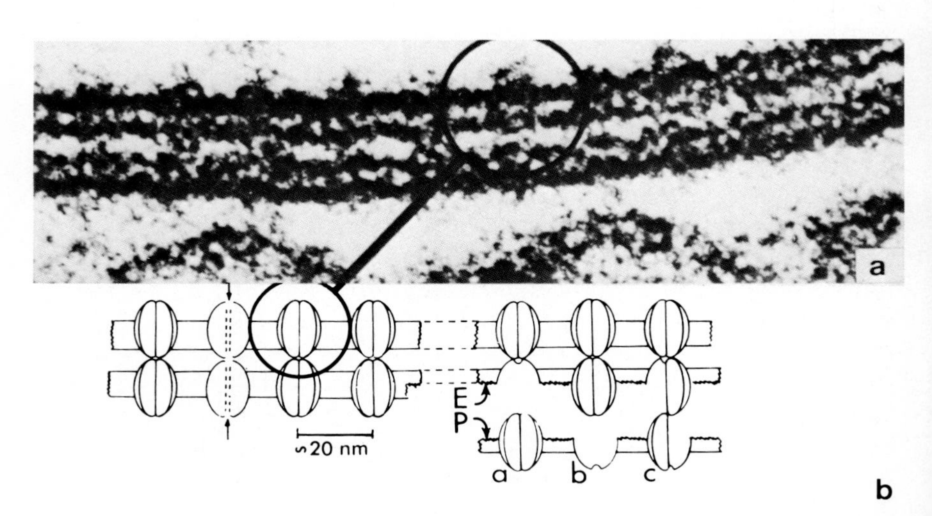


FIG. 16. (a) Cross-sectioned profile of an invertebrate (crayfish neurons) gap junction. (b) Schematic diagram of the architecture and fracture properties of the crayfish gap junctions. Intramembrane particles made up of six subunits are spanning the membrane thickness. Right, cleavage plane obtained by freeze etching. [Peracchia (1980).]

factor involved in the closure of the gap junctions is an increased accumulation of intramembrane particles, which can be seen in freeze-fracture replicas (Spray *et al.*, 1981). An increase in the free $Ca^{2+}$ content of the cell and inhibition of energy production also decrease the permeability of the cell junctions (Rose and Loewenstein, 1975a); an increase in $CO_2$ leads to the disappearance of the clusters of intramembrane particles (Lee *et al.*, 1982).

Gap junctions are also permeable to small molecules, which can thus move from one cell to the next. This was first shown by Loewenstein (reviewed in 1976) and his colleagues, who injected fluorescein into cells and studied its passage into neighboring cells. Similar experiments are generally done with the highly fluorescent dye Lucifer Yellow CH. For an analysis of electrical coupling between cells by a combination of the Lucifer Yellow and freeze-fracture replica techniques, see Andrew *et al.* (1981).

The Lucifer Yellow technique has the obvious advantages of simplicity and sensitivity, but it cannot give us information on exchanges of biologically important molecules between cells. The strategy that is generally used for this purpose is to mix together unlabeled cells (or cells whose nuclei have been labeled with [$^3$H]thymidine) and cells containing, in radioactive form, the substance of interest (nucleotide, nucleic acid, sugar, protein, etc.): coupling between neighboring cells allows the transfer of the radioactive material from cell to cell and is thus detectable by autoradiography (Fig. 17). This will happen only if the molecules are small enough to pass through the gap junctions [which can be envisaged as hydrophilic channels with a diameter of 16–20 Å in mammalian cells or as molecular sieves (Schwarzmann *et al.*, 1981; Dahl *et al.*, 1981)]. The result of all of the experiments is that nucleotides or sugars, but not RNA or proteins, can be transferred to other cells provided that these are in direct contact with the labeled cell. This phenomenon, which has been reviewed by Hooper and Subak-Sharpe (1981), is the basis of metabolic cooperation (reviewed by Hooper, 1982). If one combines certain mutants, which cannot synthesize a given metabolite due to genetic defects, with other cells with a different genetic background, the results will vary according to the cell strains and the composition of the medium. Cells unable to grow can be induced to proliferate because of metabolic cooperation between appropriate cells. Under other experimental conditions, metabolic cooperation may be harmful and even lethal. These opposite effects of metabolic cooperation result in what Hooper and Subak-Sharpe (1981) have called a "kiss of love" or a "kiss of death" between cells.

Finally, a few words must be said about the conditions required for the establishment of gap junctions and thus intercellular transfer of substances between cells. Work by Pitts and Simms (1977) and Hunter and Pitts (1981) has shown that primary cell cultures establish junctional communications without cell-type specificity. However, in some established cell lines cultivated for many generations specificity can be observed. Interestingly, gap junctions can be formed

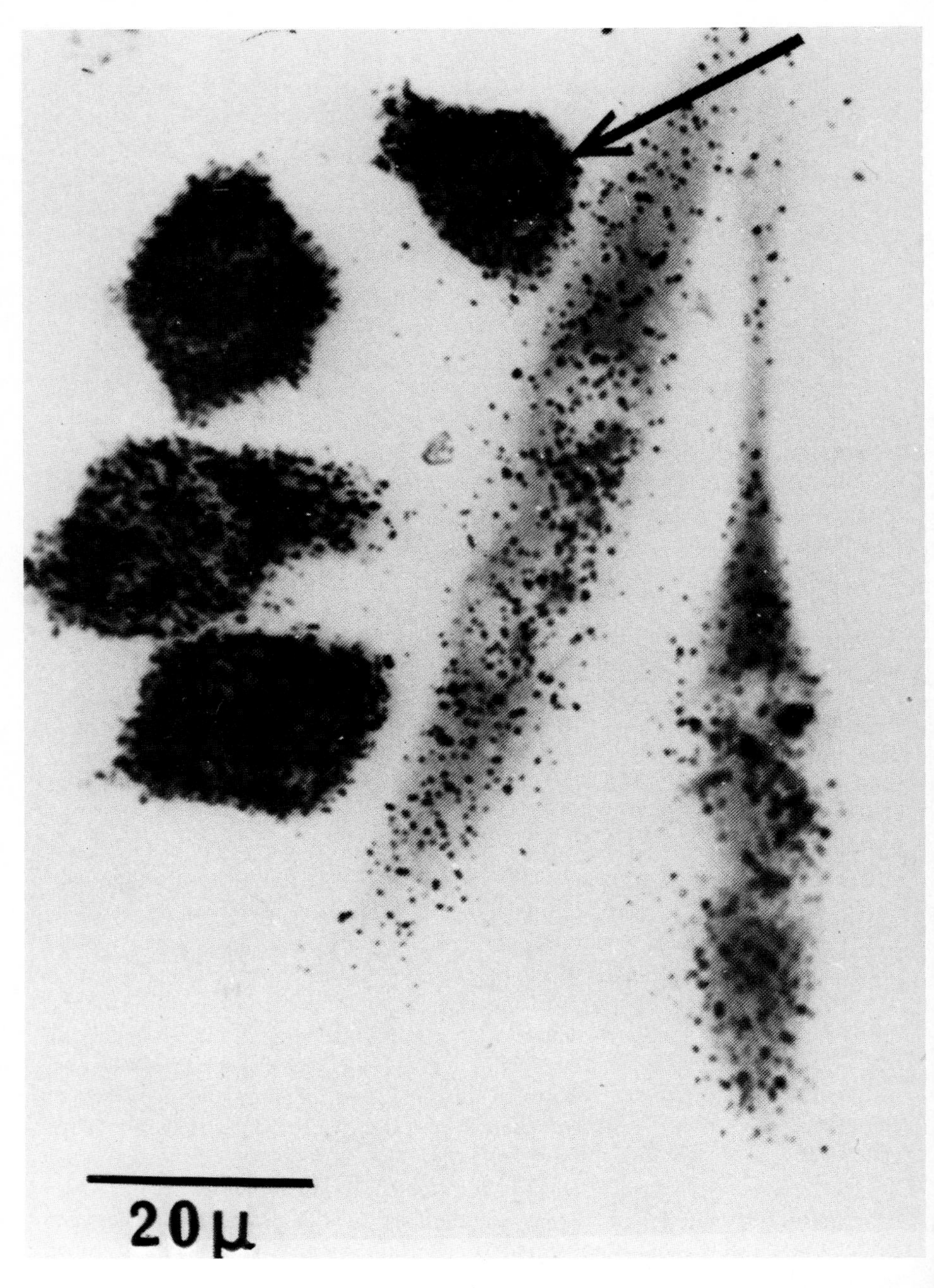


FIG. 17. Metabolic cooperation between donor and recipient cells in cocultures. Donor cells (left) have been heavily labeled with [$^{3}$H]hypoxanthine; recipient cells in contact with the labeled cells, either directly (arrow) or indirectly, show incorporation of label into nucleic acids. [Hooper and Slack (1977).]

when two anucleate cytoplasts establish contact. Their formation requires neither the presence of a cell nucleus nor energy production (Cox *et al.*, 1976; Bols and Ringertz, 1979).

Regarding the biochemical mechanisms that control junctional intracellular communications, we must single out (besides the already mentioned internal pH) the free $Ca^{2+}$ concentration and cAMP. An increase in intracellular free $Ca^{2+}$ is followed by down regulation, i.e., closure of the intercellular channels; in contrast, cAMP increases lead to up regulation by inducing the slow proliferation of channels (Loewenstein, 1982). The cAMP effects are due to increased protein phosphorylation, since addition of the catalytic subunit of adenylate kinase restores permeable junctions in cells lacking protein kinase I activity (Wiener and Loewenstein, 1983). Finally, we should mention that gap juntions may display some tissue specificity: according to Nicholson *et al.* (1983), they are made up of different proteins in liver and lens. However, injection of an antibody against a major 27-kD gap junction protein inhibits electrical coupling and dye transfer in coupled cells in culture (Hertzberg *et al.*, 1985).

### 2. *Cell Aggregation or Rejection*

The capacity of the cell membrane to recognize specific determinants on the surface of other cells is probably as old as the appearance of eukaryotic organisms during evolution. Rejection by elementary immunological mechanisms at the cellular level of tissues or organisms belonging to another species has been observed in organisms as primitive as corals and sponges. H. V. Wilson, whose classical filtration experiments date to 1907, disaggregated sponges of two species that differed in color and filtered the cells through silk cloth. Wilson first observed that the two species reaggregated to form composite small sponges. After a few hours, sorting out took place: the cells from the two different species segregated from each other, and small sponges possessing the two parental colors were then obtained. Experiments of the same type have since been done on many animal species. These experiments are possible because the cells that form a tissue can be dissociated by treatment with either trypsin or a $Ca^{2+}$-chelating agent (EDTA or EGTA). Such treatments destroy the cell coat (extracellular matrix) of glycoproteins and mucoproteins that link the cells together and weaken the cell junctions. These constituents of the cell coat, which play an essential role in cell aggregation or rejection, are reconstituted by the cells within a few hours. Many experiments on cell reaggregation have been done by embryologists interested in cellular or embryonic differentiation and by cancerologists because, as we shall see in Chapter 8, malignant transformation affects the chemical composition significantly, and thus the physiological properties of the extracellular matrix. Sponges, however, remain an attractive model.

J. Holtfreter (1939) demonstrated the significance of "tissue affinities" in amphibian gastrulation. He showed (Fig. 18) that if one closely apposes a piece

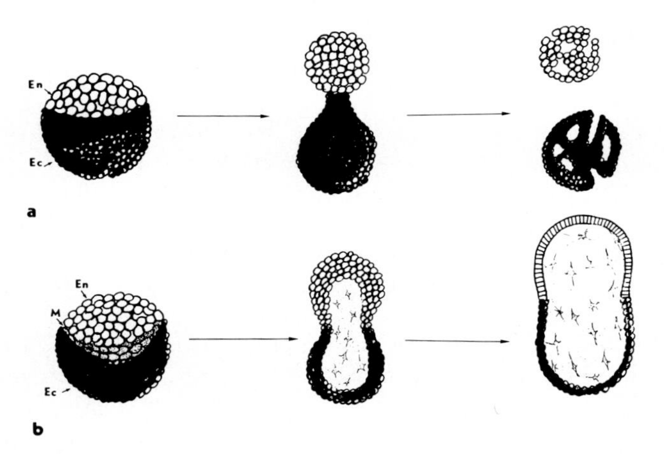


FIG. 18. Holtfreter's experiments on tissue affinities. (a) Ectoderm (Ec) and endoderm (En) display a negative tissue affinity; there is sorting out of the two fragments. (b) Mesoderm (M) has a positive tissue affinity for both ectoderm and endoderm; there is no sorting out. [P. Van Gansen, modified after Holtfreter (1939).]

of ectoderm and a piece of endoderm from young gastrulas they first stick together and then separate from each other. This rejection or sorting out demonstrates a negative tissue affinity between the two tissues. However, if a piece of mesoderm is inserted between the fragments of ectoderm and endoderm, there is no rejection. Thus, mesoderm has a positive affinity for both ectoderm and endoderm. This experiment shows that the cell surface possesses specific sites that allow recognition between cells from the various regions of the young gastrula. An "immunological" explanation of Holtfreter's (1939) results would be that some surface sites present on mesoblastic cells are complementary to sites present on the surface of both ectoderm and endoderm cells, but there is no complementarity between ectoderm and endoderm cell surface sites.

Interesting experiments have also been done on dissociated sea urchin blastulas. In 1962, Giudice dissociated blastulas from two species (*Arbacia* and *Paracentrotus*) that differ in the color of their eggs. When the dissociated cells were mixed together, composite aggregates of red *Arbacia* and white *Paracentrotus* cells were formed. However, after a few hours, sorting out took place: red and white blastulas were then observed. Thus, sea urchin blastula cells behave like the adult sponge cells in Wilson's (1907) classical experiments. The species specificity of reaggregation of sea urchin blastula cells has been confirmed by Spiegel and Spiegel (1978), who found, in addition, that adhesion between cells from the same species takes place through the intervention of microvilli and a hyaline coat material. No interaction between microvilli and the hyaline coat of adjacent cells takes place in chimeric aggregates. Experiments by Noll *et al.*

(1979) support the aforementioned immunological hypothesis: antibodies (Fab fragments) raised against membranes isolated from sea urchin blastulas inhibit the aggregation of dissociated sea urchin blastula cells. The addition of proteins present in butanol extracts reverses the inhibitory effects of the antibodies and accelerates aggregation of untreated blastula cells. However, since these proteins do not display species specificity, this result does not offer a complete explanation. The importance of surface glycoproteins for sea urchin gastrulation is further shown by the fact that tunicamycin, an inhibitor of protein glycosylation (reviewed by Elbein, 1981), modifies cell shape and adhesion between the cells and inhibits gastrulation movements (Schneider *et al.*, 1978).

Classic experiments by Moscona (1957; Moscona and Moscona, 1963) have shown that, in some cases at least, organ specificity can override species specificity in dissociation–reaggregation experiments. Embryonal retina and kidney cells from chickens and mice were dissociated and the isolated cells were mixed together, The outcome was that cells with the same organotypic specificity tended to reaggregate preferentially: chimeric renal tubules and retinas, composed of chicken and mouse cells, were obtained. However, superiority of organotypic over species specificity is not an absolute rule. For instance, adhesion between hepatocytes and heart myocytes of rats and chickens is strictly tissue specific (Albanese *et al.*, 1982).

In the early 1970s, much work was done on the effects of vegetal lectins (in particular, ConA) on the agglutination of dissociated cells. The conclusions drawn from these experiments were often contradictory [see, for instance, the papers by Moscona (1971) and Steinberg and Gepner, (1973)], but the fact that the cell surface undergoes changes during the differentiation of retina cells seems to have been established beyond doubt (Kleinschuster and Moscona, 1972).

However, interest in the effects of plant lectins on cell aggregation waned when it was discovered that animal cells release their own lectins (ligands) in the culture medium that promote cell-specific aggregation and bind to specific sugar residues. In 1972, Garber and Moscona found that "conditioned media" (i.e., media in which cells have been cultured for a certain period of time) of brain cells reaggregated brain cells specifically. This result is due to the fact that cells in culture release ligands that bind together receptors present on the surface of adjacent cells (Pessac and Defendi, 1972; Balsamo and Lilien, 1974). As shown schematically in Fig. 19, a "ligator" molecule is required for binding together the receptors present on the surface of brain or retina cells. The soluble factor, which is released from retina cells and specifically promotes their aggregation, is a glycoprotein ($M_r = 50{,}000$). The specificity seems to reside more in the protein than in the sugar moiety (Hausman and Moscona, 1975). Moscona has coined the term "cognins" for these extrinsic (soluble) bridging factors that link together the specific receptors located on the cell surface.

Models describing the possible mode of action of cognins and similar "cell

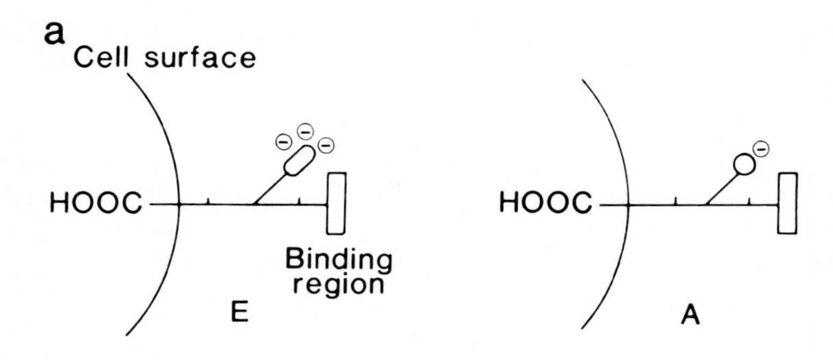


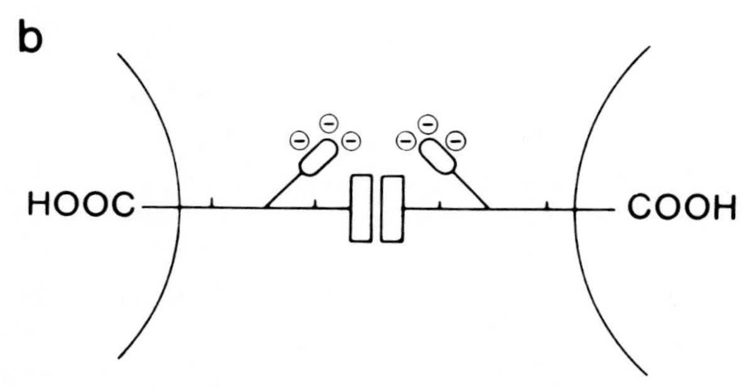


FIG. 19. Schematic model for N-CAM (neuron-specific cell adhesion molecules). (a) E and A forms of ligand molecules. The sialic acid-free binding region is represented by the rectangle and the sialic acid-rich sugar moiety by an oval; the possibility of more than one site of sugar attachment is shown by short vertical lines. The E → A conversion removes large amounts of sialic acid. (b) Adhesion of two homologous cells (neurons) by association of ligand molecules. [Edelman (1983). Copyright 1983 by the American Association for the Advancement of Science.]

adhesion molecules'' (CAM) have been proposed by Rutishauser *et al.* (1976) and discussed by Edelman and Rutishauser (1981). That aggregation between retina cells is due not only to released soluble factors but also to specific glycoproteins present on the cell surface itself was demonstrated by Hausman and Moscona (1976). In addition, the role of calcium ions in bridging the receptors has been shown for a number of cells. Work by Takeichi *et al.* (1979), Thornes and Steinberg (1981), and Brackenbury *et al.* (1981) has shown that there are two different mechanisms ($Ca^{2+}$ dependent and $Ca^{2+}$ independent) for cell aggregation. According to Takeichi *et al.* (1979), aggregation occurs only between cells that have the same $Ca^{2+}$ dependency; otherwise, rejection follows. Calcium ions were found necessary for aggregation of liver cells, but not for adhesion of brain or retina cells. Neural adhesion molecules (N-CAM) are different from liver adhesion molecules (L-CAM) (Brackenbury *et al.*, 1981). Finally, adding to a still confused situation a report by Thomas *et al.* (1981) indicates that $Ca^{2+}$-dependent and $Ca^{2+}$-independent aggregation mechanisms may coexist in retina cells and that the affinity between the cells would be modulated by extracellular proteases.

Specific CAMs are being extensively characterized in G. Edelman's laboratory. The N-CAM is made up of two closely related polypeptide chains of 140 and 170 kilodaltons, respectively. Their amino terminus extends outside the cell membrane and probably bears a binding site for another N-CAM extending from a neighboring cell. The carboxyl terminal region, at the other end of the molecule, penetrates into the plasma membrane lipid bilayer and is therefore anchored in the cell (Cunningham *et al.*, 1983). L-CAM, which is responsible for adhesion between liver cells, is an acidic glycoprotein with a molecular weight of 140,000 (Gallin *et al.*, 1983). As we shall see in Volume 2, Chapter 2, N-CAM and L-CAM have different distributions in chick embryos and undergo characteristic stage-specific changes.

That the factors that stimulate adhesion between cells reside in the cell membrane seems to be well established. For instance, isolated plasma membranes from liver cells stimulate adhesion between hepatocytes; the active factor has been identified as a glycoprotein by Schmell *et al.* (1982).

As we can see, the subject of cell adhesion, which has been reviewed in detail by Hayashi and Ishimaru (1981), is still progressing. Intrinsic membrane factors and extrinsic bridging factors (cognins, CAM) undoubtedly play an essential role in cell recognition; but, as pointed out by Marchase *et al.* (1976), other factors such as calcium ions, electrospecific attraction, hydrogen bonds between sugars, and establishment of bonds between glycoproteins by glycosyltransferases might very well also play a role in the highly complex and important process of cell recognition.

The molecular mechanisms of cell aggregation involving ligands and receptors are basically the same for sponges and for chicken retina cells. Aggregation factors and glycoproteins comparable to Moscona's cognins and Edelman's CAMs have been isolated from different species of sponges by Weinbaum and Burger (1973) and by Müller and Zahn (1973). An inhibitory aggregation factor responsible for rejection in heterologous graft experiements has been isolated by Müller *et al.* (1981). It is a glycoprotein ($M_r$ 27,000) that competes with the aggregation factor for its binding site (receptor) on the cell surface. Synthesis of this "graft rejection factor" becomes detectable 3 days after the interspecific graft has been made. Interest in dissociation–reaggregation experiments in sponges has been further intensified by the fact that these organisms are composed of various kinds of cells that have a typical morphology. After dissociation and reaggregation of a sponge, cells belonging to the various types reassociate according to complex patterns that will not be described here but can be found in a review by Van de Vÿver (1975).

## F. Cell Locomotion and Cell Shape

When cells that are tightly held together *in situ* by desmosomes or tight junctions are dissociated by treatment with trypsin or a $Ca^{2+}$-chelating agent, their shape becomes irregular and they display locomotion. These cells look like small amebas or leukocytes, and their movements after adhesion to the substratum can be followed by microcinematography. The morphology and locomotion of fibroblasts and epithelial cells have been reviewed by Vasiliev and Gelfand (1977), while the formation of protrusions of the cell surface and their role in cell migration (in particular, in relation to gastrulation and directional cell movements during morphogenesis) have been analyzed by Trinkaus (1980). The biochemical mechanisms of locomotion in large amebas have been discussed by Lansing Taylor and Condeelis (1979). They emphasized the importance of local differences in free $Ca^{2+}$ concentration and in intracellular pH for ameboid movement.

During locomotion, the cell changes its shape without increasing or decreasing its surface. The plasma membrane apparently plays only a passive role in cell locomotion, but its molecular organization may undergo local changes. It is generally believed that the main factor in cell locomotion resides in the cell surface–associated cytoskeleton—in particular, its contractile actomyosin microfilaments. However, it should be kept in mind that this is only a part of the entire cytoplasmic cytoskeleton: cooperation of all constituents of the microfilament system is probably needed to ensure locomotion in many, if not all, cells.

There are many different types of protrusions on the cell surface such as microvilli, which do not play a role in locomotion. Certain organelles are highly specialized for locomotion, namely, cilia and flagella, but they are absent from most cells. The various kinds of protuberances involved in cell locomotion, such as fibroblasts or early embryonic cells, have been classified: lamellipodia, filopodia, lobopodia, microspikes, and blebs (see Trinkaus, 1980, for details). Lamellipodia are flat and adhere to the substratum. They are only 0.1–0.4 μm thick and contain a network of actin microfilaments; their margins often present ruffles. Filopodia are very long, thin protrusions that adhere to the substratum by enlarged tips called "filopodial footpads"; their microfilaments are arranged in parallel and tend to form bundles. Microspikes are short filopodia, while the lobopodia of embryonic cells are fingerlike processes. Finally, blebs are hemispheric protrusions devoid of cytoplasmic inclusions (except ribosomes in eggs). Their role in locomotion seems to be passive rather than active since they usually lack contractile microfilaments (Tickle and Trinkaus, 1977).

Cell locomotion (except for cells that possess a specialized locomotory apparatus in the form of cilia and flagella) is usually of the ameboid type. Our understanding of the molecular mechanisms that induce the formation of cell protrusions is incomplete. However, progress in this area is likely to be rapid because of the large amount of work presently being done on cells that have been "transformed" by infection with an oncogenic virus or by treatment with a chemical carcinogen. Normal and transformed cells vary greatly in shape and locomotion. This variation will be discussed later in Volume 2, Chapter 3. The first demonstration of differences between normal and cancer cells came from the classic work of Abercrombie and Heaysman (1954) on contact inhibition of movement and of cell division. By time-lapse microcinematography studies of cultured cells, they found that when the density of a cell culture becomes high, cells stop moving after they have collided; such contacts between normal cells also lead to a cessation of proliferation. These two control mechanisms (contact inhibition of movement and contact inhibition of cell division) fail to operate properly in cancer cells: collisions between these cells do not impede cell movement and cell proliferation. Contact inhibition is now frequently called "density-dependent inhibition" of growth because it is not certain that contact between the cells causes the arrest of locomotion and cell division.

As pointed out above, it is believed that, in amebas that are large enough to be injected with substances capable of locally modifying the $Ca^{2+}$ concentration and/or the pH, locomotion (which is due to cyclic sol–gel changes) is largely controlled by these two factors (Taylor and Condeelis, 1979). It is probable that the same factors play a role in the ameboid locomotion of other cells. For instance, Dulbecco and Elkington (1975) found that calcium ions modify the morphology of fibroblasts and suggested that they increase the cGMP content. Cyclic nucleotides have, indeed, been proposed as regulators of cell movements (Willingham and Pastan, 1975). For Estensen *et al.* (1973), cAMP would be a "stop signal" and cGMP a "go signal." In the general context of our present views on the mechanisms of action of $Ca^{2+}$ and c-nucleotides on cell movement (reviewed by Sharma, 1982), these findings suggest that phosphorylation of membrane proteins plays an essential role in cell locomotion. However, calcium ions also affect cell motility by a direct action on the membrane-associated cytoskeleton. For example, Nicolson *et al.* (1976) reported that in the presence of the local anesthetic procaine, which displaces calcium bound to membranes, the cells become spherical because of a decrease in the membrane-associated microtubules and microfilaments.

That the contractile actomyosin microfilaments are directly involved in cell locomotion seems to be established beyond any doubt. Fraser and Zalik (1977) showed that cytochalasin B, which inhibits actin polymerization, prevents the formation of lobopodia in dissociated amphibian blastulas. According to Bliokh *et al.* (1980), during the spreading of fibroblasts on the substratum, only patches of the cell surface attach to the substratum; their transformation in lamellipodia is inhibited by cytochalasin B. It is important to mention here that an inhibitor of calmodulin, trifluoperazine, like cytochalasin B, inhibits the spread and migration of cultured cells. These are thus calcium-dependent, microfilament-mediated processes (Connor *et al.*, 1981). But decisive proof that movement results from the contraction of actomyosin filaments is provided by elegant experiments by Sheetz and Spudich (1983) on the very large cells of the alga *Nitella:* they dissected the alga in order to expose the longitudinal actin microfilament cytoskeleton, added small myosin-coated fluorescent beads, and observed a unidirectional movement of the beads. This movement requires ATP and is due to myosin since it is arrested if the "heads" of the myosin molecules are inactivated.

The local adhesion plaques between the cell and the substratum contain receptors for ConA, but no fibronectin (Chen and Singer, 1980). The composition of the extracellular matrix thus becomes different from that of the rest of the matrix when localized areas of the cell adhere to the substratum. Both locomotion and anchorage to the substrate require the formation of new adhesion plaques. It is probable that the plasma membrane itself undergoes molecular rearrangements during locomotion. For instance, it has been reported that regions of the cell surface where there is a continuous formation of lobopodia and ruffling lack ConA receptors (Domnina *et al.*, 1977). In contrast, the filopodial footpads are exceptionally rich in intramembrane particles (Robinson and Karnovsky, 1980).

This brief but incomplete (we have hardly mentioned the cyclosis movements of plant cells in which actomyosin microfilaments are also involved) overview of cell locomotion shows the extreme biochemical complexity of a fascinating process. Time passes quickly for those who look under a microscope at amebas crawling away from a light source.

It is likely that in the very near future, a more integrated molecular picture of cell locomotion will emerge. Sol–gel transformations of the cytoplasm, which are essential for cell motility in amebas (Allen and Taylor, 1975), almost certainly result from the sol–gel transformations of actin microfilaments that will be discussed in the next section. These changes in viscosity are $Ca^{2+}$ dependent and are regulated by the numerous proteins that bind to actin. One of them, filamin, seems to play a particularly important role in cell locomotion, since, according to the actin/filamin ratio, actin forms a gel or is liquid; gel formation apparently results from the crosslinking of actin filaments by filamin (Nunnally *et al.*, 1981). Another possibly important factor for locomotion is vinculin, which attaches the actin cytoskeleton via transmembrane connections (Trotter, 1981; Singer, 1982) to the substratum in the adhesion plaques. In a short review of vinculin, Hynes and Yamada (1982) point out the importance of this "transmembrane complex" for cell locomotion and anchorage to the substratum. This complex involves actin, vinculin, and another actin-associated protein (α-actinin) on the inner side of the cell; and fibronectin, proteoglycans, and collagen on its outer side. It is probable that the functioning of this transmembrane complex is modulated by protein phosphorylation and dephosphorylation, which could modify the conformation of the proteins involved in the complex. It is known that vinculin is phosphorylated, in a specific way, by protein kinase C. This recently discovered enzyme requires $Ca^{2+}$ and phospholipids for activity (Werth *et al.*, 1983).

It should be added that our knowledge of cell movement and adhesion is based to a very large extent on studies of cultured fibroblasts. It would be unwise to extend conclusions drawn from a single type of cell cultured under artificial conditions to other cell types present in normal tissues without serious experimental analysis.

## II. THE CYTOSKELETON

### A. General Background

It has often been suggested that the ground substance (hyaloplasm) of the cell is not really homogenous and that it possesses an invisible structure. This idea was based on experiments showing that granules that were displaced in a centrifuged cell resumed their original position upon standing.

In "Biochemical Cytology" the cytoskeleton and the endoplasmic reticulum, which had been recently discovered, were discussed together. At that time, the true constituents of the cytoskeleton (microtubules, microfilaments, and intermediate filaments) were unknown, and one had to await the development of immunocytochemical methods to learn that the tubules and filaments seen under the electron microscope formed the network of protein fibers that had been hypothesized by Monné in 1946.

The term "cytoskeleton" is perhaps not the best term that might have been chosen, because it implies a rigidity (by analogy to our own bony skeleton) that is seldom found in cells. For example, during locomotion, the shape of the cell changes continuously; during mitosis, elongate cells round up and their microtubules undergo complete reorganization. How could such changes be possible if the cells were immobilized by a rigid skeleton? The cytoskeleton should be regarded, like the cell surface and all the cell organelles, as a dynamic, not a static, structure. Since everyone uses the term "cytoskeleton," we shall do the same.

It is generally agreed that there are three major components of the cytoskeleton: the microtubules (MT), composed of tubulin and a few associated proteins, are about 25 nm in their outer diameter. they are uniform in size and straight, their length reaching several millimicrons. The microfilaments (MF), which have already been found in close association with the cell surface, are composed of actin molecules that associate with a host of other proteins, including myosin, which confers contractility on the microfilaments; They are much smaller in width (7 nm) than the microtubules and are often assembled into bundles. Finally, the intermediate filaments (IF) are heterogeneous, both in size and in width (8–15 nm) and, particularly, in chemical composition. Since the majority of the

IFs are approximately 100 Å, they are often called "10-nm filaments," which is somewhat misleading in view of their highly variable chemical composition.

MTs, MFs, and IFs constitute a three-dimensional network in the cytoplasm. Is there a higher-order assembly between these cell constituents leading to an integrated network? According to Wolosewick and Porter (1979), high-voltage electron microscopy of whole cells previously extracted with nonionic detergents indicates the existence of a three-dimensional trabecular network (Fig. 20). This cytoskeletal network is composed of filaments of 40 to 55 nm where actin predominates (Batten *et al.*, 1980). Ribosomes are associated in detergent-treated HeLa cells with the microtrabeculae, which constitute the cytoskeletal framework (Cervera *et al.*, 1981). According to a report by Schliwa *et al.*, (1981b), tiny filaments (2–3 nm in diameter and 20–30 nm in length) link together the various constituents of the three-dimensional cytoskeleton (MTs, MFs, and IFs). Schliwa *et al.* (1981a) have shown that if cells have been extracted by the detergent BRIJ 58, a microtrabecular network is seen under the electron microscope. If these cells are further extracted with another detergent, Triton X-100, only the conventional cytoskeletal network remains. Since this second treatment with Triton X-100 removes a number of proteins, it has been concluded that many cellular proteins are structure bound. It is too early to express a firm opinion about the reality of the trabecular network, which might result from technical artifacts (Ris, 1985). The existence of a three-dimensional network in the cell is conceptually very satisfactory, but further work is required before this view can be totally accepted. It has been shown by Kondo (1984) that one can observe structures similar to the microtrabecular network in red blood cells (which have no cytoskeleton) and even after "fixation" of proteins in solution. For Kondo (1984), the microtrabeculae are only markers of high-molecular-weight proteins. Such a negative view seems exaggerated but reminds us of the always possible danger of artifacts.

The hyaloplasm (cytomatrix) is imprisoned in the meshes of the cytoskeletal network. Its composition is not well known; it is probably very complex and different from what biochemists call cytosol (supernatants of centrifuged cell homogenates). Jacobson and Wojciewszyn (1984) have studied the diffusion of various chemicals in the cytoplasmic matrix and have found that it is dominated by the binding of the diffusing species to elements of the cytomatrix. There is no relationship between the speed of diffusion and the molecular weight of the diffusing substance, and there are no effects on diffusion of various agents that disrupt cytoskeletal organization.

## B. Microtubules

Microtubules have been the object of extensive studies during the last few years. Results have been summarized in two books (P. Dustin, 1984; Roberts and Hyams, 1979). A number of review articles (Snyder and McIntosh, 1976;

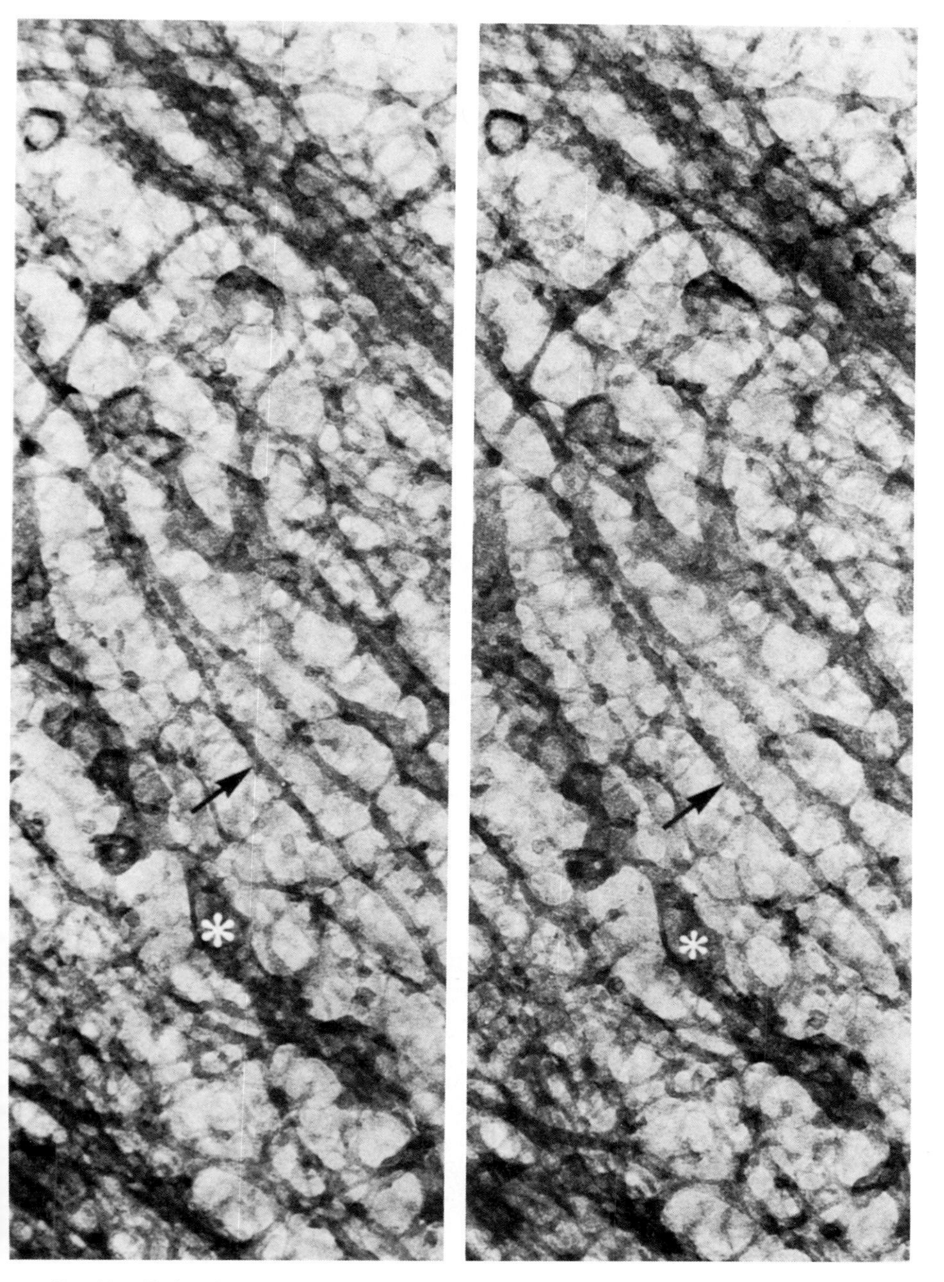

FIG. 20. Trabecular network shown by high-voltage electron microscopy in stereomicrographs of dried cells. Fusion of microtrabeculae with the surfaces of MT (arrow), the cisternae of the smooth ER (asterisk), and the MFs can be seen. [Wolosewick and Porter (1979). Reproduced from *The Journal of Cell Biology,* 1979, **82,** p. 120 by copyright permission of The Rockefeller University Press.]

Stephens and Edds, 1976; Mohri, 1976; Jacobs and Cavalier-Smith, 1977; Kirschner, 1978; Brinkley *et al.*, 1980; DeBrabander *et al.*, 1985) have been devoted to the many interesting problems raised by the MT: chemical composition, cytochemical detection, assembly and disassembly, spatial organization, etc.

Microtubules (Fig. 21) are hollow cylinders 25 nm in diameter and of variable length. Under the electron microscope, their morphology is very similar in all types of cells, animal or vegetal. Their main chemical constituent is tubulin, a heterodimer of two subunits, namely, α- and β-tubulin; each of the subunits has an $M_r$ of about 55,000 (Borisy *et al.*, 1972). Refined methods of analysis, including x-ray crystallography, have shown that the α, β heterodimers are organized into 13 protofilaments, as shown in Fig. 22 (Bryan, 1974; Amos and Klug, 1974). However, MTs with 12 or 15 protofilaments have also been observed. Tubulin has two binding sites for guanosine triphosphate (GTP). One of the tubulin-bound GTP molecules is readily exchangeable with added free GTP; the other GTP molecule is nonexchangeable and is located in the β-tubulin subunit (Weisenberg *et al.*, 1976; Geahlen and Haley, 1977).

The main property of tubulin is its capacity to undergo linear polymerization,

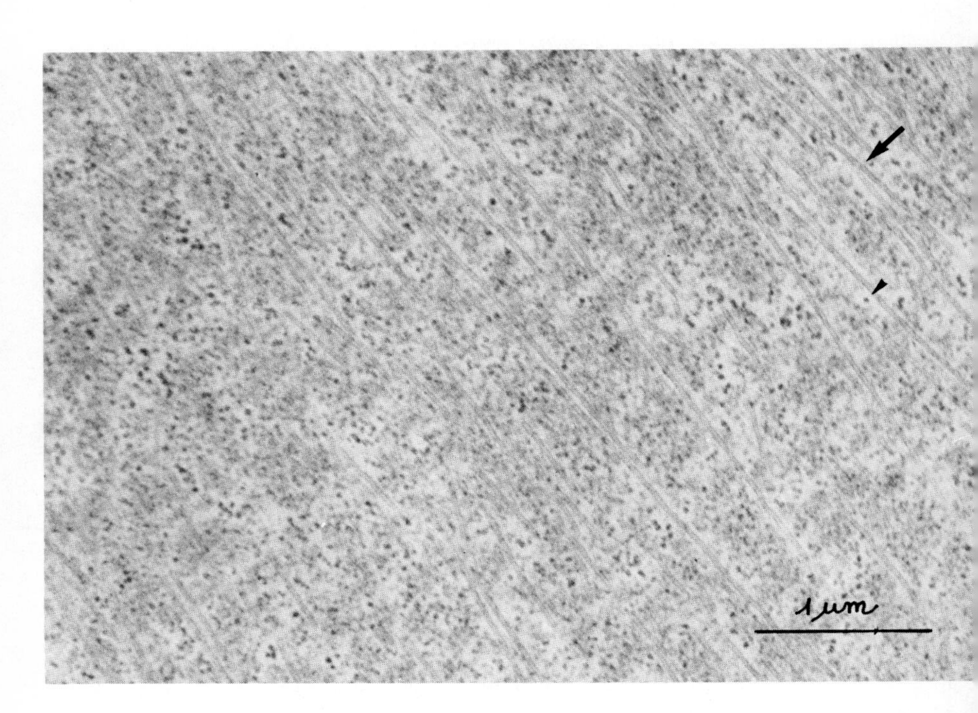


FIG. 21. Ultrathin section of the first meiotic spindle in a maturing *Xenopus* oocyte; array of MTs (arrow) and trapped ribosomes (arrowhead). [Courtesy of G. Steinert.]

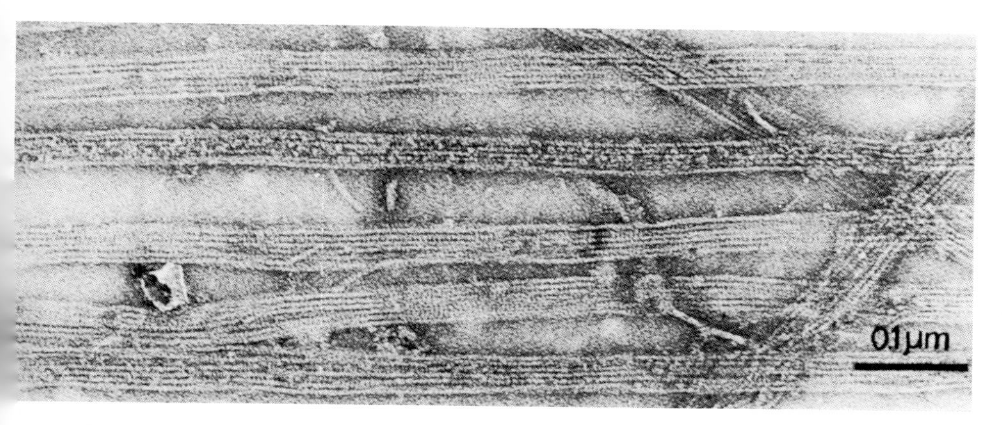


FIG. 22. Microtubules isolated from the cortical region of *Triturus* erythrocytes; the constitutive protofilaments are clearly shown by negative staining. [Bertolini and Monaco (1976).]

leading to assembly of MTs. It is now generally believed that according to the model proposed by Kirschner (1980) polymerization results from a "treadmilling" process: a flux of subunits through the polymers. A similar treadmilling mechanism also accounts for the polymerization of the actin MF. Treadmilling implies polymerization of the subunits at one end and depolymerization at the other; it is coupled with GTP hydrolysis in the case of tubulin, and ATP hydrolysis in the case of actin (reviewed by Cleveland, 1982). Other models of MT polymerization will be discussed in Chapter 5.

It is easy to obtain tubulin polymerization and MT formation under *in vitro* conditions. The source of tubulin is generally porcine brain. *In vitro* polymerization of tubulin requires only the presence of GTP and magnesium ions. It is inhibited by $Ca^{2+}$, even at low concentrations. A $Ca^{2+}$-chelating agent is, therefore, generally added to the polymerization medium. Tubulin polymerization is also inhibited by cold, by agents that oxidize reduced glutathione (Burchill *et al.*, 1978), and, in a more specific way, by substances that bind to tubulin. The number of these "MT poisons" already known is quite large and is continuously increasing. The most popular poison used for *in vitro* and *in vivo* studies is colchicine (or its derivative colcemid), whose antimitotic activity was discovered in 1934, in Brussels, by A. P. Dustin (1934). Other popular inhibitors of tubulin polymerization are the *Vinca* alkaloids (vinblastine and vincristine), podophyllotoxin, maytansine, griseofulvin, and various derivatives of benzimidazole (nocodazole, parbendazole). All these substances bind to the tubulin dimer and prevent its polymerization both *in vivo* and *in vitro*. According to Mandelbaum-Shavit *et al.* (1976), maytansine and vinblastine bind to the same site in the tubulin molecule; this site is distinct from the one binding colchicine.

Self-assembly and polymerization of tubulin dimers are possible *in vitro*, provided that the tubulin concentration is higher than 2.5 mg/ml (Herzog and Weber, 1977). This concentration can be lowered in the presence of agents that stabilize the MTs [glycerol, dimethyl sulfoxide, heavy water ($D_2O$): Lee and Timasheff, 1977] or by the addition of polycations such as polylysine (Lee *et al.*, 1978). On the other hand, polyanions, RNA in particular, inhibit tubulin polymerization *in vitro* (Bryan *et al.*, 1975). As we shall see in Volume 2, Chapter 2, polyanions and polycations may also play a regulatory role in MT assembly under *in vivo* conditions. Other interesting reagents for the stimulation of *in vitro* (and even *in vivo*) MT assembly are nonhydrolyzable derivatives of guanosine diphosphate (GDP) and GTP (Penningrath and Kirschner, 1977; Sandoval *et al.*, 1977, 1978). Microtubules polymerized in the presence of these agents are resistant to excess $Ca^{2+}$, but disappear in the cold. These observations have cast some doubt on the necessity of GTP hydrolysis for MT assembly. However, in favor of the classic interpretation is the fact that GDP inhibits *in vitro* polymerization of tubulin (Penningrath and Kirschner, 1977). Furthermore, Cote and Borisy (1981) have concluded that GTP hydrolysis is necessary for the subunit flow through the polymer during treadmilling.

In 1975, Weingarten *et al.* isolated microtubule-associated proteins (MAP) during purification of hog brain tubulin by repeated cycles of polymerization and depolymerization. A factor called "factor tau" greatly accelerates the rate of *in vitro* MT assembly. Further work showed that factor tau was made of at least four polypeptides with an $M_r$ ranging between 55,000 and 70,000. It bound to tubulin and favored the elongation of short MTs (Cleveland *et al.*, 1978). At about the same time, Murphy and Borisy (1975) found that proteins of much higher $M_r$ (300,000–350,000) also remained closely associated with brain tubulin during its purification by repeated polymerization and depolymerization cycles. These "HMW-MAPs," called $MAP_1$ ($M_r > 300{,}000$) and $MAP_2$ ($M_r$ 270,000–300,000) form periodical lateral extensions on the MTs and favor their polymerization (Fig. 23). According to Herzog and Weber (1978), the tau proteins and the HMW-MAPs have the same favorable activity on *in vitro* tubulin polymerization. $MAP_2$ has been purified by Kim *et al.* (1979), who found that one molecule of this protein binds to nine molecules of tubulin. Under the electron microscope, $MAP_2$ "decorates" the microtubules (with formation of lateral extensions).

In addition, one of its functions may be to crosslink adjacent MTs and to stabilize preexisting MTs and protect them against cold treatment. It should be added that what has been found for pig brain is not necessarily directly applicable to all cells. Although the general principles are likely to be the same, for instance, in HeLa cells, three MAPs stimulate MT assembly; their $M_r$ values (220,000, 200,000, and 125,000) are very different from those found in porcine brain MAPs (Bulinski and Borisy, 1980b).

The inhibitory effect of low concentrations (1–4 $\mu M$) of calcium ions on *in*

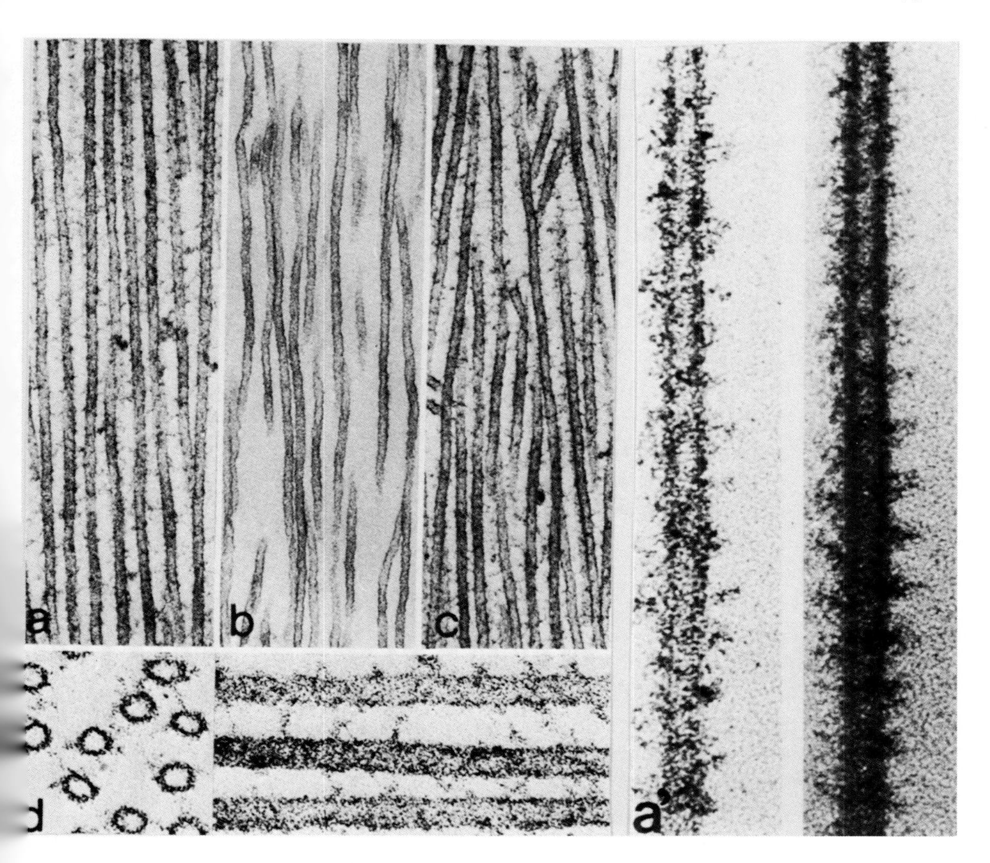


FIG. 23. (*Left*) (a) Microtubules obtained from an unfractionated MT protein; (b) obtained from purified tubulin; (c) obtained from tubulin reconstituted with HMW-MAPs; (d) transverse and longitudinal sections of MTs obtained from an unfractionated MT protein. [Murphy and Borisy (1975).] (*Right*) (a) Electron micrographs of thin sections of $MAP_2$-saturated MTs. [Kim *et al.* (1979). Reproduced from *The Journal of Cell Biology*, 1979, **80**, pp. 266–276 by copyright permission of The Rockefeller University Press.]

*vitro* MT assembly has been the subject of several investigations, since this factor probably also plays an important role *in vivo*. It has been found that a $Ca^{2+}$-activated protease specifically destroys the MAPs, in particular $MAP_2$. This, of course, negatively affects *in vitro* polymerization (Sandoval and Weber, 1978). Another possibility is the intervention of calmodulin (the main $Ca^{2+}$ receptor and modulator; reviewed by Lin, 1982) in the $Ca^{2+}$-mediated inhibition of tubulin polymerization. For instance, Runge *et al.* (1979) found that calmodulin inhibits (in the presence of $Ca^{2+}$) MT assembly and stated that this inhibition might be due to the stimulation of an MT-associated phosphodiesterase. Interesting experiments by Schliwa *et al.* (1981a) on detergent-treated, permeabilized cells have shown that addition of MAPs or trifluoperazine (a calmodulin inhibitor) protects microtubules against depolymerization by 1–4 $\mu M$ $Ca^{2+}$.

Other factors that may control MT assembly are phosphorylation of the MAPs and addition (or removal) of a tyrosine residue to α-tubulin. It has been shown by Sheterline (1977) and by Vallee *et al.* (1981) that a cAMP-dependent protein kinase is associated with the MTs and that this enzyme phosphorylates one of the HMW-MAPs. More recent work by Jameson and Caplow (1981) indicates that such phosphorylation of the MAPs decreases MT assembly; this could explain the role played by cAMP in MT formation in living cells, which will be examined later.

In 1977, Raylin and Flavin discovered a new enzyme (tyrosine ligase) that adds one tyrosine molecule to α-tubulin in the presence of ATP and removes it in the presence of ADP. The significance (if any) of tyrosine addition or removal for tubulin polymerization remains unknown.

We are now in the era of specific gene sequencing; the α- and β-tubulin genes are under scrutiny using modern methods of molecular biology. In 1978, Bryan *et al.* showed that different genes coded for the mRNAs of tubulins. Valenzuela *et al.* (1981) synthesized and sequenced the complementary DNAs (cDNAs) corresponding to the α- and β-tubulin mRNAs. From the cDNA base sequences, they deduced the amino acid sequences of both α- and β-tubulins. This analysis showed, among other things, that the tyrosine residue present in the terminal position of the α-tubulin chain was encoded by the α-tubulin gene; the role of tyrosine ligase is thus to remove this tyrosine residue, not to add it. The cDNAs of tubulin mRNAs from sea urchins, *Drosophila,* chicken, and mammals possess common sequences demonstrating that at least part of the molecule has been highly conserved during evolution. A peptide map analysis by Little *et al.* (1981) has disclosed that it is β-tubulin that has been best conserved. In contrast, α-tubulins present in eggs and sperm of sea urchins are different from each other. However, there are great similarities between the α-tubulins of sperm from three sea urchin species and from the echiuroid worm *Urechis*. Finally, the question of the number of tubulin gene copies in the genome has been tackled by Kirschner (1980), who cloned the cDNA copies of the α- and β-tubulin mRNAs of the brain. He concluded, in his preliminary report, that there are four α- and four β-genes, distributed on at least three different chromosomes in mammals. He also found that factor tau and the HMW-$MAP_2$ are very widespread and have been highly conserved during evolution. That there are many "isoforms" of tubulin is now quite clear; in the liver of mouse embryos, four α- and three β-tubulins were separated (Denoulet *et al.*, 1982). The presence in the genomes of both *Drosophila* and mammals of multiple DNA sequences coding for α- and β-tubulins proves that the two tubulin subunits exist in multiple forms. Readers interested in the organization of the tubulin genes might consult Cleveland's (1983) short review with profit.

This type of work should ultimately lead to a better understanding of the role played by MTs in living cells, although the link between DNA sequences and MT assembly and disassembly in a living cell still appears to be distant.

What do we know about MT organization in cells? A major contribution is that of Klaus Weber *et al.* (1975a,b), who studied the intracellular localization of tubulin with fluorescent antibodies. They discovered that MTs form a network in the cytoplasm and that there is little tubulin in the nucleus (Fig. 24). The cytoplasmic network often shows a greater density in a region close to the nucleus; this region is believed to be associated with the centriole. The MT network vanishes, as one would expect from what has been said, if the cells are treated with colchicine or left in the cold. Sherline and Schiavone (1977), using fluorescent antibodies against MAPs, found that these proteins are distributed in filaments that extend from the nucleus to the plasma membrane. This filamentous network is broken down by colchicine treatment, demonstrating that MAPs are actually associated with MTs in living cells. This was confirmed, in a variety of cells, by Connolly *et al.* (1978) and by Connolly and Kalnins (1980) for the HMW-MAPs; however, fluorescent anti-tau antibodies did not stain the MT network in all cells. This MT network does not show major differences in

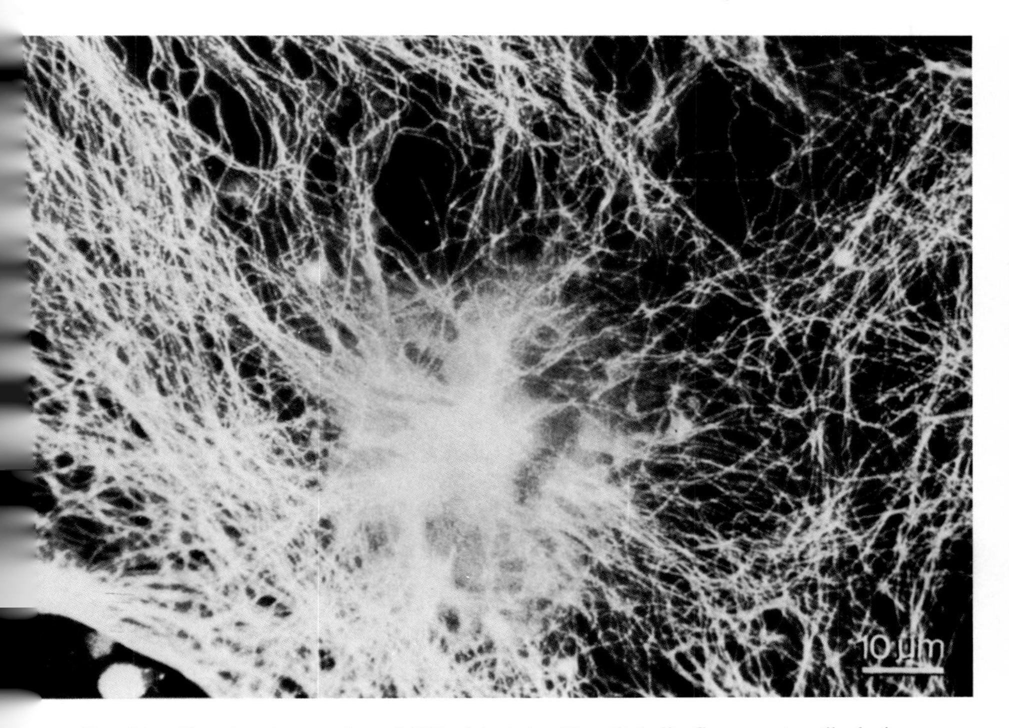


FIG. 24. Cytoplasmic complex of MTs detected with antitubulin fluorescent antibody in a fibroblast. [Osborn and Weber (1976).]

transformed (malignant) and normal cells. Differences in the organization of the MT network in the two kinds of cells, which had been reported by others, are due to artifacts that result from the fact that transformed cells do not spread as evenly on the substratum as normal cells (De Mey *et al.*, 1978).

*In vivo* polymerization of tubulin molecules requires the intervention of microtubule organizing centers (MTOC), which would, as in crystal seeds, act as nucleation centers for soluble, unpolymerized tubulin dimers. The nature and mode of action of the MTOCs are not yet perfectly clear. Frankel (1976) suggested that MTs grow around a centriole, but more recent work by Spiegelman *et al.* (1979) on cells that had been subjected to a colchicine pulse and allowed to recover showed that there are as many as 5–10 MT initiation sites in fibroblasts; 10–30 MTs grow around each site. However, epithelial cells have a single MTOC that might be the centriole, and this seems to be the rule in interphase cells, according to Brenner and Brinkley (1981).

It is generally believed that the main function of the MT network is to maintain the shape of the cells. However, according to Eichorn and Peterkofsky (1979), experimentally produced changes in the ratio of free tubulin dimers and of polymerized MTs does not necessarily modify cell shape. Another possible function of MT in association with IFs might be the positioning of the nucleus and the centrioles in the cell; this results from experiments on syncytia obtained by fusion of several cells (Wang *et al.*, 1979a,b). Besides the inhibitors of tubulin polymerization, an agent that is proving useful in this analysis is taxol, which has opposite effects: it promotes the elongation of preexisting MTs and induces the spontaneous nucleation of new MTs (Schiff *et al.*, 1979; Schiff and Horwitz, 1980). The effects of this interesting substance on cell division and egg cleavage will be examined later.

Work with colchicine, antitubulin antibodies, taxol, etc. emphasizes the major importance of the MT network for the maintenance of normal cell morphology and physiology. For instance, all anti-MT agents produce the fragmentation and disorganization of the Golgi apparatus (Wehland *et al.*, 1983; Pavelka and Ellinger, 1983); both its localization and its integrity thus depend on the MTs. Geuens *et al.* (1983) and David-Pfeuty (1983) found that MTs and IFs form a complex. Redistribution of the MTs after removal of taxol or nocodazole is always followed by redistribution of the IFs. Similar results have been reported by Maro *et al.* (1983) and by Herman *et al.* (1983), who conclude that the cytoplasmic MT complex controls the distribution of the cell organelles, the IFs and the actin MFs: if so, the MT network would really deserve to be named the cytoskeleton. This conclusion is reinforced by the finding that the plasma membrane, the mitochondrial membrane, and the membrane that surrounds secretory granules all possess tubulin as an integral component: all these membranes have binding sites for tubulin, and it would not be surprising if the same was found for IFs (Bernier-Valentin *et al.*, 1983). There is growing evidence that MAP 1 and MAP 2 act as connecting links between the various components (MT, MF, and IF) of the cyskeleton (reviewed by Wicke, 1985) and that cell organelles are translocated along the MTs in an ATP-dependent manner (short review by Schroer and Kelly, 1985).

Another interesting problem is the *control of tubulin synthesis* in living cells. The size of the unpolymerized tubulin pool is probably of crucial importance, since Cleveland *et al.* (1983) found that injection of tubulin into Chinese hamster ovary cells selectively inhibited tubulin synthesis; overall protein synthesis remained unaffected. When such injections are performed on dividing sea urchin eggs, the injeted tubulin molecules are incorporated into the aster MTs unless they have reached their maximal size. Thus, in dividing cells, microinjected tubulin can participate in tubulin polymerization under *in vivo* conditions. It is also known that depolymerization of the MTs by colchicine and some other antitubulin drugs selectively decreases tubulin synthesis. As a result of the increased concentration of tubulin dimers in the cell, there is a fast decrease in both α- and β-tubulin mRNAs in the treated cells. This decrease is due to instability (half-life of 1–2 hr) of these mRNAs and to decreased transcription of the tubulin genes. Thus, the size of the monomer pool may regulate the rate of tubulin mRNA transcription (Ben Ze'ev *et al.,* 1979; Cleveland *et al.,* 1981).

Our present knowledge of the factors that control MT assembly and disassembly in intact cells remains scanty, but there is littlė doubt that what has been established with the *in vitro* model systems remains true for living cells. Colchicine and the other inhibitors of tubulin polymerization—taxol, stabilizing agents such as $D_2O$, GTP analogs, high $Ca^{2+}$, cold treatment, etc.—exert the same effects *in vivo* and *in vitro* whenever it is possible to use them on living cells. Among the regulatory factors that control MT assembly are the cyclic nucleotides (Kram and Tomkins, 1973). According to Willingham and Pastan (1975) and Hsie *et al.* (1975), treatments that increase the cAMP content of the cells (for instance, addition of dibutyrul-cAMP, which penetrates into the cells better than cAMP, or treatment with theophyllin, which inhibits the hydrolysis of cAMP by phosphodiesterase) favor MT assembly *in vivo.* Another factor that influences MT assembly and disassembly is ATP. It is surprising that we had to wait until 1981 to discover that a decrease in ATP content (induced by inhibitors of oxidative phosphorylation) slows down the disassembly of MTs in the presence of colchicine; an increase in ATP content accelerates this disassembly (Bershadsky and Gelfand, 1981, 1983). Likewise, De Brabander *et al.* (1981a) found that inhibition of oxidative metabolism impedes MT disassembly and stabilizes free MTs. A possible explanation of these unexpected findings is that a decrease in ATP production could lead to a decrease in GTP, which is believed to play an important role in Kirschner's (1980) treadmilling model of MT elongation.

### C. Microfilaments

Actin-containing contractile MFs (4–5 nm in diameter) have already been mentioned in the discussion of the membrane-associated cytoskeleton, which plays a fundamental role in capping, endocytosis, and cell locomotion. They can be detected with antisera or by combination with heavy meromyosin (the head region of myosin, which binds to actin and has the ATPase activity). The heavy meromyosin molecules bound to actin can be seen under the electron microscope, and the structurally modified MFs are said to be "decorated."

Actin is by no means the only consistuent of the MFs, but it is the major one from the quantitative viewpoint. In fact, nonmuscle cells (like muscle cells) contain a host of "contractile proteins" (reviewed by Gröschel-Stewart, 1980; Schliwa, 1981; Weeds, 1982; Craig and Powell, 1980). These proteins, like actin, can be detected by immunofluorescence; in this way, it has been shown that they are associated with actin within the cell, where they form a network (Fig. 25) that is distinct from the MT network. This can be demonstrated by treating living cells with MT- or MF-selective poisons (colchicine or cytochalasin B or D, respectively) and staining with the appropriate fluorescent antisera; colchicine destroys the MT network without greatly affecting the distribution of the MFs, and cytochalasin B or D has the opposite effect. However, there are multiple and complex interactions between MTs and MFs (Schliwa *et al.*, 1982).

FIG. 25. Tropomyosin associated with the MFs in a fibroblast. Detection by immunofluorescence. [Lazarides and Burridge (1975). Copyright by M.I.T.]

Unpolymerized actin ($M_r \sim 42{,}000$) is a globular protein called "G-actin"; the polymerized, filamentous form is called "F-actin." One molecule of ATP or ADP is bound to each actin molecule; *in vitro* polymerization of G-actin in F-actin is readily obtained in the presence of $Ca^{2+}$ or $Mg^{2+}$ and is accompanied by the hydrolysis of the bound ATP. *In vivo,* the G$\rightarrow$F-actin polymerization is believed to take place by the treadmilling mechanism proposed by Kirschner (1980) for both MTs and MFs, in which GTP and ATP, respectively, would play a distinct role. Calmodulin seems to be associated with the actin filaments and is probably also involved in *in vivo* polymerization, as well as the MAPs which bind to actin and calmodulin in addition to tubulin (Sobue *et al.,* 1985).

Organisms provided with a heart and with skeletal and smooth muscles possess six different actin species. Nonmuscle cells, in which we are mainly interested here, have only two of these six species; they are called "β-" and "δ-actins" and differ only by four amino acid residues (Bryan *et al.,* 1981). The six actins of *Drosophila* are coded by a family of six closely related actin genes (Storti and Rich, 1976; Fyrberg *et al.,* 1980).

As already mentioned, cytochalasin B binds to G-actin and inhibits its polymerization by preventing the addition of actin monomers at the tip of the growing MF (MacLean-Fletcher and Pollard, 1980). The use of this drug has greatly increased our understanding of the many functions of the MFs, especially those associated with the plasma membrane (see Wessells *et al.,* 1971, for a review of early work on the biological effects of the cytochalasins). More recently, a substance isolated from the venenous mushroom *Amanita phalloides* and called "phalloidin" has also proved useful for the analysis of the role played by the MFs, especially in locomotion. In complete contrast to cytochalasin B, it favors the polymerization of G-actin and stabilizes the F-actin network. If it is injected into cells or amebas, it inhibits locomotion and multiplication (Wehland *et al.,* 1977, 1978). Thus, both cytochalasin B and phalloidin inhibit cell locomotion, which, therefore, closely depends on the reversibility of G-actin polymerization and depolymerization and on the accompanying $Ca^{2+}$-mediated sol$\rightleftharpoons$gel changes.

Actin is remarkable in that it has the capacity to bind many proteins that are present in muscles, taking part in contractility. This was first shown by Lazarides (1975) and by Lazarides and Burridge (1975), who discovered that two typical muscle proteins (tropomyosin and α-actinin) are present in all cells and are associated with actin. As in muscles, tropomysin allows interaction between actin and myosin, and is thus involved in the contraction of the MFs. α-Actinin is required for the formation of bundles of MFs, particularly in stress fibers (Fig. 26) that are involved in the adhesion of fibroblasts to their substratum. In these cables of MFs, actin is associated with myosin, α-actinin, and tropomyosin (or a similar protein present in smooth muscles called "filamin") (Badley *et al.,* 1980).

Interestingly, if fluorescent actin or tropomyosin is injected into cells or

FIG. 26. α-Actinin associated with the MFs in a fibroblast. Detection by immunofluorescence. [Lazarides and Revel (1979).]

amebas, it is quickly incorporated into the MFs, where it participates in assembly and disassembly (Taylor and Wang, 1978; Kreis *et al.,* 1979; Wehland and Weber, 1980; Glacy, 1983). After injection of fluorescent α-actinin isolated from smooth muscles (*in vitro* α-actinin binds to F-actin and produces the crosslinking and gelation of the actin MFs) into fibroblasts, there is a preferential incorporation in the stress fibers that are ''decorated'' at regular intervals. They display a ''regular striped arrangement'' like that of the sarcomeres in muscles, are contractile in the presence of ATP, and contain actomyosin (Feramisco and Blose, 1980; Kreis and Birchmeier, 1980). In fact, according to Schloss and Goldman (1980), MFs have a beaded appearance in mammalian cells, thus resembling dwarf skeletal muscles, They contain actin, myosin, α-actinin, and filamin, like our smooth muscles, and, in addition, several kinds of tropomyosins. They also possess a high-molecular-weight, actin-binding protein. Sanger *et al.* (1983) have proposed a model for the molecular organization of stress fibers in epithelial cells and fibroblasts, where the distribution of the long actin fibers and the shorter, rod-like α-actinin and tropomyosin molecules is basically the same as in striated muscle. A paper by Niederman *et al.* (1983) sheds some light on the possible role of the actin-binding protein. The authors found that mixtures of actin and an actin-binding protein from macrophages self-assemble *in vitro* to form a three-dimensional network similar to the peripheral cytoskeleton of mobile cells. But how injected actin, tropomyosin, or α-actinin correctly finds its way through the cell and inserts in the correct place, and how this "recruitment" process occurs with such accuracy, are still unknown. The cell must possess a system of "policemen" or red and green lights to regulate the traffic of molecules. What we know about this system will be discussed later.

Myosin is essential for MF contractility, since actin is not contractile per se; myosin is required to pull two actin filaments in opposing directions, thus producing contraction. This process requires energy derived from ATP hydrolysis by actomyosin ATPase. Microfilaments however, do not have to exert as strong a tension as muscles do on tendons and skeleton. It is, therefore, not surprising that in nonmuscle cells the actin/myosin ratio is greater than 100 (Crawford *et al.*, 1980), instead of about 0.5 in myofibrils.

It was originally thought that MFs were composed only of actin, myosin, filamin, α-actinin, and tropomyosin. Yet, in fibroblasts moving on a solid substrate, the distribution of these essential constituents of the MFs was not as constant as one would have expected. While actin, myosin, and filamnin are generally associated in the same bundles of MFs, the situation seems to differ in microspikes and ruffles of moving fibroblasts, where there is little myosin (Heggeness *et al.*, 1977). This suggests that the molecular composition of the MFs is not constant, varying in different parts of the cell during locomotion.

There have been many newcomers to the family of actin-binding proteins (actin is a very generous host indeed!). A protein called "profilin" inhibits actin polymerization by combining with G-actin in a 1 : 1 complex called "profilactin" (Blickstad *et al.*, 1980; Mockrin and Korn, 1980). Villin is a $Ca^{2+}$-dependent nucleation factor for actin assembly that is necessary for unidirectional assembly. An increase in the villin/actin ratio increases the number of MFs, but they become shorter in length. These changes are modulated by the free $Ca^{2+}$ concentration (Glenney *et al.*, 1981a,b). According to Jockusch and Isenberg (1981), both vinculin and α-actinin bind to actin, but with opposite effects: α-actinin favors gelation, whereas vinculin inhibits it. These two proteins could thus play an important regulatory role in cell locomotion. Finally, β-actinin accelerates actin polymerization, according to Maruyama and Sakai (1981).

In an excellent review on the immunocytochemical detection of contractile proteins, Gröschel-Stewart (1980) listed, besides actin, the following associated proteins: pancreatic DNase I and profilin, which maintain G-actin in its unpolymerized form; α-actinin, an "actin-binding protein," filamin, gel-actin, which favors its polymerization and the aggregation of the F-actin filaments in bundles; and myosin and tropomyosin, which play the role of regulatory proteins in MF assembly and disassembly. However, this list is not yet complete. In a paper, Maruyama and Sakai (1981) also listed proteins that affect actin polymerization. In addition to DNase I and profilin, an F-actin depolymerizing factor and a so-called PI factor can bind to F-actin monomers and prevent their polymerization, and β-actinin, like cytochalasin B, can bind to filamentous actin. $Ca^{2+}$-sensitive proteins (fragmin, gelsolin, villin) fragment the actin filaments without net depolymerization. Other lists of actin-binding proteins have been published by Weeds (1982) and by Craig and Powell (1980). At least 15 different proteins may associate with cytoplasmic actin. Six of them are $Ca^{2+}$-insensitive crosslinking factors that induce the formation of actin gels; four proteins are $Ca^{2+}$-sensitive modulators that control the rigidity of actin gels; one of them, villin, changes from an F-actin bundling factor into an F-actin severing factor according to the free calcium ion concentration. Finally, there are five $Ca^{2+}$-insensitive modulators that inhibit gelation and, like cytochalasin B, prevent the formation of F-actin by polymerization. Vinculin does not bind to actin, but to sites between the microfilaments and the plasma membrane. This array of proteins that bind to actin with opposite effects and different $Ca^{2+}$ sensitivities is needed to produce the sol–gel changes required for cell motility and for modification of the cell shape. One could predict that this already impressive list of actin-binding proteins is not yet complete. Newcomers to this family include acumentin in macrophages (Southwick and Hartwig, 1982), a 36,000-dalton, actin-binding protein of *Physarum* (Ogihara and Tonomura, 1982), a very widespread actinogelin (Mimura and Asano, 1982), and depactin, which in starfish oocytes cuts the actin MFs into fragments. (Mabuchi, 1983). According to a recent review, mainly devoted to the regulation of the microfilaments by tropomyosin, about 30 actin-modulating proteins have been identified so far (Payne and Rudnick, 1984). In a recent discussion, Pollard (1984) concluded that there are probably more than 60 actin-binding proteins, but that they belong to a limited number of families and that they can be subdivided into three different classes. Some proteins bind to actin monomers and affect actin polymerization; others bind along actin filaments and cross-link them; the last group of proteins cross-links actin filaments to other structures such as myosin, spectrin, microtubule-associated proteins, etc. Maruta *et al.* (1984) reported the presence of three different actin-capping proteins in *Physarum*. These proteins are unexpectedly very similar to actin itself, except that they do not undergo polymerization. It will be interesting to determine whether these actin-like proteins are encoded by genes other than the actins.

This wealth of regulatory proteins clearly indicates that accurate polymerization and depolymerization of MFs are of vital importance for the cell. A curious finding might interest cell biologists. If amebas or HeLa cells are treated with high concentrations of dimethyl sulfoxide (DMSO), actin bundles can be detected in the cell nucleus (Fukui and Katsumaru, 1979). Further analysis of this unexpected phenomenon has confirmed that after treatment with 10% DMSO, the stress fibers disappear and their actin is translocated into the nucleus. However, there is no translocation into the nucleus of the other main components of the stress fibers, tropomyosin, α-actinin, and myosin (Sanger *et al,*, 1980). These observations have been made under unphysiological conditions in which the cells are under great stress. However, they may be of interest in furthering our understanding of MF architecture and nuclear membrane permeability, which will be discussed in Volume 2, Chapters 1 and 2.

### D. Intermediate Filaments

As previously mentioned, IFs are much more variable in size and chemical composition than MTs and MFs, in which one protein molecule (tubulin or actin) so heavily predominates that it imposes its morphology on the cytoskeletal structures to which it belongs. The common-denomination IFs are composed of many cytoskeletal structures that, from the biochemical viewpoint, have little or nothing in common.

Lazarides (1981) provided an excellent mini-review (only two pages) of the present status of the IF question. There are four major classes of IF with a width of about 100 Å: epiderm cells contain cytokeratin filaments; neurons have three distinct categories of neurofilaments composed of three proteins of $M_r$ 210,000, 160,000, and 68,000; glial cells possess a 50,000-dalton glial fibrillary acidic protein; and finally, cells of mesodermal origin have IF composed of either vimentin (52,000 daltons) or desmin (50,000 daltons). These last two proteins are present in a great number of cells, either alone or in combination (Fig. 27). Finally, smooth and skeletal muscles, but not heart muscle or fibroblasts, have IF composed of a larger protein, synemin ($M = 230{,}000$). Interestingly, during differentiation of myoblasts, the vimentin/desmin ratio decreases; IFs of the common vimentin type are progressively replaced by more specialized IFs in which desmin (the major subunit of the IF in smooth muscles) predominates. This is an example of the many molecular changes that occur during muscle differentiation. It should be added that several proteins present in the IFs, including vimentin, desmin, and the three proteins present in the neurofilaments, can be phosphorylated by cAMP-dependent protein kinases. This could induce changes in protein conformation and control the formation or disappearance of IFs in a variety of cell types.

Newcomers to the IF family are plectin and decamin. Plectin crosslinks the elements of the cytoskeleton (Wiche and Baker, 1982); decamin IF are bundles of 2- to 3-nm protofilaments (Steven *et al.*, 1982).

In discussing intercellular adhesions and desmosomes, we noted that the tonofilaments that reinforce the contacts between adjacent epithelial cells are made up of cytokeratins (Franke *et al.*, 1981a,b). There is an entire family of

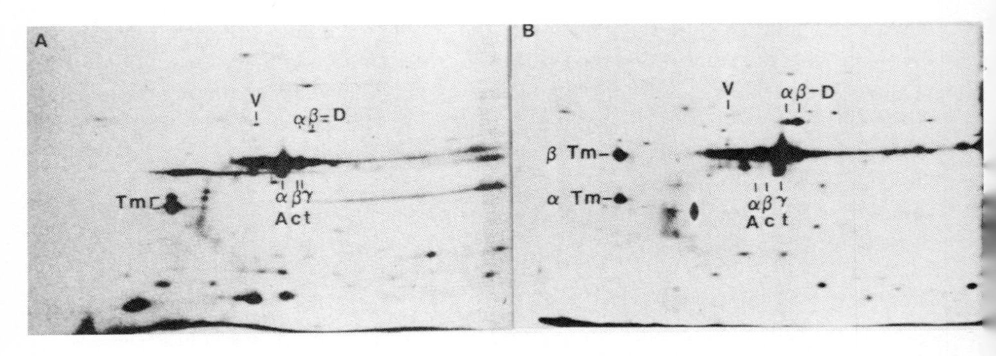


Intermediate-filament Nomenclature

| Filament type | Major synonym | Tissue (cell) | Molecular weight (× 10⁻³) | Reference |
|---|---|---|---|---|
| α-Keratins | Cytokeratins<br>Prekeratins<br>Epidermins | Epithelium | 68 63, 60,<br>58, 54, 52 | Matoltsy (1965)<br>Franke *et al.* (1978a) |
| Vimentin | Fibroblastic<br>Intermediate<br>Filament protein (F-IFP)<br>Decamin<br>Lentin<br>58-kd Protein of fibroblast | Mesenchyme | 52, 57, 54, 58 | Franke *et al.* (1978b) |
| Desmin | Mesosin<br>Skeletin<br>100-Å filament<br>Subunit protein | Muscle | 50, 55<br>85 | Lazarides and Hubbard (1976)<br>Small and Sobieszek (1977) |
| Neurofilament protein | Filarin<br>68-kd Core protein | Neurons | 68, 160, 210 | Schmitt (1968) |
| Glial filament protein | Glial fibriliary protein (GFP)<br>Glial fibriliary acidic protein | Glial | 51 | Huneeus and Davison (1970)<br>Eng *et al.* (1971)<br>Liem *et al.* (1978)<br>Goldman *et al.* (1978) |

FIG. 27. Translation of skeletal and smooth muscle mRNA *in vitro*. The [$^{35}$S] methionine-labeled translation products of skeletal (A) and smooth (B) muscle poly(A)$^+$ RNA have been electrophoresed on two-dimensional isoelectric focusing-sodium dodecyl sulfate gels. The positions of actin (Act), desmin (D), vimentin (V), and tropomyosin (Tm) are indicated. [O'Connor *et al.* (1981). Table from Brinkley (1982).]

these cytokeratins of $M_r$ 48,000–65,000 (10–20 members encoded by different structural genes). Parenchymatous cells of liver and pancreas have IFs that are morphologically like epidermal tonofilaments, but their cytokeratin differs from the prekeratin of epidermal cells (Franke *et al.,* 1981b).

In general, a single type of IF protein is expressed in a given cell type, and these proteins may therefore be taken as useful markers of cell differentiation; they might also be useful for a classification of tumors based on more precise grounds than purely morphological criteria (reviewed by Osborn and Weber, 1982). However, simultaneous expression of vimentin and cytokeratin IFs is not a rare event: epithelial cells (Henderson and Weber, 1981), cells of the parietal endoderm in chick embryos (Lane *et al.*, 1983), and human metastatic carcinoma cells (Ramaekers *et al.*, 1983) coexpress vimentin and cytokeratin. Dräger (1983) observed the coexistence of neurofilaments and vimentin IFs in mouse retina neurons. It is probable that, in epithelial cells, the vimentin and cytokeratin networks are associated. Injection into the cells of an antibody raised against either vimentin on cytokeratin induces the collapse of the two networks (Klymskowsky, 1982). When epithelial cells express only vimentin, expression of their cytokeratin genes is somehow repressed; elegant experiments in which keratin mRNA has been injected into such cells have shown that the injected cells synthesize keratin and assemble it in IFs (Kreis *et al.*, 1983). The absence of keratin in the uninjected cells thus resulted from inactivity (repression) of the corresponding genes.

We still know little about the functions of the IFs in living cells. This is due to lack of suitable inhibitors and to molecular heterogeneity. It is more difficult to inject an antivimentin or anticytokeratin antibody into a few cells than to add colchicine, taxol, or cytochalasin to a whole cell culture. However, a few suggestions about the role played by the IFs in the cell economy have been made. One of their functions might be the anchorage of the nucleus in the center of the cell (Lehto *et al.*, 1978; Wang *et al.* 1979a,b; Henderson and Weber, 1980); some of the vimentin filaments form a "cage" around the nucleus in colcemid-treated cells (Fig. 28). Vimentin IFs might also play a role in capping: when lymphocytes are induced to form a cap, there is a redistribution of the vimentin IFs; the fact that one observes the cocapping of actin, myosin, tubulin, and vimentin confirms that there are physical or functional connections between the three main cytoskeletal networks (Dellagi and Brouet, 1982); of course, the accumulation of vimentin in the caps might well be a passive process and does not necessarily imply an active role of the IFs in capping. Lane *et al.* (1983) have pointed out that in chick embryos, vimentin is expressed only in cells in which cell-to-cell contacts are reduced (endothelial and mesenchymal cells). In agreement with this view is the finding of Ramaekers *et al.* (1983) that solid tumors possess only cytokeratin, while both cytokeratin and vimentin are expressed in ascites or pleural fluids. All these observations show that, as expected, cytokeratins play an important role in cell-to-cell adhesion in epithelia and a few other cell types. But the physiological role of vimentin remains obscure. A paper by Venetianer *et al.* (1983) brings a glimpse of light in darkness: they studied a clone of dedifferentiated hepatoma cells that no longer synthesized cytokeratins and found that the arrest of cytokeratin synthesis had no consequences for growth and differentiation; they suggested that the IF might be related to the specialized functions of differentiated cells. Of potential interest for future research in this field is the isolation by Nelson and Traub (1981, 1983) of a $Ca^{2+}$-requiring protease that specifically hydrolyzes vimentin and desmin; this enzyme, like the IFs themselves, is associated with the detergent-resistant cytoskeleton; since it

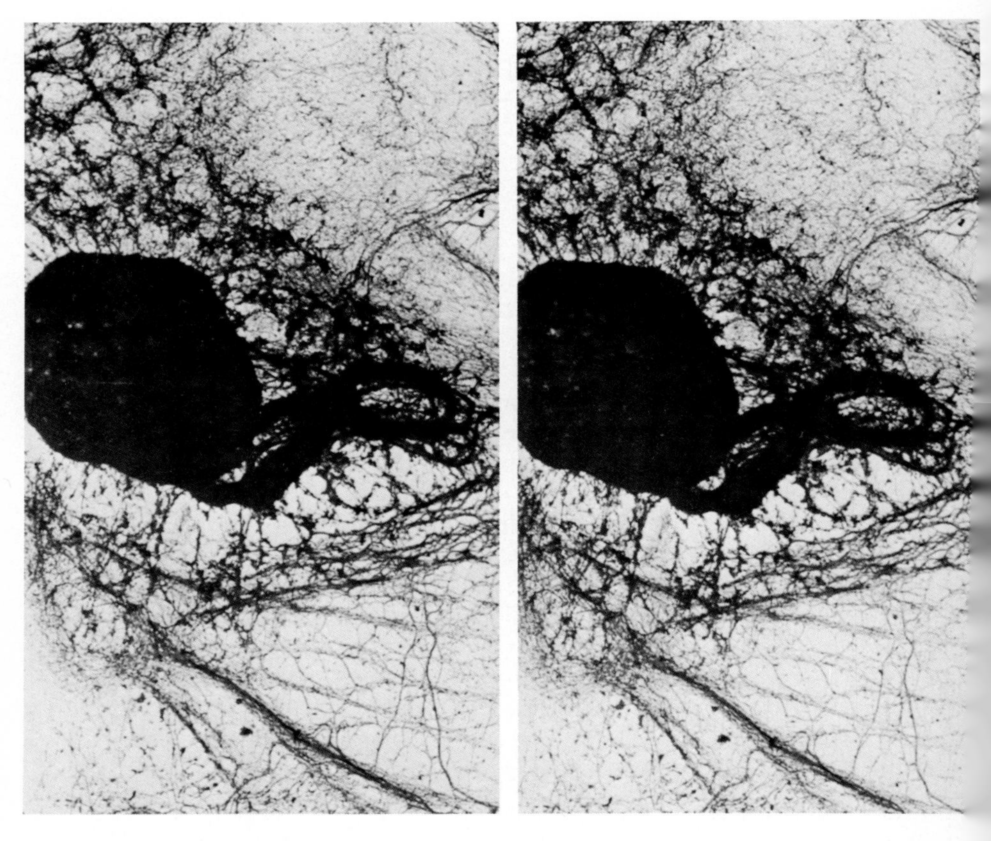

FIG. 28. Intermediary filaments form a reticulated "cage" around the cell nucleus in colcemid-treated cells. [Henderson and Weber (1981).]

produces only limited proteolysis of its substrates, it might play a role in the control of the still enigmatic vimentin IF functioning.

Before closing this section, one should stress the important fact that the cytoskeleton is not a static, but a dynamic structure. Microtubules, microfilaments, and probably intermediate filaments continuously undergo polymerization and depolymerization of their subunits. In addition, there are close interactions between the various components of the cytoskeleton. For instance, Celis *et al.* (1984) have recently shown that in monkey epithelial kidney cells, the cytokeratin and vimentin intermediate filaments interact with both microtubules and microfilaments. This was demonstrated by experiments where the cells were treated with colchicine and cytochalasin. Interesting details about these questions can be found in a recent symposium published in the *J. Cell Biol.* (**99,** (1, 2), 1984).

## III. SINGLE MEMBRANE–BOUND CYTOPLASMIC ORGANELLES: THE GERL SYSTEM

### A. General Background

The cytoplasm imprisoned in the meshes of the cytoskeletal network contains many organelles that differ widely in their ultrastrucutre, chemical composition, and function. Some of them, which display the capacity to replicate in a semiautonomous way and thus have genetic continuity, will be discussed in Section IV; mitochondria, chloroplasts, and centrioles with their appendages (cilia or flagella). The organelles that will be examined here are characterized morphologically by the presence of a single membrane that separates them from the structureless hyaloplasm (the cytosol of the biochemists). This membrane is basically similar to the plasma membrane and is thus composed of a lipid bilayer with associated proteins. These proteins possess both similarities and differences, which are observed when one compares the various cell membrane fractions isolated by differential centrifugation of homogenates (Elder and Morré, 1976). For Morré, there is a continuity between the plasma membrane and the nuclear membrane. Between the two is the endoplasmic reticulum (ER). This continuous membrane undergoes local changes in chemical composition. The single membranes surrounding all of the cytoplasmic cell organelles to be discussed here appear to have a common origin, but this unitary view is not shared by all investigators. Fujiki *et al.* (1982) have pointed out that the proteins of the different organelles differ widely. These organelles include the ER, the Golgi bodies, lysosomes and related particles or vacuoles, the peroxisomes, and, in plants, the glyoxysomes, and secretion granules of gland cells (see Fig. 2, Chapter 1, for a schematic representation of these cytoplasmic organelles). Both their structure and functions are different. ER is the site of synthesis and transport of secretory proteins that, after undergoing modifications (glycosylation in particular) in the Golgi region, are stored in the secretion granules and exported by exocytosis through the cell surface. The lysosomes are the digestive organelles of the cell. Peroxisomes and glyoxysomes are the sites of oxidative processes that differ from those that take place in the mitochondria.

From this brief overview, it is clear that there is a continuous movement of proteins exported from the ER to the plasma membrane via Golgi elements. Novikoff (reviewed in 1976) integrated the lysosomes into this scheme. Using cytochemical studies at the ultrastructural level, he showed that Golgi elements, ER, and lysosomes are part of a common intracellular system—the GERL—that is characterized by the presence of acid phosphatases. Peroxisomes are believed to originate from the ER, as observed from electron micrographs.

## B. The Endoplasmic Reticulum

### *1. Historical Background*

As we saw in Chapter 1, the ER story began with the discovery by Albert Claude of microsomes in the homogenates of chick embryos (1943). These small RNA-rich granules (50–200 nm) are present in all cells (animal or vegetal) and contain lipids and proteins. They were considered by Brachet and Jeener (1944), and quite accurately, as the probable agents of cellular protein synthesis. This was proved true about 10 years later. Under the electron microscope, microsomes appear as vesicles limited by a membrane covered by very small granules. Treatment with bile salts [deoxycholate (DOC), for instance] disintegrates the microsomal membrane without dissolving the small granules (Palade and Siekevitz, 1956). Thus the microsomes, isolated by high-speed centrifugation of homogenates, are composed of two constituents: a DOC-sensitive lipoprotein membrane constituent and the small RNA-rich granules, which are now called "ribosomes."

Electron microscopic studies by Porter (1954, 1955) and Palade (1955), at about the same time, showed that the structureless hyaloplasm contained a very delicate network, the endoplasmic reticulum. It was a network of canalicules uniting the plasma membrane to the nuclear membrane, thus allowing the intracellular distribution of substances present in the lumen. When the canalicules anastomosed, they enlarged and formed cisternas (Fig. 29). Simultaneously—and in the same laboratory—Palade discovered (1955) that parts of the ER were covered by small (10 to 15 nm) granules that were particularly abundant in all protein-synthesizing cells. It was soon discovered that "Palade's small granules" were very rich in RNA and were indeed instrumental in protein synthesis. These structures are now universally called "ribosomes," but it should be recalled that Palade saw them under the electron microscope in fixed cells before biochemists isolated them from DOC-treated microsomes. Thus, microsomes are mainly fragments of the ER. Furthermore, since the entire ER is not covered with ribosomes, one has to distinguish between rough and smooth ER (see Fig. 29; see also Fig. 31). Only the first bears ribosomes; the smooth ER is essentially composed of membranes and is free of ribosomes. All ribosomes are not associated with ER membranes; many ribosomes, particularly in protein-synthesizing cells, lie free in the hyaloplasm.

An excellent description of the rough and smooth ER has been provided by Porter (1961); we have already mentioned Novikoff's review (1976) based on cytochemical observations. A more recent review, oriented more toward the biochemical aspects, has been published by Svardal and Pryme (1980).

### *2. Chemical Composition and Role of ER Membranes*

Improved methods of differential centrifugation have allowed the separation of rough and smooth ER membranes (Beaufay *et al.*, 1974). Analysis of these

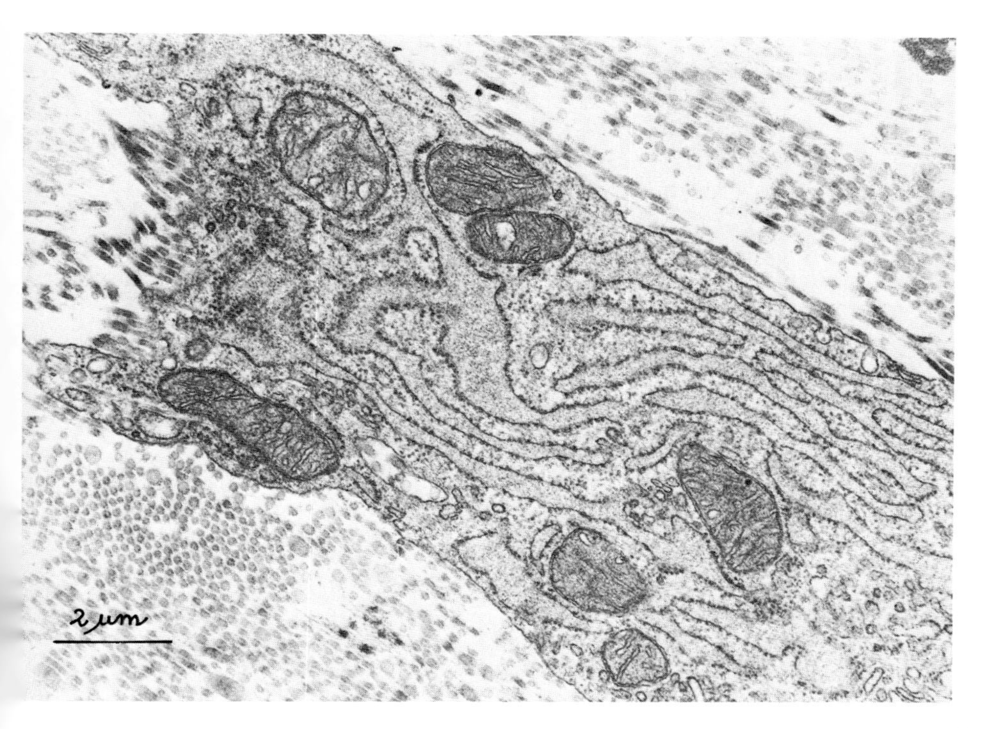


FIG. 29. Fibroblast rough ER, with cisternae dilated to various extents (skin of a young mouse). [Courtesy of N. Van Lerberghe.]

membranes, which have phospholipids and proteins as their main constituents, has not disclosed profound differences between the two. Paiement *et al.* (1980) in Beaufay's laboratory have discovered the intriguing fact that microsomal membranes fuse together upon addition of GTP, but the molecular basis of this phenomenon and its possible occurrence in the living cell remain unknown.

Most of the biochemical work on ER membranes has been done on material isolated from rat liver. The preferred marker enzymes for ER membranes are glucose-6-phosphatase and NADH-cytochrome *c* reductase, which are abundant in liver microsomal membranes (Fig. 30). However, it is not absolutely certain that these two enzymes are reliable markers for microsomal membranes isolated from all tissues and organs, since they are related to two specialized functions of the liver: glycogen breakdown and detoxification, by oxidation, of toxic organic chemicals.

There is good evidence, largely based on ulttrastructural studies, that the membranes of the smooth ER are involved in the synthesis, transport, and accumulation of lipids and glycogen (see Porter, 1961).

In contrast to cells that synthesize proteins and have an abundant rough ER,

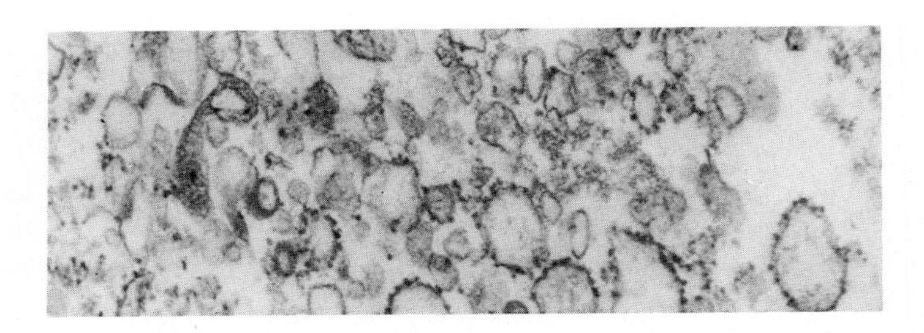

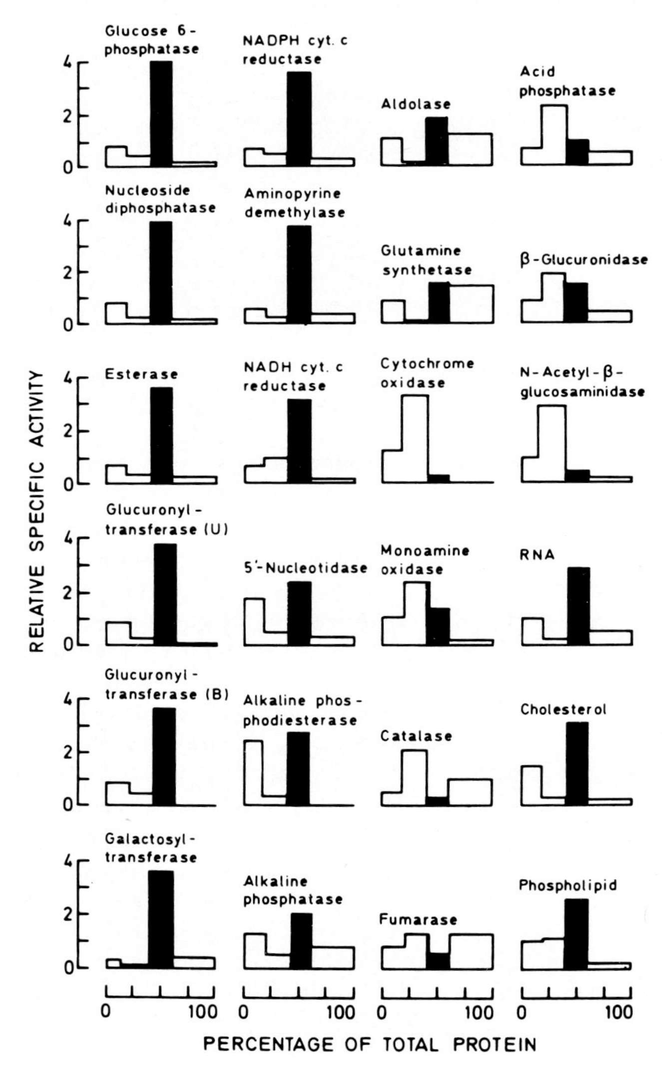
Glucose 6-phosphatase
NADPH cyt. c reductase
Aldolase
Acid phosphatase
Nucleoside diphosphatase
Aminopyrine demethylase
Glutamine synthetase
β-Glucuronidase
Esterase
NADH cyt. c reductase
Cytochrome oxidase
N-Acetyl-β-glucosaminidase
Glucuronyl-transferase (U)
5'-Nucleotidase
Monoamine oxidase
RNA
Glucuronyl-transferase (B)
Alkaline phos-phodiesterase
Catalase
Cholesterol
Galactosyl-transferase
Alkaline phosphatase
Fumarase
Phospholipid
RELATIVE SPECIFIC ACTIVITY
0
2
4
0
100
PERCENTAGE OF TOTAL PROTEIN

smooth ER membranes are accumulated in cells that synthesize steroid hormones (testosterone, for instance, in the Sertoli cells of the testis) or carotenoids (in the retina) (Fig. 31). It has been shown by Drochmans (1963) that glycogen granules accumulate and form clusters in close association with smooth ER membranes in the liver. These morphological observations are corroborated with biochemical findings showing the association of such enzymes as cholesterolesterase, vitamin A esterase, a fatty acid oxidation system, and, as we have seen, glucose-6-phosphatase with ER membranes. According to a report by Witters *et al.* (1981), microsomes have a high acetyl-CoA carboxylase activity and could thus play an important role in fatty acid synthesis.

Another very important function of ER membranes in the liver is detoxification of organic compounds, which can take place by various biochemical pathways (glucuronoconjugation, for instance); the most important of these processes is oxidation of aromatic hydrocarbons (arylhydrocarbons). The enzyme system that catalyzes the oxidation of organic molecules in the ER membrane has been called "mixed-function oxidases" or the "monooxygenase system" because it can oxidize a large variety of chemically unrelated organic compounds. Interestingly, the activity of the mixed-function oxidase system is low in normal liver and is greatly increased by the presence of an aromatic oxidizable substrate (barbiturates, for instance); this provides an example of substrate induction of enzymatic activity in mammalian cells. Furthermore, the number of enzyme inducers (i.e., possible substrates for the enzyme) is so large that activation of the monooxygenase system can almost be compared to the multiplicity of immunological responses toward a huge number of antigens. The microsomal oxidizing system for aromatic hydrocarbons (often called "arylhydrocarbon hydroxylase") is particularly important, since among its substrates are carcinogenic polycyclic hydrocarbons present in, among other places, tobacco tar and smoke. Microsomes from liver can inactivate carcinogenic hydrocarbons by oxidation; unfortunately, the reverse usually occurs: arylhydrocarbon hydroxylase very often transforms a noncarcinogenic hydrocarbon into a carcinogenic one. This is the reason why, in the so-called Ames (1979) test for screening mutagenicity and putative carcinogenicity of organic compounds (this test is based on the induction of mutations in bacteria), it is necessary to add liver microsomes to the chemical that is being tested; innocuous organic chemicals are often transformed by the microsomal enzymatic machinery into strongly mutagenic substances.

From a biochemical viewpoint, the major component of the monooxygenase system is a special cytochrome, called "cytochrome *P*-450"; it is reduced by

---

FIG. 30. Distribution pattern of marker enzymes and other constituents in four fractions obtained by differential centrifugation of a rat liver homogenate. (*Left to right*) Nuclei, mitochondria + lysosomes, microsomes, (black areas) and supernatant. [Redrawn from Beaufay *et al.* (1974.] (*Top*) Section through a microsomal pellet. [Courtesy of A. Vokaer and A. Schram.]

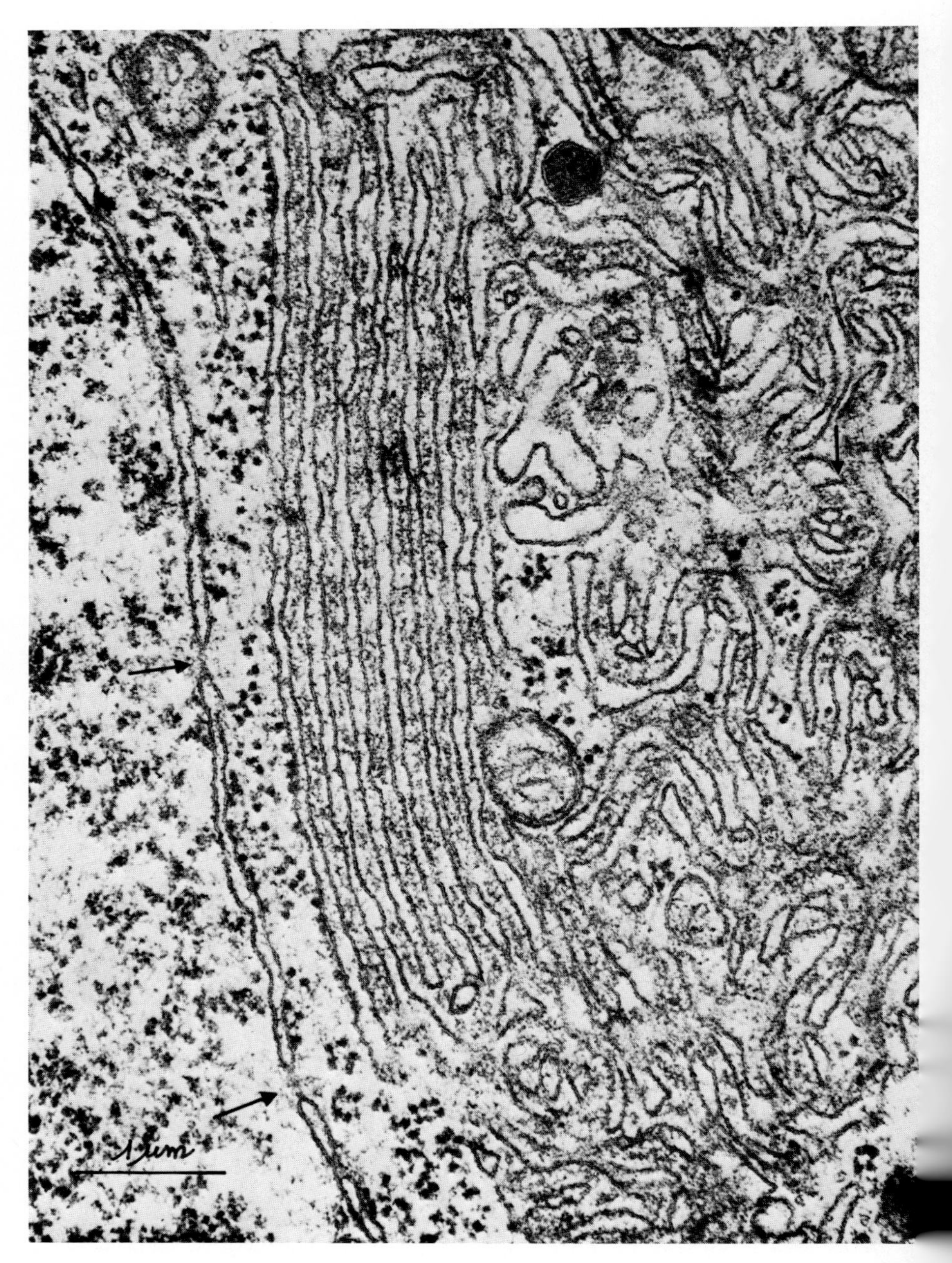


FIG. 31. Endoplasmic reticulum in the "perikaryon" of the rat lateral geniculate nucleus. Arrows show nuclear membrane pore complexes. [Karlson (1966).]

NADPH–cytochrome *P*-450-reductase and cytochrome $b_5$, which are also located in microsomal membranes. Cytochrome *P*-450 (so called because it has a maximum of light absorption at 450 mm when it is in the reduced form) can bind to substrate and molecular oxygen. Since cytochrome *P*-450 catalyzes the metabolism of a huge number of endogenous or exogenous substrates, it should exist in multiple forms. This is indeed the case: according to a report by Guengerich *et al.* (1981), differences in cytochrome *P*-450 can be detected when one compares human liver microsomes in two individuals. It is also known that the multiple forms of cytochrome *P*-450 are encoded by a family of closely related, but not identical, genes (Ramperz and Walz, 1983).

*3. Ribosomes*

Despite the author's interest (see Chapter 1) in the role played by cytoplasmic RNA (at least 80% of this RNA is ribosomal in most cells) in protein synthesis, very little will be said here about the ribosomes and their intervention in protein synthesis. Since this subject has been treated in detail in many books dealing with biochemistry or molecular biology, a detailed discussion here would be superfluous.

Very briefly, eukaryotic ribosomes have a sedimentation constant of 80 S and are formed from two subunits of unequal size (sedimentation constants of 60 and 40 S, respectively). Ribosomes are composed of approximately 50% RNA and 50% proteins. The rRNAs present in the large (60 S) subunit are the 28 S RNA (with a small 5.8 S fragment that is attached noncovalently and can be released upon heating) and a very small (120-nucleotide) 5 S RNA. The small 40 S subunit has a single 18 S rRNA. The approximate $M_r$ of the two major 28 and 18 S rRNAs are 1,200,000 and 600,000 daltons, respectively. As we shall see, they are coded by different genes located in the nucleolus; the 5 S genes are localized on the chromosomes. The biogenesis of the ribosomes, which has been reviewed by Hadjiolov (1981), will be discussed in the chapters dealing with the nucleolus and with oogenesis. There are about 40 distinct proteins in the large subunit and 30 in the small one. One of these ribosomal proteins, called the "S 5 protein," is of current interest because it is easily phosphorylated by protein kinases. There is, as yet, no evidence that phosphorylation and dephosphorylation of this protein modulate the rate of protein synthesis in the cell. If one compares ribosomes extracted from a variety of eukaryotes, one finds a large number of homologous sequences in their rRNAs. This shows that they have been highly conserved during evolution. Ribosomal proteins are more variable and have thus evolved to a greater extent than the rRNAs.

It is well known that protein synthesis takes place on polyribosomes (or *polysomes*) where 80 S ribosomes associate with an mRNA coding for a given protein. The number of ribosomes associated in the polysomal chains depends on the size of the mRNA, which is correlated with the size of the protein that is being synthesized. Outside the polyribosomes, the ribosomes are dissociated and form a pool of free subunits. Besides mRNA, transfer RNAs (tRNAs) charged with the appropriate amino acids are bound, in polysomes, to the ribosomes. Formation of the initiation complex (reviewed by Thomas *et al.*, 1981) requires GTP, methionine tRNA ($tRNA^{Met}$), an initiation codon in mRNA, 80 S ribosomes, and three protein factors; elongation of the growing polypeptide chain requires other protein factors (one of them, called "translocase," hydrolyzes GTP into GDP and inorganic phosphate, producing the energy required for elongation). Two termination factors and a termination codon on the mRNA are necessary for the arrest of the process and the release of the completed polypeptide chain. The protein-synthesizing machinery is so complex that we cannot go into greater detail here.

Specific inhibitors of cytoplasmic protein synthesis are important tools for the cell biologist. There is an entire array of such inhibitors. The most popular are cycloheximide, which inhibits the elongation step (it "freezes" the polysomes), and puromycin, which releases incomplete polypeptide chains and dissociates the polysomes.

Many cells possess both free and membrane-bound polysomes. Oocytes and eggs, in particular, possess a very high proportion of free ribosomes; their ribosomes can even undergo crystallization (Fig. 32) if protein synthesis is maintained at low levels by a cold environment. This occurs, for instance, in lizard oocytes during hibernation (Taddei, 1972); their crystallized ribosomes are inactive in protein synthesis under *in vivo* conditions, but they are fully active when they are introduced in a system for *in vitro* protein synthesis (Taddei *et al.*, 1973). Ribosome crystallization thus allows the oocyte to store part of its protein-synthesizing machinery when it is not required. According to an x-ray analysis, RNA forms a dense core and proteins are mainly located at the periphery of crystallized ribosomes (Kühlbrandt and Unwin, 1982).

In the rough ER, the ribosomes are attached to the membrane by their 60 S subunit (Baglioni *et al.*, 1971). Why do ribosomes bind only to the rough ER and not to the smooth ER? The specificity of binding is inherent not in the ribosomes but in the membrane. Kreibich *et al.* (1978, 1982) have shown that membranes from the rough ER contain two proteins, called "ribophorins I and II," which can bind to the ribosomes; these proteins are absent from the smooth membranes. The two ribophorins are transmembrane glycoproteins of 65,000 and 63,000 daltons; membranes from the rough ER (previously stripped from their ribosomes), in contrast to the smooth ER membranes, have high-affinity binding sites for ribosomes; the ribophorins are associated with these sites, which do not exist in smooth ER membranes (Marcantino *et al.*, 1982).

There are differences in the mRNA populations associated with free and membrane-bound ribosomes. The mRNAs coding for secretory proteins are associated with the rough ER; free ribosomes are linked to other mRNAs coding for

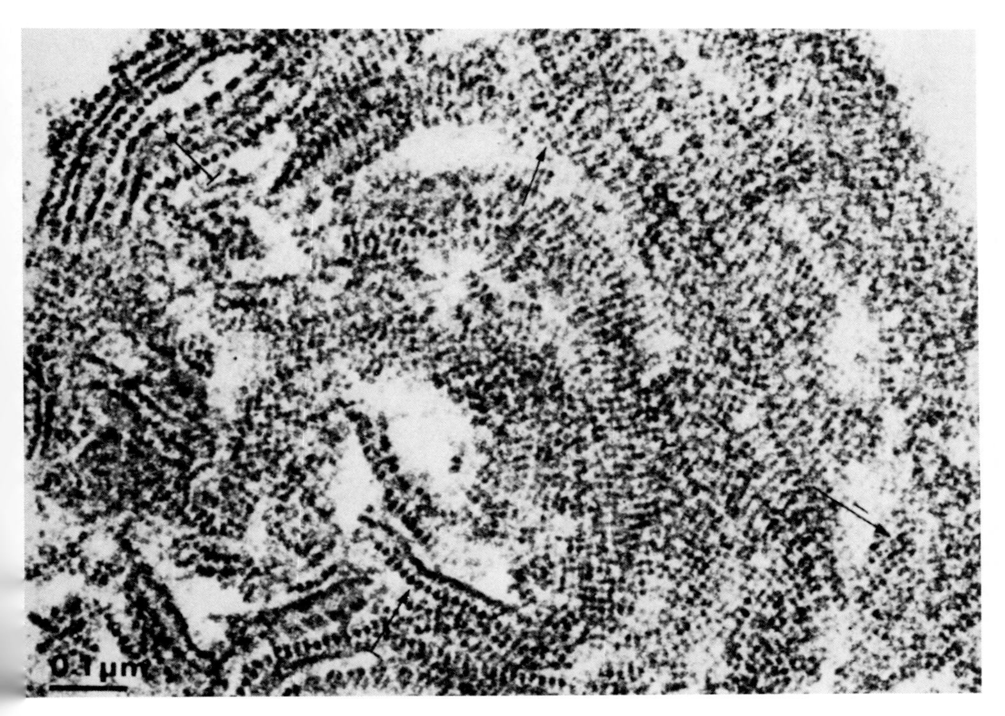


FIG. 32. Electron micrograph of crystalline ribosomes (ribosomal pellet from lizard oocytes). [Taddei *et al.* (1973).]

other proteins, which are generally released in the hyaloplasm (Clissold *et al.*, 1981).

This leads us to another interesting problem: the role of the ER in the synthesis and translocation of secretory proteins. The first point that should be made is that for many proteins (and, in particular, for those that should be excreted from the cells), the product of mRNA translation in the polysomes is not the protein itself, but an inactive precursor that contains an additional peptide sequence. This sequence must be split by an appropriate proteolytic enzyme in order to activate the protein. These precursors of enzymes and hormones are called "proproteins" [for instance, proalbumin, procathepsin D (a lysosomal enzyme), proinsulin, etc.]. The splitting of the additional peptide occurs after the secreted protein has moved out of the ER and the Golgi complex. In the case of the much studied exocrine pancreas, the transformation of the inactive zymogens present in the secretory granules into active enzymes has been discussed at length by Neurath and Walsh (1976). Clearly, the pancreatic cell could not survive if enzymes such as trypsin, chymotrypsin, amylase, nucleases, etc. were set loose and allowed to diffuse freely into their hyaloplasm. There is a double protection against such a catastrophe: the presence of a membrane around the secretory granules and the inactivity of the enzymes when they are in the form of zymogens (proenzymes).

How are proteins synthesized on polysomes attached to the rough ER finally excreted? This problem was first studied in the pancreas by Palade (reviewed in 1975), who used autoradiography (after pulses at various time intervals with tritiated amino acids) at the ultrastructural level. His kinetic analysis of protein synthesis and protein transport in the pancreas showed that exportable proteins (like the pancreatic enzymes) are synthesized by polysomes attached to the ER; such proteins accumulate and move forward in the canaliculi (and thus move into the lumen) of the ER. They are first stored in its cisternas, then in the Golgi region, and finally exocytosed at the cell surface. This general pathway is now universally accepted.

At the molecular level, two problems remain: (1) how do the proteins synthesized on polysomes located on the external cytoplasmic surface of the ER membranes penetrate into their lumen, and (2) how are the secretory products transported once they are inside the ER lumen? A satisfactory answer to the first question has been given by G. Blobel and his colleagues (Sabatini and Blobel, 1970; Blobel and Dobberstein, 1975a,b; Blobel, 1977). His "signal hypothesis" (Fig. 33) assumes that the polysomes associated with the cytoplasmic face of the ER synthesize secretary preproproteins, which are provided with a hydrophobic tail (the signal peptide) composed of a number of amino acid residues. This hydrophobic tail leading sequence would penetrate into the lipid bilayer and

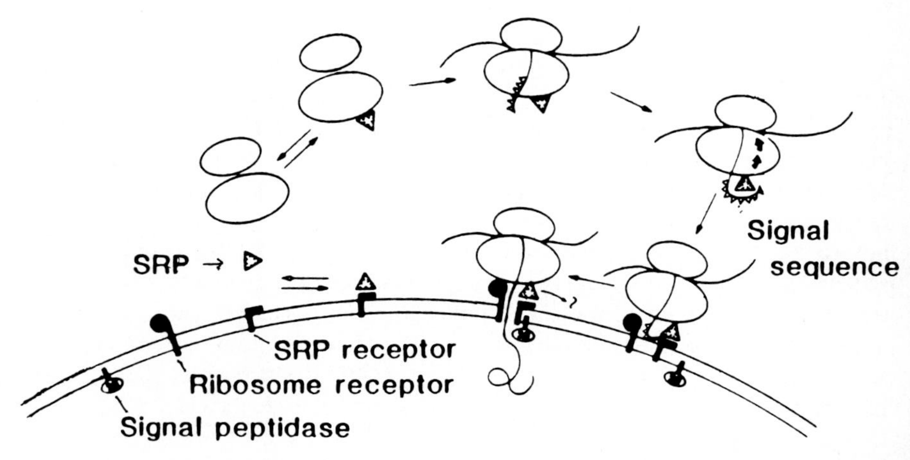


FIG. 33. Model of protein translocation. The signal recognition particle that acts as a transient third unit of a ribosome in linking protein translation with the SRP is composed of 7 S RNA and six proteins. [Walter *et al.* (1982).] Accumulation occurs in the ER cisternae.

conduct the proprotein through the ER interface during its synthesis: protein synthesis and transport would thus be linked together. At the luminal surface of the ER membrane, a signal peptidase would specifically remove the signal peptide, leaving the preprotein in the ER lumen. An important point of Blobel's hypothesis is that the binding of the proteins to the ER membrane and their transport to the lumen are closely linked to their translation.

More recent work has also strongly favored the view that the ER membrane possesses specific receptors for the signal sequences—signal recognition protein (Prehn *et al.*, 1980, 1981; Meyer and Dobberstein, 1980; Walter and Blobel, 1980; Jackson *et al.*, 1980. Bendzko *et al.*, 1982; Anderson *et al.*, 1982; Gilmore *et al.*, 1982). Extraction of microsomes by salt suppresses the capacity of the microsomal membranes to translocate the nascent secretory proteins vectorially across their lipid bilayer. However, the signal peptide receptor present on the cytoplasmic face of the ER is responsible only for binding, and the transport of the nascent protein through the ER membrane requires ribosomal synthetic activity (Prehn *et al.*, 1981).

The signal recognition protein (SRP) of microsomal membranes from dog pancreas has been isolated and characterized by Walter *et al.* (1981) and Walter and Blobel (1981a,b). It is an 11 S complex that has the remarkable property of inhibiting the *in vitro* translation of the mRNAs coding for secretory proteins, but not that of other mRNAs; for instance, the translation of globin mRNA is unaffected by addition of SRP. It thus binds selectively to ribosomes engaged in the synthesis of secretory proteins. Chain translocation across the ER membranes is mediated by specific proteins and is thus a receptor-mediated event. SRP reacts with the amino terminal signal peptide of the nascent chain and this inhibits further translation.

An interesting paper by Meyer *et al.* (1982) shows that this process is even more complicated than the preceding discussion indicates. The present situation, in this dynamic field, is summarized in Fig. 33. Synthesis of secretory proteins begins on free ribosomes. After polymerization of the 60–70 amino acids which form the signal (leading) sequence, and its emergence, protein synthesis is stopped by (Walter and Blobel, 1981a,b). This SRP is a 250,000-dalton complex to which a 7 S RNA is associated, forming a signal recognition particle; (Walter and Blobel, 1982); its function is to selectively inhibit the translation of nascent proteins. Thus, the 11 S signal recogniton particle is made up of six nonidentical polypeptide chains and one molecule of 7 S RNA. This small cytoplasmic RNA is absolutely required for protein translocation across the ER and for the selective arrest of secretory protein synthesis; however, it is inactive in the absence of the SRP proteins. If one reconstitutes the signal recognition particles by mixing 7 S RNA with the six SRP proteins, 11 S particles are obtained; they are fully active in protein translocation across the protein integration into ER membranes (Walter and Blobel, 1983). However, more recent recombination experiments by Siegel and Walter (1985) have shown that if 2 out of the 6 SRP proteins are missing, protein translocation can occur in the absence of an arrest of polypeptide chain elongation; this arrest is thus not a prerequisite for translocation.

The ER possesses another specific membrane protein discovered by Meyer *et al.* (1982), who called it "docking protein"; its function is to release the block of protein synthesis induced by SRP. The result is that the initiated ribosomal complex makes contact with the correct membrane, i.e., the membrane that contains the docking protein. When the block in protein synthesis is lifted by contact with this docking protein, translocation tnrough the membrane, coupled with a continuation of translation, can take place. The function of the SRP is to allow the penetration of the secretory protein in the ER; otherwise the protein would be released into the cytoplasm (cytosol, hyaloplasm) with disastrous results, as in the case of pancreatic enzymes (proteases, nucleases, amylase).

To summarize, two factors of the translocation machinery are of major importance: the 11 S signal recognition particle made up of six distinct proteins and one molecule of 7 S RNA, and the docking protein; it is a 72,000-dalton integral protein of the ER. The SRP decodes the information contained in the signal sequences of nascent secretory proteins (preproproteins), lysosomal and membrane proteins. Reaction between SRP and its receptor allows the resumption of protein synthesis (polypeptide chain elongation); the translocated protein is then integrated into the lipid bilayer of the ER membrane. The interaction between ribosome, SRP, and SRP receptor is transitory and requires the intervention of the above-mentioned ribophorins I and II (Walter *et al.*, 1982). According to a more recent review by Walter *et al.* (1984), the 7 S RNA (now called SL7 RNA) is the structural backbone of the signal recognition particle. This cytoplasmic RNA displays some homology with highly repeated interspersed DNA sequences present in the genome (the so-called *Alu* sequences, to be discussed in Chapter 4). The SRP receptor (docking protein) is an integral membrane protein of the endoplasmic reticulum.

It should be added that glycosylation of proteins (membrane and secretory glycoproteins) may be a cotranslational or posttranslational event: *N*-glycosylation (addition of a polymannose chain to ovalbumin, for instance) takes place in the rough ER during the synthesis of the polypeptide chain; *O*-glycosylation, leading to mucinlike molecules, is a posttranslational event that takes place in the Golgi cisternae (reviewed by Hanover and Lennarz, 1981).

The translocation of proteins through membranes has been discussed by Davis and Tai (1980), Waksman *et al.* (1980), Wickner (1980), and Sabatini *et al.* (1982). They all point out that the signal hypothesis is not necessarily the only valid one: another possibility is the "membrane-triggered folding hypothesis," which emphasizes self-assembly and the role of changing protein conformation during the transfer from an aqueous compartment into a membrane. This alternate hypothesis is based on the fact that some secretory proteins have no terminal signal sequence, but have instead a hydrophobic amino acid sequence in the middle of the molecule. For Engelman and Steitz (1981), secretion of proteins across membranes begins with the spontaneous (thus, not driven by protein synthesis) penetration of the hydrophilic portion of the bilayer by a helical hairpin. Finally, one should mention that, according to Hortin and Boime (1981), complete removal of the presequence of preprolactin does not prevent its secretion. Thus, while Blobel's signal hypothesis rests on an impressive accumulation of facts, one cannot exclude the coexistence of other mechanisms.

In regard to our second question, the transport of the secretory proteins after their penetration into the ER lumen, we have only one—but promising—theory at our disposal. A paper by Chiu and Phillips (1981) shows that the membranes of the ER are degraded and synthesized at the same rate as the membrane-associated secretory products. This would produce a unidirectional membrane flow due to synthesis at one end and degradation at the other; this flow would allow the transport of the secretory product.

We should also mention that Blobel (1980) has extended his hypothesis to the problem of the intracellular localization of proteins; it would be determined by "topogenic sequences." Signal sequences would initiate the translocation of specific proteins through membranes, stop–transfer sequences would arrest their translocation, sorting sequences would regulate the traffic of proteins between donor and receiver membranes and, finally, insertion sequences would be needed for the unilateral integration of proteins into lipid bilayers. This process is still hypothetical, but necessary if we wish to understand the integration of specific proteins (for instance, $Na^+/K^+$–ATPase into the plasma membrane) or, as pointed out in Section II,C, the correct recruitment by stress fibers of injected actin, tropomyosin, or α-actinin. We have made an analogy between the traffic of proteins in the cell and the traffic of cars in a city. If Blobel (1980) is correct, a traveling protein requires several passports, visas, etc., in order to reach its goal. Is the cell thus run, like so many countries today, by a combination of police and bureaucracy? There is strong evidence, reviewed by Olden *et al.* (1982), that the tags that direct glycoproteins to specific cell organelles are their carbohydrate moieties. We shall return to this question later.

### C. Annulate Lamellae

Cells, especially oocytes or eggs, often display curious figures that have been called "annulate lamellae" (reviewed by Kessel, 1983) (Fig. 34). They are a piling-up of dilated membranes presenting, at regular intervals, denser structures sometimes called "fenestrae." Annulate lamellae have a superficial resemblance to the nuclear membrane (which will be examined in Chapter 6). It has, therefore, been generally assumed that they arise from a delamination of the nuclear membrane. We do not think that this is very likely; it seems more probable that annulate lamellae are a degenerate form of the ER (Scheer and Franke, 1972; Steinert-Meulemans, 1980). Their size increases when cells or oocytes are placed under suboptimal conditions, as was the case for the *Xenopus* oocyte in which the huge annulate lamellae shown in Fig. 34 were found. There is very often a continuity between annulate lamellae and ER, suggesting that the two are closely related. Treatment of cells with inhibitors of MT assembly (De Brabander and Borgers, 1975; Maul, 1977) often induces the formation of annulate lamellae. Interestingly, in fibroblasts treated with cytochalasin B, vinblastine-

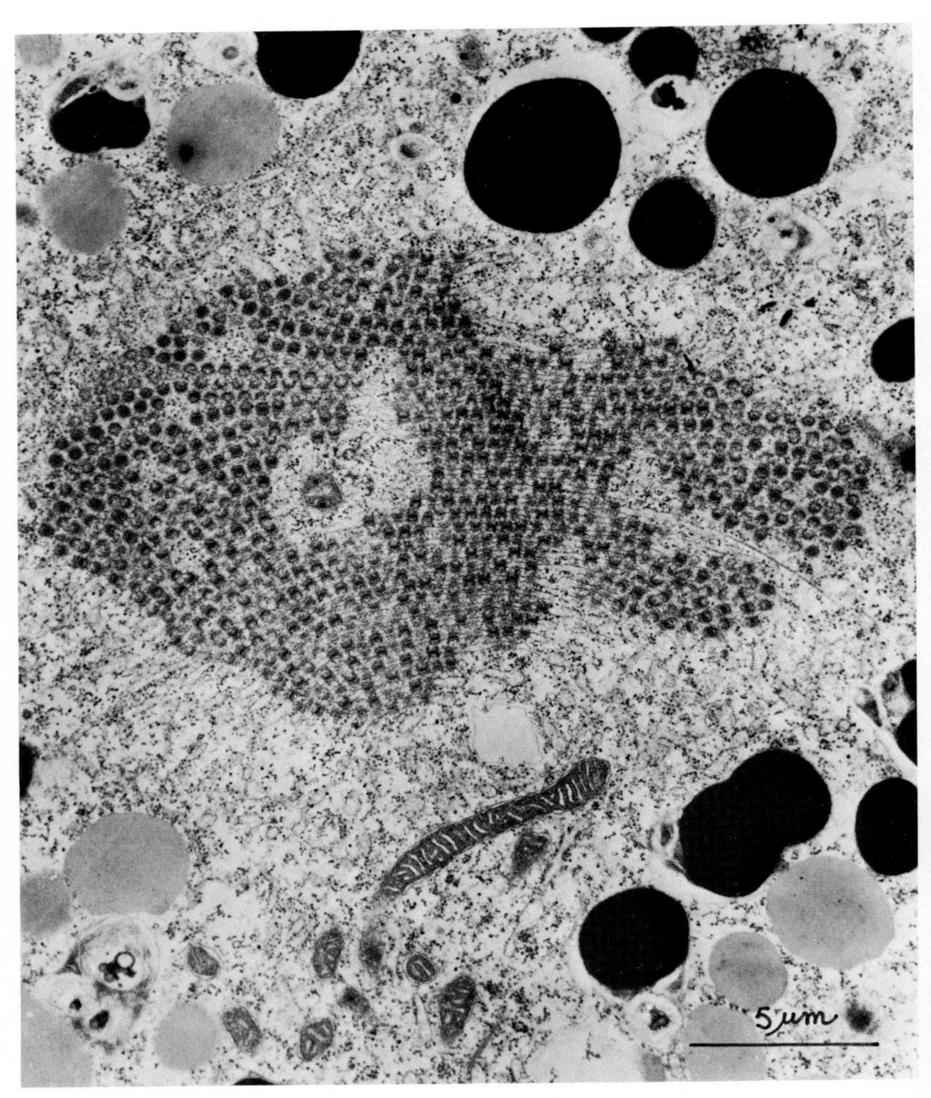


FIG. 34. Annulate lamellae in a full-grown *Xenopus* oocyte. [Courtesy of G. Steinert.]

induced annulate lamellae persist for 2 days in anucleate cytoplasts; they disappear during mitosis and reappear afterward (Maul, 1977). Of course, annulate lamellae will remain somewhat mysterious as long as they are not isolated and characterized: cytochemistry only shows that they are made of RNA and proteins of a still-unidentified nature.

### D. Golgi apparatus (Golgi complex)

The "internal reticular apparatus" described by C. Golgi (1898) in neurons stained with a silver impregnation method has long been considered an artifact, but electron microscopy in the 1950–1955 period clearly demonstrated its presence in all cells (see the discussion of this important question by G. Palade, 1975). The Golgi apparatus is not, however, a reticulum since the main Golgi elements, the dictyosomes, are not linked together by fibers or canaliculi, but are dispersed in the cytoplasm. For this reason, the name "Golgi complex" (reviewed by Tartakoff 1982a,b, 1983a,b) is now often preferred to the older "Golgi apparatus." The functions of the Golgi complex have long remained mysterious and are not yet entirely clear. Cytochemistry at the ultrastructural level and biochemical studies on isolated Golgi structures have greatly, but not completely, clarified the significance of the Golgi complex. The main conclusion that can be drawn from all of these studies is that the Golgi complex is an obligate intermediary between ER, on one side, and lysosomes, secretory granules, and the cell surface, on the other (Novikoff's GERL concept). It plays an essential role in intracellular transport and in the glycosylation and eventual exocytosis of the products of ER activity (reviewed by Mellman, 1982).

The dictyosomes, as shown in Fig. 35, are stacks of cisternae and a system of attached tubules and secretory vesicles; the tubules connect the secretory vesicles with the central saccules (Morré and Ovtracht, 1981). They have both a convex and a concave face and, therefore, display a distinct polarity. The concave face is closest to the nucleus; the convex face, close to the ER, is called the *cis,* proximal or forming phase; the concave face is the *trans,* distal, or maturing phase. The cisternae end in saccules surrounded by vesicles; these Golgi vesicles are smaller on the convex side, where they are called "transition vacuoles," than on the concave side, where they constitute the secretory vesicles. It has been proposed by Rothman (1981) that the *cis* and *trans* faces constitute two distinct organelles and that their function is to "purify" the proteins exported from the ER by a type of distillation process. These exported proteins would enter through the *cis* face and exit the Golgi through the *trans* face, where glycosylases are accumulated. If the Golgi complex is indeed a kind of distribution apparatus, it really deserves to be called again the "Golgi apparatus." There is growing evidence that the proximal and distal faces differ in function and chemical composition (Dunphy *et al.,* 1981); for instance, the cisternae of the two faces bind different lectins that are specific for distinct carbohydrate residues (Tartakoff and Vassali, 1983).

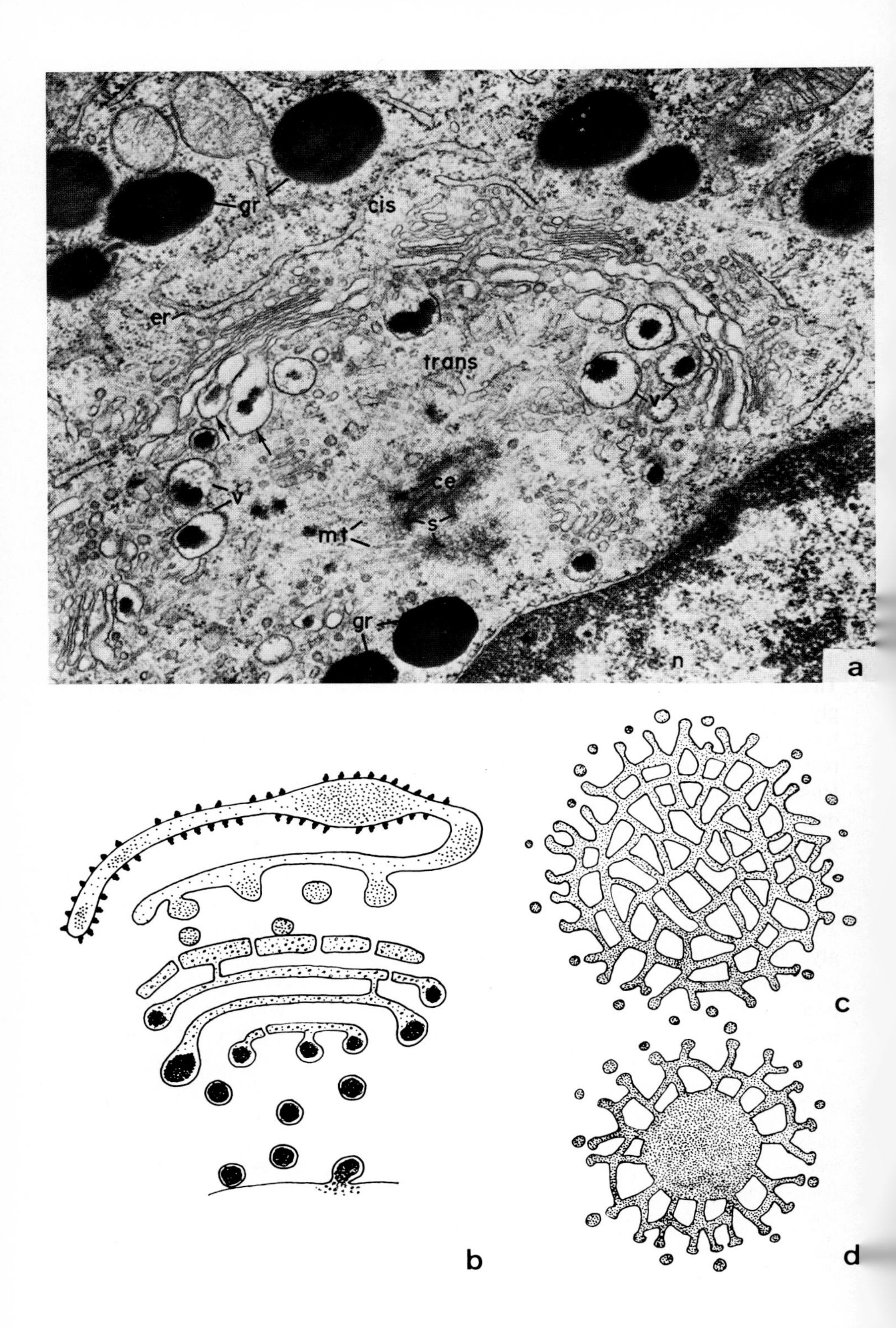

gr
cis
er
trans
v
ce
s
mt
v
gr
n
a
b
c
d

It is believed that there is a continuous membrane flow from the proximal to the distal end, where secretory vacuoles are continuously released into the cytoplasm. These vacuoles are in close contact with the smooth ER and are also involved in the formation of the lysosomes, as first proposed by Novikoff when he developed the GERL concept. GERL is a hydrolase-rich region of the ER, situated at the *trans* face of the Golgi, from which lysosomes seem to arise. It was once believed that the dictyosomes were a reserve of smooth membranes that could be used to provide new membrane components to the ER and the plasma membrane. Currently, it seems more likely that there is a continuous flow of membranes starting from the ER, reaching the dictyosomes, and ending in lysosomal and plasma membranes (Morré, 1977). However, a dynamic interpretation of morphological pictures is by no means easy when one is dealing with a system as labile as the Golgi complex. The recent discovery by Lipsky and Pagano (1985) of a specific fluorescent vital dye for the Golgi complex might allow a better understanding of the dynamics of this complex.

The Golgi complex can be isolated by differential centrifugation of homogenates; it is composed of about 60% protein and 40% lipid. The complexity of the protein pattern decreases from ER to Golgi and from Golgi to plasma membrane, suggesting that a membrane flow takes place in that direction. Among the proteins of the Golgi complex are several enzymes (phosphatases, nucleotidases, thiamine pyrophosphatase) that have been widely used as morphological markers. The enzyme markers used in most biochemical studies on homogenates are the glycosyltranserases which are involved in the synthesis of the glycoproteins. It has been shown that sialyltransferase and the neuraminic acid of the sialylglycoproteins are located on the luminal side of the Golgi membranes (Carey and Hirschberg, 1981); galactosyltransferase and nucleoside diphosphatase are found only on the *trans* (concave) side of the dictyosomes (Roth and Berger, 1982). The first enzyme transfers galactose on secretory proteins within the lumina of the Golgi saccules (Bergeron *et al.*, 1982). A specific α-mannosidase is present in purified preparations from the Golgi complex (Dewald and Touster, 1973). This suggests that the Golgi apparatus plays a very important role in the glycosylation of the proteins. However, this role is not exclusive, since it is known, as we have seen, that the major part of protein glycosylation already occurs in the ER, together with the synthesis of the protein. Further glycosylation in the Golgi apparatus seems to be essential for the recognition of a protein by its

---

FIG. 35. Dictyosomes. (a) Golgi region of a developing leukocyte (promyelocyte). ce, Centriole; s, centriolar satellite; v, vacuoles. Arrows show condensing secretory products (lysosomal enzymes and peroxidase). gr, dense azurophil granules; mt, microtubules; trans, transmost cisternae. [Farquhar and Palade (1981). Reproduced from *The Journal of Cell Biology*, 1981, **17,** 375 by copyright permission of The Rockefeller University Press.] (b) Longitudinal section. (c) Apical view of cisternae (ergastoplasmic face). (d) Apical view of cisternae (secretory face). [Original drawings by P. Van Gansen.]

target site inside the cell. As we shall see, this is probably true of the lysosomal enzymes.

The Golgi apparatus is also deeply involved in the synthesis of mucopolysaccharides in animal cells and of the cell wall in plants (Whaley *et al.*, 1972; Morré, 1977). Many dictyosomes are accumulated at the tip of a growing alga, as shown in Fig. 36. Similar observations can be made on dividing plant cells: dictyosomes accumulate where a new cell wall forms.

It has been shown that the monovalent ion-ionophore monensin (Tartakoff, 1982, 1983b) inhibits the secretion of collagen and fibronectin by fibroblasts; the block takes place at the Golgi level, since the Golgi vesicles are greatly enlarged (Uchida *et al.*, 1980). That monensin inhibits the migration of Golgi vesicles filled with glycoprotein toward the plasma membrane has also been shown by Vladutiu and Rattazzi (1980). These findings show that an adequate $Na^+/K^+$ equilibrium (monensin imposes on the cell a high $Na^+$ and a low $K^+$ concentration) is required for the proper functioning of the Golgi complex, which plays a central role in intracellular traffic. Monensin slows down carbohydrate maturation of the glycoproteins in the Golgi distal compartment (where sulfatation of proteoglycans also takes place). It seems that by increasing the $Na^+$ intake, it

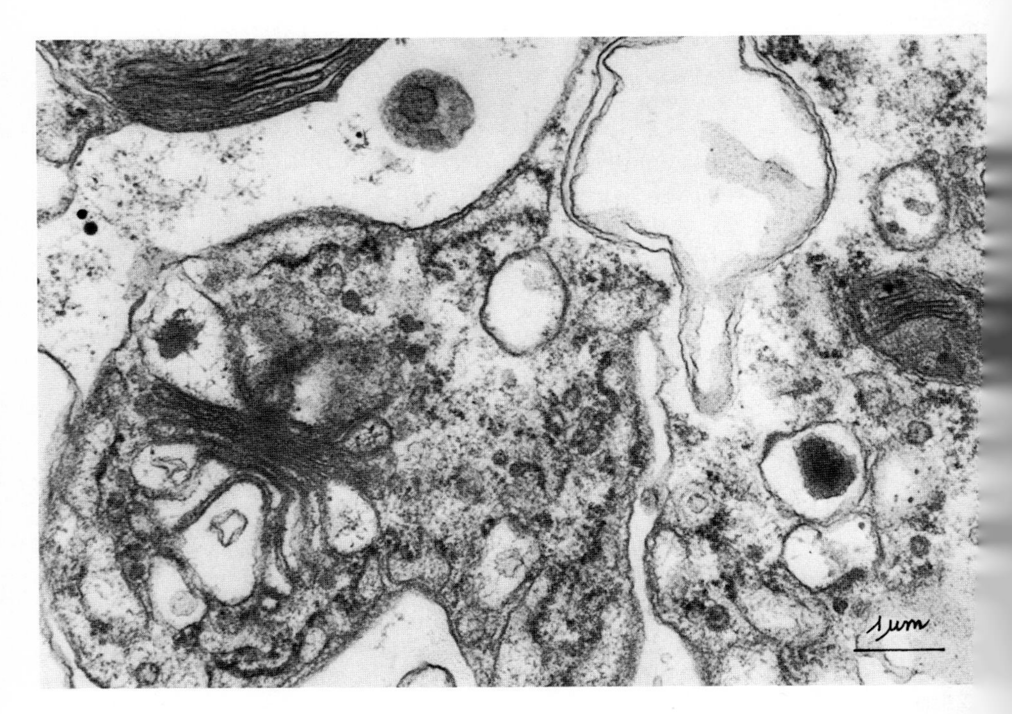


FIG. 36. Dictyosomes at the apical (growing) end of the unicellular alga *Acetabularia mediterranea*. [Courtesy of M. Boloukhère.]

raises the intracellular pH of the Golgi cisternae (and of the lysosomes); certain work has indeed shown that the Golgi membranes contain an electrogenic ATP-driven $H^+$ pump (Glickman *et al.*, 1983).

Today, the Golgi complex is believed to be a sorting center for proteins; it is polarized, and its face proximal to the rough ER receives the products of its synthetic activity; small Golgi vesicles may shuttle back and forth to the ER. The distal face adds terminal sugars, since only the distal cisternae possess galactosyl-, fucosyl-, and sialyltransferases. While the membranes of the ER remove the signal peptides of preproteins and add to them a high mannose tail (cotranslational processing), the Golgi adds terminal sugars to asparagine-linked oligosaccharides (posttranslational processing) (Rogers *et al.*, 1982). There is thus, in the Golgi complex, a physical separation of the enzymes involved in the biosynthesis and processing of the oligosaccharides that constitute glycoproteins (Deutscher *et al.*, 1983).

The author remembers when Golgi apparatus was considered a fixation and staining artifact. Today he still finds it difficult to clearly represent the complex functions of a labile, elusive system that fully deserves the name "Golgi complex." Its main function seems to be the distribution in the cell of proteins whose glycosylation has been completed in the Golgi complex itself. The main pathway seems to be Golgi dictyosomes → Golgi secretory elements → plasma membrane (Evans *et al.*, 1980). As already mentioned, the Golgi complex is also involved in the production of the lysosomes and the related peroxisomes, a discussion of which is now quite appropriate.

### E. Lysosomes

The exciting story of the discovery of the lysosomes and their subsequent characterization by de Duve and his co-workers (Hers, Berthet, Beaufay, Wattiaux, Baudhuin and many others) has been related by de Duve himself in his Nobel Prize lecture (1975). Very aptly, it is entitled "Exploring the Cell with a Centrifuge."

Around 1950, de Duve was working on the enzymes present in mitochondria. It had already been realized that the mitochondrial fraction that could be isolated by the then available techniques of differential centrifugation was not homogeneous. Over the mitochondrial pellet was a "fluffy layer" composed of "light mitochondria." This heterogeneous pellet contained both oxidative and hydrolytic enzymes. Patient and intelligent experimenting allowed de Duve to demonstrate that the oxidative enzymes, but not the hydrolases, are constituents of true mitochondria. The so-called light mitochondria have nothing to do with them; they contain a number of hydrolases with an acidic pH optimum. These hydrolytic enzymes are in a latent form. Freezing or treatment with the detergent Triton X-100 simultaneously solubilizes all the acid hydrolases, showing that they are imprisoned inside a proteolipidic membrane. These particles—more exactly, vacuoles—were called "lysosomes" by de Duve *et al.* (1955). The

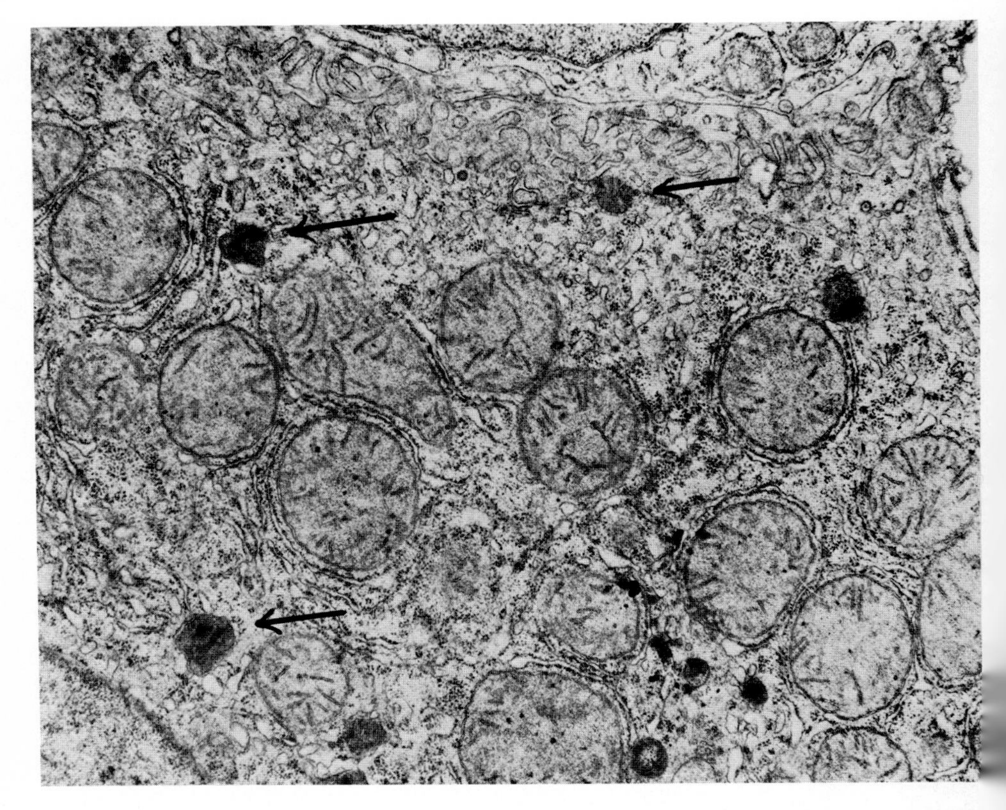

FIG. 37. Lysosomes (arrows) in a rat liver cell. [Courtesy of P. Baudhuin.]

enzymes are held together inside a membrane as in Fig. 37; their release in the cell sap would be a catastrophe and could only lead to death. This is why de Duve first considered the lysosomes as "suicide bags." Fortunately for our cells, it turned out that the lysosomes are more resistant than many other organelles (mitochondria, for instance) when the cells are placed under adverse conditions (x-ray irradiation, heat shocks, etc.). This finding led de Duve to a new—and correct—interpretation of the role of the lysosomes in the cell: they are the cellular equivalents of the digestive vacuoles of protozoa. As we shall see, their function is to digest foreign material that has penetrated in the cell by pinocytosis or phagocytosis; they also digest degenerating cell organelles, for instance, old mitochondria; thanks to its lysosomes, the cell is always kept clean.

The lysosomal enzymes released simultaneously by detergent treatment were, at first, acid phosphatase, RNase, DNase, and cathepsin (Wattiaux and de Duve, 1956). Currently, about 50 different enzymes (all of them hydrolases with an acidic pH optimum) have been found in lysosomes (see Allison, 1974, for a list of these enzymes). They are able to digest a very large number of biological substances. Of these enzymes, acid phosphatase deserves special mention. This enzyme has two advantages: its activity can easily be measured by biochemical methods, and its intracellular localization can be detected by cytochemical techniques (at both the light and EM levels). Therefore, acid phosphatase is the classic enzyme marker for the lysosomes (Fig. 38), and its intracellular localization has been studied in a wide variety of cells. All the lysosomal enzymes are glycoproteins with asparagine-linked, mannose-rich oligosaccharides.

Cytochemical and ultrastructural studies have shown that the lysosomes, even in the same organ (mammalian liver, for instance, which remains a favorite material for the study of the lysosomes), vary considerably in size, electron density, and acid phosphatase content. Work in de Duve's laboratory demonstrated that the lysosomes are identical to the "dense bodies" found by electron microscopy (Fig. 37). Large-scale separation of dense bodies by differential

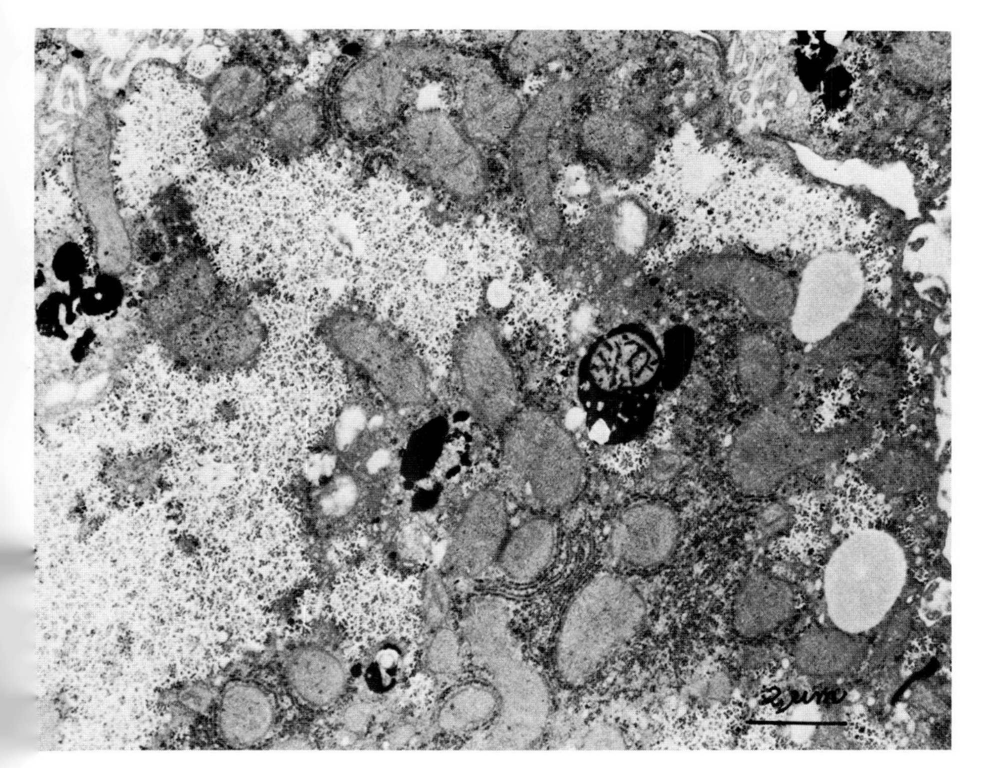

FIG, 38. Cytochemical detection, at the ultrastructural level, of acid phosphatase in the lysosomes of rat liver cells. [Courtesy of G. Steinert.]

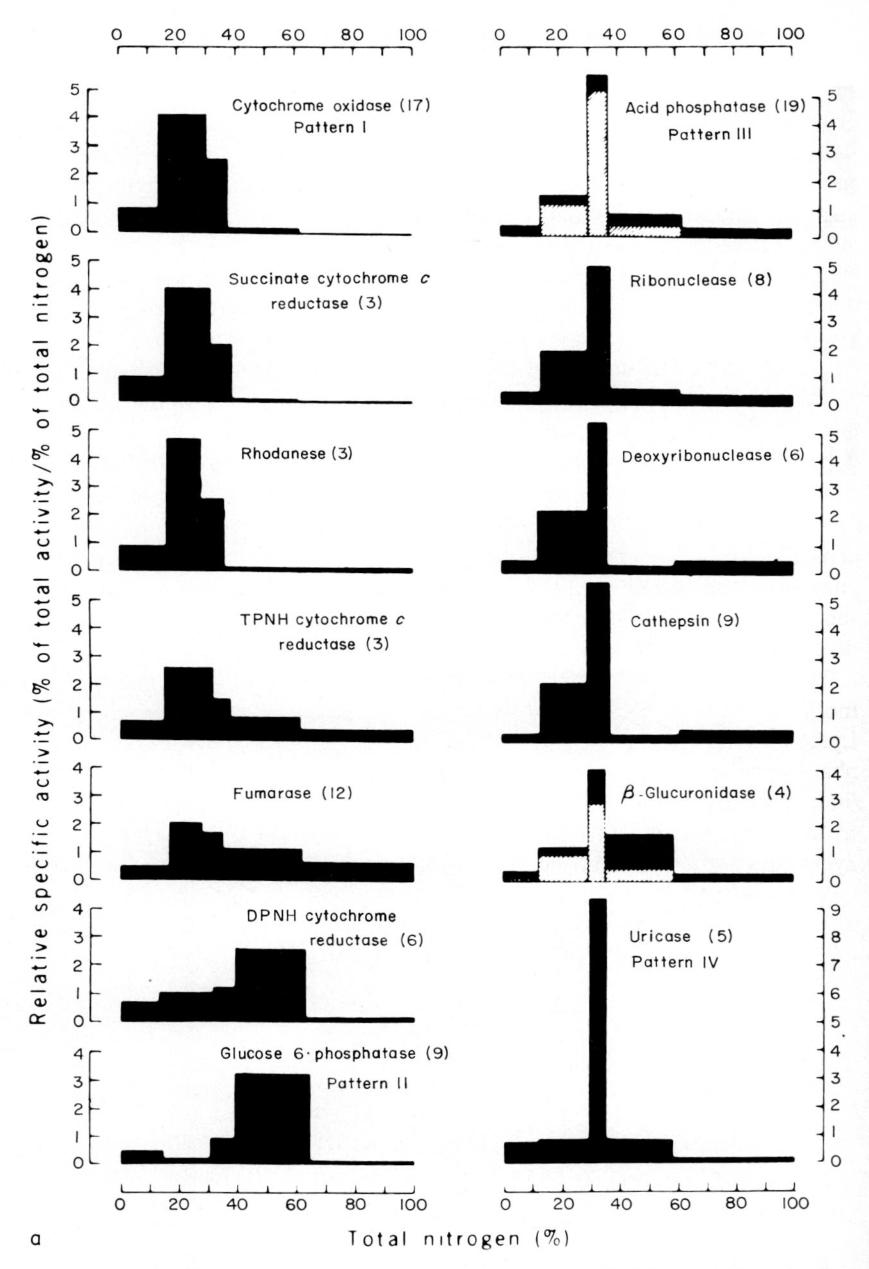


FIG. 39. (a) Identification, by measurement of marker enzyme activity, of various fractions obtained by differential centrifugation of a rat liver homogenate. Pattern I, mitochondria (three enzymes); pattern II, microsomes; pattern III, lysosomes; pattern IV, peroxisomes. Brackets show the number of determinations [de Duve (1975). Copyright 1975 by the American Association for the Advancement of Science.] (b) Purified fraction of liver lysosomes seen under the electron microscope. [Courtesy of Prof. P. Bandhuin.]

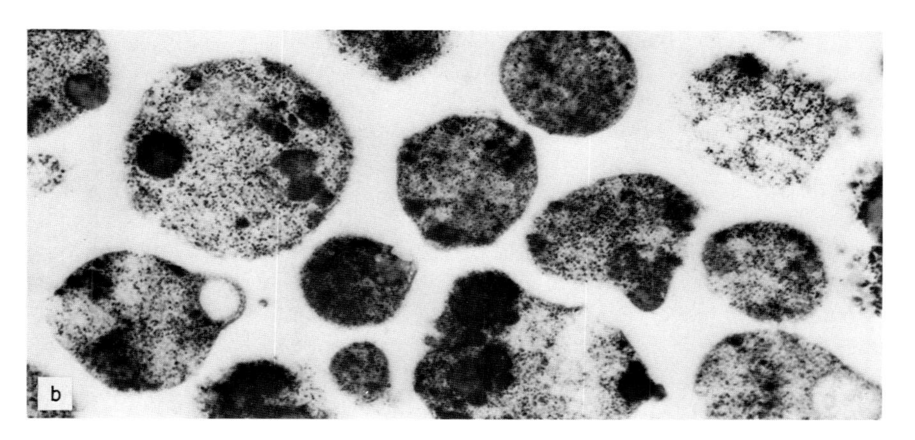


FIG. 39. (*Continued*)

centrifugation and subsequent characterization by electron microscopy and by enzymatic analysis (Fig. 39) provided the final proof that the lysosomes were separate entities that had nothing in common with the mitochondria (Leighton *et al.* 1968).

The morphological and chemical heterogeneity of lysosomes and their polymorphism result from their history, which is shown diagrammatically in Fig. 40. Lysosomes result from the fusion of a phagosome with a primary lysosome; the phagosome is a vesicle resulting from endocytosis of the solid (phagocytosis) or liquid (pinocytosis) material that will ultimately be digested by the lysosome hydrolytic enzymes. Primary lysosomes have a more complex origin. The enzymes are synthesized in the rough ER (although an exogenous origin of the

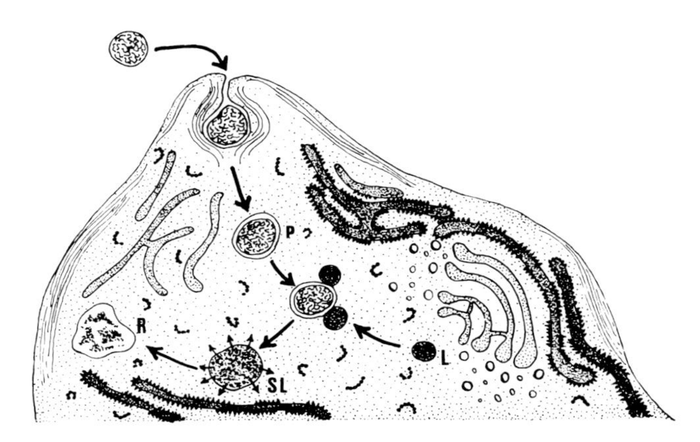


FIG. 40. Schematic representation of endocytosis. A phagosome (or pinosome, P) fuses with a primary lysosome (L) originating from the Golgi bodies to become a secondary lysosome (SL). R, residual body. [Original drawing by P. Van Gansen.]

lysosomal enzymes has been reported in endothelial and smooth muscle cells by Hasilik *et al.*, 1981); the membrane of the lysosomes originates from the Golgi region, while that of the phagosome is derived from the plasma membrane. The result of the fusion between a phagosome and a primary lysosome is the formation of a secondary lysosome (also called a ''heterophagosome'' or ''digestive vacuole''), where digestion of the endocytosed material takes place. If digestion is complete, the soluble breakdown products move through the lysosomal membrane into the hyaloplasm, where they can be further metabolized. If digestion is incomplete, the lysosome becomes filled with solid residues that are eliminated by a defecation process; the resulting electron-opaque lysosomes are called ''residual bodies.'' Finally, in the cytolysosomes (reviewed by Glaumann *et al.*, 1981) (or autophagosomes), degenerated cell organelles are destroyed (Fig. 41). Curiously, the formation of autophagosomes, but not the digestion of foreign proteins by liver lysosomes, is inhibited specifically by 3-methyladanine (Seglen and Gordon, 1982). The formation of cytolysosomes, containing mitochondria and fragments of the ER, can be induced by vinblastine (Marzella *et al.*, 1982).

Lysosomes are ubiquitous, but are particularly abundant and active in cells specializing in phagocytosis, such as polynuclear neutrophils or macrophages. Their proteolytic enzymes (of the cathepsin type) certainly play an important role in protein degradation, particularly when the proteins have been altered by the introduction of abnormal amino acids (fluorophenylalanine, for instance). However, it seems unlikely that all protein turnover takes place in the lysosomes, since there are also proteases in the cell sap (hyaloplasm). Nevertheless, the lysosomes can help the cell to correct possible errors made by the complicated machinery used for protein synthesis.

Lysosomal enzymes, in order to be active, require a low intralysosomal pH. According to Ohkuma and Poole (1978), this is 4.7–4.8 (as compared to about 7.2 in the surrounding cytoplasm); an active mechanism, requiring energy, is obviously needed to drive protons through the lysosomal membrane and to maintain the pH difference. Indeed, Ohkuma and Poole (1978) found that nigericin, which mediates an electroneutral $K^+/H^+$ exchange, increases the lysosomal pH. More recently, Schneider (1981) showed that addition of $Mg^{2+}$-ATP decreases the intralysosomal pH by about 1 pH unit and suggested that the acid pH of the lysosomes might be due to an ATP-driven proton pump. This would explain why, according to Mego and Farb (1978), degradation of labeled ovalbumin by lysosomes requires energy that would serve to maintain a low pH inside the lysosomes. More recent work by Ohkuma *et al.* (1982), Harikumar and Reeves (1983), and Schneider (1983) has provided evidence for the presence, in lysosomes, of an $H^+$-ATPase (an electrogenic proton pump driven by $Mg^{2+}$-ATP).

So-called lysosomotropic drugs have stimulated a good deal of interest. They are primary amines (ammonia, methylamine, and the antimalarial drug chloro-

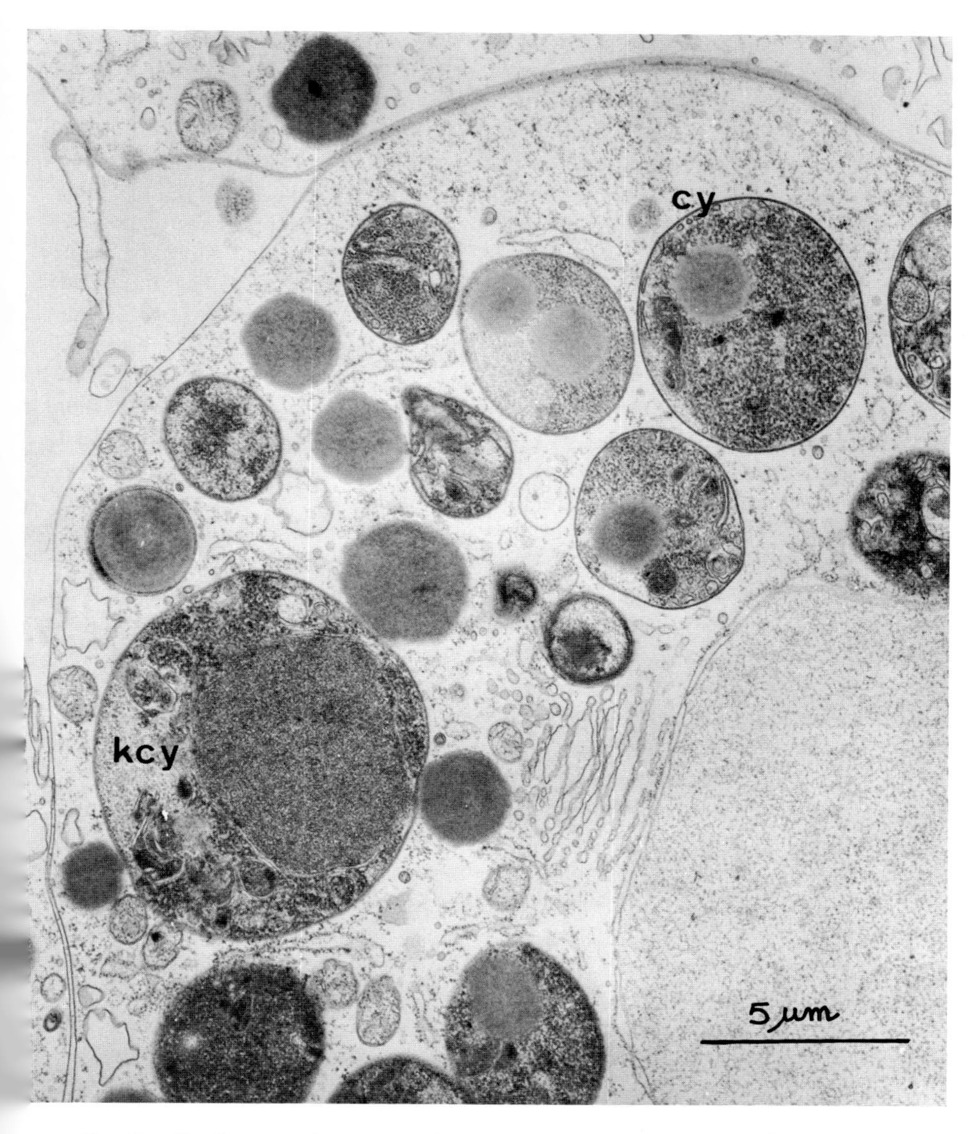


FIG. 41. Cytolysosomes in a spong archaeocyte. cy, cytolysosome; kcy, cytolysosome with an ingested nucleus. [Courtesy of L. De Vos.]

quine) that penetrate easily into the lysosomes and increase their pH. In particular, chloroquine is believed to accumulate selectively in the lysosomes (Harder *et al.*, 1981). Although definite conclusions cannot yet be drawn, it seems that primary amines inhibit fusion between lysosomes and phagosomes (Gordon *et al.*, 1980), but not endocytosis and internalization of hormone–receptor complexes (King *et al.*, 1980; Yarden *et al.*, 1981). What seems well established is that they inhibit the degradation of hormones that, like insulin or the epidermal growth factor (EGF), have favorable effects on cell proliferation; after treatment with chloroquine, there is an accumulation of EGF (King *et al.*, 1981; Savion *et al.*, 1981; Yarden *et al.*, 1981) and of insulin (Tsai and Seeman, 1981) in the cells. It seems that, as one would expect from the effects of the lysosomotropic drugs on lysosomal pH, their main effect is to arrest the degradation by the lysosomal enzymes of endocytosed proteins. How hormone–receptor complexes located in coated pits can avoid meeting lysosomes was discussed earlier when we presented the receptosome concept of Willingham and Pastan; they provided compelling evidence for the view that internalization of exogenous proteins takes place through the coated-pit receptosomes, the Golgi-GERL, and the lysosome pathway (Willingham *et al.*, 1981d; Pastan and Willingham, 1981).

Another question that is likely to arouse considerable interest in the near future is: by which mechanisms are the lysosomal enzymes recruited and selectively accumulated in the lysosomes? It appears fhat these enzymes need a "passport" or "ticket" to allow their recognition and that carbohydrate chains present in lysosomes constitute the recognition signal. In 1974, Bishayee and Bachhawat observed that all the lysosomal enzymes known at that time were glycoproteins that bound ConA. More recently, however, Hasilik (1980) pointed out that the lysosomal enzymes were modified by glycosylation during their synthesis and that many of them bore a mannose phosphate group. This group, according to a report by Waheed *et al.* (1981), is *N*-acetylglucosamine phosphomannose. Work by Gabel *et al.* (1982) has shown that the phosphomannosyl residues of the lysosomal enzymes are necessary for binding the enzymes to mannose 6-phosphate receptors present in the lysosomal membranes and for their translocation into the lysosomes. There is no binding of the lysosomal enzymes to the mannose 6-phosphate receptors if the *N*-acetylglucosamine residues have not been removed. However, lysosomal enzymes may accumulate in the lysosomes of cells that are deficient in mannose 6-phosphate receptors (Gabel *et al.*, 1983): it seems that the mannose 6-phosphate "passport" is not always required for entry into the lysosomes.

Brown and Farquhar (1983) give some information about the intracellular localization of the mannose 6-phosphate receptor: detection by immunocytochemistry shows that it is accumulated in the *cis*-Golgi cisternae, the lysosomes, and the endosomes; it is hardly present in the ER. This suggests that sorting of the proteins that have been synthesized in the rough ER takes place in the *cis*-

Golgi cisternae, and that one of their functions is to send the lysosomal enzymes to their final destination.

Another question has remained puzzling until now: how can lysosomal enzymes, once they have reached their destination, escape immediate degradation by the lysosomal cathepsins? A paper by Hoogeveen *et al.* (1983) shows that, at least for three lysosomal enzymes, a dual protective mechanism operates: aggregation in a multimeric enzyme complex of high molecular weight and binding to a 32,000-dalton protective protein.

Progress in this field will be awaited with interest, since it may help us to understand the still baffling topological specificity of the GERL system. A paper by Rosenfeld *et al.* (1982) marks progress in this area: the authors found that lysosomal enzymes are synthesized on polysomes bound to the rough ER, cotranslationally glycosylated, and transferred into its lumen. The neosynthesized enzymes have a dual destination: 60% are processed proteolytically; the final proteolytic cleavage takes place in the lysosomes themselves, as shown by the fact that processing is inhibited by chloroquine; the remaining 40% are secreted without proteolytic processing, but some of their oligosaccharides are modified during their passage through the Golgi apparatus. All the unglycosylated hydrolases synthesized by cells treated with tunicamycin are exported without undergoing proteolytic processing; the authors concluded that modified sugar residues serve as sorting-out signals that send the hydrolases to their lysosomal destination.

Finally, a few words should be said about the importance of the lysosomes in cell pathology. Release of lysosomal enzymes may occur in many pathological conditions (rhumatoid arthritis, gout, myocardial infarction, etc.), but for the cell biologist, the most interesting of the "lysosomal diseases" are undoubtedly the storage diseases (Hers and Van Hoof, 1973). Due to an autosomal recessive mutation, a lysosomal enzyme might be missing in homozygotes, and its substrate will therefore accumulate in lysosomes. For instance, in glycogenosis type II, which was first studied by Hers and Van Hoof, liver and muscle lysosomes are filled with glycogen; the disease is due to the congenital lack of α-glucosidase, an enzyme that degrades glycogen to glucose. Introduction of the missing enzyme into the cells should cure the disease, but this is currently wishful thinking, at least at the level of the whole organism. The problem is simpler at the cellular level: Cori *et al.* (1983) succeeded in correcting a genetically caused enzyme defect (a deficiency in liver glucose-6-phosphatase due to deletion of one of the chromosomes) by fusing the deficient hepatocytes with normal cells. Cell transplantation procedures might be useful for treating hereditary deficiencies in lysosomal enzymes. It has been shown that certain lysosomal enzymes (but not all) can be transferred from lymphocytes to fibroblasts (Olsen *et al.*, 1983). Attempts to use liposomes as vectors for the enzyme are now being made, but it is too early to judge the actual possibilities of this approach. Lysosomes (and liposomes) are also interesting for cancer chemotherapy (Trouet, 1978). The drugs currently used for this purpose can be bound to DNA or proteins and then injected; these complexes would be degraded in the lysosomes, allowing a slow release of the drug with the advantages of longer action and lower toxicity. Intensive and important work on the use of lysosomes in medicine is now being done in the Institute of Cell Pathology, founded and directed in Brussels by C. de Duve.

### F. Lysosome-Related Organelles

We shall now consider briefly a miscellaneous collection of cell organelles that bear a close or distant similarity to the lysosomes.

Very similar, if not identical, to the lysosomes are the protein resorption droplets (phagosomes) of the kidney, which have been studied by Straus (1954) and Straus and Oliver (1955). Injected proteins (for instance, horseradish peroxidase, which can easily be detected cytochemically) are reabsorbed in the proximal segment of the nephron, with the subsequent formation of large droplets. Since the injected protein is ultimately degraded, the phagosomes of Straus can be considered as giant, specialized lysosomes.

Another type of very large lysosomes are the digestive vacuoles of Protozoa, as shown by the classic studies of Holter (1954), who compared the distribution of various enzymes in homogenates of and in intact amebas, which had been cut into two halves after centrifugation. His elegant experiments allowed him to separate *in vivo* the vesicles containing the acid hydrolases from the mitochondria at a time when the lysosomes were still considered light mitochondria of the fluffy layer.

Other lysosome-like organelles are the multivesicular bodies (Fig. 42) that can be stained vitally with dyes such as toluidine blue or acridine orange and give a positive acid phosphatase reaction. The dyes penetrate by endocytosis and are concentrated in vacuoles that are thus homologous to phagosomes. In invertebrate eggs, these vacuoles have been called "meta-granules" (Dalcq *et al.*, 1956) because they stain red in eggs treated with toluidine blue. Such metachromatic staining is usually due to the presence of acid mucopolysaccharides in the vacuoles. The meta-granules, which have the ultrastructure of multivesicular bodies, seem to play a role in yolk formation during oogenesis and in cleavage after fertilization.

A highly specialized lysosome is the acrosome (Fig. 43) of the spermatozoa from many species; as we shall see in Volume 2, Chapter 2, it plays an important role in fertilization. The acrosome is a vacuole located at the proximal end of the spermatozoon, containing an acrosome granule composed of unpolymerized G-actin. Like the lysosomes, the acrosome originates from the Golgi region and contains hydrolytic enzymes, including the typical acid phosphatase. Physiologically, the most important of these enzymes is acrosin, a trypsinlike protease that

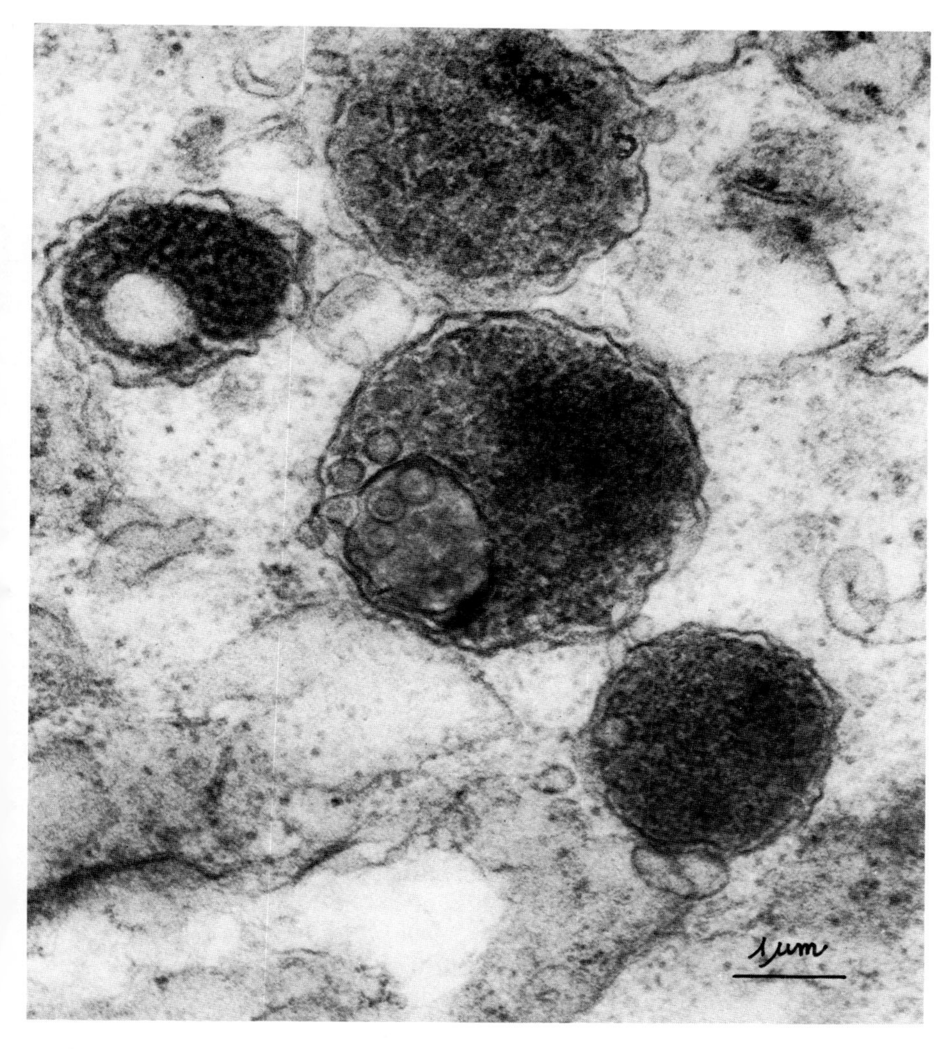


FIG. 42. Multivesicular bodies in a blastomere of the mollusk *Ilyanassa obsoleta*. [Courtesy of Y. Gérin.]

after its release from the acrosome, breaks down the membranes that surround and protect the egg. Acrosin formation results from the transformation of an inactive pro-acrosin by an autocatalytic process (Mukerji and Meizel, 1975); according to Stambaugh and Smith (1978), acrosin is associated with tubulin in the acrosome. As one can see, the acrosome displays several of the basic proper-

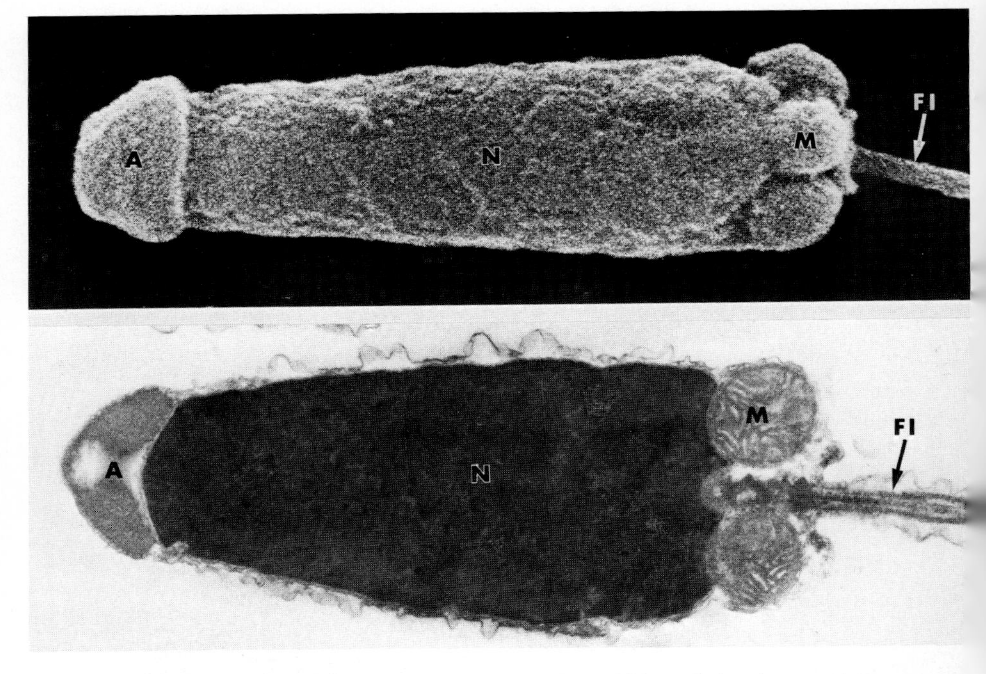


FIG. 43. Spermatozoon of the mollusk *Dentalium*. (*Bottom*) Transmission electron microscopy. (*Top*) Scanning electron microscopy. A, acrosome; N, nucleus; Fl, flagellum; M, mitochondrion. [Courtesy of Drs. Dufresne-Dubé *et al.* (1983).]

ties of the lysosomes but differs from them in other respects (the presence of G-actin and tubulin, in particular).

Whether yolk platelets in eggs are related to the lysosomes has been and remains the subject of a good deal of discussion. As already mentioned, yolk platelets result from the receptor-mediated endocytosis of vitellogenin, a precursor of the yolk phosphoproteins. The presence in oocytes of very large microvilli (as in the kidney) allows the formation of large phagosomes where the phosphoproteins undergo precipitation and even crystallization in some species. In sea urchin eggs, the yolk platelets have a very strong acid phosphatase activity (Schuel, 1978) and in this respect resemble lysosomes. On the other hand, investigations in our laboratory (Decroly *et al.*, 1979; Steinert and Hanocq, 1979) have shown that yolk platelets isolated from full-grown *Xenopus* oocytes have no detectable acid phosphatase activity; the enzyme is present in smaller particles, which differ, however, from the classic mammalian liver lysosomes in several ways. On the other hand, according to Vernier and Sire (1977), the yolk platelets of trout oocytes are authentic secondary lysosomes, and are provided with three typical lysosomal enzymes. There are obviously considerable differences among yolk platelets from eggs of different species, and the question of their relationships with the lysosomes has not yet been answered satisfactorily.

### G. Peroxisomes and Glyoxysomes

In 1969, de Duve isolated a new type of cell organelle and showed that it is identical to the "microbodies" observed in the liver and kidney by electron microscopists. These spherical microbodies (Fig. 44) have a diameter of about 0.6–0.7 μm and are limited by a single membrane. Characteristic of microbodies is the presence of an electron-dense core or "nucleoid," which sometimes undergoes crystallization. These organelles, after their isolation by de Duve (1969), were renamed "peroxisomes" by him because they contain a number of enzymes involved in the direct oxidation (cyanide-insensitive) of a number of substrates to produce hydrogen peroxide ($H_2O_2$). The peroxisomes also contain catalase, the enzyme responsible for the breakdown of $H_2O_2$ in water and oxygen. It is probable that the main function of the peroxisomes is to eliminate $H_2O_2$, a very toxic compound. The enzymatic equipment of the peroxisomes—

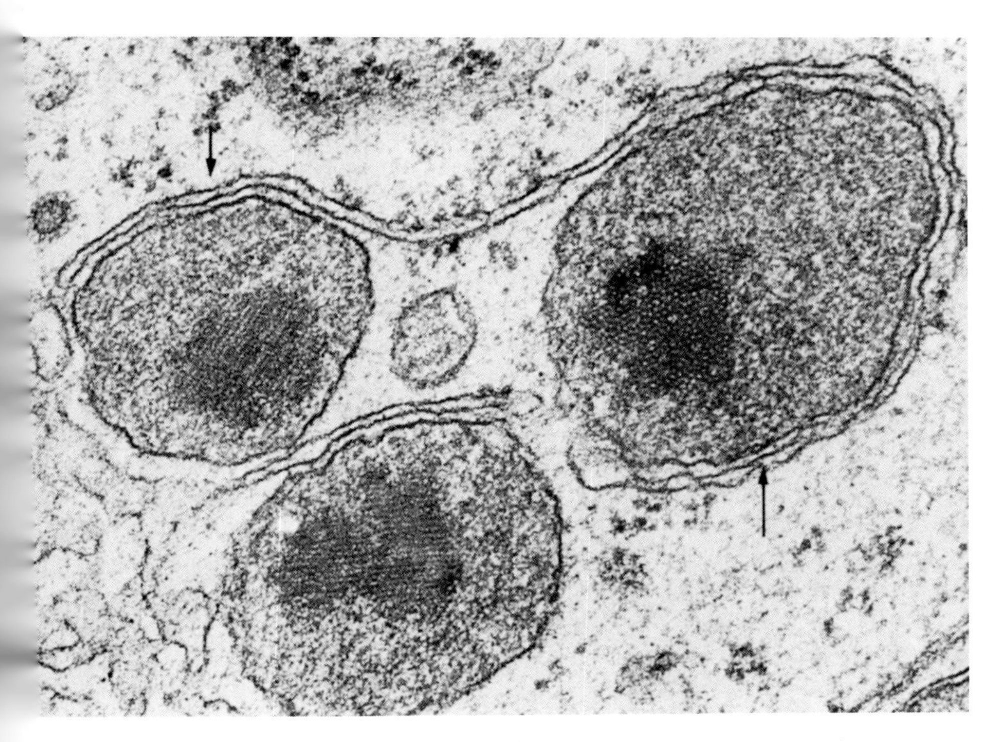

Fig. 44. Peroxisomes in a rat liver cell. [Schnitka (1966).] Arrows indicate smooth ER membranes that probably give rise to peroxisome microbodies.

in contrast to that of the lysosomes—varies from tissue to tissue, but catalase is always present. Among the enzymes present in the peroxisomes are D-amino acid oxidase, polyamine oxidase, and uricase, the crystal that is often observed in the matrix of the peroxisomes. Thus, the function of the peroxisomes (reviewed by Lord, 1980, and Lord and Roberts, 1980) is to allow the cell to eliminate waste products, such as uric acid or D-amino acids, and to dispose of the $H_2O_2$ produced by their oxidation by molecular oxygen. The toxicity of $H_2O_2$ is due to the fact that it easily forms highly reactive free radicals that attack nucleic acids and proteins. This topic will be reviewed later in a discussion of cancer and cell aging. Peroxisomes also possess a system for the cyanide-insensitive oxidation of long-chain fatty acids (Ishii *et al.*, 1980), but oxidation of palmityl-CoA is quantitatively more important in mitochondria than in peroxisomes (Foerster *et al.*, 1981). Peroxisomes might also play a role in fatty acid synthesis, since acetyl-CoA synthetase is present on their cytoplasmic side (Mannaerts *et al.*, 1982).

The role of the peroxisomes in biosynthetic processes has been somewhat neglected in the past. Heymans *et al.* (1983) have studied, from the biochemical viewpoint, patients suffering from the so-called Zellweger syndrome, in which peroxisomes are absent in liver and kidney. They found that these patients are almost unable to synthesize a particular class of phospholipids, the plasmalogens. The key enzymes for plasmalogen synthesis are thus localized in the peroxisomes. Since plasmalogens are found in all cell membranes, it is not surprising that the hereditary Zellweger syndrom is lethal and that the patients die in infancy.

In plant cells, the equivalents of the peroxisomes are the glyoxysomes (reviewed by Lord, 1980, and Lord and Roberts, 1980), which are responsible for photorespiration. Glycolic acid produced by chloroplasts in the light is oxidized by a glyoxysomal glycolic acid oxidase with the production of $H_2O_2$; the latter is destroyed, as in peroxisomes, by catalase. The glyoxysomes also allow the conversion of fats to carbohydrates during the growth of seedlings through the so-called glyoxylate cycle, which is described in most biochemistry textbooks.

What is known today about the biosynthesis of the microbodies (peroxisomes and glyoxysomes) was reviewed by Kindl in 1982. Their proteins are synthesized on free polyribosomes and are posttranslationally imported in the growing organelles. Assembly then takes place in the organelles by the acquisition of prosthetic groups (catalase is an example), either by processing or by oligomerization. The membrane is believed to originate from the smooth ER, as can be seen in Fig. 44.

### H. Secretory Granules

There is little to add to what has already been said in this chapter. As shown and summarized by Palade (1975), pancreatic enzymes are synthesized by the ribosomes associated with the rough ER, penetrate into the lumen of the ER

[Blobel's (1977) signal sequences], accumulate in ER cisternae, and agglomerate in the Golgi region, where membrane-bound secretory granules are elaborated. Their membrane fuses with the plasma membrane and exocytosis, which is a $Ca^{2+}$-, ATP-dependent process requiring the intervention of the membrane-associated microfilaments, takes place. At a late stage of this process, the inactive zymogens are transformed into active enzymes (reviewed by Neurath and Walsh, 1976) by limited proteolysis (see, for instance, the classic trypsinogen–trypsin transformation). Of historical interest is the fact that detection *n situ* of the pancreatic enzymes is one of the first (if not the first) applications of immunological methods to cytology. In 1954, Marshall applied the immunofluorescence method that had been devised by Coons (1952) to the pancreas. He found that the zymogen granules actually contain chymotrypsinogen and carboxypeptidase in easily detectable amounts; the nucleases have a more ubiquitous distribution in pancreatic cells, being present both in secretory granules and in the rest of the cytoplasm.

## IV. SEMIAUTONOMOUS CELL ORGANELLES

### A. General Background

In this section, the heterogeneous population of cell organelles, which are endowed with genetic continuity and thus have the potential capacity to self-replicate, will be reviewed.

The classic examples of semiautonomous organelles are, of course, the mitochondria and the chloroplasts, which have their own DNA. The genetic information present in mitochondrial or chloroplastic DNA is insufficient to ensure full independence for these organelles. They must receive, through the nuclear gene-coded cytoplasmic machinery of protein synthesis, many proteins that they are unable to synthesize by themselves. The relationships between the nucleus and the chloroplasts are best known in the alga *Acetabularia;* they will be examined in Volume 2, Chapter 1.

It is perhaps farfetched to place in this category of semiautonomous cell organelles the centrioles and basal bodies. However, there is no doubt that centrioles can give rise to daughter centrioles when the cell is preparing for mitotic division. It is also certain that centrioles and basal bodies are morphologically and functionally very similar, if not identical. Cilia and flagella, which have been included in this section, are not endowed with genetic continuity, but are so closely linked to the basal bodies that it is convenient to consider them as appendages of these bodies.

### B. Mitochondria

These cell organelles, which play the major role in energy production in animal cells, have been known since the days of Benda (1902). They can be stained with specific dyes on fixed preparations; in living cells, mitochondria can be seen under phase contrast, vitally stained with Janus green or Rhodamin 123 and dyes that bind to mitochondrial DNA (ethidium bromide DAPI), or cytochemical reactions for oxidases (Nadi reaction) or dehydrogenases (tetrazolium reduction).

Phase contrast microscopy combined with microcinematography has shown that in the living cell mitochondria undergo deformations; they move inside the cytoplasm and sometimes undergo fragmentation by constriction, leading to an increase in number (Frédéric and Chèvremont, 1953). Equational division, characteristic of chromosomes during mitosis, has never been observed. The ultrastructure of the mitochondria is basically the same in all eukaryotes, but there are appreciable quantitative differences in size, shape, and fine structure among tissues. As shown in Figs. 45 and 46, mitochondria possess a double membrane. The two membranes are separated by an intermembrane space; the inner compartment is the mitochondrial matrix. While the outer membrane is smooth, the inner membrane displays infoldings called "mitochondrial crests" (cristae) that greatly increase its surface. The crests are covered with small (8.5-nm), regularly spaced particles called "elementary particles."

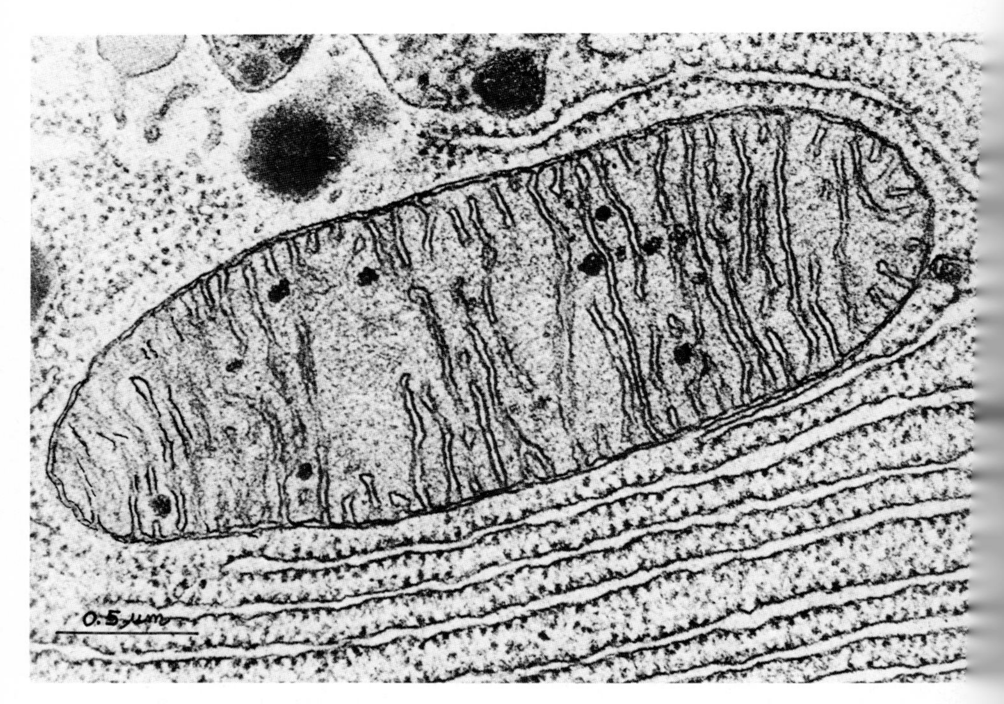


FIG. 45. Mitochondria in an exocrine cell of the pancreas (transmission electron microscopy). [Original photograph.]

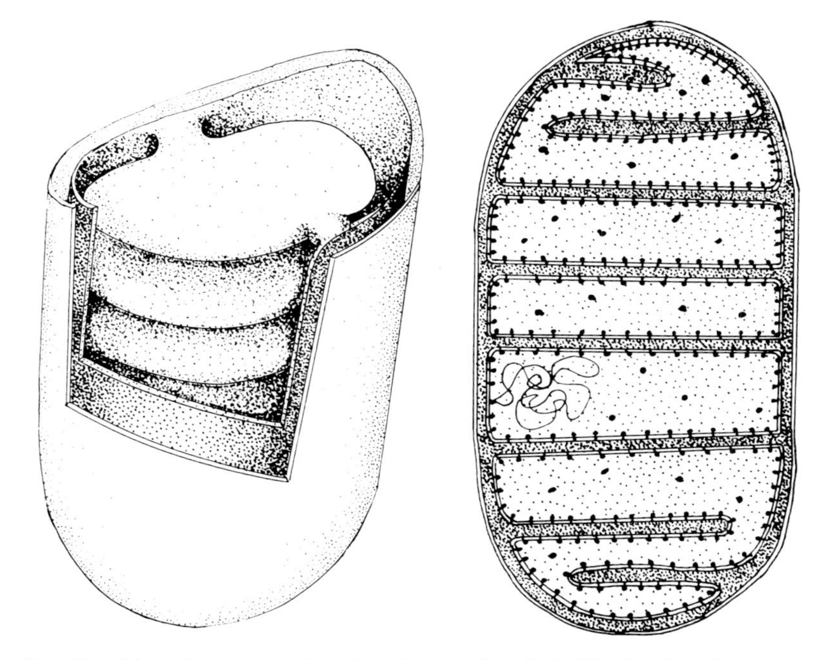

FIG. 46. Schematic representation of a mitochondrion. (*Left*) Mitochondrial double membrane has been partially removed. (*Right*) Longitudinal section through a mitochondrion. The granules seen on the cristae (oxysomes) are believed to have ATPase activity. A circular mitochondrial DNA molecule is attached to the inner mitochondrial membrane. [Original drawing by P. Van Gansen.] For clarity, the space between the outer and inner membranes has been exaggerated.

When the mitochondria are not "energized," i.e., when they are not functioning normally, a condensation of their inner compartment can be observed. This occurs when cells are placed under anaerobic conditions or treated with inhibitors of oxidative phosphorylation (see the review on form and function *in vivo* of mitochondria by Smith and Ord, 1983).

There are 700–1600 mitochondria in a liver cell; they often appear to be associated with MTs but not with MFs; whether they associate with IFs is not yet clear (Ball and Singer, 1982; Summerhayes *et al.*, 1983). Complexes between rough ER and mitochondria have been isolated from broken cells by Meier *et al.* (1981) and Cascarano *et al.* (1982).

Mitochondria, as we already know from lysosomes, can be isolated by differential centrifugation of homogenates; the enzyme markers are cytochrome oxidase and succinic dehydrogenase, which play a prominent role in biological oxidations. Isolated mitochondria easily undergo osmotic swelling, which can be prevented by the addition of ATP + $MgCl_2$. Their main components are proteins and phospholipids. In addition, they possess small amounts of nucleic acids (both DNA and RNA), the nature and function of which will be discussed here.

By treatment with detergents (digitonin and lubrol), it is possible to separate the outer mitochondrial membrane from the rest of the mitochondrion, which is called a "mitoplast." The classic marker enzymes are monomine oxidase for the outer mitochondrial membrane, adenylate kinase for the intermembrane space, cytochrome oxidase and succinic dehydrogenase for the internal membrane, and malic dehydrogenase for the matrix. Chemical analysis of the outer and inner membranes has shown that the former contains about twice as many lipids as the latter; with 40% lipids, the outer mitochondrial membrane is similar in overall composition to the various cell membranes we have already discussed. The inner mitochondrial membrane, on the other hand, is distinctly different from all other cell membranes in many respects: the presence of crests and elementary particles, low lipid content, and, most important of all, the presence of the machinery required for oxidative phosphorylation, i.e., energy production. The mitochondrial matrix contains the soluble enzymes of the Krebs tricarboxylic cycle, two or three "nucleoids" composed of mitochondrial DNA, small ribosomes called "mitoribosomes," and often dense granules with a high calcium content. Their probable role is $Ca^{2+}$ sequestration and release. More details about the structure and overall composition of mitochondria can be found in the review or books of Novikoff (1961), Lehninger (1964), Munn (1974), Kuroiwa (1982), and Rosamond (1982).

The main function of the mitochondria is to provide energy to the cell by ATP production via electron transport and oxidative phosphorylation. A discussion of the mechanisms of the coupling between respiration and phosphorylation of ADP to ATP is best left to biochemists. (Lehninger, 1964, 1982; Boyer *et al.*, 1977; Ernster, 1977; Green, 1977). The subject is still a hot one: there are still proponents of the direct chemical coupling and conformational coupling theories, despite the fact that the chemiosmotic theory is the "official" one since the award of a Nobel Prize to P. Mitchell. His Nobel lecture was published in 1979 and gives a clear and understandable account of his theory. The source of energy, according to Mitchell, is a transmembrane electrochemical proton gradient. The proton motive force $\Delta$ p results from two gradients through the mitochondrial membrane: one is a chemical gradient that is the difference in pH ($\Delta$ pH) through the membrane of the mitochondria, between the matrix, and the surrounding cytoplasm; the other component of $\Delta p$ is an electrical potential difference ($\Delta\psi$) between the outer and inner sides of the membrane. The proton motive force $\Delta$ $p$/or $\Delta\ \mu H^+$) is equal to $\Delta\psi - RT/F\ \Delta$ pH. Electrochemical gradients would be produced by electron transport; the energy contained in the two gradients would then be used by a membrane ATPase, allowing it to work in the unusual direction of ATP synthesis ($ADP + P_i \rightarrow ATP$). Mitochondrial energy production can be inhibited at various points of the electron transport chain by a number of drugs:

oligomycin, antimycin, rotenone, and cyanide affect various steps of electron transfer; uncouplers (which increase respiration and decrease ATP production) such as dinitrophenol are proton ionophores, which drive protons into the mitochondria and acidify its matrix; potassium ionophores (which exchange $K^+$ against $H^+$) are useful for testing the two components of the proton motive force; valinomycin collapses the potential difference $\Delta\psi$ and usually uncouples oxidation and phosphorylation; the electroneutral $K^+$-ionophore nigericin abolishes the chemical gradient ΔpH and seldom acts as an uncoupler.

We have already mentioned the presence of calcium-containing dense granules in the mitochondrial matrix. Such calcium stores are interesting for cell biologists, in view of the importance of the free $Ca^{2+}$ level for cell communication, MT and MF assembly, locomotion, and eno- and exocytosis. Rose and Loewenstein (1975b) showed that if calcium is injected into a cell and if its localization is followed by the light emission of aequorin, it does not diffuse from the injection point, being sequestered in the mitochondria and microsomes. Sequestration of $Ca^{2+}$ is linked to energy production, since the injected $Ca^{2+}$ diffuses freely in KCN-treated cells; indeed, $Ca^{2+}$ accumulation in mitochondria and microsomes requires respiration (Lucas *et al.*, 1978). Accumulation of excess $Ca^{2+}$ in isolated mitochondria has adverse results: oxidative phosphorylation is inhibited unless ATP and $MgCl_2$ are added to isolated mitochondria or to permeabilized cells (Villalobo and Lehninger, 1980). It seems that the all-important maintenance of a proper free $Ca^{2+}$ content in the cell depends largely on the $Ca^{2+}$ influx/efflux ratio in the mitochondria. This ratio is controlled by the intramitochondrial $Ca^{2+}$ content and is adjusted by $Ca^{2+}$ sequestration into the microsomes (Becker *et al.*, 1980). Calcium influx in the mitochondria is mediated by an endogenous $Ca^{2+}$ ionophore (calciphorin, a small molecule of 3000 daltons) while $Ca^{2+}$ efflux results from the activity of a calcium pump (an $Mg^{2+}$–activated ATPase), according to Jeng and Shamoo (1980). That mitochondria play a major role in $Ca^{2+}$ binding is substantiated by the fact that fluorescent camodulin, which may be taken as a calcium buffer, binds much more strongly to mitochondria than to MFs and MTs in fibroblasts (Pardue *et al.*, 1980). The complicated mechanisms of $Ca^{2+}$ transport in mitochondria (which may involve exchange with either three $Na^+$ or two $H^+$ for $Ca^{2+}$ efflux), have been reviewed by Nicholls and Crompton (1980) and by Barrett (1981). As was mentioned above, excess intramitochondrial-free $Ca^{2+}$ has effects on oxidative phosphorylation; when mitochondria store $Ca^{2+}$, they are performing a primarily "selfish job": no more than 0.1% of mitochondrial calcium should be free, and this low concentration has to be kept within narrow limits in order to ensure normal oxidation levels in tissues (Denton and McCormack, 1980). Only under conditions of severe stress do the mitochondria release part of their bound calcium.

While accumulation and release of $Ca^{2+}$ from a mitochondrial store are important for cell economy, it should be pointed out that there are other nonmitochondrial stores of $Ca^{2+}$ in the cell; the use of suitable inhibitors allows us to distinguish between the two: while the uptake of $Ca^{2+}$ in the mitochondria is inhibited by antimycin and oligomycin (which interfere with cellular oxidation), vanadate is a suitable inhibitor of $Ca^{2+}$ uptake in the nonmitochondrial storage compartment. There is now strong evidence that release of $Ca^{2+}$ from this nonmitochondrial store is, as we have already seen, mediated by the breakdown of phosphoinositides and the production of inositol triphosphate (Streb *et al.*, 1983). It is generally accepted that this store is accumulated in endoplasmic reticulum vesicles and cisternae (Somlyo *et al.*, 1985), but one cannot exclude the possibility that the cell possesses other stores of $Ca^{2+}$ than mitochondria and endoplasmic reticulum.

In "Biochemical Cytology" (1957), there was no discussion about mitochondrial DNA, its transcription and translation (reviewed by Borst, 1977; Saccone and Quagliariello, 1975; Attardi, 1985). However, the experiments of Boris Ephrussi (1949, 1953) and Ephrussi and Slonimski (1955) on the *petites* mutations in yeasts were mentioned. Ephrussi found that treatment of yeast cells with acriflavin (ethidium bromide, which is more efficient than acriflavin, is now often preferred) leads to a cytoplasmic mutation. Cells lose their respiratory enzymes and form colonies of much smaller size than normal. Genetic analysis clearly showed the cytoplasmic localization of the mutation; mitochondria were immediately suspected as being the mutated cells organelles because the respiratory enzymes were lost. This remarkable case of cytoplasmic heredity was discovered because growth in yeasts was possible since fermentation can occur in the absence of respiration. In obligatory aerobic cells or organisms, the mutation would have been lethal and, thus, would not have been detected. Electron microscopy showed that the *petites* have mitochondria, though less developed than in normal yeast cells (Yotsuyanagi, 1955); their ultrastructure is somewhat simplified. Similar observations were made later on a respiratory cytoplasmic mutant called *poky* in *Neurospora* and on trypanosomes treated with ethidium bromide.

This work raised the question of the eventual presence of DNA in mitochondria. Work done in the 1960s in a number of laboratories (see the review by Borst, 1977) led to the isolation of mitochondrial DNA from many sources. It was found that mitochondria always contain circular (as in bacteria) DNA molecules (Fig. 47); their contour length varies between 20 and 46 μm in yeasts and *Neurospora,* but is only about 5 μm in higher eukaryotes (*Xenopus,* humans), with an $M_r$ of $<10^6$. In the *petites* mutants, the size of mitochondrial DNA is variable, but the molecules are always shorter than in normal yeast cells. In general, they have a higher adenine + thymine content. These findings have led to the conclusion that the *petites* mutations result from numerous deletions in mitochondrial DNA (Morimoto *et al.*, 1975). Elegant hybridization experiments (reciprocal crosses between two different *Xenopus* species and analysis by mo-

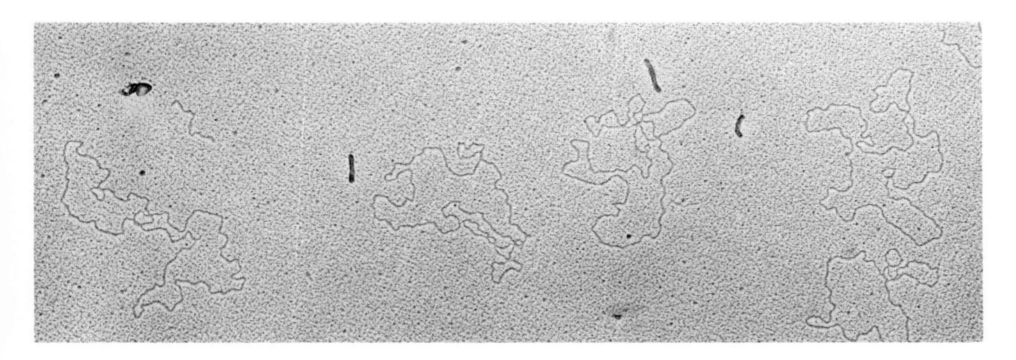

FIG. 47. Circular mitochondrial DNA molecules isolated from *Xenopus* oocytes. [Courtesy of F. Hanocq.]

lecular hybridization of the mitochondrial DNA present in the offspring) led Dawid and Blackler (1972) to several important conclusions: (1) All the mitochondria present in the adult derive from those of the egg. (2) There is no sequence homology between mitochondrial and nuclear DNAs; the nucleus does not contain copies of mitochondrial DNA, which is an independent entity responsible for cytoplasmic heredity. As we shall see, this last conclusion is no longer valid today. Mitochondrial DNAs have strongly diverged during evolution; trypanosomes that are morphologically indistinguishable can be differentiated by analysis of their mitochondrial DNAs (Steinert and Van Assel, 1974). According to Upholt and Dawid (1977), it is even possible to show that the mitochondrial DNAs of two dogs are not identical.

The mitochondrial DNA of trypanosomes deserves mention as a curiosity of nature. Trypanosomes have a single large mitochondrion that contains a Feulgen-positive body (Fig. 48). This body was called ''kinetoplast'' by the old protozoologists, because they believed that it was related to the flagellum and its motility. In fact, it is a strange network of mitochondrial DNA molecules. Work by Steinert and Van Assel, Borst, Riou, Wolstenholme, Simpson, and others (reviewed by Kallinikova 1981) has shown that kinetoplastic DNA (kDNA) is a network of $40 \times 10^6$ daltons composed of a large number of intermingled minicircles of 0.8 μm contour length. Each minicircle has $1.5 \times 10^6$ daltons of kDNA (1–2 kilobases, or kb). The network is composed of between 3000 and 25,000 (according to species) of these circles; in addition, the network comprises around 45 maxicircles made up of about 10 to 15 times larger DNA molecules (Fig. 49). The minicircles, according to Borst *et al.* (1980), would be more responsible than the maxicircles for the aforementioned species differences in the composition of kDNA. According to the latest available report (Barrois *et al.*, 1981), kDNA is made up of 3000 heterogeneous coiled minicircles of 1012 base pairs (bp) and 50 supercoiled maxicircles of 23,000 bp. Using restriction endonucleases, analysis of the molecular organization of kDNA shows that it contains

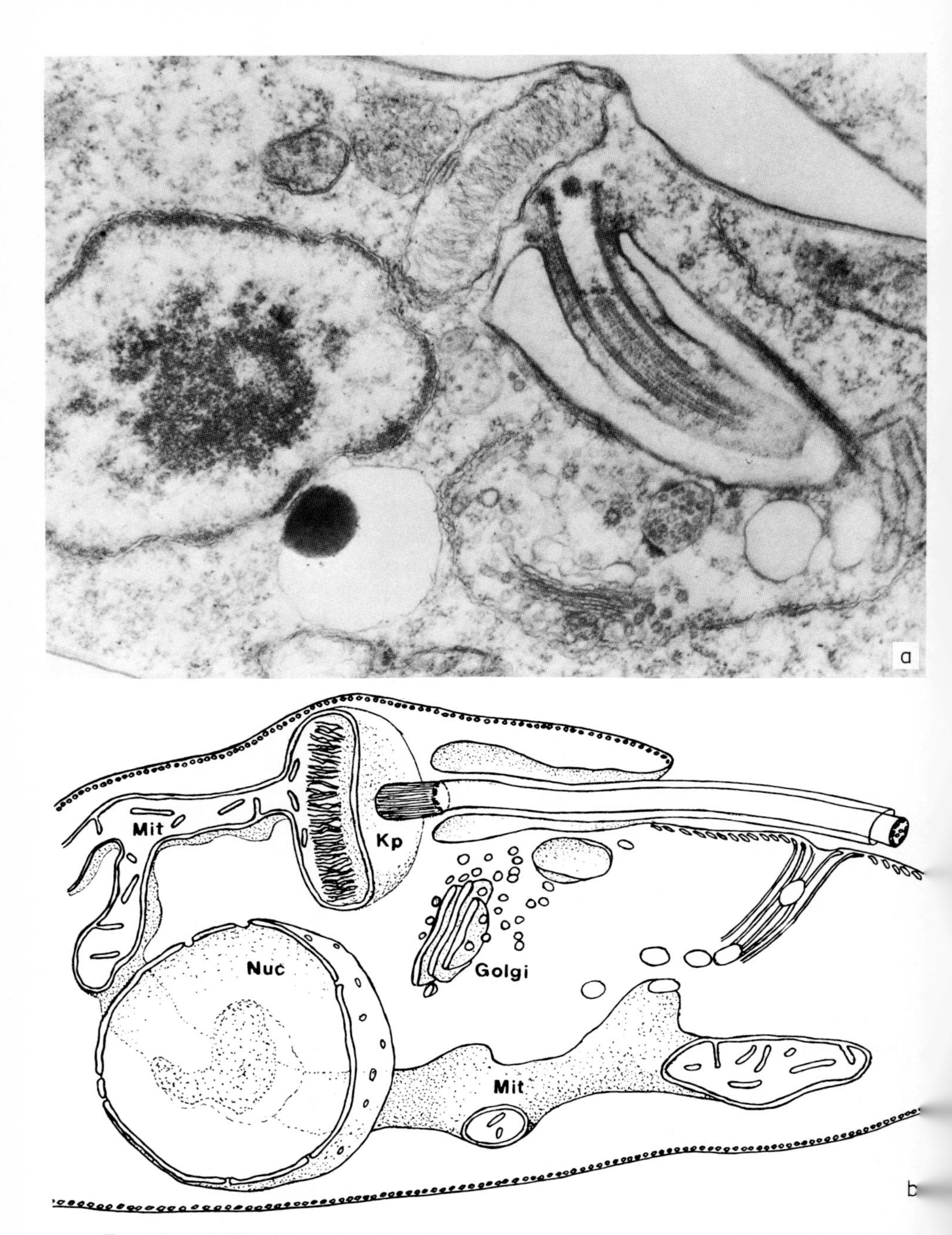


FIG. 48. (A) Ultrathin section through a trypanosome (*Trypanosoma mega*). (B) Schematic representation of the mitochondrial system in trypanosomids. Kp, kinetoplast; Nuc, nucleus; mit, mitochondria. [Courtesy of M. Steinert.]

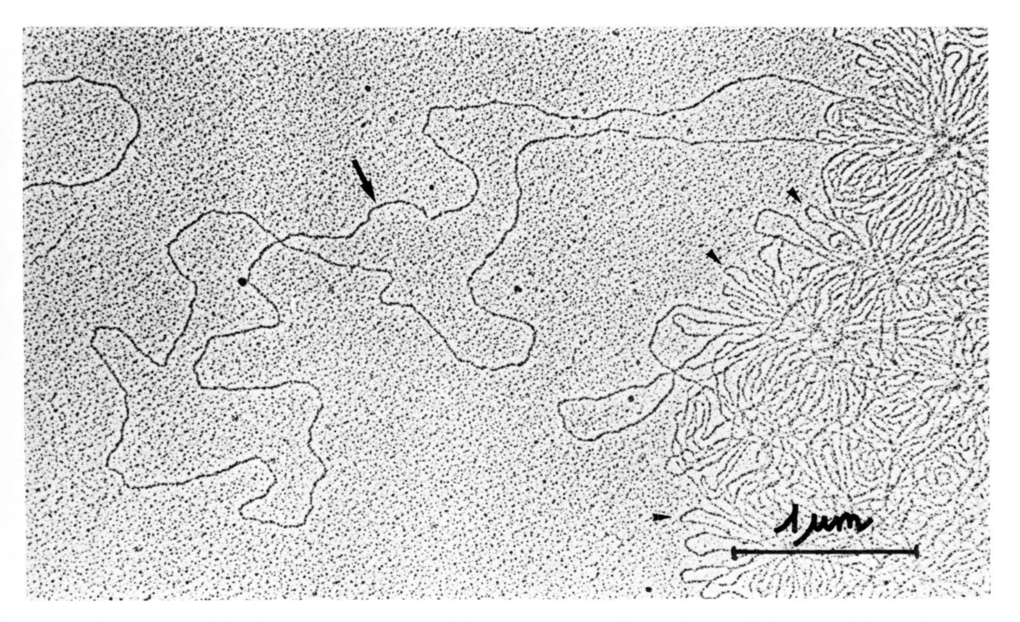


FIG. 49. Mini- (arrowheads) and maxicircles (arrow) in kinetoplastic DNA isolated from the trypanosome *Crithidia luciliae*. [Courtesy of M. Steinert.]

many termination codons. Thus, the coding potentiality of kDNA would be restricted to oligopeptides of 20 amino acids transcribed on the maxicircles. However, this inference is not necessarily true, since it is not known whether kDNA transcription uses the same genetic code as nuclear DNA; the reasons for this reservation will become apparent when we discuss mitochondrial transcription and translation. According to Stuart and Gelvin (1982), the kDNA maxicircles are largely, but not completely, transcribed; there is no, or very little, transcription of the minicircles. Finally, according to Eperon *et al.* (1983), the major transcripts of kDNA are very small rRNAs (12 and 9 S).

These reports widen, if possible, the kDNA mystery: why did trypanosomes accumulate so much DNA in their mitochondrion, and why are the coding potentialities of this DNA so limited? What is the function of the noncoding minicircles? It should be added that akinetoplastic strains of trypanosomes, which have lost their kDNA by treatment with ethidium bromide or berenil, are perfectly viable. The minicircle network can be uncatenated by treatment with the bacterial enzyme DNAgyrase. Conversely, Shlomai and Zadok (1983) have found that trypanosomes possess a DNA topoisomerase activity (see the next chapter) that interlocks the minicircles in huge catenanes; this enzyme, which requires $Mg^{2+}$ and ATP for activity, is probably responsible for the formation of the kinetoplastic network.

When mitochondria multiply by fission or fragmentation, their DNA undergoes replication by mechanisms similar to those discovered by Cairns (1963) for DNA replication in the circular chromosome of *Escherichia coli*. Characteristic of mitochondrial replication are its sensitivity to ethidium bromide and the presence of displacement loops (D-loops), which can be seen under the electron microscope. They are short nascent DNA strands maintained at one origin of replication (Fig. 50), and result from the asynchronous replication of DNA where the two strands have different origins. Replication begins by initiation of DNA synthesis at the origin of the heavy strand; synthesis of the light strand does not begin before the heavy strand has been elongated to the origin of the light strand (Doda *et al.*, 1981). The subject has been reviewed by Clayton (1982). According to Tapper and Clayton (1981), human mitochondrial DNA has two distinct replication origins related to the position of the mitochondrial rRNA and the tRNA genes, which will soon be discussed. In the mitochondria of *Xenopus* oocytes, a 12,500-dalton protein binds preferentially to the single-stranded fiber of the D-loop and presumably stabilizes it (Barat and Mignotte, 1981). It should be added that mitochondrial DNA is not "naked" in the nucleoids, since these

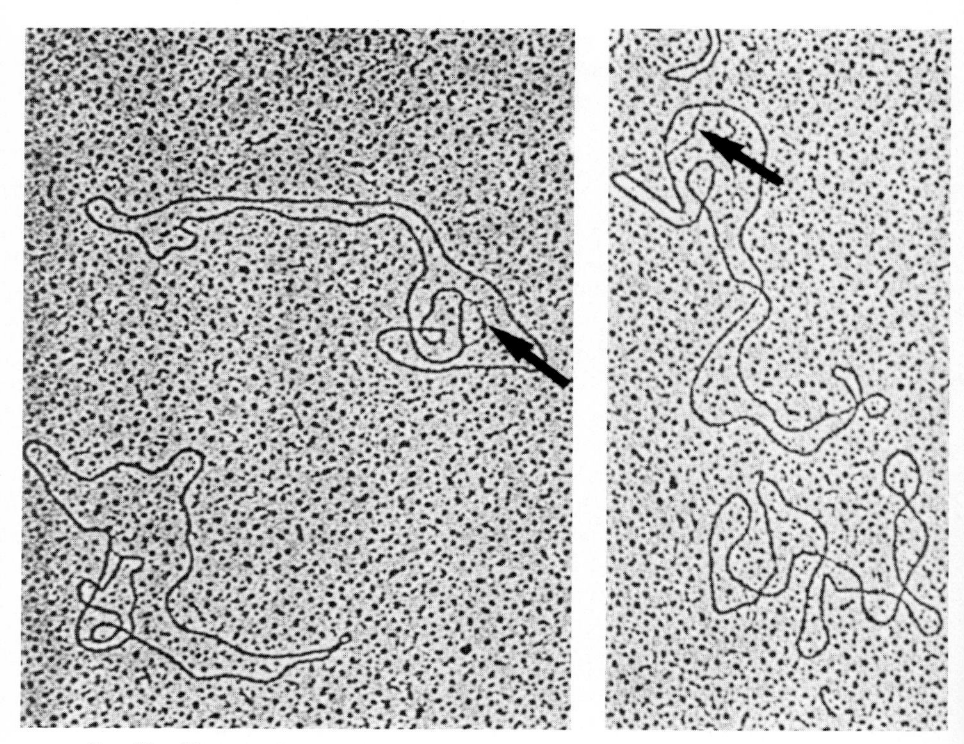

FIG. 50. Electron micrographs of closed circular mitochondrial DNA molecules. D-loops are indicated by arrows. [Kasamatsu *et al.* (1971).]

bodies contain twice as much protein than DNA and, according to Rickwood *et al.* (1981), are attached to the inner mitochondrial membrane by RNA-containing structures. Finally, it is generally believed that replication is catalyzed by a γ-DNA polymerase located in the mitochondria themselves (Adams and Kalf, 1980).

We come now to mitochondrial DNA transcription and translation. In 1970, Swanson and Dawid reported that mitochondria isolated from *Xenopus* oocytes contain ribosomes (the so-called mitoribosomes. These are smaller than cytoplasmic ribosomes. Their sedimentation constant 55–60 S is lower than that of the bacterial 70 S ribosomes due to a higher protein/RNA ratio. Like the cytoplasmic ribosomes, mitoribosomes contain two major types of RNA, but their low sedimentation constants (21 and 13 S in *Xenopus*) distinguish them from cytoplasmic rRNAs. It was later shown that mitochondrial DNA ($11.7 \times 10^6$ daltons) of *Xenopus* contains genes coding for the two mitochondrial rRNAs and for 15 mitochondrial tRNAs, which differ from their cytoplasmic counterparts (Dawid, 1972). Later, Ojala and Attardi (1974) and Hirsch and Penman (1974) discovered that, in addition to rRNA and tRNA, mitochondrial RNA is composed of eight different mRNAs that have a polyadenylic acid "tail" [poly ($A^+$)-RNA]; they suggested that these eight mRNAs are coded for by mitochondrial DNA. At the same time, it was pointed out by Leister and Dawid (1974) that the mitoribosomes of *Xenopus* contain between 40 and 44 proteins and that, in view of its small size, mitochondrial DNA could not code for more than one-sixth of this set of proteins. It should be added that the mitoribosomes of yeasts and *Neurospora* have a higher sedimentation constant than those of *Xenopus;* the value of these mitoribosomes is 70 S, the same as found for bacteria.

Further work during the 1974–1979 period extended these results to mitochondria from a variety of organisms. This work showed, in addition, that transcription of mitochondrial DNA can be selectively inhibited by rifampicin.

An unexpected and interesting finding was that key mitochondrial enzymes, such as cytochrome oxidase and ATPase (both are made up of several subunits) result from a coordinated synthesis by mitochondrial and cytoplasmic ribosomes; some of their subunits are synthesized on mitoribosomes and others on cytoplasmic ribosomes (see, for instance, Koch, 1976). The most unexpected finding came more recently. In 1980, Martin *et al.* and Heckman *et al.* discovered that the "universal" genetic code does not apply to mitochondria isolated from yeast and *Neurospora*. In bacteria and in the polyribosomes of eukaryotic cells, the triplet codon UGA is a "stop" codon; it interrupts the reading of the mRNA base sequences by the ribosomes, terminates the elongation of the growing polypeptide chain, and contributes to its release. In mitochondria, the UGA codon is recognized by a mitochondrial $tRNA^{tryp}$. This means that when UGA is read, a tryptophan residue is inserted into the polypeptide chain; growth of the latter continues. More recently, de Bruijn (1983) has discovered other peculiarities of the mitochondrial genetic code: in *Drosophila* mitochondrial DNA, the AGA triplet codes for serine instead of arginine, and the ATAA sequence [which, in the nucleus, is a signal for the addition of a poly(A) tail to many mRNAs] is used for the initiation of translation.[1]

Another breakthrough came with findings of Anderson *et al.* (1981). They succeeded in completely sequencing the 16,569 bp of the human mitochondrial genome. This allowed them to accurately map, in the circular mitochondrial DNA molecule, the position of the 16 S and 12 S rRNA genes of 22 tRNA genes (these tRNAs differ from the tRNAs coded by the nuclear genes), the genes for three subunits (I, II, III) of cytochrome oxidase, the genes for subunit 6 of mitochondrial ATPase, the genes for cytochrome *b* and the genes for eight other as yet unidentified mitochondrial proteins. This base analysis showed that, in contrast to the situation prevailing in the much larger yeast mitochondria, as shown in Slonimski's laboratory by Lazowska *et al.* (1980), there are very few noncoding bases between the human mitochondrial genes. Anderson *et al.* (1981) were also able to decipher the entire mitochondrial genetic code, in which UGA is read tryptophan and the termination codons are AGA and AGC. These termination codons are not, as usual, coded by the DNA, but are created by polyadenylation of mRNAs. Since both strands of human mitochondrial DNA are transcribed, evolution has resulted in an extraordinary economy in genetic material. In contrast, in the much longer mitochondrial DNA of yeast cells, the genes are separated by nontranscribed spacers; in addition, there are intervening sequences (introns) in the middle of the genes themselves. The possible biological and molecular functions of these mitochondrial DNA introns have been the subject of interesting hypotheses by Slonimski (1980) that will be discussed in the next chapter. There are no introns in human mitochondrial DNA (Ojala *et al.*, 1980); it has, therefore, been suggested that the entire DNA "heavy" strand is transcribed as a single transcript (polycistronic transcription, as in bacteria). This large transcript would then be cleaved at the correct places by endonucleases; the necessary punctuation would be afforded by tRNA sequences (Ojala *et al.*, 1981).

The work of Anderson *et al.* (1981) has been extended to the mouse (Bibb *et al.*, 1980) and bovine (S. Anderson *et al.*, 1982) mitochondrial genomes; both have now been completely sequenced. The mouse mitochondrial genome is composed of 16,925 bp; the genes for the 16 S and 12 S rRNAs, for 22 different tRNAs, for three subunits of cytochrome oxidase, one ATPase subunit, and eight unidentified proteins have been located. There are no (or perhaps very short) spacers. The D-loop has 879 nucleotides containing the origins of both heavy-strand and light-strand replication, and is the only region in mitochondrial DNA that does not code proteins. Bovine mitochondrial DNA has 16,358 nucleotides

[1]It has been recently found that in ciliates the stop codons UAA and UAG code for glutamic acid or glutamine (Caron and Meyer, 1985; Preer *et al.*, 1985; Halftenbein, 1985). There are thus exceptions to the universality of the genetic code.

and the same general organization as its human counterpart. Although there is between 63 and 79% homology between the two, there is little homology in the D-loop region, which is variable in length.

We have mentioned that work using up-to-date methodology has shown that the classic concept that there is no sequence homology between mitochondrial and nuclear DNAs can no longer be accepted. Indeed, it came as a surprise when van den Boogaart *et al.* (1982) reported that sequences homologous to yeast mitochondrial DNA can be detected in nuclear DNA; these findings have been confirmed by Farrelly and Butow (1983). This is not a peculiarity of lower eukaryotes, which, as we have seen, have unusually large mitochondrial DNA molecules; similar findings have been reported for maize (Kemble *et al.*, 1983), locusts (Gellissen *et al.*, 1983), see urchins (Jacobs *et al.*, 1983), and even normal rat liver (Hadler *et al.*, 1983). The phenomenon appeared to be very widespread, if not universal. Closer analysis has shown that, in sea urchins, typical mitochondrial DNA sequences, coding for mitochondrial 16 S RNA and for the cytochrome oxidase I subunit, hybridize to genomic DNA (Jacobs *et al.*, 1983). The authors conclude that, during evolution, a germ line transposition of part of the mitochondrial genome into nuclear DNA took place; this event was followed by rearrangements of the DNA sequences and single nucleotide substitutions. In rat liver, the mitochondrial DNA sequences inserted into nuclear DNA are shorter than 3 kb; they correspond to the portion of the mitochondrial genome that contains the D-loop and part of the rRNA genes (Hadler *et al.*, 1983). As the authors point out, the D-loop, which is species specific and is the site of mitochondrial DNA replication, is a good candidate for an interaction with the nuclear genome.

The unescapable conclusion is that small fragments of the mitochondrial genome have escaped from the mitochondria, penetrated into the cell nucleus, and been integrated in the nuclear genome. When did this event happen? Does it still happen today? Farrelly and Butow (1983) think that the transfer of mitochondrial DNA into the nucleus started during the pre-Cambrian period (this is not easy to prove experimentally) and that it is still continuing. That this transfer might still occur today is suggested by observations by Kemble *et al.* (1983) on maize. They found that, in this plant, sequences homologous to mitochondrial DNA are found not only in the nucleus but also in DNA-containing viruslike particles (episomes); these particles might be the carriers of mitochondrial DNA sequences to the nucleus. More work in this new field will certainly lead to exciting discoveries.

Clearly, few polypeptides are synthesized on mitoribosomes under the direction of mRNAs transcribed on mitochondrial DNA. However, these polypeptides are very important since lack of their synthesis would result in the lack of activity for enzymes of such fundamental importance for energy production as cytochromoxidase or mitochondrial ATPase. Other mitochondrial proteins are synthesized on the mitoribosomes, but the required information comes from mRNAs which have been synthesized in the nucleus. Finally, the majority of the mitochondrial proteins [including all the proteins of the outer mitochondrial membrane, according to Gellerfors and Linden (1981)] are synthesized on cytoplasmic ribosomes programmed by mRNAs of nuclear origin; they must move from the cytoplasm into the mitochondria.

Cell biologists must distinguish between the proteins that are synthesized on mitochondrial and on cytoplasmic ribosomes. We have mentioned that mitochondrial DNA and RNA synthesis can be halted, in a specific way, by ethidium bromide and rifampicin, respectively. Luckily, we have also at our disposal a specific inhibitor of mitochondrial protein synthesis: the antibiotic chloramphenicol, which inhibits mitochondrial protein synthesis without affecting cytoplasmic protein synthesis. Its effects are thus exactly the opposite of those of cycloheximide, which affects only cytoplasmic protein synthesis.

A considerable amount of work, done mainly in P. Slonimski's laboratory for many years, has shown that it is possible to isolate many mitochondrial DNA mutants, resistant to antibiotics or inhibitors of the respiratory electron chain. Similar work on mammalian cells in culture has allowed the isolation of cells able to grow in the presence of chloramphenicol. Interestingly, fusion of cytoplasts (anucleate cytoplasm) from such chloramphenicol-resistant cells with normal cells produces cybrids that are, at least for several generations, chloramphenicol-resistant (Bunn *et al.*, 1974). These elegant experiments provide direct evidence that chloramphenicol resistance is inherited by the cytoplasm, probably through mitochondrial DNA.

We have seen that the mitochondrial proteins may have different origins, and this raises interesting questions. How can these proteins of endogenous or exogenous origins assemble together in a harmonious way? How does the complex structure of the mitochondria (with their crests, etc.) arise? At the molecular level, how can an enzymatically active cytochrome oxidase or ATPase be built up with some subunits synthesized in the cytoplasm and others in the mitochondrion itself? How can the mitochondria select the proteins that they will utilize for their own purpose from among the innumerable proteins that surround them in the hyaloplasm? Our answers to these questions are still very limited. What is known is that the proteins that are synthesized on the mitoribosomes under the direction of mitochondrial DNA are strongly hydrophobic, which should greatly facilitate their incorporation into the lipid bilayer of the internal mitochondrial membrane. Regarding the penetration into the mitochondria of proteins synthesized in the cytoplasm, we know that these proteins are synthesized on polysomes in the form of precursors that are larger (longer by 2000–3000 daltons) than the final products; this is followed by energy–dependent proteolytic processing of the precursor with simultaneous uptake of the protein (Schatz, 1979; Lewin *et al.*, 1980; Neupert and Schatz, 1981; reviewed by Ades, 1982). This vectorial energy-dependent processing differs in one important respect from what was seen when the transport of secretory proteins into the ER was discussed. In mitochondria, the transport is linked to ATP production and not to protein synthesis. In the ER protein, synthesis and transport are associated (Blobel, 1977). According to Anderson (1981), the proteolytic processing of the mitochondrial protein precursors removes a basic peptide necessary for the recognition of the protein by the "mitochondrial uptake apparatus." There is, indeed, some evidence that mitochondria possess distinct receptors for the different cytoplasmic proteins (Zimmerman *et al.,* 1981). Great progress in this interesting field is expected in the next few years.

Work on the subject has disclosed that all mitochondrial proteins of cytoplasmic origin do not derive from higher-molecular-weight precursors and subsequent proteolytic cleavage. This is the case for cytochrome C, in which incorporation into mitochondria requires only the cleavage of the terminal methionine residue (Matsuura *et al.,* 1981). There is good evidence for the view that insertion into the external mitochondrial membrane of proteins synthesized on cytoplasmic polysomes does not require energy; they bind to receptors located on this outer membrane. But insertion into the inner compartment of the mitochondrion requires energy and, in most cases, proteolytic processing (Gasser and Schartz, 1983; Zwizinski *et al.,* 1983; Riezman *et al.,* 1983). Proteolytic processing of a precursor results from the action of a matrix neutral protease, an enzyme that displays remarkable specificity; it hydrolyzes neither the mature protein nor nonmitochondrial proteins (Cerletti *et al.,* 1983).

In summary, mitochondrial proteins encoded by the nuclear genes are synthesized on cytoplasmic polysomes and released in the cytosol; they then bind to receptors located on the outer mitochondrial membrane and are translocated by an energy-dependent process; in most cases, but not all, the amino extension is cleaved by a protease present in the mitochondrial matrix. Finally, refolding and assembly of the protein take place (Schatz and Butow, 1983).

In a recent review dealing with the entry of proteins into mitochondria, Hay *et al.* (1984) stress the following points: mitochondrial proteins are synthesized as precursors that bind to receptors located on the mitochondrial outer membrane; translocation through this membrane is energy-dependent and is followed by the proteolytic processing of the translocated precursors; and the last stage is assembly of the processed proteins into functional units. However, all mitochondrial proteins do not conform to this general scheme. For instance, cytochrome *c* enters into the intermembrane space, and this requires only the presence of heme and a mitochondrial-specific receptor. Neither energy nor proteolysis is required. The B subunit of the mitochondrial $F_1$-ATPase, which moves into the mitochondrial matrix, requires the presence of a receptor, energy, and proteolytic activity to reach its goal. Cytochrome $b_2$ moves first into the matrix and then back into the intermembrane space. This requires a receptor, energy, a matrix protease, and an intermembrane space protease. Finally, porin is inserted into the outer mitochondrial membrane without a requirement for energy and proteolysis. Transport of proteins into mitochondria is, as one can see, a complex and highly selective process.

At this point, a number of questions should be asked. Did mitochondria originate from bacteria? Should they be considered symbiotic organisms? The author would be hard pressed to answer these questions, since the answers are unknown. After all, none of us had the opportunity to witness the invasion of a primitive eukaryote bacterium (if it ever happened). Soon after the end of World War II, the author was asked by Jacques Monod whether he believed that mitochondria are bacteria. André Lwoff, a witness to this scene, whimsically smiled and said, "This is a very dangerous question; be careful." The author answered "no"; the advent of electron microscopy showed that he was right, since no bacteria have the well-developed crests so characteristic of mitochondria. However, the possibility that mitochondria were once bacteria should not be dismissed for that reason alone. From the molecular viewpoint, the two have many things in common: circular DNA, probably polycistronic transcription, reproduction by fission, etc. If mitochondria have evolved from bacteria that invaded a primitive eukaryote, they have undergone numerous changes during prolonged symbiotic life. Their genetic information is very limited, they have their own genetic code, their ribosomes are smaller than those of bacteria, and most of their proteins come from their host. In a review by Mahler (1983), it was pointed out that the organizations of the bacterial and mitochondrial genomes are very different. For this reason, Mahler does not believe in the prokaryotic origin of the mitochondria and suggests that they appeared autogenously in proto-eukaryotic cells. Unfortunately, no one knows what these hypothetical proto-karyotic cells looked like. However, chloroplasts, which will be discussed in the next section, look very much like primitive unicellular green algae. They display so many similarities to mitochondria that the symbiotic theory is very attractive. In a detailed review, Gray and Doolittle (1982) came to similar conclusions. The case for the symbiotic theory is stronger for these organelles than for mitochondria.

## C. Chloroplasts

Despite the tremendous importance of this subject (there would be no life without photosynthesis), relatively little will be said about it in this book, which deals mainly with animal cells. As has been done in this volume for mitochondria, the major biochemical function of the chloroplasts, photosynthesis, will be left to biochemistry textbooks.

Plant cells contain many types of plastids (leukoplasts, chromoplasts, chloroplasts, etc.) that are believed to originate from undifferentiated proplastids. Interconversion of one type of plastid into another can certainly take place. The whole subject of the origin and interconversion of plastids has been reviewed in great detail by Schnapf (1980).

Chloroplasts, which are several micrometers long, can easily be seen under the light microscope as ovoid green particles. Chloroplasts (Figs. 51 and 52) are surrounded, like the mitochondria, by a double membrane. Within the chloroplasts are stacks of lamellae or sacks, called "thylakoids," with an average thickness of 25 nm. Inner membranes, thus, are very well developed in chloroplasts, as they are in mitochondria (in chloroplasts, mitochondrial crests are

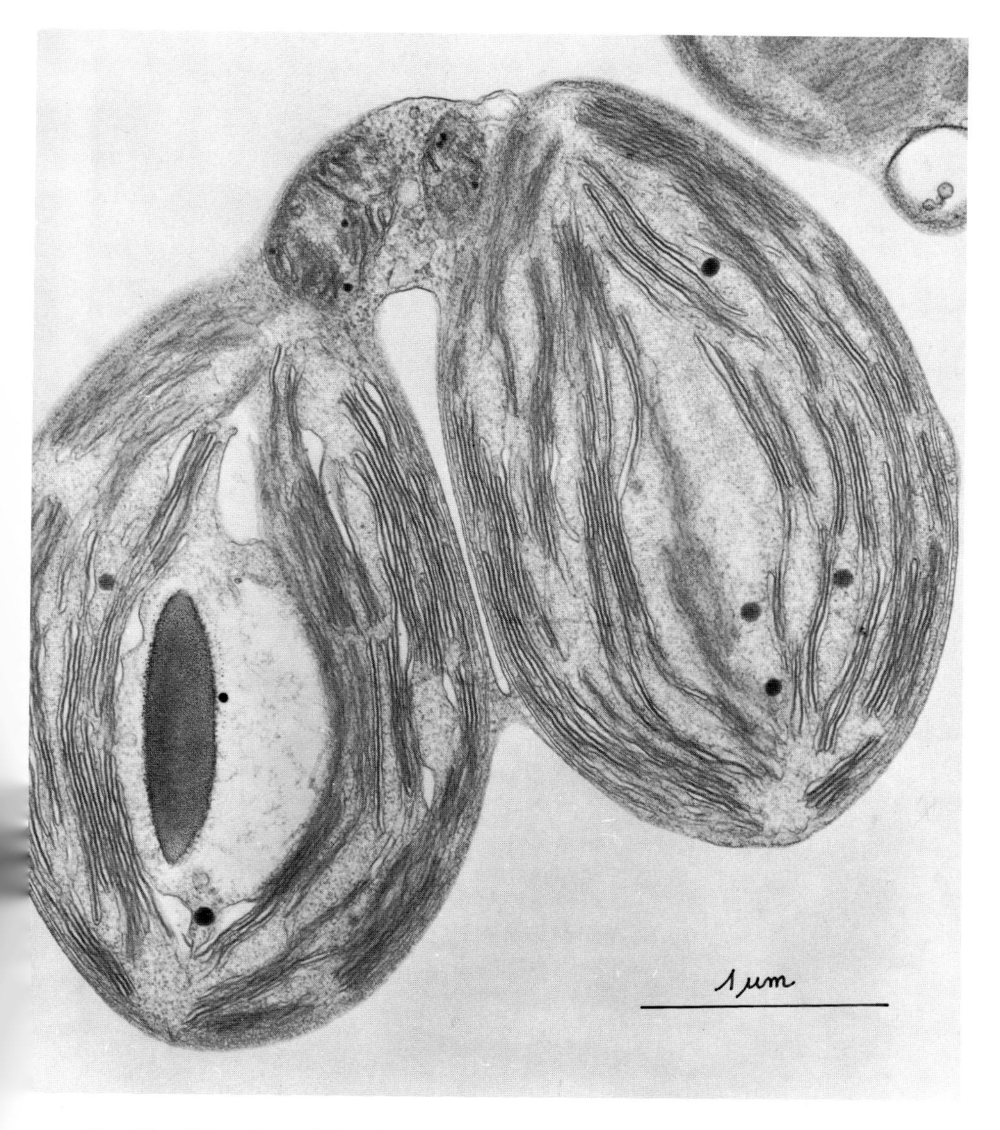


FIG. 51. Chloroplasts of the alga *Acetabularia mediterranea*, showing the thylakoids. The chloroplast at the left contains a starch granule. [Courtesy of M. Geuskens.]

replaced by thylakoids). In higher plants, stacks of thylakoids form dense cylinders called "grana," in which the machinery for photosynthesis is concentrated (Fig. 53). In algae, as in the *Acetabularia* chloroplasts represented in Fig. 51, the organization in grana is much less conspicuous. Thus, one speaks of "pseudograna." The structureless matrix of the chloroplast, called the "stroma" con-

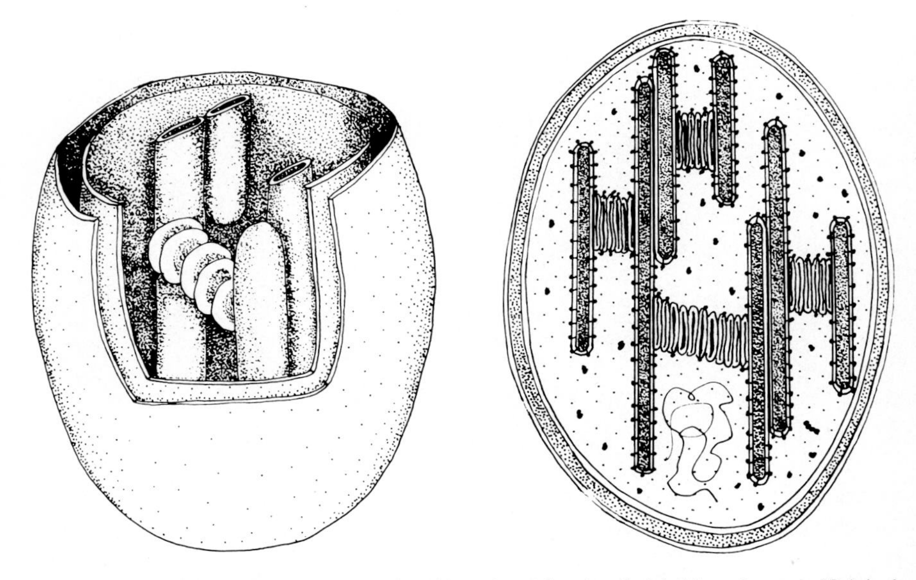

FIG. 52. Schematic representation of a chloroplast, showing thylakoids and grana. [Original drawing by P. Van Gansen.]

tains many ribosomes, chloroplastic DNA, and about 50% of the chloroplastic proteins in solution. The other 50% are part of the thylakoids and are thus insoluble. The thylakoids contain the chlorophylls, plastoquinone, and carotenoid pigments, assembled in photosystems I and II, which are required for photosynthesis. In addition to these lipids, proteins involved in electron transport and maintenance of the structure are associated with the photosystems. The thylakoids are surrounded by a membrane that is basically similar to the plasma membrane (with a lipid bilayer containing many intramembrane particles of various sizes). The major protein of the stroma is the key enzyme of the Calvin cycle, ribulose-1,5-bisphosphate carboxylase, which is responsible for the initial step of $CO_2$ fixation. This large enzyme is mentioned here not only because of its importance in photosynthesis, but also because it is composed of eight large subunits synthesized by the chloroplast itself and of eight small subunits of cytoplasmic origin. This is a striking example of the cooperation between a semiautonomous organelle and cytoplasm that has already been seen with mitochondrial cytochrome oxidase and ATPase.

There are many similarities between chloroplasts and mitochondria. This was illustrated by Ephrussi's work on the *petites,* which demonstrated the role played by the mitochondria in cytoplasmic heredity in yeasts. The existence of cytoplasmic heredity in plants presumably due to chloroplasts, has been known for many years. The classical examples are maternal heredity in *Epilobium* (Michaelis, 1951) and the cytoplasmic factors causing male sterility in maize (Rhoades, 1950). More recently, a considerable amount of work has been done on *Chlamy-*

*domonas* by Sager and her colleagues (1981), demonstrating convincingly the role of chloroplastic DNA in heredity.

Like mitochondria, chloroplasts divide by fission (Fig. 54). Their multiplication is not dependent on the presence of chlorophyll, as shown by studies on the effects of streptomycin on various plants and, in particular, on the unicellular alga *Euglena* (Provasoli *et al.*, 1951; De Deken-Grenson, 1955). Streptomycin-treated algae do not form chlorophyll, and it seemed as if the antibiotic had "cured plants of their chloroplasts." However, the conclusion of these studies was that streptomycin inhibits the growth and differentiation of proplastids without preventing their multiplication. Readers interested in the control of chloroplast division should read the extensive review of Possingham and Lawrence (1983). They will see there how chloroplasts acquire a dumbbell shape and then divide by constriction and fission. Except in unicellular algae, there is no correlation between cell division and chloroplast division.

The key material for cytoplasmic heredity in plants and for chloroplast multiplication is, of course, chloroplastic DNA (reviewed by Rochaix, 1985). As in mitochondria, chloroplastic DNA is composed of circular molecules; in higher plants (tobacco, mustard), its $M_r$ is around $100 \times 10^6$, almost 10 times greater than that of mitochondrial DNA in *Xenopus* or humans (Kolodner *et al.*, 1976; Link *et al.*, 1981). There are no repeated sequences among the 160kb pairs of chloroplastic DNA. It replicates, as in bacteria and mitochondria, by the Cairns

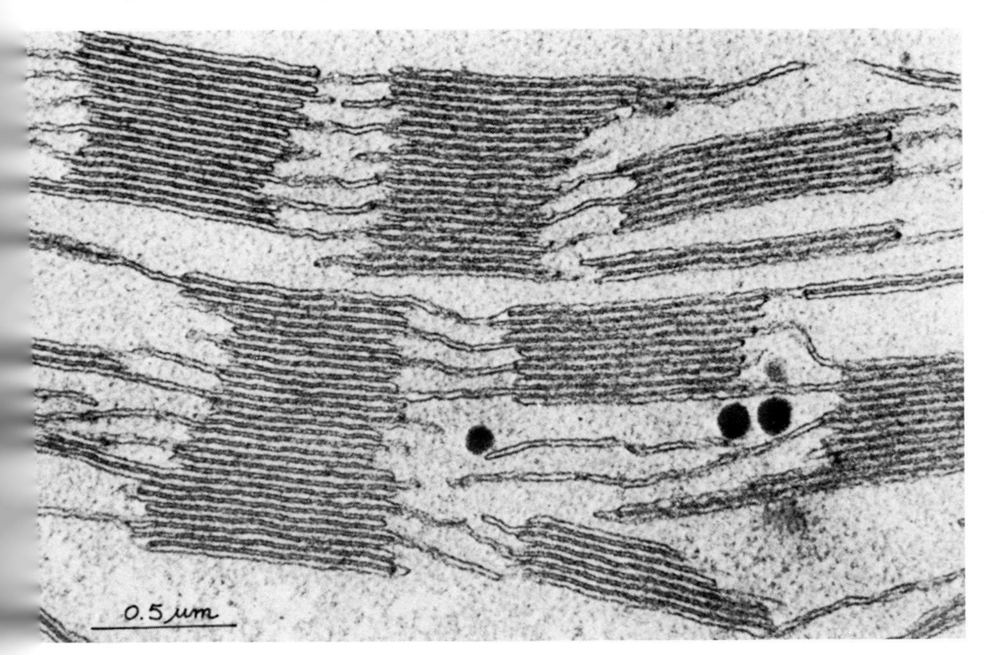


FIG. 53. Grana in the chloroplasts of a higher plant. Note the differences in structure with the less-differentiated chloroplasts of an alga shown in Fig. 51. [Miller (1979).]

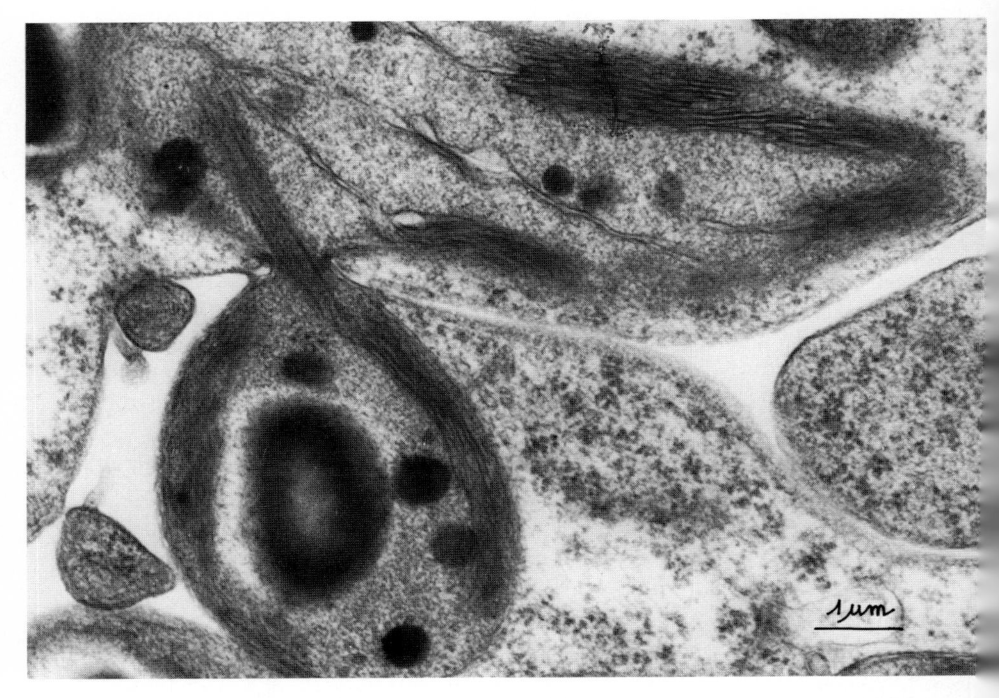


FIG. 54. A dividing chloroplast in *Acetabularia mediterranea*. [Courtesy of M. Bouloukhère.]

circle mechanism (Kolodner and Tewari, 1975). Again, as in mitochondria, ethidium bromide is a specific inhibitor of chloroplastic DNA replication (Flechtner and Sager, 1973). It seems that, in algae, the size of the multiple DNA copies present in a single chloroplast may vary. In *Euglena,* the average size of the chloroplastic DNA molecules is 44 μm, but digestion with specific endonucleases shows that there are differences in size between various copies (Jenni *et al.,* 1981). In *Acetabularia,* the chloroplasts contain both minicircles of 4.2 μm and linear molecules greater than 200 μm. The latter probably result from the breakage of very long circular molecules, possibly of 700–800 μm contour length (with an $M_r$ larger than $10^9$ (Green, 1976; Green *et al.,* 1977; Padmanabhan and Green, 1978). DNA replication in chloroplasts is catalyzed, as in mitochondria, by a γ-like DNA polymerase (Sala *et al.,* 1980). The same authors have pointed out further similarities between chloroplastic and mitochondrial DNAs: the presence of D-loops during replication and the absence of the beaded structure due, in chromatin, to associated histones (see the next chapter). Another similarity between the DNAs of the two semiautonomous cell organelles is that their replication, as shown by autoradiography after thymidine incorporation, is not synchronous with nuclear DNA replication. Chloroplast DNA replication can take place in isolated chloroplasts; all that is needed to promote [$^3$H]thymidine incorporation into isolated pea chloroplasts is a sufficient amount of light (Mills and Baumgartner, 1983). This shows that the chloroplasts possess all of the machinery (precursors and enzymes) required for their own DNA replication.

We have seen that mitochondrial DNA sequences are present in the nuclear DNA of many animal and plant cells. Similar findings have been made for chloroplastic DNA by Timmis and Scott (1983). They found that, in spinach leaves, DNA sequences homologous to chloroplastic DNA are present in nuclear DNA and are integrated at specific sites in the nuclear genome. If DNA sequences have moved during evolution from both mitochondria and chloroplasts into the nuclei, could similar exchanges take place between the DNAs of the two kinds of organelles? Yes, as shown by the work of Stern and Lonsdale (1982) and Lonsdale *et al.* (1983). They found that, in maize, mitochondrial DNA contains a segment homologous to chloroplastic DNA; the chloroplastic DNA fragment inserted into mitochondrial DNA contains sequences coding for the large subunit of the important enzyme ribulose-1,6-bisphosphate carboxylase; even the flanking sequences of the gene, which might play a regulatory role, are inserted into the mitochondrial genome. Recombination between chloroplastic and mitochondrial DNA sequences has thus taken place. For such mobile sequences, Stern and Lonsdale (1982) have coined the term "promiscuous" sequences; they point out that DNA promiscuity must have been a frequent event during evolution.

This work raises intriguing questions that will certainly be answered in the not very distant future: do all of the mitochondria present in maize cells possess the promiscuous DNA sequences of chloroplastic origin, or did this event take place in only a part of the mitochondrial population? Is the chloroplastic DNA fragment that is inserted into the mitochondrial genome transcribed by mitochondrial RNA polymerase? We know that other subunits of ribulose-1,6-bisphosphate carboxylase are synthesized in the cytoplasm under the control of nuclear genes. Do these subunits penetrate into both maize mitochondria and chloroplasts, or are they rejected from mitochondria because they lack specific receptors on their membrane? Is the presence of chloroplastic DNA in the mitochondrial genome a relic of evolution, or does this incorporation of chloroplastic DNA material into mitochondrial DNA still occur today? If so, what are the frequency and the mechanisms of this process? Answers to these questions might provide fresh evidence for the frequently held view that chloroplasts once were unicellular algae that, during evolution, became endosymbiontic organelles.

Chloroplasts contain 70 S ribosomes (like bacteria and yeast mitochondria) that can form polysomes after binding with mRNAs of endogenous (i.e., coded by chloroplastic DNA) or exogenous (i.e., coded by nuclear genes) origin. Isolated chloroplasts can synthesize proteins after addition of labeled amino acids (as well as nucleic acids and chloroplastic pigments), as shown by Goffeau and

Brachet (1965) and by Shephard and Bidwell (1973); they function perfectly in photosynthesis, producing the energy required for protein synthesis. In fact, according to a report by Green (1980), chloroplasts isolated from the alga *Acetabularia* synthesize between 20 and 24 different proteins; 13 to 15 of them are bound to chloroplastic membranes. As in mitochondria, protein synthesis is inhibited by chloramphenicol but not by cycloheximide. However, in contrast to mitochondria, protein synthesis by isolated chloroplasts is entirely light dependent. This difference between the two organelles is to be expected in view of their different mechanisms for energy production. Another difference between chloroplasts and mitochondria is that the mRNAs of the former lack a polyadenylic tract (Wheeler and Hartley, 1975). In both cases, RNA synthesis is catalyzed by a rifampicin-sensitive RNA polymerase.

It should be apparent from this summary that the informational content of chloroplastic DNA (only 0.3% of that of nuclear DNA in *Chlamydomonas*) is insufficient to ensure full autonomy to the chloroplasts. However, it can explain a number of cases of cytoplasmic heredity (cf. Granick, 1961; Bogorad, 1977; Sager *et al.*, 1981, for more details). According to a report by von Wettstein (1980), more than 30 proteins in chloroplastic membranes are implicated in photosynthesis; it is already known that 9 of them are coded by chloroplastic DNA and 7 by nuclear DNA. As for the mitochondria, we are again faced with the problem of the correct integration of all these proteins in the chloroplasts, but here we know nothing yet about its mechanisms. The important question of the interactions between chloroplastic and nuclear genes will be reexamined when we review nucleocytoplasmic interactions in the giant unicellular alga *Acetabularia* (Volume 2, Chapter 1).

### D. Centrioles and Basal Bodies (Kinetosomes)

Many years ago, light microscopy demonstrated the presence of small granules in the center of the asters during mitotic division and at the basis of the cilia or flagella. The former are the centrioles. Since they are often composed of adjacent granules, such pairs of centrioles are sometimes called "diplosomes." Around the diplosomes is an amorphous osmiophilic material called the "pericentriolar cloud." Diplosomes and the cloud together form a centrosome, which plays a very important role as an MTOC during cell division in animal cells (Chapter 5).

Protozoologists and cytologists observed, at the bases of cilia and flagella, granules that looked very similar to the centrioles of dividing cells; they called them "kinetosomes" or "parabasal bodies." The appellation "basal bodies" is now preferred for two reasons. First, the term "kinetosome" is misleading because it incorrectly implies a role of the granule in ciliary or flagellar movement; in fact, Linderman and Rikmenspoel showed, in 1972, that the "kinetosome" is not necessary for the motility of bull sperm flagella. There is now ample evidence that this is, in fact, not the exception but the rule. Second, the term "basal body" is more appropriate than "parabasal body" because electron microscopy has shown that there is a direct continuum between this body and the corresponding cilium or flagellum (see the next section).

In 1897, the French cytologist L. F. Henneguy (1898) and the Czech protozoologist von Lenhossék (1898) emphasized the homology between centrioles and basal bodies. The Henneguy–Lenhossék theory was fully confirmed when electron microscopy showed that the two organelles had exactly the same ultrastructure (De Harven and Bernhard, 1956). As shown in Fig. 55, centrioles or basal bodies are cylinders of $0.2 \times 0.5$–$2\ \mu m$. They have no membrane, but their walls contain nine triplets of MTs in a cartwheel arrangement. The three tubules that form one of the triplets are labeled A, B, and C. The triplets are surrounded by an electron-dense material. Fibers connecting the A subfibers of each triplet with the center of the cylinder can sometimes be seen.

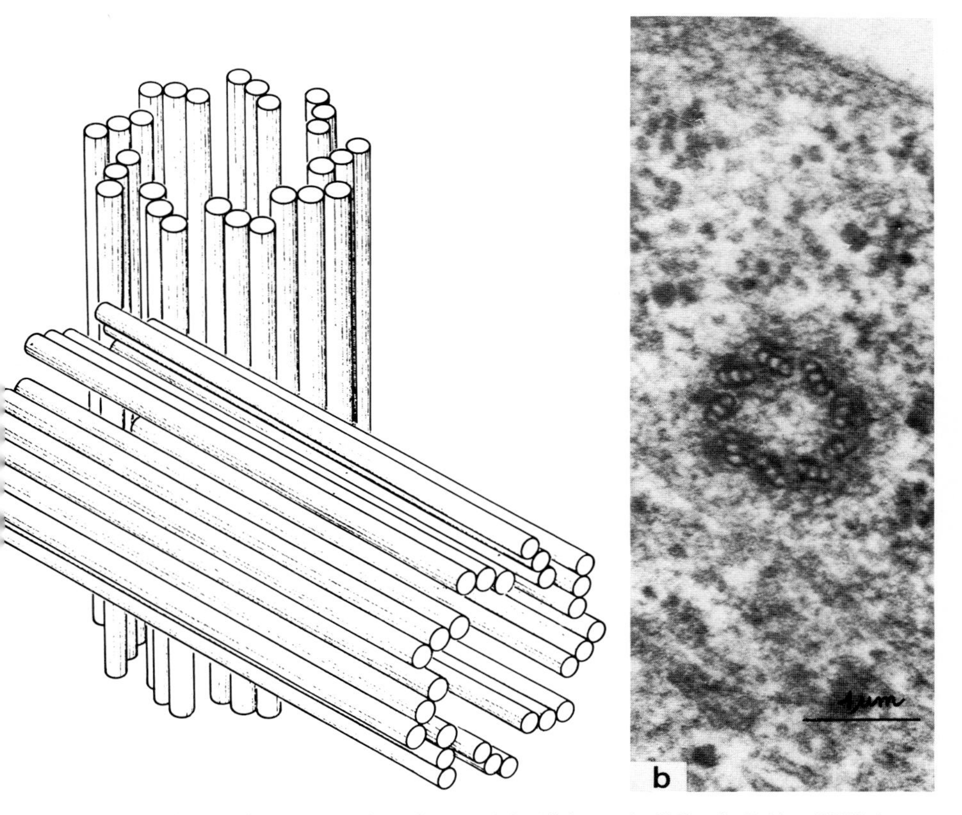


FIG. 55. (a) Schematic representation of a centriole (diplosome). [Albrecht-Buhler (1978).] (b) Transverse section through a centriole in a chick embryo blastomere. [Courtesy of P. E. Messier.]

That centrioles replicate during cell division has been known for many years. That the basal bodies of ciliates are also endowed with genetic continuity has been repeatedly emphasized by André Lwoff (1950). However, replication of the centrioles is not the result of binary fission or budding. Electron microscopy shows almost invariably the presence, near the basis of the centriole, of a "daughter centriole" or "procentriole" (Fig. 56). These daughter centrioles are never in direct contact with their "mothers," which excludes multiplication by budding. Instead, they grow progressively from their distal end. According to Phillips and Rattner (1976), the formation of the procentrioles, but not their elongation, requires protein synthesis. On the other hand, regeneration of cilia in deciliated sea urchin larvae requires RNA synthesis (Merlino *et al.*, 1978). It is very likely that, in eggs at any rate, centrioles can assemble *de novo* from dispersed precursor molecules; this can occur even in the absence of the cell nucleus (Kato and Sugiyama, 1971; Miki-Noumura, 1977; Moy *et al.*, 1977). This will be discussed in Volume 2, Chapter 2.

The correctness of the Henneguy–Lenhossék theory is further demonstrated by the fact that, in somatic cells, centrioles can give rise to cilia when culture conditions do not allow the cells to divide. Thus, centrioles never rest, and are

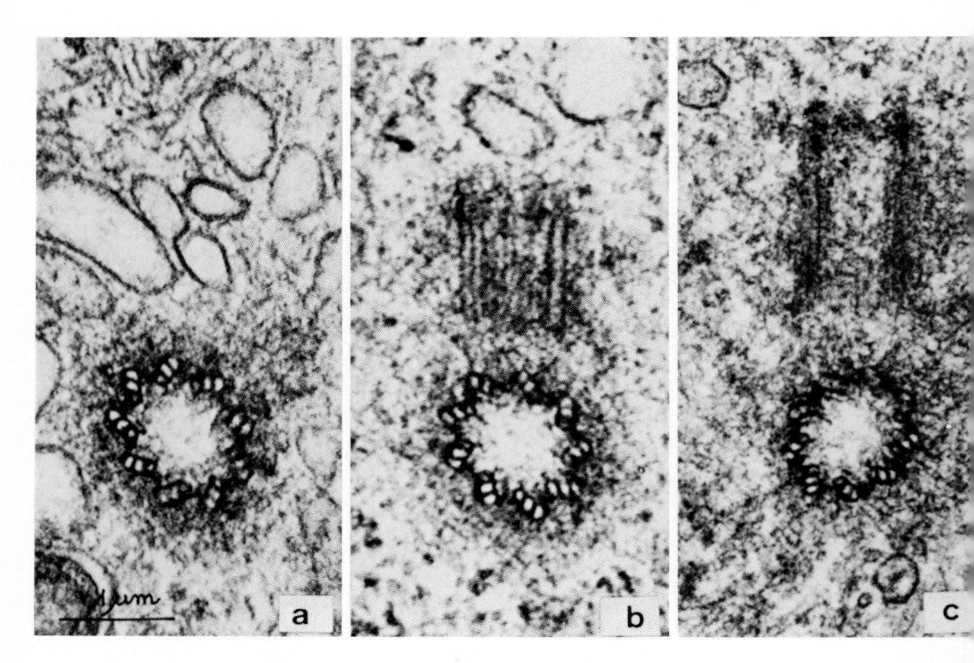


FIG. 56. Centriolar elongation during the cell cycle. (a) Early $G_1$ phase; (b) S phase; (c) prophase. [Rattner and Phillips (1973). Reproduced from the *Journal of Cell Biology*, 1973, **57**, pp. 359–372 by copyright permission of The Rockefeller University Press.]

always active. For instance, when fibroblasts are preparing for DNA synthesis and mitotic division, they often build up a cilium; deciliation follows when the cell is ready to divide (Tucker *et al.*, 1979). In dense cultures of hepatocytes, where cell division seldom takes place, there is an increase in the number of cilia; centrioles that can no longer be used for mitosis become basal bodies (Mori *et al.*, 1979). According to Tucker *et al.* (1983), all mammalian cells form a nonmotile primary cilium; since it shortens in the presence of $Ca^{2+}$ and the bivalent ion ionophore A23187, Tucker *et al.* (1983) suggest that the function of the primary cilium might be to control $Ca^{2+}$ flux; this remains one of many possible hypotheses. The primary cilium may be a sensory receptor, but we have no facts for or against this possibility. The primary cilia regress when the DNA of quiescent cells, after addition of growth factors, begins to replicate (Tucker *et al.*, 1979).

Unfortunately, we know very little about the chemical makeup of centrioles and basal bodies. Early cytochemical studies suggested that they contain DNA but this has not been substantiated. This is not surprising, now that we know that these organelles do not replicate in the same manner as chromosomes, chloroplasts, and mitochondria, which absolutely need DNA for their replication. Formation of daughter centrioles apparently results from processes more similar to crystallization after addition of a seed; centrioles might be the nucleating center not only for MT assembly but also for their own duplication. The presence of RNA in the centrosomal "cloud," if not in the centrioles themselves, is probable, as we shall see when we discuss cell division. It has been reported by Mabuchi and Mabuchi (1973) and by Anderson (1977) that centrioles of sea urchin eggs and basal bodies of cilia from the chick oviduct have a $Ca^{2+}$, $Mg^{2+}$-activated ATPase. However, a more recent report by Anderson and Floyd (1980) states that basal bodies isolated from chick oviduct cilia lack ATPase activity and have no myosin or desmin; in contrast, they contain 20% tubulin (different from the cytoskeletal tubulin and localized in the nine MT triplets), a major protein of $M_r$ 17,000 and a few higher-molecular-weight proteins. ($M_r$ between 90,000 and 180,000). That basal bodies are essentially proteinaceous is further shown by the fact that in the flagellate *Polytomella* striated "rootlets" homologous to basal bodies attached to the basal body contain four different proteins (Stearns and Brown, 1979).

Clearly, much more work is needed before we understand the chemical nature of the centrioles and basal bodies, and thus the exact mechanism of their replication and function.

## E. Cilia and Flagella

Few things provide as much pleasure to a cell biologist as watching, under phase contrast microscopy, the elegant movement of a *Paramecium* or a sea urchin blastula or the hectic agitation of a swarm of spermatozoa.

The ultrastructure of these motile appendages to the basal bodies is now so well known that a detailed description is no longer needed. As shown in Figs. 57 and 58, cilia and flagella (reviewed by Gibbons, 1981) are made up of a core structure, called the "axoneme," surrounded by a membrane, which is continuous with the plasma membrane. The axoneme is characterized by nine doublets of outer microtubules that form a cylinder surrounding a single pair (two singlets) of central microtubules. This 9 + 2 tubule organization is the rule, but as with all rules, there are exceptions as we shall see. The subfibers that form the outer doublets are called A and B; subfiber A is a complete MT, whereas subfiber B is incomplete and is C-shaped. Subfibers A, in each of the nine doublets, carry outer and inner arms that, as we shall see, play an essential role in cilia or flagella motility. To subfibers A are also attached interdoublet links and radial spokes, projections that converge toward the central pair of MTs. This complicated structure of the axoneme is almost constant. In contrast, the interactions between the axoneme and the surrounding membrane display great variations. Structures such as a "ciliary crown" at the apex of the cilium bridges between the axoneme and the membrane, and the presence of particles arranged in a "necklace" or in "plaques" at the base of cilia or flagella, are, according to a review by Dentler (1981), highly variable. More constant is the presence of a ciliary plate separating the cilium from its basal body. The role of these various structures in ciliary or flagellar motility is not yet known. According to a paper by Brokaw *et al.* (1982), the function of the radial spokes would be to convert a symmetric bending pattern into an asymmetric one.

The fundamental characteristic of cilia and flagella is, of course, their motility (reviewed by Gibbons, 1977, 1981; Mohri, 1976; Blum and Hines, 1979). Since the pioneering studies of Summers and Gibbons (1971) and Brokaw (1972), everyone agrees that the bending of flagella results from the sliding of the outer MTs (Fig. 59) (sliding filaments theory). Two important factors were discovered at the same time: calcium fluxes control the locomotion of ciliates (Eckert, 1972) and the role of ATP in inducing contractions. In 1973, Brokaw and Gibbons separated axonemes of sea urchin spermatozoa by treating their isolated flagella with the detergent Triton X-100. This technique allowed them to show that addition of ATP induces the contraction of the denuded flagella. ATP must be hydrolyzed in order to provide energy for MT sliding. This is done by a special ATPase, called "dynein," which is located in the arms attached to the A subfibers of the outer doublets [now called "dynein arms" (Gibbons, 1965)]. Dynein is a very widespread high-molecular-weight ATPase that exists in two different forms (Mabuchi *et al.*, 1976). It is activated by calcium and magnesium ions (reviewed by Warner and Mitchell, 1980, who call dynein "the motility coupled ATPase for microtubule-based filament mechanisms"). It has been reported that the dyneins located in the inner and outer arms are immunologically different (Ogawa *et al.*, 1982).

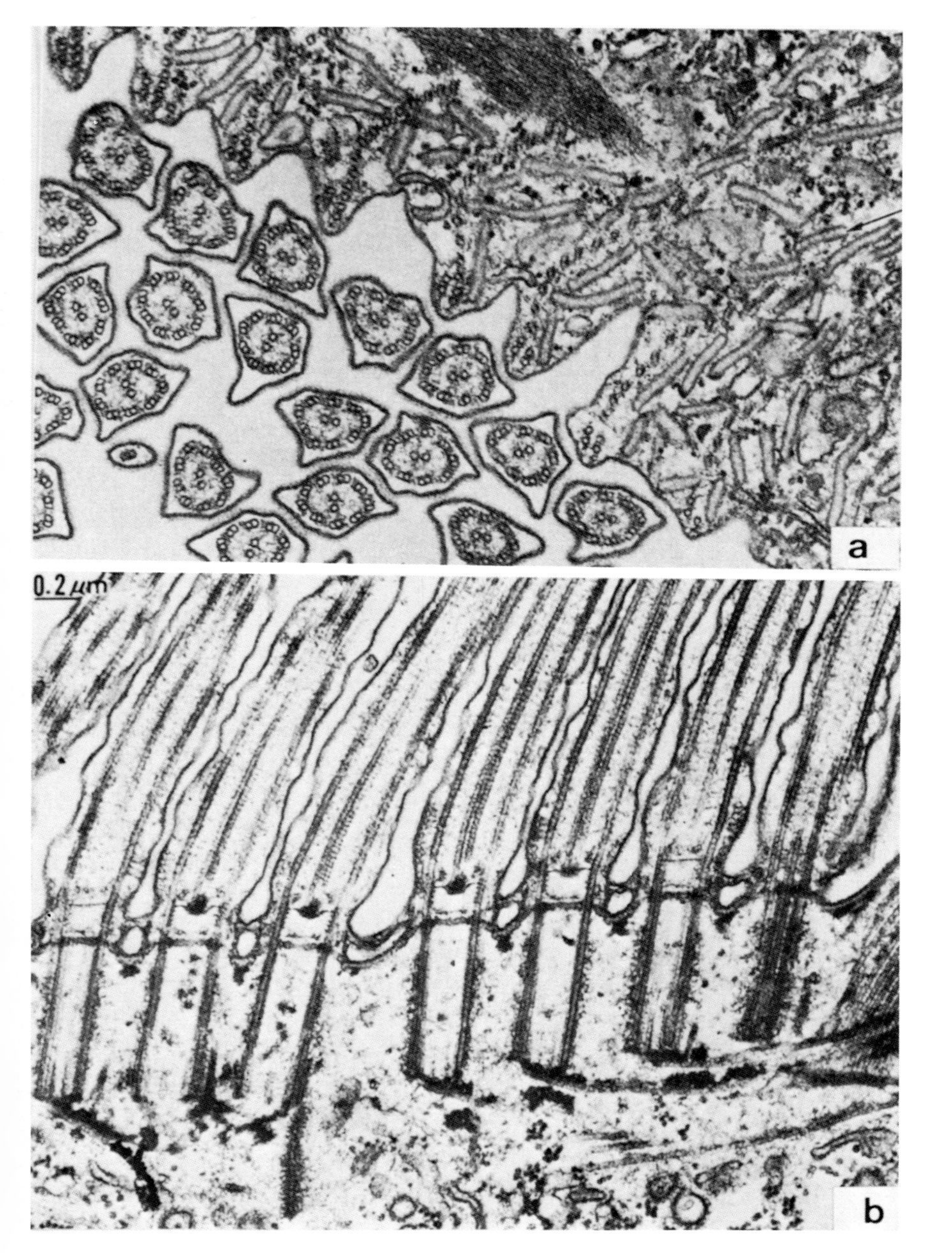


FIG. 57. Cilia in the pharynx of the ciliate *Paramecium*. (a) Transverse section; (b) longitudinal section. [Ehret and McArdle (1974).]

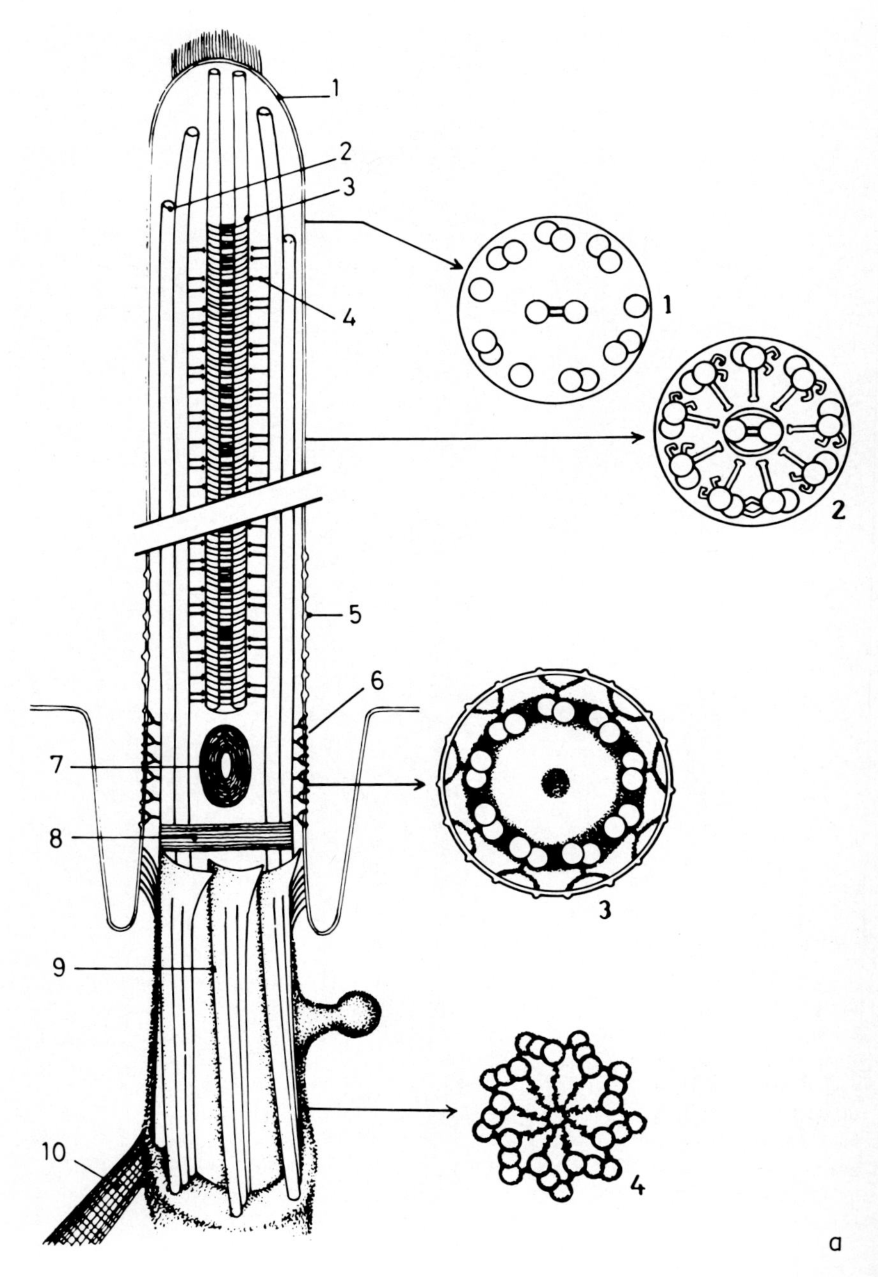


FIG. 58a.

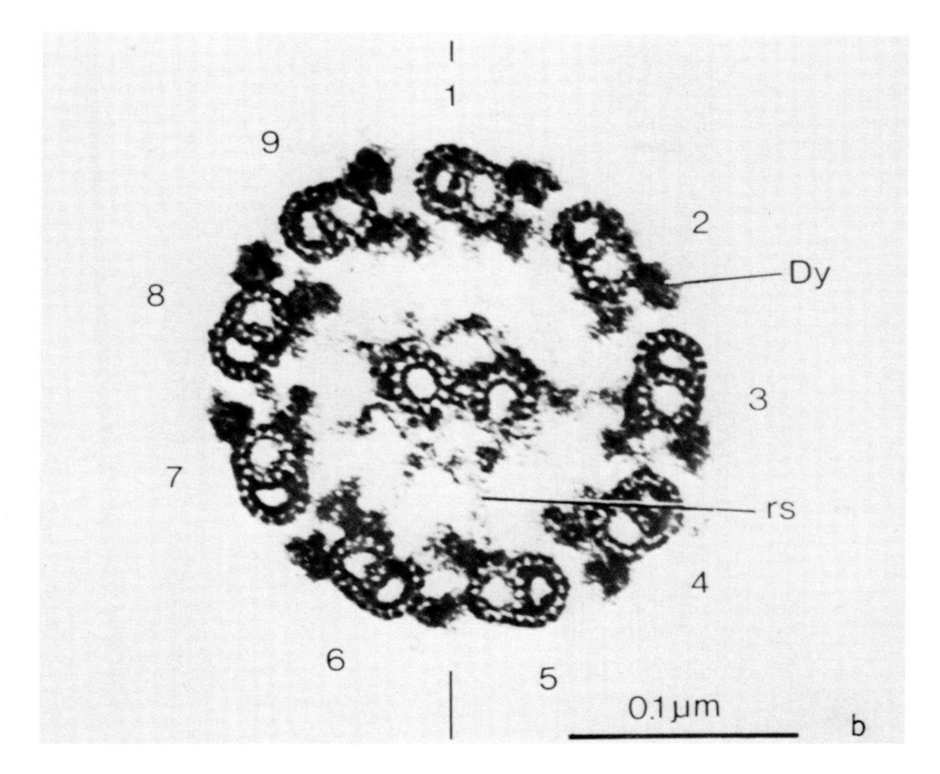


FIG. 58b.

FIG. 58. (a) Diagrammatic representation of a cilium. (*Left*) Longitudinal section: (1) ciliary membrane with its apical differentiation; (2) ciliary doublets, the A MT being longer than the B; (3) the central pair, with complex bonding between the two MTs; (4) bridges (spokes) extending from the central pair toward the peripheral doublets; these are grouped by three, and are unequally spaced; (5) granular membrane differentiations of the cilium; (6) ''champagne-glass'' units linking the doublets to the membrane; (7) ciliary vesicle; (8) basal plate; (9) alar sheets surrounding the basal body's triplets; (10) striated ciliary rootlet. (*Right*) Cross sections: (1) nine doublets and the central pair of MTs; (2) dynein links between the doublets, the radial spokes, and links between the two central MTs; (3) complex differentiations at the ciliary basis, with a dense material surrounding the MT and connected to the champagne-glass structures; (4) basal body triplets with radial links. [Courtesy of Prof. P. Dustin.] (b) Transverse section of a cilium, showing the nine doublets and central pair of microtubules. Dy, dynein arms; rs, radial spokes.

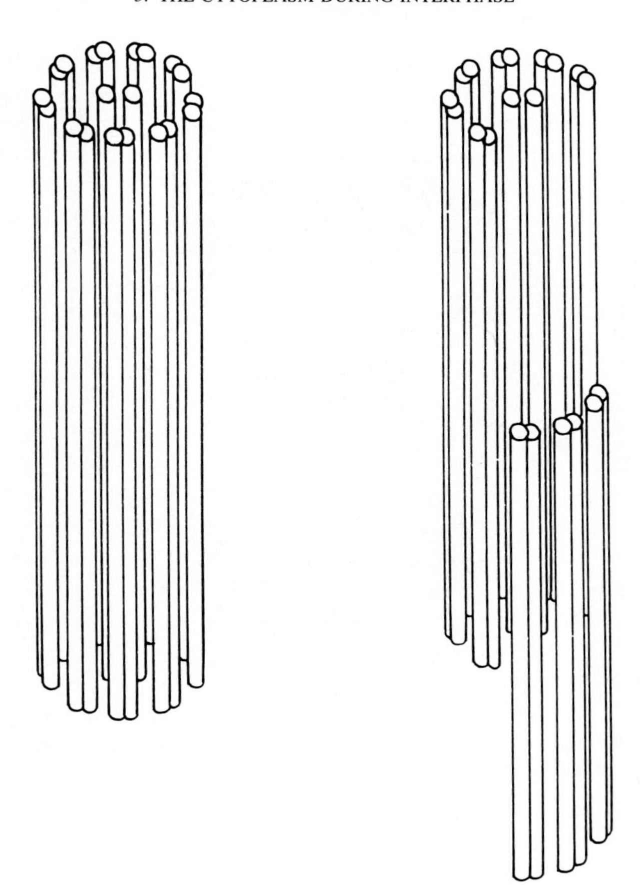

FIG. 59. Schematic representation of the sliding filament model of ciliary beating; sliding of the outer doublets allows lengthening and shortening of the cilium. [Dustin (1984).]

That dynein activity is absolutely required for flagellar and ciliary motility has already been demonstrated by three different sets of observations. In humans, patients suffering from the so-called Kartagener's syndrome are infertile because their immotile or poorly motile (Jouannet *et al.*, 1983) spermatozoa lack dynein arms (Afzelius *et al.*, 1975). Antidynein serum inhibits the contraction by ATP of demembranated flagella (Okuno *et al.*, 1976). Finally, more recent work by Gibbons *et al.* (1978) and Cande and Wolniak (1980) has shown that dynein activity is inhibited by 0.5–4 $\mu M$ vanadate, and that such treatment with vanadate completely arrests the motility of denuded cilia or flagella (from sea urchin larvae or sperm) that had been reactivated by ATP addition. All these experiments reinforce Brokaw's conclusion (1975) that flagellar movements result from the sliding of the MTs of the outer fibers due to dynein and crossbridges between the A and B MTs (Fig. 59). In a model presented more recently by Rikmenspoel (1982), sliding of filaments operates in the tubulin–dynein system of axonemes in both cilia and flagella; there would be attachment and detachment of the dynein crossbridges during effective stroke and recovery stroke, respectively. However, Mitchell and Rosenbaum (1985) recently described a *Chlamydomones* mutant which lacks dynein and nevertheless swims.

As mentioned above, a critical factor in flagellar or ciliary motility is the free $Ca^{2+}$ concentration in or around the axoneme. An increase in this concentration above a critical level, by addition of the divalent ion inophore A23187 in the presence of $CaCl_2$, for instance, arrests the beating of the cilia in mussel gills (M. F. Walter and Satir, 1978). Another factor involved is cAMP, which activates the motility of demembranated sea urchin spermatozoa (Brokaw, 1983); initiation of flagellar motility in *Chlamydomonas* requires the cAMP-dependent phosphorylation of a 55,400-dalton protein (Kakar *et al.*, 1983). According to Hyams and Borisy (1978), the direction of movement in *Chlamydomonas* flagella depends on the $Ca^{2+}$ concentration of the outer medium. At calcium concentrations lower than $10^{-6}$ *M* the flagella display forward movement; at $10^{-3}$ *M*, they stop beating. Similarly, denuded sea urchin sperm flagella reactivated with ATP display asymmetric beating at 0.005 *M* calcium and circular movement at high or low calcium concentrations (Brokaw *et al.*, 1974; Brokaw, 1979). Blum *et al.* (1980) have found that the $Ca^{2+}$-binding protein calmodulin increases dynein activity 10 times in the cilia of the ciliate *Tetrahymena*. Thus, the conclusion that the direction of ciliary beating is $Ca^{2+}$ dependent seems unavoidable; calmodulin is involved, since its inhibitors reverse the direction of flagellar wave propagation (Marchese-Ragona *et al.*, 1983). However, one cannot exclude the idea that other ions than $Ca^{2+}$ also play a role in ciliary motion. According to Nichols and Rikmanspoel (1978a), microinjection of $Mg^{2+}$ stimulates the motility of the flagella in *Chlamydomonas,* while microinjection of $Ca^{2+}$ has the opposite effect. In *Euglena,* injection experiments of divalent ions–chelating agents (EDTA or EGTA)—indicate that a decrease in the internal $Ca^{2+}$ concentration does not stimulate flagellar movement. On the other hand, $Mg^{2+}$ activates dynein and stimulates motility (Nichols and Rikmenspoel, 1978b). According to the same authors, an increase in $Mg^{2+}$ might also induce a membrane potential. That such a factor is probably involved in ciliary movement is suggested by our own unpublished experiments on sea urchin larvae treated with excess KCl or with the $K^+$ ionophore valinomycin. Treatments that affect the membrane potential produce a temporary, spontaneously reversible paralysis of the larvae. Nevertheless, the main regulatory factor in ciliary or flagellar locomotion is certainly free $Ca^{2+}$. Its concentration must be maintained within narrow limits; in mammalian spermatozoa, this is done by a calcium pump ($Ca^{2+}$-dependent $Mg^{2+}$-ATPase) and a quantitatively more important $Na^+/Ca_2{}^+$ exchange system (Bradley and Forrester, 1980).

There is some evidence, reviewed by Tash and Means (1983), that cAMP and protein phosphorylation are also involved in flagellar motility. These authors propose that cAMP activates a cAMP-dependent kinase that phosphorylates sperm proteins; calmodulin would mediate many of the effects exerted by $Ca^{2+}$ on motility: $Ca^{2+}$-calmodulin would pump calcium ions out of the flagellum and would control dynein ATPase and myosin light chain kinase.

In summary, the force required for motility results from the relative sliding of the outer doublets; the dynein arms interact with a neighboring outer doublet and induce the sliding. Free $Ca^{2+}$ plays an important regulatory role in the process. This appears to be highly satisfactory; but what are the two central singlets doing? The answer is much less clear at the present time. In the past, it was often assumed that the axis between the two singlets might direct the orientation of the beat, but there is no good evidence for this view.

According to Omoto and Kung (1980), the central pair of MTs rotates by 360° counterclockwise with each ciliary beat cycle. Its function would be to determine which of the outer MTs should actively slide. The motive force would reside in the continuous rotation in one direction of the central pair of MTs (Omoto and Witman, 1981). This conclusion is not accepted by Tamm and Tamm (1981), who, using a similar methodology, concluded that the $Ca^{2+}$-dependent ciliary binding does not result from rotation and orientation of the central pair of MTs. Finally, a paper by Marchese-Ragona and Holwill (1980) adds to the confusion. They found strains of the trypanosomid *Crithidia* where the flagellum has only one central singlet and displays normal motility nevertheless. The function of the two central MTs in cilia and flagella thus remains a mystery. The suggestion that centrioles and basal bodies rotate like the engines of a jet plane (Bornens, 1978) is attractive, but is still hypothetical.

Comparative studies on the ultrastructure of cilia and flagella in the animal kingdom have disclosed unsuspected and very interesting features: for instance, the axonemes of eel spermatozoa lack the outer dynein arms, the radial spoke complex, and the central doublet of MTs (Baccetti *et al.*, 1979; Gibbons *et al.*, 1983); nevertheless their flagella beat at a frequency that is the highest so far recorded for eukaryotic flagella. Still more surprising is the fact that some gregarines have functional cilia provided with only three outer doublets of MTs (Prensier *et al.*, 1980). It is impossible not to share the conclusion drawn humorously by Gibbons (1983) at the end of a symposium on cilia and flagella: little more than dynein arms and MTs is required for undulatory beating.

Finally, the biochemical composition of cilia and flagella deserves mention. In 1975, Weber *et al.* (1975a,b) demonstrated that antibodies against external doublets of sea urchin sperm flagella crossreact with the cytoskeletal MTs of mammalian cells. There are thus antigenic similarities between tubulins of sea urchin spermatozoa and mammalian MTs. However, biochemical work by Bibring *et al.* (1976) has shown that, in sea urchins, there is a heterogeneity of α-tubulins when one compares sperm flagella, cilia of blastulas, and mitotic apparatus MTs. That the biochemical makeup of flagella is probably extremely complex is indicated by a paper by Piperno *et al.* (1981). Working on flagella isolated from *Chlamydomonas,* they detected the presence of at least 170 different proteins in the axoneme; no fewer than 17 of them were missing in a mutant in which the radial spokes were absent. The molecular organization of cilia and flagella is thus exceedingly complex, and a great deal of interesting work still lies ahead for cell biologists interested in this fascinating field.

## REFERENCES


Abercrombie, M., and Heaysman, J. E. M. (1954). *Exp. Cell Res.* **6,** 293.

Adams, W. J., Jr., and Kalf, G. F. (1980). *Biochem. Biophys. Res. Commun.* **95,** 1875.

Ades, I. Z. (1982). *Mol. Cell. Biochem.* **43,** 113.

Afzelius, B. A., Eliasson, R., Johnsen, Ø., and Lindholmer, C. (1975). *J. Cell Biol.* **66,** 225.

Albanese, J., Kuhlenschmidt, M. S., Schmell, E., Slife, C. W., and Roseman, S. (1982). *J. Biol. Chem.* **257,** 3165.

Albertini, D. F., Berlin, R. D., and Oliver, J. M. (1977). *J. Cell Sci.* **26,** 57.

Ali, I. V., and Hynes, R. O. (1978). *Cell* **14,** 439.

Albrecht-Buhler, G. (1978). *Sci. Amer.* **238**(4), 75.

Allen, R. D., and Taylor, D. L. (1975). *In* "Molecules and Cell Movements" (S. Inoue and R. E. Stephens, eds.), p. 239. Raven Press, New York.

Allison, A. C. (1974). "Lysosomes," Vol. 58 (Oxford Biol. Readers). Oxford Univ. Press, London and New York.

Ames, B. N. (1979. *Science* **204,** 587.

Amos, L. A., and Klug, A. (1974). *J. Cell Sci.* **14,** 523.

Anderson, D. J., Walter, P., and Blobel, G. (1982). *Cell Biol.* **93,** 501.

Anderson, L. (1981). *Proc. Natl. Acad. Sci. U.S.A.* **78,** 2407.

Anderson, R. G. W. (1977). *J. Cell Biol.* **74,** 547.

Anderson, R. G. W., and Floyd, A. K. (1980). *Biochemistry* **19,** 5625.

Anderson, S., Bankier, A. T., Barrell, B. G., de Bruijn, H. L., Coulson, A. R., Drouin, J., Eperon, I. C., Nierlich, D. P., Roe, B. A., Sanger, F., Schreier, P. H., Smith, A. J. H., Staden, R., and Young, I. G. (1981). *Nature (London)* **290,** 457.

Anderson, S., de Bruijn, M. H. L., Coulson, A. R., Eperon, I. C., Sanger, F., and Young, I. G. (1982). *J. Mol. Biol.* **156,** 683.

Andrew, R. D., MacVicar, B. A., Dudek, F. E., and Hatton, G. I. (1981). *Science* **211,** 1187.

Aplin, J. D., and Hughes, R. C. (1983). *Biochim. Biophys. Acta* **694,** 375.

Ash, J. F., Louvard, D., and Singer, S. J. (1977). *Proc. Natl. Acad. Sci. U.S.A.* **74,** 5584.

Attardi, G. (1985). *Intl. Rev. Cytol.* **93,** 93.

Avnur, Z., and Geiger, A. (1981). *Cell* **25,** 121.

Baccetti, B., Burrini, A. G., Dalla, R., and Pallini, V. (1979). *J. Cell Biol.* **80,** 334.

Badley, R. A., Woods, A., Carruthers, L., and Rees, D. A. (1980). *J. Cell Sci.* **43,** 379.

Baglioni, C., Bleiberg, I., and Zanderer, M. (1971). *Nature (London), New Biol.* **232,** 8.

Ball, E. H., and Singer, S. J. (1982). *Proc. Natl. Acad. Sci. U.S.A.* **79,** 123.

Balsamo, J., and Lilien, J. (1974). *Nature (London)* **251,** 522.

Barat, M., and Mignotte, B. (1981). *Chromosoma* **82,** 583.

Barrett, G. J. (1981). *Trends Biochem. Sci.* **6,** 322.

Barrois, M., Riou, G., and Galibert, F. (1981). *Proc. Natl. Acad. Sci. U.S.A.* **78,** 3323.

Batten, B. E., Aalberg, J. J., and Anderson, E. (1980). *Cell* **21,** 885.

Beaufay, H., Amar-Costesec, A., Thinès-Sempoux, D., Wibo, M., Robbi, M., and Berthet, J. (1974). *J. Cell Biol.* **61,** 213.
Becker, G. L., Fiskum, G., and Lehninger, A. L. (1980). *J. Biol. Chem.* **255,** 9009.
Benda, C. (1902). *Ergeb. Anat. Entwicklungsgesch.* **12,** 743.
Bendzko, P., Prehn, S., Pfeil, W., and Rapoport, T. A. (1982). *Eur. J. Biochem.* **123,** 121.
Ben-Ze'ev, A., Farmer, S. R., and Penman, S. (1979). *Cell* **17,** 319.
Bergeron, J. J. M., Rachubinski, R. A., Sikstrom, R. A., Posner, B. I., and Paiment, J. (1982). *J. Cell Biol.* **92,** 139.
Bernier-Valentin, F., Aunis, D., and Rousset, B. (1983). *J. Cell Biol.* **97,** 209.
Berridge, M. J. (1984). *Biochem. J.* **220,** 345.
Berridge, M. J., and Irvine, R. F. (1984). *Nature* **312,** 315.
Berridge, M. J., Heslop, J. P., Irvine, R. F., and Brown, K. D. (1984). *Biochem. J.* **222,** 195.
Bershadsky, A. D., and Gelfand, V. I. (1981). *Proc. Natl. Acad. Sci. U.S.A.* **78,** 3610.
Bershadsky, A. D., and Gelfand, V. I. (1983). *Cell Biol. Int. Rep.* **7,** 173.
Bertolini, B., and Monaco, G. (1976). *J. Ultrastr. Res.* **54,** 59.
Besterman, J. M., and Low, R. B. (1983). *Biochem. J.* **210,** 1.
Bibb, M. J., Van Etten, R. A., Wright, C. T., Walberg, M. W., and Clayton, D. A. (1980). *Cell* **26,** 167.
Bibring, T., Baxandall, J., Denslow, S., and Walker, B. (1976). *J. Cell Biol.* **69,** 301.
Bishayee, S., and Bachhawat, B. K. (1974). *Biochim. Biophys. Acta* **334,** 378.
Blickstad, I., Erikksson, S., and Carlsson, L. (1980). *Eur. J. Biochem.* **109,** 317.
Bliokh, Z. L., Domnina, L. V., Ivanova, O. Y., Pletjushkina, O. Y., Svitkina, T. N., Smolyaninov, V. A., Vasiliev, J. M., and Gelfand, I. M. (1980). *Proc. Natl. Acad. Sci. U.S.A.* **77,** 5919.
Blobel, G. (1977). *In* "International Cell Biology" (B. Brinkley and K. Porter, eds.), p. 318. Rockefeller Univ. Press, New York.
Blobel, G. (1980). *Proc. Natl. Acad. Sci. U.S.A.* **77,** 1496.
Blobel, G., and Dobberstein, B. (1975a) *J. Cell Biol.* **67,** 835.
Blobel, G., and Dobberstein, B. (1975b) *J. Cell Biol.* **67,** 852.
Bloom, W. S., School, W., Feageson, E., Ores, C., and Puszkin, S. (1980). *Biochim. Biophys. Acta* **598,** 447.
Blum, J. J., and Hines, M. (1979). *Rev. Biophys.* **12,** 103.
Blum, J. J., Hayes, A., Jamieson, G. A., Jr., and Vanaman, T. C. (1980). *J. Cell Biol.* **87,** 386.
Bogorad, L. (1977) *In* "International Cell Biology" (B. R. Brinkley and K. R. Porter, eds.), p. 175. Rockefeller Univ. Press, New York.
Bols, N. C., and Ringertz, N. R. (1979). *Exp. Cell Res.* **120,** 15.
Borisy, G. G., Olmsted, J. B., and Klugman, R. A. (1972). *Proc. Natl. Acad. Sci. U.S.A.* **69,** 2890.
Bornens, M. (1978). *C. R. Hebd. Seances Acad. Sci.* **287,** 1417.
Borst, P. (1977). *In* "International Cell Biology" (B. R. Brinkley and K. R. Porter, eds.), p. 237 Rockefeller Univ. Press, New York.
Borst, P., Fase-Fowler, F., Hoeijmakers, J. H. J., and Frasch, A. C. C. (1980). *Biochim. Biophys. Acta* **610,** 197.
Bosmann, H. B. (1977). *Int. Rev. Cytol.* **50,** 1.
Bourguignon, L. Y. W., and Singer, S. J. (1977). *Proc. Natl. Acad. Sci. U.S.A.* **74,** 5031.
Boyer, P. D., Chance, B., Ernster, L., Mitchell, P., Racker, E., and Slater, E. C. (1977). *Annu. Rev. Biochem.* **46,** 955.
Brachet, J., and Jeener, R. (1944). *Enzymologia* **13,** 196.
Brackenbury, R., Rutishauser, U., and Edelman, G. M. (1981). *Proc. Natl. Acad. Sci. U.S.A.* **78,** 387.
Bradley, M. P., and Forrester, I. T. (1980). *FEBS Lett.* **121,** 15.
Brenner, S. L., and Brinkley, B. R. (1981). *Cold Spring Harbor Symp. Quant. Biol.* **46,** 241.
Bretscher, A., and Weber, K. (1980a). *Cell* **20,** 839.
Bretscher, A., and Weber, K. (1980b). *J. Cell Biol.* **86,** 335.

Bretscher, A., Osborn, M., Wehland, J., and Weber, K. (1981). *Exp. Cell Res.* **135,** 213.
Bretscher, M. S., and Pearse, B. M. F. (1984). *Cell* **38,** 3.
Bretscher, M. S., Thomson, J. N., and Pearse, B. M. F. (1980). *Proc. Natl. Acad. Sci. U.S.A.* **77,** 4156.
Brinkley, B. R. (1982). *Cold Spring Harbor Symp.* **46**(2), 1029.
Brinkley, B., Fistel, S., Marcum, J. M., and Pardue, R. L. (1980). *Int. Rev. Cytol.* **63,** 59.
Brokaw, C. J. (1972). *Science* **178,** 455.
Brokaw, C. J. (1975). *Proc. Natl. Acad. Sci. U.S.A.* **72,** 3102.
Brokaw, C. J. (1979). *J. Cell Biol.* **82,** 401.
Brokaw, C. J. (1983). *J. Cell Biol.* **97,** 196a.
Brokaw, C. J., and Gibbons, I. R. (1973). *J. Cell Sci.* **13,** 1.
Brokaw, C. J., Josselin, R., and Bobrow, L. (1974). *Biochem. Biophys. Res. Commun.* **58,** 795.
Brokaw, C. J., Luck, J. L., and Huang, B. (1982). *J. Cell Biol.* **92,** 722.
Brown, M. S., Anderson, R. G. W., and Goldstein, J. L. (1983). *Cell* **32,** 663.
Brown, S. S., Malinoff, H. L., and Wicha, M. S. (1983). *Proc. Natl. Acad. Sci. U.S.A.* **80,** 5927.
Brown, W. J., and Farquhar, M. G. (1983). *J. Cell Biol.* **97,** 431a.
Brownell, A. G., Bessem, C. C., and Slavkin, H. C. (1981). *Proc. Natl. Acad. Sci. U.S.A.* **78,** 3711.
Bryan, J. (1974). *Fed. Proc. Fed. Am. Soc. Exp. Biol.* **33,** 152.
Bryan, J., Nagle, B. W., and Doenges, K. H. (1975). *Proc. Natl. Acad. Sci. U.S.A.* **72,** 3570.
Bryan, R. N., Cutter, G. A., and Hayashi, M. (1978). *Nature (London)* **272,** 81.
Bryan, R. N., Bassinger, J., and Hayashi, H. (1981). *Dev. Biol.* **81,** 349.
Bulinski, J. C., and Borisy, G. G. (1980a). *J. Cell Biol.* **87,** 792.
Bulinski, J. C., and Borisy, G. G. (1980b). *J. Biol. Chem.* **255,** 11570.
Bunn, C. L., Wallace, D. C., and Eisenstadt, B. (1974). *Proc. Natl. Acad. Sci. U.S.A.* **71,** 1681.
Burchill, B. R., Oliver, J. M., Pearson, C. B., Leinbach, E. D., and Berlin, C. D. (1978). *J. Cell Biol.* **76,** 439.
Burgess, D. R. (1982). *J. Cell Biol.* **95,** 853.
Burgess, G. M., Claret, M., and Jenkinson, D. H. (1979). *Nature (London)* **279,** 544.
Cairns, J. (1963). *J. Mol. Biol.* **6,** 208.
Campbell, A. K., Daw, R. A., Hallett, M. B., and Luzio, J. P. (1981). *Biochem. J.* **194,** 551.
Cande, W. Z., and Wolniak, S. M. (1980). *J. Cell Biol.* **79,** 573.
Carafoli, E. (1975). *Mol. Cell. Biochem.* **8,** 133.
Carey, D. J. and Hirschberg, B. (1981). *J. Biol. Chem.* **25,** 989.
Caron, F., and Meyer, E. (1985). *Nature* **313,** 185.
Carraway, C. A. C., Corrado, F. J., IV, Fogle, D. D., and Carraway, K. L. (1980). *Biochem. J.* **191,** 45.
Cascarano, J., Montisano, D. F., Pickett, C. B., and James, T. W. (1982). *Exp. Cell Res.* **139,** 39.
Celis, J. E., Small, J. V., Mose Larsen, P., Fey, S. J., De Mey, J., and Celis, A. (1984). *Proc. Natl. Acad. Sci. U.S.A.* **81,** 1117.
Cerletti, N., Böhni, P. C., and Suda, K. (1983). *J. Biol. Chem.* **258,** 4944.
Cervera, M., Dreyfuss, G., and Penman, S. (1981). *Cell* **23,** 113.
Chapman-Andresen, C., and Holter, H. (1955). *Exp. Cell Res., Suppl.* **3,** 52.
Chen, W. T., and Singer, S. J. (1980). *Proc. Natl. Acad. Sci. U.S.A.* **77,** 7318.
Chiu, R., and Phillips, A. H. (1981). *J. Biol. Chem.* **256,** 3103.
Christensen, H. N. (1979). *Adv. Enzymol.* **49,** 41.
Christensen, H. N. (1984). *Biochem. Biophys. Acta* **779,** 255.
Chuang, D. M. (1981). *J. Biol. Chem.* **256,** 8291.
Claude, A. (1943). *Biol. Symp.* **10,** 3.
Clayton, D. A. (1982). *Cell* **28,** 693.
Cleveland, D. W. (1982). *Cell* **28,** 689.
Cleveland, D. W. (1983). *Cell* **34,** 330.

Cleveland, D. W., Hwo, S. Y., and Kirschner, M. W. (1978). *J. Mol. Biol.* **116,** 207.
Cleveland, D. W., Lopata, M. A., Sherline, P., and Kirschner, M. W. (1981). *Cell* **25,** 537.
Cleveland, D. W., Pittenger, M. F., and Feramisco, J. R. (1983). *Nature* (*London*) **305,** 738.
Clissold, P. M., Mason, P. J., and Bishop, J. O. (1981). *Proc. Natl. Acad. Sci. U.S.A.* **78,** 3697.
Cobb, M. H., and Rosen, O. M. (1984). *Biochim. Biophys. Acta* **738,** 1.
Connolly, J. A., and Kalnins, V. I. (1980). *Exp. Cell Res.* **127,** 341.
Connolly, J. A., Kalnins, V. I., Cleveland, B. W., and Kirschner, M. W. (1978). *J. Cell Biol.* **76,** 781.
Connor, C. G., Brady, R. C., and Brownstein, B. L. (1981). *J. Cell. Physiol.* **108,** 299.
Coons, A. H. (1952). *Symp. Soc. Exp. Biol.* **6,** 166.
Cori, C. F., Gluecksohn-Waelsch, S., Shaw, P. A., and Robinson, C. (1983). *Proc. Natl. Acad. Sci. U.S.A.* **80,** 6611.
Cote, R. H., and Borisy, G. G. (1981). *J. Mol. Biol.* **150,** 577.
Cox, R. P., Krauss, M. R., Balis, M. E., and Dancis, J. (1976). *J. Cell Biol.* **71,** 693.
Craig, S. W., and Powell, L. D. (1980). *Cell* **22,** 739.
Crawford, N., Chahal, H., and Jackson, P. (1980). *Biochim. Biophys. Acta* **626,** 218.
Cunningham, B. A., Hoffman, S., Rutishauser, V., Hemperly, J. J., and Edelman, G. M. (1983). *Proc. Natl. Acad. Sci. U.S.A.* **80,** 3116.
Cuppoletti, J., Mayhew, E., and Jung, C. Y. (1981). *Biochim. Biophys. Acta* **642,** 392.
Dahl, G., Azarnia, R., and Werner, R. (1981). *Nature* (*London*) **289,** 683.
Dalcq, A., Pasteels, J., and Mulnard, J. (1956). *Bull. Cl. Sci., Acad. R. Belg.* **62,** 771.
Danielli, J. F. (1954). *Colston Pap.* **7,** 1.
Davis, B. D., and Tai, P. C. (1980). *Nature* (*London*) **283,** 433.
David-Pfeuty, T. (1983). *Eur. J. Cell Biol.* **30,** 93.
Dawid, I. B. (1972). *J. Mol. Biol.* **63,** 201.
Dawid, I. B., and Blackler, A. W. (1972). *Dev. Biol.* **29,** 152.
De Brabander, M., and Borgers, M. (1975). *J. Cell Sci.* **19,** 331.
De Brabander, M., De Mey, J., Geuens, G., Nuydens, R., Aerts, F., Willebrords, R., and Moeremans, M. (1985). *Hormones Cell Reg.* **9,** 85; INSERM European Symposium (J. E. Dumont, B. Hamprecht, and J. Nuney, eds.). Elsevier.
De Brabander, M., Geuens, G., Nuydens, R., Willebrords, R., and De Mey, J. (1981a). *Cell Biol. Int. Rep.* **5,** 913.
De Brabander, M., Geuens, G., Nuydens, R., Willebrords, R., and De Mey, J. (1981b). *Cold Spring Harbor Symp. Quant. Biol.* **46,** 227.
de Bruijn, M. H. L. (1983). *Nature* (*London*) **304,** 234.
Decroly, M., Goldfinger, M., and Six-Tondeur, N. (1979). *Biochim. Biophys. Acta* **587,** 567.
De Deken-Grenson, M. (1955). *Biochim. Biophys. Acta* **17,** 35.
de Duve, C. (1969). *Proc R. Soc. London, Ser. B* **173,** 71.
de Duve ,C. (1975). *Science* **189,** 186.
de Duve, C., Pressman, B. C., Gianetto, R. J., Wattiaux, R., and Appelmans, F. (1955). *Biochem. J.* **60,** 604.
De Harven, E., and Bernhard, W. (1956). *Z. Zellforsch. Mikrosk. Anat.* **45,** 378.
De Kruiff, B., Gerritsen, W. J., Oerlemans, A., Demel, R. A., and Van Deenen, L. L. M. (1974). *Biochim. Biophys. Acta* **339,** 30.
de Laat, S. W., Barts, P. W. J., and Bakker, M. I. (1976). *J. Membr. Biol.* **27,** 109.
Dellagi, K., and Brouet, J. C. (1982). *Nature* (*London*) **298,** 284.
De Mey, J., Joniau, M., De Brabander, M., Moens, W., and Geuens, J. (1978). *Proc. Natl. Acad. Sci. U.S.A.* **75,** 1339.
Denoulet, P., Jeantet, C., and Gros, F. (1982). *Biochem. Biophys. Res. Commun.* **105,** 806.
Dentler, W. L. (1981). *Int. Rev. Cytol.* **72,** 1.
Denton, R. M., and McCormack, J. G. (1980). *FEBS Lett.* **119,** 1.
De Rosier, D. J., and Censullo, R. (1981). *J. Mol. Biol.* **146,** 77.

Deutscher, S. L., Creek, K. E., Merion, M., and Hirschberg, C. B. (1983). *Proc. Natl. Acad. Sci. U.S.A.* **80,** 3938.
Dewald, B., and Touster, O. (1973). *J. Biol. Chem.* **248,** 7223.
Di Caprio, R. A., French, A. S., and Sanders, E. J, (1974). *Biophys. J.* **14,** 387.
Dickson, R. B., Willingham, M. C., and Pastan, I. (1981a). *J. Biol. Chem.* **256,** 3454.
Dickson, R. B., Willingham, M. C., and Pastan, I. (1981b). *J. Cell Biol.* **89,** 29.
Dickson, R. B., Beguinot, L., Hanover, J. A., Richert, N. D., Willingham, M. C., and Pastan, I. (1983). *Proc. Natl. Acad. Sci. U.S.A.* **80,** 5335.
Doda, J. N., Wright, C. T., and Clayton, D. A. (1981). *Proc. Natl. Acad. Sci. U.S.A.* **78,** 6116.
Domnina, L. V., Pletyushkina, O. Y., Vasiliev, J. M., and Gelfand, I, M. (1977). *Proc. Natl. Acad. Sci. U.S.A.* **74,** 2865.
Dräger, U. C. (1983). *Nature (London)* **303,** 169.
Drochmans, P. (1963). *Biochem. Soc. Symp.* **23,** 127.
Dufresne-Dubé, L., Picheral, B., and Guerrier, P. (1983). *J. Ultrastruct. Res.* **83,** 242.
Dulbecco, R., and Elkington, J. (1975). *Proc. Natl. Acad. Sci. U.S.A.* **72,** 1584.
Dunphy, W. G., Fries, E., Urbani, L. J., and Rothman, J. E. (1981). *Proc. Natl. Acad. Sci. U.S.A.* **78,** 7453.
Dustin, A. P. (1934). *Bull. Cl. Sci., Acad. R. Belg.* **14,** 587.
Dustin, P. (1984). "Microtubules," 2nd ed. Springer-Verlag, Berlin and New York.
Eckert, R. (1972). *Science* **176,** 473.
Edelman, G. M. (1976). *Science* **192,** 218.
Edelman, G. M. (1983). *Science* **219,** 450.
Edelman, G. M., and Rutishauser, V. (1981). *J. Supramol. Struct. Cell. Biochem.* **16,** 259.
Edelman, G. M., Yahara, I., and Wang, J. L. (1973). *Proc. Natl. Acad. Sci. U.S.A.* **70,** 1442.
Ehret, C. F., and McArdle, E. W. (1974). *In* "Paramecium, A Current Survey" (W. J. Van Wagtendonk, ed.), p. 263. Elsevier.
Eichhorn, J. H., and Peterkofsky, B. (1979). *J. Cell Biol.* **82,** 572.
Elbein, A. D. (1981). *Trends Biochem. Soc.* **6,** 219.
Elder, J. H., and Morré, D. J. (1976). *J. Biol. Chem.* **251,** 5054.
Engelman, D. M., and Steitz, T. A. (1981). *Cell* **23,** 411.
Eperon, I. C., Janssen, J. W. G., Hoeijmakers, J. H. J., and Borst, P. (1983). *Nucleic Acids Res.* **11,** 105.
Ephrussi, B. (1949). "Unités biologiques douées de continuité génétique," p. 165. C.N.R.S., Paris.
Ephrussi, B. (1953). "Nucleocytoplasmic Relations in Micro-organisms." Oxford Univ. Press, London and New York.
Ephrussi, B., and Slonimski, P. P. (1955). *Nature (London)* **176,** 1207.
Ernster, L. (1977). *Annu. Rev. Biochem.* **46,** 981.
Estensen, R. D., Hill, H. R., Quie, P. G., Hogan, N., and Goldberg, N. D. (1973). *Nature (London)* **245,** 458.
Evans, R. M., and Fink, L. M. (1977). *Proc. Natl. Acad. Sci. U.S.A.* **74,** 5341.
Evans, W. H. (1980). *Biochim. Biophys. Acta* **604,** 27.
Evans, W. H., Flint, N. A., and Vischer, P. (1980). *Biochem. J.* **192,** 903.
Farquhar, M. G., and Palade, G. E. (1963). *J. Cell Biol.* **17,** 375.
Farrelly, F., and Butow, R. A. (1983). *Nature (London)* **301,** 296.
Fehlmann, M., Carpentier, J. L., Van Obberghen, E., Freychet, P., Thamm, P., Saunders, D., Brandenburg, D., and Orci, L. (1982). *Proc. Natl. Acad. Sci. U.S.A.* **79,** 5921.
Feramisco, J. R., and Blose, S. H. (1980). *J. Cell Biol.* **86,** 608.
Flechtner, V. R., and Sager, R. (1973). *Nature (London), New Biol.* **241,** 277.
Fleit, H., Conklyn, M., Stebbins, R. D., and Silber, R. (1975). *J. Biol. Chem.* **250,** 8889.
Foerster, E. C., Fährenkemper, T., Rabe, U., Graf, P., and Sies, H. (1981). *Biochem. J.* **196,** 705.
Forgac, M., Cantley, L., Wiedenmann, B., Altstiel, L., and Branton, D. (1983). *Proc. Natl. Acad. Sci. U.S.A.* **80,** 1300.

Fox, T. O., Sheppard, J. R., and Burger, M. M. (1971). *Proc. Natl. Acad. Sci. U.S.A.* **68,** 244.
Franke, W. W., Weber, K., Osborn, M., Schmid, E., and Freudestein, C. (1978). *Exp. Cell Res.* **116,** 429.
Franke, W. W., Denk, H., Kalt, R., and Schmid, E. (1981a). *Exp. Cell Res.* **131,** 299.
Franke, W. W., Schiller, D. L., Moll, R., Winter, S., Schmid, E., Engelbrecht, I., Denk, H., Krepler, R., and Platzer, B. (1981b). *J. Mol. Biol.* **153,** 933.
Franke, W. W., Schmid, E., Grund, C., Müller, H., Engelbrecht, I., Moll, R., Stadler, J., and Jarasch, E. D. (1981c). *Differentiation* **20,** 217.
Franke, W. W., Winter, S., Grund, C., Schmid, E., Schiller, D. L., and Jarasch, E. D. (1981d). *J. Cell Biol.* **90,** 116.
Frankel, F. R. (1976). *Proc. Natl. Acad. Sci. U.S.A.* **73,** 2798.
Fraser, B. R., and Zalik, S. E. (1977). *J. Cell Sci.* **27,** 227.
Frédéric, J., and Chèvremont, M. (1953). *Arch. Biol.* **63,** 109.
Fujiki, Y., Fowler, S., Shio, H., Hubbarb, A. L., and Lazarow, P. B. (1982). *J. Cell Biol.* **93,** 103.
Fukui, Y., and Katsumaru, H. (1979). *Exp. Cell Res.* **120,** 451.
Fyrberg, E. A., Kindler, K. L., and Davidson, N. (1980). *Cell* **19,** 365.
Gabel, C. A., Goldberg, D. E., and Kornfeld, S. (1982). *J. Cell Biol.* **95,** 536.
Gabel, C. A., Goldberg, D. E., and Kornfeld, S. (1983). *Proc. Natl. Acad, Sci. U.S.A.* **80,** 775.
Gallin, W. J., Edelman, G. M., and Cunningham, B. A. (1983). *Proc. Natl. Acad. Sci. U.S.A.* **80,** 1038.
Galloway, C. J., Dean, G. E., Marsh, M., Rudnick, G., and Mellman, I. (1983). *Proc. Natl. Acad. Sci. U.S.A.* **80,** 3334.
Garber, B., and Moscona, A. A. (1972). *Dev. Biol.* **27,** 235.
Gasser, S. M., and Schatz, G. (1983). *J. Biol. Chem.* **258,** 3427.
Geahlen, R. L., and Haley, B. E. (1977). *Proc. Natl. Acad. Sci. U.S.A.* **74,** 4375.
Geiger, B. (1983). *Biochim. Biophys. Acta* **737,** 305.
Geiger, B., Tokuyasu, K. T., Dutton, A. H., and Singer, S. J. (1980). *Proc. Natl. Acad. Sci. U.S.A.* **77,** 4127.
Gellissen, G., Bradfield, J. Y., White, B. N., and Wyatt, G. R. (1983). *Nature (London)* **301,** 631.
Geuens, G., De Brabander, M., Nuydens, R., and De Mey, J. (1983). *Cell Biol. Int. Rep.* **7,** 35.
Gibbons, B. H., Gibbons, I. R., and Baccetti, B. (1983). *J. Submicrosc. Cytol.* **15,** 15.
Gibbons, I. R. (1965). *Arch. Biol.* **76,** 317.
Gibbons, I. R. (1977). *In* "International Cell Biology" (B. R. Brinkley and K. R. Porter, eds.), p. 348. Rockefeller Univ. Press, New York.
Gibbons, I. R. (1981). *J. Cell Biol.* **91,** 107s.
Gibbons, I. R. (1983). *J. Submicrosc. Cytol.* **15,** 243.
Gibbons, I. R., Cosson, M. P., Evans, J. A., Gibbons, B. H., Houck, B., Martinson, K. H., Sale, W. S., and Tang, W. J. Y. (1978). *Proc. Natl. Acad. Sci. U.S.A.* **75,** 2220.
Gilmore, R., Blobel, G., and Walter, P. (1982). *J. Cell Biol.* **95,** 463.
Giudice, G. (1962). *Develop. Biol.* **5,** 402.
Glacy, S. D. (1983). *J. Cell Biol.* **97,** 1207.
Glaumann, H., Ericsson, J. L. E., and Marzella, L. (1981). *Int. Rev. Cytol.* **73,** 149.
Glenney, J. R., and Glenney, P. (1983). *Cell* **34,** 503.
Glenney, J. R., and Glenney, P. (1984). *Eur. J. Biochem.* **144,**529.
Glenney, J. R., and Weber, K. (1980). *J. Biol. Chem.* **255,** 10551.
Glenney, J. R., Jr., Bretscher, A., and Weber, K. (1980). *Proc. Natl. Acad. Sci. U.S.A.* **77,** 6458.
Glenney, J. R., Jr., Geisler, N., Kaulfus, P., and Weber, K. (1981a). *J. Biol. Chem.* **256,** 8156.
Glenney, J. R., Jr., Kaulfus, P., and Weber, K. (1981b). *Cell* **24,** 471.
Glickman, J., Croen, K., Kelley, S., and Al-Awquari, Q. (1983). *J. Cell Biol.* **97,** 1303.
Glynn, I. M., and Karlish, S. J. D. (1975). *Annu. Rev. Physiol.* **37,** 13.
Godfraind-De Becker, A., and Godfraind, T. (1980). *Int. Rev. Cytol.* **67,** 141.
Goffeau, A., and Brachet, J. (1965). *Biochim. Biophys. Acta* **95,** 302.

Goldstein, J. L., Anderson, R. G. W., and Brown, M. S. (1979). *Nature* (*London*) **279,** 679.
Golgi, C. (1898). *Arch. Ital. Biol.* **30,** 60.
Goodenough, D. A. (1974). *J. Cell Biol.* **61,** 557.
Gorbsky, G., and Steinberg, M. S. (1981). *J. Cell Biol.* **90,** 243.
Gordon, A. H., D'Arcy Hart, P., and Young, M. R. (1980). *Nature* (*London*) **286,** 79.
Gospodarowicz, D., Delgado, D., and Vlodavsky, I. (1980). *Proc. Natl. Acad. Sci. U.S.A.* **77,** 4094.
Gospodarowicz, D., Greenburg, G., Foidart, J. M., and Savion, N. (1981). *J. Cell Physiol.* **107,** 171.
Gotlib, L. J., and Searls, D. B. (1980). *Biochim. Biophys. Acta* **602,** 207.
Granick, S. (1961). *In* "The Cell" (J. Brachet and A. E. Mirsky, eds.), Vol. 2, p. 489. Academic Press, New York.
Gray, M. W., and Doolittle, W. F. (1982). *Microbiol. Rev.* **46,** 1.
Green, B. R. (1976). *Biochim. Biophys. Acta* **447,** 156.
Green, B. R., Muir, B. L., and Padmanabhan, U. (1977). *In* "Progress in *Acetabularia* Research" (C. F. L. Woodcock, ed.), p. 107. Academic Press, New York.
Green, B. R. (1980). *Biochim. Biophys. Acta* **609,** 107.
Green, D. E. (1977). *Trends Biochem. Sci.* **2,** 113.
Gröschel-Stewart, U. (1980). *Int. Rev. Cytol.* **65,** 193.
Guengerich, F. P., Wang, P., Mason, P. S., and Mitchell, M. B. (1981). *Biochemistry* **20,** 2370.
Hadjiolov, A. A. (1981). *Subcell. Biochem.* **7,** 1.
Hadler, H. I., Dimitrijevic, B., and Mahalingam, R. (1983). *Proc. Natl. Acad. Sci. U.S.A.* **80,** 6495.
Halftenbein, N. E. (1985). *Nucl. Ac. Res.* **13,** 415.
Hanover, J. A., and Lennarz, W. J. (1981). *Arch. Biochem. Biophys.* **211,** 1.
Harder, A., Pakalapati, G., and Debuch, H. (1981). *Biochem. Biophys. Res. Commun.* **99,** 9.
Harikumar, P., and Reeves, J. P. (1983). *J. Biol. Chem.* **258,** 10403.
Harris, H. (1983). *Nature* (*London*) **302,** 106.
Hasilik, A. (1980). *Trends Biochem. Sci.* **5,** 237.
Hasilik, A., Voss, B., and von Figura, K. (1981). *Exp. Cell Res.* **133,** 23.
Hausman, R. E., and Moscona, A. A. (1975). *Proc. Natl. Acad. Sci. U.S.A.* **72,** 916.
Hausman, R. E., and Moscona, A. A. (1976). *Proc. Natl. Acad. Sci. U.S.A.* **73,** 3594.
Hay, R., Böhni, P., and Gasser, S. (1984). *Biochem. Biophys. Acta* **779,** 65.
Hayashi, H., and Ishimaru, Y. (1981). *Int. Rev. Cytol.* **70,** 139.
Heckman, J. E., Sarnoff, J., Alzner-De Weerd, B., Yin, S. and BajBhandary, U. L. (1980). *Proc. Natl. Acad. Sci. U.S.A.* **77,** 3159.
Heggeness, M. H., Wang, K., and Singer, S. J. (1977). *Proc. Natl. Acad. Sci. U.S.A.* **74,** 3883.
Helenius, A., Mellman, I., Wall, D., and Hubbard, A. (1983). *Trends Biochem. Sci.* **8,** 245.
Henderson, D., and Weber, K. (1980). *Exp. Cell Res.* **129,** 441.
Henderson, D., and Weber, K. (1981). *Exp. Cell Res.* **132,** 297.
Henis, Y. I., and Elson, E. L. (1981). *Proc. Natl. Acad. Sci. U.S.A.* **78,** 1072.
Henneguy, L. F. (1898). *Arch. Anat. Microsc. Morphol. Exp.* **1,** 481.
Herman, B., and Albertini, D. F. (1983). *Nature* (*London*) **304,** 738.
Herman, B., Langevin, M. A., and Albertini, D. F. (1983). *Eur. J. Cell Biol.* **31,** 34.
Hers, H. G., and Van Hoof, F. (1973). "Lysosomes and Storage Diseases." Academic Press, New York.
Hertzberg, E. L., Spray, D. C., and Bennett, M. V. L. (1985). *Proc. Natl. Acad. Sci. U.S.A.* **82,** 2412.
Herzog, W., and Weber, K. (1977). *Proc. Natl. Acad. Sci. U.S.A.* **74,** 1860.
Herzog, W., and Weber, K. (1978). *Eur. J. Biochem.* **92,** 1.
Heymans, H. S. A., Schutgens, R. B. H., Tan, R., vanden Bosch, H., and Borst, P. (1983). *Nature* (*London*) **306,** 69.

Hirokawa, N., Keller, T. C. S., Chasan, R., and Mooseker, M. S. (1983). *J. Cell Biol.* **96,** 1325.
Hirsch, M., and Penman, S. (1974). *J. Mol. Biol.* **83,** 131.
Hogan, B. (1981). *Nature (London)* **290,** 737.
Holter, H. (1954). *Proc. Roy. Soc.* **142,** 140.
Holter, H., and Marshall, J. M. (1954). *C.R. Trav. Lab. Carlsberg, Ser. Chim.* **29,** 7.
Holtfreter, J. (1939). *Arch. Exp. Zellforsch. Besonders Gewebezuecht.* **23,** 169.
Hoogeveen, A. T., Verheijhen, F. W., and Galjaard, H. (1983). *J. Biol. Chem.* **258,** 12143.
Hooper, M. L. (1982). *Biochim. Biophys. Acta* **651,** 85.
Hooper, M. L., and Slack, C. (1977). *Dev. Biol.* **55,** 251.
Hooper, M. L., and Subak-Sharpe, J. H. (1981). *Int. Rev. Cytol.* **69,** 45.
Hortin, G., and Boime, I. (1981). *Cell* **24,** 453.
Hsie, A. W., Kawashima, K., O'Neill, J. P., and Schröder, C. H. (1975). *J. Biol. Chem.* **250,** 984.
Hunter, G. K., and Pitts, J. (1981). *J. Cell Sci.* **49,** 163.
Hüttner, I., and Peters, H. (1978). *J. Ultrastr. Res.* **64,** 303.
Hyams, J. S., and Borisy, G. G. (1978). *J. Cell Sci.* **33,** 235.
Hynes, R. O. (1973). *Proc. Natl. Acad. Sci. U.S.A.* **70,** 3170.
Hynes, R. O. (1982). *Cell* **28,** 437.
Hynes, R. O., and Yamada, K. M. (1982). *J. Cell Biol.* **95,** 369.
Ishii, H., Horie, S., and Suga, T. (1980). *J. Biochem. (Tokyo)* **87,** 1855.
Jackson, R. C., Walter, P., and Blobel, G. (1980). *Nature (London)* **286,** 174.
Jacobs, H. T., Posakony, J. W., Grula, J. W., Roberts, J. W., Xin, J. H., Britten, R. J., and Davidson, E. H. (1983). *J. Mol. Biol.* **165,** 609.
Jacobs, M., and Cavalier-Smith, T. (1977). *Biochem. Soc. Symp.* **42,** 193.
Jacobson, K., and Wojcieszyn, J. (1984). *Proc. Natl. Acad. Sci. U.S.A.* **81,** 6747.
Jameson, L., and Caplow, M. (1981). *Proc. Natl. Acad. Sci. U.S.A.* **78,** 3413.
Jeng, A. Y., and Shamoo, A. E. (1980). *J. Biol. Chem.* **255,** 6897.
Jenni, B., Fasnacht, M., and Stutz, E. (1981). *FEBS Lett.* **125,** 175.
Jockusch, B. M., and Isenberg, G. (1981). *Proc. Natl. Acad. Sci. U.S.A.* **78,** 3005.
Joseph, S. K. (1984). *Trend Biochem. Sci.* **9,** 420.
Jouannet, P., Escalier, D., Serres, C., and David, G. (1983). *J. Submicrosc. Cytol.* **15,** 67.
Kakar, S. S., Means, A. R., and Tash, J. S. (1983). *J. Cell Biol.* **97,** 197a.
Kallinikova, V. D. (1981). *Int. Rev. Cytol.* **69,** 105.
Karlson, U. (1966). *J. Ultrastr. Res.* **16,** 429.
Kartenbeck, J., Schmid, E., Müller, H., and Franke, W. W. (1981). *Exp. Cell Res.* **133,** 191.
Kasamatsu, H., Robberson, D. L., and Vinograd, J. (1973). *Proc. Natl. Acad. Sci. U.S.A.* **68,** 2252.
Kasuga, M., Fujita-Yamaguchi, Y., Blithe, D. L., and Kahn, C. R. (1983). *Proc. Natl. Acad. Sci. U.S.A.* **80,** 2137.
Kato, K. H., and Sugiyama, M. (1971). *Dev. Growth Differ.* **13,** 359.
Keen, J. H., Willingham, M. C., and Pastan, I. (1981). *J. Biol. Chem.* **256,** 2538.
Keller, T. C. S., III, and Mooseker, M. S. (1982). *J. Cell Biol.* **95,** 943.
Kemble, R. J., Mans, R. J., Gabay-Laughman, S., and Laughman, J. R. (1983). *Nature (London)* **304,** 744.
Keski-Oja, J., Todaro, G. J., and Vaheri, A. (1981). *Biochim. Biophys. Acta* **673,** 323.
Kessel, R. G. (1983). *Int. Rev. Cytol.* **82,** 181.
Kilberg, M. S. (1982). *J. Membr. Biol.* **69,** 1.
Kim, H., Binder, L. I., and Rosenbaum, J. L. (1979). *J. Cell Biol.* **80,** 266.
Kindl, H. (1982). *Int. Rev. Cytol.* **80,** 193.
King, A. C., Hernaez-Davis, L., and Cuatrecasas, P. (1981). *Proc. Natl. Acad. Sci. U.S,A.* **77,** 3283.
Kirchhausen, T., and Harrison, S. C. (1981). *Cell* **23,** 755.
Kirschner, M. W. (1978). *Int. Rev. Cytol.* **54,** 1.
Kirschner, M. W. (1980). *J. Cell Biol.* **86,** 330.

Kirschner, M. W. (1980). *Eur. J. Cell Biol.* **22,** 285.

Kistler, J., and Bullivant, S. (1980). *J. Ultrastruct. Res.* **72,** 27.

Kleinman, H. K., Klebe, R. J., and Martin, G. R. (1981). *J. Cell Biol.* **88,** 473.

Kleinschuster, S. J., and Moscona, A. A. (1972). *Exp. Cell Res.* **70,** 397.

Klingenberg, M. (1981). *Nature (London)* **290,** 449.

Klymkowsky, M. W. (1982). *EMBO J.* **1,** 161.

Koch, G. (1976). *J. Biol. Chem.* **251,** 6097.

Kolodner, R. D., and Tewari, K. K. (1975). *Nature (London)* **256,** 708.

Kolodner, R., Tewari, K. K., and Warner, R. C. (1976). *Biochim. Biophys. Acta* **447,** 144.

Kondo, H. (1984). *J. Ultrastr. Res.* **87,** 124.

Kram, R., and Tomkins, G. M. (1973). *Proc. Natl. Acad. Sci. U.S.A.* **70,** 1659.

Kreibich, G., Ulrich, B. L., and Sabatini, D. D. (1978). *J. Cell Biol.* **77,** 464.

Kreibich, G., Ojakian, G., Rodriguez-Boulan, E., and Sabatini, D. D. (1982). *J. Cell Biol.* **93,** 111.

Kreis, T. E., and Birchmeier, W. (1980). *Cell* **22,** 555.

Kreis, T. E., Winterhalter, K. H., and Birchmeier, W. (1979). *Proc. Natl. Acad. Sci. U.S.A.* **76,** 3814.

Kreis, T. E., Geiger, B., Schmid, E., Jorcano, J. L., and Franke, W. W. (1983). *Cell* **32,** 1125.

Kühlbrandt, W., and Unwin, P. N. T. (1982). *J. Mol. Biol.* **156,** 431.

Kuroiwa, T. (1982). *Int. Rev. Cytol.* **75,** 1.

Lakshminarayanaiah, N. (1979). *Subcell. Biochem.* **6,** 401.

Lane, E. B., Hogan, B. L. M., Kurkinen, M., and Garrels, J. I. (1983). *Nature (London)* **303,** 701.

Langer, P. R., Waldrop, A. A., and Ward, D. D. (1981). *Proc. Natl. Acad. Sci. U.S.A.* **78,** 6633.

Larkin, J. M., Brown, M. S., Goldstein, J. L., and Anderson, R. G. W. (1983). *Cell* **33,** 273.

Läuger, P. (1980). *J. Membr. Biol.* **57,** 163.

Lazarides, E. (1975). *J. Cell Biol.* **65,** 549.

Lazarides, E. (1981). *Cell* **23,** 649.

Lazarides, E., and Burridge, K. (1975). *Cell* **6,** 289.

Lazarides, E., and Nelson, W. J. (1982). *Cell* **31,** 505.

Lazarides, E., and Revel, J. P. (1979). *Sci. Am.* **240**(5), 97.

Lazowska, J., Jacq, C., and Slonimski, P. P. (1980). *Cell* **22,** 333.

Lee, J. C., and Timasheff, S. N. (1977). *Biochemistry* **16,** 1754.

Lee, J. C., Tweedy, N., and Timasheff, S. N. (1978). *Biochemistry* **17,** 2783.

Lee, W. M., Cran, D. G., and Lane, N. J. (1982). *J. Cell Sci.* **57,** 215.

Lehninger, A. L. (1964). ''The Mitochondria.'' Benjamin, New York.

Lehninger, A. L. (1982). ''Principles of Biochemistry.'' Worth Publ., New York.

Lehto, V. P., Virtanen, I., and Kurki, O. (1978). *Nature (London)* **272,** 175.

Leighton, F., Poole, B., Beaufay, H., Baudhuin, P., Coffey, J. W., Fowler, S., and de Duve, C. (1968). *J. Cell Biol.* **37,** 482.

Leister, D. E., and Dawid, I. B. (1974). *J. Biol. Chem.* **249,** 5108.

Levitzki, A., and Helmreich, E. J. M. (1979). *FEBS Lett.* **101,** 213.

Lewin, A., Gregor, I., Mason, T. L., Nelson, N., and Schatz, G. (1980). *Proc. Natl. Acad. Sci. U.S.A.* **77,** 3998.

Lewis, W. (1931). *Bull. Johns Hopkins Hosp.* **49,** 17.

Lin, Y. M. (1982). *Mol. Cell. Biochem.* **45,** 101.

Linderman, C. B., and Rikmenspoel, R. (1972). *Science* **175,** 337.

Link, G., Chambers, S. E., Thompson, J. A., and Falk, H. (1981). *Mol. Gen. Genet.* **181,** 454.

Lipsky, N. G., and Pagano, R. E. (1985). *Science* **228,** 745.

Little, M., Ludueña, R. F., Langford, G. M., Asnes, C. F., and Farrell, K. (1981). *J. Mol. Biol.* **149,** 95.

Loewenstein, W. R. (1976). *Cold Spring Harbor Symp. Quant. Biol.* **40,** 49.

Loewenstein, W. R. (1982). *Physiol. Rev.* **61,** 829.

Lonsdale, D. M., Hodge, T. P., Howe, C. J., and Stern, D. B. (1983). *Cell* **34,** 1007.
Lord, J. M. (1980). *Subcell. Biochem.* **7,** 171.
Lord, J. M., and Roberts, L. M. (1980). *Trends Biochem. Sci.* **5,** 271.
Lucas, M., Schmid, G., Kromas, R., and Löffler, G. (1978). *Eur. J. Biochem.* **85,** 609.
Lwoff, A. (1950). "Problems of Morphogenesis in Ciliates." Wiley, New York.
Mabuchi, I. (1983). *J. Cell Biol.* **97,** 1612.
Mabuchi, I., Shimizu, T., and Mabuchi, Y. (1976). *Arch. Biochem. Biophys.* **176,** 564.
Mabuchi, Y., and Mabuchi, I. (1973). *Exp. Cell Res.* **82,** 271.
McClain, D. A., D'Eustachio, P., and Edelman, G. M. (1977). *Proc. Natl. Acad. Sci. U.S.A.* **74,** 666.
MacLean-Fletcher, S., and Pollard, T. B. (1980). *Cell* **20,** 329.
Mahler, H. R. (1983) *Int. Rev. Cytol.* **82,** 1.
Majerus, P. W., Neufeld, E. J., and Wilson, D. B. (1984). *Cell* **37,** 701.
Maltoltsy, A. G., Maltoltsy, M. N., and Cliffel, P. J. (1981). *Biochim. Biophys. Acta* **668,** 160.
Mandelbaum-Shavit, F., Wolpert-DeFilippes, M. K., and Johns, D. G. (1976). *Biochem. Biophys. Res. Commun.* **72,** 47.
Mannaerts, G. P., Van Veldhoven, P., Van Broekhoven, A., Vandebroek, G., and DeBeer, L. J. (1982). *Biochem. J.* **204,** 17.
Maruta, H., Knoerzer, W., Hinssen, H., and Isenberg, G. (1984). *Nature* **312,** 424.
Marcantonio, E. E., Grebenau, R. C., Sabatini, D. D., and Kreibich, G. (1982). *Eur. J. Biochem.* **124,** 217.
Marchase, R. B., Vosbeck, K., and Roth, S. (1976). *Biochim. Biophys. Acta* **457,** 385.
Marchese-Ragona, S. P., and Holwill, M. E. J. (1980). *Nature (London)* **287,** 867.
Marchese-Ragona, S. P., Mellor, J. S., and Holwill, M. E. J. (1983). *J. Submicrosc. Cytol.* **15,** 43.
Maro, B., Sauron, M. E., Paulin, D., and Bornens, M. (1983). *Biol. Cell.* **47,** 243.
Marshall, J. R., Jr. (1954). *Exp. Cell Res.* **6,** 240.
Martin, N. C., Pham, H. D., Underbrink-Lyon, K., Miller, D. L., and Donelson, J. E. (1980). *Nature (London)* **285,** 579.
Maruyama, K., and Sakai, H. (1981). *J. Biochem. (Tokyo)* **89,** 1337.
Marzella, L., Ahlberg, J., and Glaumann, H. (1982). *J. Cell Biol.* **93,** 144.
Mastro, A. M., and Rozengurt, E. (1976). *J. Biol. Chem.* **251,** 7899.
Matsuura, S., Arpin, M., Hannum, C., Margoliash, E., Sabatini, D. D., and Morimoto, T. (1981). *Proc. Natl. Acad. Sci. U.S.A.* **78,** 4368.
Maul, G. G. (1977). *Exp. Cell Res.* **104,** 233.
Means, A. R., Lagace, L., Guerriero, V., Jr., and Chafouleas, J. G. (1982). *J. Cell. Biochem.* **20,** 317.
Mego, J. L., and Farb, F. M. (1978). *Biochem. J.* **172,** 233.
Meier, P. J., Spycher, M. A., and Meyer, U. A. (1981). *Biochim. Biophys. Acta* **646,** 283.
Mellman, I. S. (1982). *Nature (London)* **299,** 301.
Mellman, I. S., Steinman, R. M., Unkeless, J. C., and Cohn, Z. A. (1980). *J. Cell Biol.* **86,** 712.
Merlino, G. T., Chamberlain, J. P., and Kleinsmith, L. J. (1978). *J. Biol. Chem.* **253,** 7078.
Meyer, D. I., and Dobberstein, B. (1980). *J. Cell Biol.* **87,** 503.
Meyer, D. I., Krause, E., and Dobberstein, B. (1982). *Nature (London)* **297,** 647.
Michaelis, P. (1951). *Cold Spring Harbor Symp. Quant. Biol.* **16,** 121.
Miki-Noumura, T. (1977). *J. Cell Sci.* **24,** 203.
Mills, W. R., and Baumgartner, B. J. (1983). *FEBS Lett.* **163,** 124.
Miller, K. R. (1979). *Sci. Amer.* **241**(4), 102.
Mimura, N., and Asano, A. (1982). *J. Cell Biol.* **93,** 899.
Mitchell, D. R., and Rosenbaum, J. L. (1985). *J. Cell Biol.* **100,** 1228.
Mitchell, P. (1979). *Science* **206,** 1148.
Mockrin, S. C., and Korn, E. D. (1980). *Biochemistry* **19,** 5359.

Mohri, H. (1976). *Biochim. Biophys. Acta* **456,** 85.
Monné, L. (1946). *Experientia* **2,** 153.
Mookerjee, B. K., Cuppoletti, J., Rampal, A. L., and Jung, C. Y. (1981). *J. Biol. Chem.* **256,** 1290.
Mori, Y., Akedo, H., Tanigaki, Y., Tanaka, K., and Okada, M. (1979). *Exp. Cell Res.* **120,** 435.
Morimoto, R., Lewin, A., Hsu, H. J., Rabinowitz, M., and Fukuhara, H. (1975). *Proc. Natl. Acad. Sci. U.S.A.* **72,** 3868.
Morré, D. (1977). *In* "International Cell Biology (B. R. Brinkley and K. R. Porter, eds.), p. 293. Rockefeller Univ. Press, New York.
Morré, D. J., and Ovtracht, L. (1981). *J. Ultrastruct. Res.* **74,** 284.
Moscona, A. A. (1957). *Proc. Natl. Acad. Sci. U.S.A.* **43,** 184.
Moscona, A. A. (1971). *Science* **171,** 905.
Moscona, M. H., and Moscona, A. A. (1963). *Science* **142,** 1070.
Moy, G. W., Brandriff, B., and Vacquier, V. D. (1977). *J. Cell Biol.* **73,** 788.
Mukerji, S. K., and Meizel, S. (1975). *Arch. Biochem. Biophys.* **168,** 720.
Müller, W. E., and Zahn, R. K. (1973). *Exp. Cell Res.* **80,** 95.
Müller, W. E. G., Bernd, A., Zahn, R. K., Kurelec, B., Dawes, K., Müller, I., and Uhlenbruck, G. (1981). *Eur. J. Biochem.* **116,** 573.
Munn, E. A. (1974). "The Structure of Mitochondria." Academic Press, New York.
Murphy, D. B., and Borisy, G. G. (1975). *Proc. Natl. Acad. Sci. U.S.A.* **72,** 2696.
Nandi, P. K., Pretorius, H. T., Lippoldt, R. E., Johnson, M. L., and Edelhoch, H. (1980). *Biochemistry* **19,** 5917,
Nelson, W. J., and Traub, P. (1981). *Eur. J. Biochem.* **116,** 51.
Nelson, W. J., and Traub, P. (1983). *Mol. Cell. Biol.* **3,** 1146.
Nelson, W. J., Colaço, C. A. L. S., and Lazarides, E. (1983a). *Proc. Natl. Acad. Sci. U.S.A.* **80,** 1626.
Neupert, W., and Schatz, G. (1981). *Trends Biochem. Sci.* **6,** 1.
Neurath, H., and Walsh, K. A. (1976). *Proc. Natl. Acad. Sci. U.S.A.* **73,** 3825.
Nicholls, D. G., and Crompton, M. (1980). *FEBS Lett.* **111,** 261.
Nichols, K. M., and Rikmenspoel, R. (1978a). *J. Cell Sci.* **29,** 233.
Nichols, K. M., and Rikmenspoel, R. (1978b). *Exp. Cell Res.* **116,** 333.
Nicholson, B. J., Takemoto, L. J., Hunkapiller, M. W., Hood, L. E., and Revel, J. P. (1983). *Cell* **32,** 967.
Nicolson, G. L. (1972). *Nature (London) New Biol.* **239,** 193.
Nicolson, G. L. (1976). *Biochim. Biophys. Acta* **457,** 57.
Nicolson, G. L., Smith, J. R., and Poste, G. (1976). *J. Cell Biol.* **68,** 395.
Niederman, R., Amrein, P. C., and Hartwig, J. (1983). *J. Cell Biol.* **96,** 1400.
Nishizuka, Y. (1984). *Nature* **308,** 693.
Noll, H., Matranga, V., Cascino, D., and Vittorelli, L. (1979). *Proc. Natl. Acad. Sci. U.S.A.* **76,** 288.
Novikoff, A. B. (1961). *In* "The Cell" (J. Brachet and A. E. Mirsky, eds.), Vol. 2, p. 299. Academic Press, New York.
Novikoff, A. B. (1976). *Proc. Natl. Acad. Sci. U.S.A.* **73,** 2781.
Nunnally, M. H., Powell, L. D., and Craig, S. W. (1981). *J. Biol. Chem.* **256,** 2083.
O'Connor, C. M., Asai, D. J., Flytzanis, C. N., and Lazarides, E. (1981). *Mol. Cell Biol.* **1,** 303.
Ogawa, K., Negishi, S., and Obika, M. (1982). *J. Cell Biol.* **92,** 706.
Ogihara, S., and Tonomura, Y. (1982). *J. Cell Biol.* **93,** 604.
Ohkuma, S., and Poole, B. (1978). *Proc. Natl. Acad. Sci. U.S.A.* **75,** 3327.
Ohkuma, S., Moriyama, Y., and Takano, T. (1982). *Proc. Natl. Acad. Sci. U.S.A.* **79,** 2758.
Ojala, D., and Attardi, G. (1974). *J. Mol. Biol.* **82,** 151.
Ojala, D., Merkel, C., Gelfand, R., and Attardi, G. (1980). *Cell* **22,** 393.
Ojala, D., Montoya, J., and Attardi, G. (1981). *Nature (London)* **290,** 470.

Okada, C. Y., and Rechsteiner, M. (1982). *Cell* **29,** 33.
Okuno, M., Ogawa, K., and Mohri, H. (1976). *Biochem. Biophys. Res. Commun.* **68,** 901.
Olden, K., Parent, J. B., and White, S. L. (1982). *Biochim. Biophys. Acta* **650,** 209.
Oliver, J. M., Gelfand, E. W., Pearson, C. B., Pfeiffer, J. R., and Dosch, H. M. (1980). *Proc. Natl. Acad. Sci. U.S.A.* **77,** 3499.
Olsen, I., Muir, H., Smith, R., Fensom, A., and Watt, D. J. (1983). *Nature (London)* **306,** 75.
Omoto, C. K., and Kung, C. (1980). *J. Cell Biol.* **87,** 33.
Omoto, C. K., and Witman, G. B. (1981). *Nature (London)* **290,** 708.
Osborn, M., and Weber, K. (1976). *Exp. Cell Res.* **103,** 331.
Osborn, M., and Weber, K. (1980). *Exp. Cell Res.* **130,** 484.
Osborn, M., and Weber, K. (1982). *Cell* **31,** 303.
Otto, J. J, (1983). *J. Cell Biol.* **97,** 1283.
Ovchinnikov, Y. A. (1979). *Eur. J. Biochem.* **94,** 321.
Overton E. (1895). *Vierteljahresschr. Naturforsch. Ges. Zürich* **40,** 159.
Owada, M. K., Hakura, A., Iida, K., Yahara, I., Sobue, K., and Kakiuchi, S. (1984). *Proc. Natl. Acad. Sci. U.S.A.* **81,** 3133.
Padmanabhan, V., and Green, B. R. (1978). *Biochim. Biophys. Acta* **521,** 67.
Paiement, J., Beaufay, H., and Godelaine, D. (1980). *J. Cell Biol.* **86,** 29.
Palade, G. E. (1955). *J. Biophys. Biochem. Cytol.* **1,** 567.
Palade, G. E. (1975). *Science* **189,** 347.
Palade, G. E., and Siekevitz, P. (1956). *J. Biophys. Biochem. Cytol.* **2,** 171.
Pardue, R. L., Kaetzel, M. A., Hahn, S. H., Brinkley, B. R., and Dedman, J. R. (1980). *Cell* **23,** 533.
Pastan, I. H., and Willingham, M. C. (1981). *Science* **214,** 504.
Pastan, I. H., and Willingham, M. C. (1983). *Trends Biochem. Sci.* **8,** 250.
Pavelka, M., and Ellinger, A. (1983). *J. Cell Biol.* **97,** 737.
Payne, M. R., and Rudnick, S. E. (1984). *Trends Biochim. Sci.* **9,** 361.
Pearse, B. M. F. (1978). *J. Mol. Biol.* **126,** 803.
Pearse, B. M. F. (1980). *Trends Biochem. Sci.* **5,** 131.
Penningrath, S. M., and Kirschner, M. W. (1977). *J. Mol. Biol.* **115,** 643.
Peracchia, C. (1980). *Int. Rev. Cytol.* **66,** 81.
Pessac, B., and Defendi, V. (1972). *Nature (London), New Biol.* **238,** 13.
Peters, R., Brünger, A., and Schulten, K. (1981). *Proc. Natl. Acad. Sci. U.S.A.* **78,** 962.
Petersen, O. W., and Van Deurs, B. (1983). *J. Cell Biol.* **96,** 277.
Pfeffer, S. R., Drubin, D. G., and Kelly, R. B. (1983). *J. Cell Biol.* **97,** 40.
Phillips, P. G., Furmanski, P., and Lubin, M. (1974). *Exp. Cell Res.* **86,** 301.
Phillips, S. G., and Rattner, J. B. (1976). *J. Cell Biol.* **70,** 9.
Pierce, M., Turley, E. A., and Roth, S. (1980). *Int. Rev. Cytol.* **65,** 2.
Pinto da Silva, P., and Kachar, B. (1982). *Cell* **28,** 441.
Piperno, G., Huang, B., Ramanis, Z., and Luck, D. J. L. (1981). *J. Cell Biol.* **88,** 73.
Pitts, J. D., and Simms, J. W. (1977). *Exp. Cell Res.* **104,** 153.
Plattner, H. (1981). *Cell Biol. Int. Rep.* **5,** 435.
Pollard, T. D. (1984). *Nature* **312,** 403.
Porter, K. R. (1954). *J. Histochem. Cytochem.* **2,** 346.
Porter, K. R. (1955). *Fed Proc. Fed. Am. Soc. Exp. Biol.* **14,** 673.
Porter, K. R. (1961) *In* "The Cell" (J. Brachet and A. E. Mirsky, eds.), Vol. 2, p. 621. Academic Press, New York.
Possingham, J. V., and Lawrence, M. E. (1983). Intern. *Rev. Cytol.* **84,** 1.
Preer, J. R. Jr., Preer, L. B., Rudman, B. M., and Barnett, A. J. (1985). *Nature* **314,** 188.
Prehn, S., Tsmaloukas, A., and Rapoport, T. A. (1980). *Eur. J. Biochem.* **107,** 185.

Prehn, S., Nürnberg, P., and Rapoport, T. A. (1981). *FEBS Lett.* **123,** 79.
Prensier, G., Vivier, E., Goldstein, S., and Schrével, J. (1980). *Science* **207,** 1493.
Provasoli, L., Hutner, S. H., and Pintner, I. J. (1951). *Cold Spring Harbor Symp. Quant. Biol.* **16,** 113.
Quinn, P. J. (1981). *Prog. Biophys. Mol. Biol.* **38,** 1.
Ramaekers, F. C. S., Osborn, M., Schmid, E., Weber, K., Bloemendal, H., and Franke, W. W. (1980). *Exp. Cell Res.* **127,** 309.
Ramaekers, F. C. S., Haag, D., Kant, A., Moesker, O., Jap, P. H. K., and Vooijs, G. P. (1983). *Proc. Natl. Acad. Sci. U.S.A.* **80,** 2618.
Ramperz, A., and Walz, F. G., Jr. (1983). *Proc. Natl. Acad. Sci. U.S.A.* **80,** 6542.
Rattner, J. B., and Phillips, S. G. (1973). *J. Cell Biol.* **57,** 359.
Raviola, E., Goodenough, D. A., and Raviola, G. (1980). *J. Cell Biol.* **87,** 273.
Raylin, D., and Flavin, M. (1977). *Biochemistry* **16,** 2189.
Reaven, E., and Axline, S. G. (1973). *J. Cell Biol.* **59,** 12.
Rebhun, L. I. (1977). *Int. Rev. Cytol.* **49,** 1.
Rhoades, M. M. (1950). *Proc. Natl. Acad. Sci. U.S.A.* **36,** 634.
Rickwood, D., Chambers, J. A. A., and Barat, M, (1981). *Exp. Cell Res.* **133,** 1.
Riezman, H., Hay, R,, Witte, C., Nelson, N., and Schatz, G. (1983). *EMBO J.* **2,** 1113.
Rikmenspoel, R. (1982). *J. Theor. Biol.* **96,** 617.
Ris, H. (1985). *Cell Biol.* **100,** 1475.
Roberts, K., and Hyams, J. S., eds. (1979). "Microtubules." Academic Press, London.
Robinson, J. M., and Karnovsky, M. J. (1980). *J. Cell Biol.* **87,** 562.
Rochaix, J. D. (1985). *Int. Rev. Cytol.* **93,** 57.
Rodbell, M. (1980). *Nature (London)* **284,** 17.
Rogers, G., Gruenebaum, J., and Boime, I. (1982). *J. Biol. Chem.* **257,** 4179.
Rosamond, J. (1982). *Biochem. J.* **202,** 1.
Rose, B., and Loewenstein, W. R. (1975a). *Nature (London)* **254,** 250.
Rose, B., and Loewenstein, W. R. (1975b). *Science* **190,** 1204.
Rosenfeld, M. G., Kreibich, G., Popov, D., Kato, K., and Sabatini, D. D. (1982). *J. Cell Biol.* **93,** 135.
Roth, J., and Berger, E. G. (1982). *J. Cell Biol.* **93,** 223.
Roth, R. A., and Cassell, D. J. (1983). *Science* **219,** 299.
Rothman, J. E. (1981). *Science* **213,** 1212.
Rubin, K., Johansson, S., Höök, M., and Öbrink, B. (1981). *Exp. Cell Res.* **135,** 127.
Runge, M. S., Hewgley, P. B., Puett, D., and Williams, R. C., Jr. (1979). *Proc. Natl. Acad. Sci. U.S.A.* **76,** 2561.
Ruoslahti, E., Pierschbacher, M., Hayman, E. G., and Engwall, E. (1982). *Trends Biochem. Sci.* **7,** 188.
Russell, T., and Pastan, I. (1973). *J. Biol. Chem.* **248,** 5835.
Rutishauser, U., Thiery, J. P., Brackenbury, R., Sela, B. A., and Edelman, G. M. (1976). *Proc. Natl. Acad. Sci. U.S.A.* **73,** 577.
Sabatini, D. D., and Blobel, G. (1970). *J. Cell Beiol.* **45,** 146.
Sabatini, D. D., Kreibich, G., Morimoto, T., and Adesnik, M. (1982). *J. Cell Biol.* **92,** 1.
Saccone, C., and Quagliariello, E. (1975). *Int. Rev. Cytol.* **43,** 125.
Sager, R., Grabowy, C., and Sano, H. (1981). *Cell* **24,** 41.
Sala, F., Amileni, A. R., Parisi, B., and Spadari, S. (1980). *Eur. J. Biochem.* **112,** 211.
Salisbury, J. L., Condeelis, J. S., and Satir, P. (1980). *J. Cell Biol.* **87,** 132.
Sandoval, I. V., and Weber, K. (1978). *Eur. J. Biochem.* **92,** 463.
Sandoval, I. V., MacDonald, E., Jameson, J. L., and Cuatrecasas, P. (1977). *Proc. Natl. Acad. Sci. U.S.A.* **74,** 4881.

Sandoval, I. V., Jameson, J. L., Niedel, J., MacDonald, E., and Cuatrecasas, P. (1978). *Proc. Natl. Acad. Sci. U.S.A.* **75,** 3178.
Sanger, J. W., Sanger, J. M., Kreis, T. E., and Jockusch, B. M. (1980). *Proc. Natl. Acad. Sci. U.S.A.* **77,** 5268.
Sanger, J. W., Sanger, J. M., and Jockusch, B. M. (1983). *J. Cell Biol.* **96,** 961.
Savion, N., Vlodavsky, I., and Gospodarowicz, D. (1981). *J. Biol. Chem.* **256,** 1149.
Schatz, G., and Butow, R. A. (1983). *Cell* **32,** 316.
Scheer, U., and Franke, W. W. (1972). *Planta* **107,** 145.
Schiff, P. B., and Horwitz, S. B. (1980). *Proc. Natl. Acad. Sci. U.S.A.* **77,** 1561.
Schiff, P. B., Fant, J., and Horwitz, S. G. (1979). *Nature (London)* **277,** 665.
Schliwa, M. (1981). *Cell* **25,** 587.
Schliwa, M., Euteneuer, V., Bulinski, J. C., and Izant, J. G. (1981a). *Proc. Natl. Acad. Sci. U.S.A.* **78,** 1037.
Schliwa, M., Van Blerkom, J., and Porter, K. R. (1981b). *Proc. Natl. Acad. Sci. U.S.A.* **78,** 4329.
Schliwa, M., Pryzwansky, K. B., and Van Blerkom, J. (1982). *Philos. Trans. R. Soc. London, Ser. B* **299,** 199.
Schloss, J. A., and Goldman, R. D. (1980). *J. Cell Biol.* **87,** 633.
Schmell, E., Slife, C. W., Kuhlenschmidt, M. S., and Roseman, S. (1982). *J. Biol. Chem.* **257,** 3171.
Schneider, D. L. (1981). *J. Biol. Chem.* **256,** 3858.
Schneider, D. L. (1983). *J. Biol. Chem.* **258,** 1833.
Schneider, E. G., Nguyen, H. T., and Lennarz, W. J. (1978). *J. Biol. Chem.* **253,** 2348.
Schnepf, E. (1980). *Results Prob. Cell Differ.* **10,** 1.
Schnitka, T. K. (1966). *J. Ultrastruct. Res.* **16,** 598–625.
Schroer, T. A., and Kelly, R. B. (1985). *Cell* **40,** 729.
Schuel, H. (1978). *Gamete Res.* **1,** 299.
Schwarzmann, G., Wiegandt, H., Rose, B., Zimmerman, A., Ben-Haim, D., and Loewenstein, W. R. (1981). *Science* **213,** 551.
Seglen, P. O., and Gordon, P. B. (1982). *Proc, Natl. Acad. Sci. U.S.A.* **79,** 1889.
Sharma, R. K. (1982). *Prog. Nucleic Acid Res. Mol. Biol.* **27,** 233.
Sheetz, M. P., and Spudich, J. A. (1983). *Nature (London)* **303,** 31.
Shephard, D. C., and Bidwell, R. G. (1973), *Protoplasma* **76,** 289.
Sherline, P., and Schiavone, K. (1977). *Science* **198,** 1038.
Sheterline, P. (1977). *Biochem. J.* **168,** 533.
Shia, M. A., and Pilch, P. F. (1983). *Biochemistry* **22,** 717.
Shlomai, J., and Zadok, A. (1983). *Nucleic Acids Res.* **11,** 4019.
Shotwell, M. A., Jayme, D. W., Kilberg, M. S., and Oxender, D. L. (1981). *J. Biol. Chem.* **256,** 5422.
Siegel, V., and Walter, P. J. (1985) **100,** 1913.
Singer, I. I. (1982). *J. Cell Biol.* **92,** 398.
Singer, I. I., and Paradiso, P. R. (1981). *Cell* **24,** 481.
Singer, S. J., and Nicolson, G. L. (1972). *Science* **175,** 720.
Skou, J. C, (1974). *In* "Perspectives in Membrane Biology" (S. Estrada-O and C. Gitler, eds.), p. 263. *In* "Perspectives in Membrane Biology." Academic Press, New York.
Slonimski, P. P. (1980). *C. R. Hebd. Seances, Acad. Sci., Ser. D* **290,** 331.
Smith, R. A., and Ord, M. J. (1983). *Int. Rev. Cytol.* **83,** 63.
Snyder, J. A., and McIntosh, J. R. (1976). *Annu. Rev. Biochem.* **45,** 699.
Sobue, K., Tanaka, T., Ashino, N., and Kakiuchi, S. (1985). *Biochim. Biophys. Acta* **845,** 366.
Somlyo, A. P., Bond, M., and Somlyo, A. V. (1985). *Nature* **314,** 623.
Sogin, D. C., and Hinkle, P. C. (1980). *Proc. Natl. Acad. Sci. U.S.A.* **77,** 5725.

Southwick, F. S., and Hartwig, J. H. (1982). *Nature (London)* **296,** 303.
Sowers, A. E., and Hackenbrock, C. R. (1981). *Proc. Natl. Acad. Sci. U.S.A.* **78,** 6246.
Spiegel, M., and Spiegel, E. (1978). *Exp. Cell Res.* **117,** 261.
Spiegelman, B. M., Lopata, M. A., and Kirschner, M. W. (1979). *Cell* **16,** 239.
Spray, D. C., Harris, A. L., and Bennett, M. V. L. (1981). *Science* **211,** 712.
Staehelin, L. A., and Hull, B. E. (1978). *Sci. Amer.* **238**(5), 140.
Stambaugh, R., and Smith, M. (1978). *J. Exp. Zool.* **203,** 135.
Starling, D., Duncan, R., and Lloyd, J. B. (1983). *Cell Biol. Int. Rep.* **7,** 593.
Stearns, M. S., and Brown, D. L. (1979). *Proc. Natl. Acad. Sci. U.S.A.* **76,** 5745.
Steinberg, M. S., and Gepner, I. A. (1973). *Nature (London), New Biol.* **241,** 249.
Steinert, G., and Hanocq, J. (1979). *Biol. Cell.* **34,** 117.
Steinert, M., and Van Assel, S. (1974). *Biochem. Biophys. Res. Commun.* **61,** 1249.
Steinert-Meulemans, G. (1980). Doctoral Thesis, Université libre de Bruxelles.
Steinman, R. M., Mellman, I. S., Muller, W. A., and Cohn, Z. A. (1983). *J. Cell Biol.* **96,** 1.
Stephens, R. E., and Edds, K. T. (1976). *Phys. Rev.* **56,** 709.
Stern, D. B., and Lonsdale, D. M. (1982). *Nature (London)* **299,** 698.
Steven, A. C., Wall, J., Hainfeld, J., and Steinert, P. M. (1982). *Proc. Natl. Acad. Sci. U.S.A.* **79,** 3101.
Stone, D. K., Xie, X. S., and Racker, E. (1983). *J. Biol. Chem.* **258,** 4059,
Storti, R. V., and Rich, A. (1976). *Proc. Natl. Acad. Sci. U.S.A.* **73,** 2346.
Straus, W. (1954). *J. Biol. Chem.* **207,** 745.
Straus, W., and Oliver, J. (1955). *J. Exp. Med.* **102,** 1.
Streb, H., Irvine, R. F., Berridge, M. J., and Schultz, I. (1983). *Nature (London)* **306,** 67.
Stuart, K. D., and Gelvin, S. B. (1982). *Mol. Cell Biol.* **2,** 245.
Sulakhe, P. V., and St. Louis, P. J. (1980). *Prog. Biophys. Mol. Biol.* **35,** 135.
Summerhayes, I. C., Wong, D., and Chen, L. B. (1983). *J. Cell Sci.* **61,** 87.
Summers, K., and Gibbons, I. R. (1971). *Proc. Natl. Acad. Sci. U.S.A.* **68,** 3092.
Svardal, A, M., and Pryme, I. F. (1980). *Subcell. Biochem.* **7,** 117.
Swanson, R. F., and Dawid, I. B. (1970). *Proc. Natl. Acad. Sci. U.S.A.* **66,** 117.
Taddei, C. (1972). *Exp. Cell Res.* **70,** 285.
Taddei, C., Gambino, R., Metafora, S., and Monroy, A. (1973). *Exp. Cell Res.* **78,** 159.
Takeichi, M., Ozaki, H. S., Tokunaga, K., and Okada, T. S. (1979). *Dev. Biol.* **70,** 195.
Tamm, S. L., and Tamm, S. (1981). *J. Cell Biol.* **89,** 495.
Tapper, D. P., and Clayton, D. A. (1981). *J. Biol. Chem.* **256,** 5109.
Tartakoff, A. M. (1982a). *Trends Biochem. Sci.* **7,** 174.
Tartakoff, A. M. (1982b). *Philos. Trans. R. Soc. London,* **300,** 173.
Tartakoff, A. M. (1983a). *Intern. Rev. Cytol.* **85,** 221.
Tartakoff, A. M. (1983b). *Cell* **32,** 1026.
Tartakoff, A. M., and Vassalli, P. (1983). *J. Cell Biol.* **97,** 1243.
Tash, J. S., and Means, A. R. (1983). *Biol. Reprod.* **28,** 75.
Taylor, D. L., and Condeelis, J. S. (1979). *Int. Rev. Cytol.* **56,** 57.
Taylor, D. L., and Wang, Y. L. (1978). *Proc. Natl. Acad. Sci. U.S.A.* **75,** 857.
Terranova, V. P., Rohrbach, D. H., and Martin, G. R. (1980). *Cell* **22,** 719.
Thomas, A. A. M., Benne, R., and Voorma, H. A. (1981). *FEBS Lett.* **128,** 177.
Thomas, W. A., and Steinberg, M. S. (1981). *Dev. Biol.* **81,** 106.
Tickle, C., and Trinkaus, J. P. (1977). *J. Cell Sci.* **26,** 139.
Timmis, J. N., and Scott, N. S. (1983). *Nature (London)* **305,** 65.
Toh, B. H., and Hard, G. C. (1977). *Nature (London)* **269,** 695.
Trinkaus, J. P, (1980). *In* "Tumor Cell Surfaces and Malignancy" (R. Hynes and F. Fox, eds.), p. 887. Alan R. Liss, Inc., New York.

Trotter, J. A. (1981). *Exp. Cell Res.* **132,** 235.
Trouet, A. (1978). *Eur. J. Cancer* **14,** 105.
Tsai, J. S., and Seeman, M. (1981). *Biochim. Biophys. Acta* **673,** 259.
Tucciarone, L. M., and Lanclos, K. D. (1981). *Biochem. Biophys. Res. Commun,* **99,** 221.
Tucker, R. W,, Pardee, A. B., and Fujiwara, K. (1979a). *Cell* **17,** 527.
Tucker, R. W., Meade-Cobun, K. S., Jayaraman, S., and More, N. S. (1983). *J. Submicrosc. Cytol.* **15,** 139.
Turin, L., and Warner, A. (1977). *Nature (London)* **270,** 56.
Tycko, B., Keith, C. H., and Maxfield, F. R. (1983). *J. Cell Biol.* **97,** 1762.
Uchida, N., Smilowitz, H., Ledger, P. W., and Tanzer, M. L. (1980). *J. Biol. Chem.* **255,** 8638.
Unwin, P. N. T., and Zampighi, G. (1980). *Nature (London)* **283,** 545.
Upholt, W. B., and Dawid, I. B. (1977). *Cell* **11,** 571.
Valenzuela, P., Quiroga, M., Zaldivar, J., Rutter, W. J., Kirschner, M. W., and Cleveland, D. W. (1981). *Nature (London)* **289,** 650.
Vallee, R.B., Dibartolomeis, M.J. and Theurkauf, W.E. (1981).*J. Cell Biol.* **90,** 568.
van Deenen, L. L. M. (1981). *FEBS Lett.* **123,** 1.
van den Boogaart, P., Samallo, J., and Agsteribbe, E. (1982). *Nature (London)* **298,** 187.
Van de Vijver, G. (1975). *Cur. Top. Dev. Biol.* **10,** 123.
Van Dyke, R. W., Steer, C. J., and Scharschmidt, B. F. (1984). *Proc. Natl. Acad. Sci. U.S.A.* **81,** 6747.
Vartio, T., and Vaheri, A. (1983). *Trends Biochem. Sci.* **8,** 442.
Vasiliev, J. M., and Gelfand, I. M. (1977). *Int. Rev. Cytol.* **50,** 159.
Venetianer, A., Schiller, D. M., Magin, T., and Franke, W. W. (1983). *Nature (London)* **305,** 730.
Vernier, J, M., and Sire, M. F. (1977). *Biol. Cell.* **29,** 99.
Villalobo, A., and Lehninger, A. L. (1980). *J. Biol. Chem.* **255,** 2457.
Vladutiu, G. D., and Rattazzi, M. C. (1980). *Biochem. J.* **192,** 813.
von Lenhossék, M. (1898). *Verh. Anat. Ges.* **12,** 106.
von Wettstein, D. (1980). *In* ''International Cell Biology'' (H. G. Schweiger, ed.), p. 250. Springer-Verlag, Berlin and New York.
Waheed, A., Hasilik, A., and von Figura, K. (1981). *J. Biol. Chem.* **256,** 5717,
Waksman, A,, Hubert, P., Crémel, G., Rendom, A., and Burgun, C. (1980). *Biochim. Biophys. Acta* **604,** 249.
Wallace, R. A., and Jared, D. M. (1976). *J. Cell Biol.* **69,** 345.
Walter, M. F., and Satir, P. (1978). *J. Cell Biol.* **79,** 110.
Walter, P., and Blobel, G. (1980). *Proc. Natl. Acad. Sci. U.S.A.* **77,** 7112.
Walter, P., and Blobel, G. (1981a). *J. Cell Biol.* **91,** 551.
Walter, P., and Blobel, G. (1981b). *J. Cell Biol.* **91,** 557.
Walter, P., and Blobel, G. (1982). *Nature (London)* **299,** 691.
Walter, P., and Blobel, G. (1983). *Cell* **34,** 525.
Walter, P., Ibrahimi, I., and Blobel, G. (1981). *J. Cell Biol.* **91,** 545.
Walter, P., Gilmore, R., Müller, M., and Blobel, G. (1982). *Philos. Trans. R. Soc. London, Ser. B* **300,** 225.
Walter, P., Gilmore, R., and Blobel, G. (1984). *Cell* **38,** 5.
Wang, E., Connolly, J. A., Kalnins, V. I., and Choppin, P. W. (1979a). *Proc. Natl. Acad. Sci. U.S.A.* **76,** 5719.
Wang, E., Cross, R. K., and Choppin, P. W. (1979b). *J. Cell Biol.* **83,** 320.
Warner, F. D,, and Mitchell, D. R. (1980). *Int. Rev. Cytol.* **66,** 1.
Wattiaux, R., and de Duve, C. (1956). *Biochem. J.* **63,** 606.
Weber, K., Bibring, T., and Osborn, M. (1975a). *Exp. Cell Res.* **95,** 111.
Weber, K., Pollack, R., and Bibring, T. (1975b). *Proc. Natl. Acad. Sci. U.S.A.* **72,** 459.

Weeds, A. (1982). *Nature (London)* **296,** 811.
Wehland, J. and Weber, K. (1980). *Exp. Cell Res.* **127,** 397.
Wehland, J., Osborn, M., and Weber, K. (1977). *Proc. Natl. Acad. Sci. U.S.A.* **74,** 5613.
Wehland, J., Stockem, W., and Weber, K. (1978). *Exp. Cell Res.* **115,** 451.
Wehland, J., Willingham, M. C., Dickson, R., and Pastan, I. (1981). *Cell* **25,** 105.
Wehland, J., Henkart, M., Klausner, R., and Sandoval, I. V. (1983). *Proc. Natl. Acad. Sci. U.S.A.* **80,** 4286.
Weinbaum, G., and Burger, M. M. (1973). *Nature (London)* **244,** 510.
Weingarten, M. D., Lockwood, A. H., Hwo, S. Y., and Kirschner, M. W. (1975). *Proc. Natl. Acad. Sci. U.S.A.* **72,** 1858.
Weisenberg, R. C., Deery, W. O., and Dickinson, P. J. (1976). *Biochemistry* **15,** 4248.
Werth, D. K., Niedel, J. E., and Pastan, I. (1983). *J. Biol. Chem.* **258,** 11423.
Wessells, N. K., Spooner, B. S., Asch, J. F., Bradley, M. O., Luduena, M. A., Taylor, E. L., Wrenn, J. T., and Yamada, K. M. (1971). *Science* **171,** 135.
Whaley, W. G., Dauwalder, M., and Kephart, J. E. (1972). *Science* **175,** 596.
Wheeler, A. M., and Hartley, M. B. (1975). *Nature (London)* **257,** 66.
Wiche, G. (1985). *Trends Biochem. Sci.* **10,** 67.
Wiche, G., and Baker, M. A. (1982). *Exp. Cell Res.* **138,** 15.
Wiche, G., Krepler, R., Artlieb, U., Pytela, R., and Denk, H. (1983). *J. Cell Biol.* **97,** 887.
Wickner, W. (1980). *Science* **210,** 861.
Wiener, E. C., and Loewenstein, W. R. (1983). *Nature (London)* **305,** 433.
Willingham, M. C., and Pastan, I. (1975). *J. Cell Biol.* **67,** 146.
Willingham, M. C., and Pastan, I. (1980). *Cell* **2,** 67.
Willingham, M, C., Keen, J. H., and Pastan, I. H. (1981a). *Exp. Cell Res.* **132,** 329.
Willingham, M. C., and Pastan, I. (1983). *Proc. Natl. Acad. Sci. U.S.A.* **80,** 5617.
Willingham, M. C., and Pastan, I. (1984). *Intern. Rev. Cytol.* **92,** 51.
Willingham, M. C., and Pastan, I. (1985). *Trends Biochem. Sci.* **10,** 190.
Willingham, M. C., Pastan, I. H., Sahagian, G. G., Jourdian, G. W., and Neufeld, E. F. (1981b). *Proc. Natl. Acad. Sci. U.S.A.* **78,** 6967.
Willingham, M. C., Yamada, S. S., Davies, P. J. A., Rutherford, A. V., Gallo, M. G., and Pastan, I. H. (1981c). *J. Histochem. Cytochem.* **29,** 17.
Willingham, M. C, Rutherford, A. V., Gallo, M. G., Wehland, J., Dickson, R. B., Schlegel, R., and Pastan, I. H. (1981d). *J. Histochem. Cytochem.* **29,** 1003.
Wilson, H. V. (1907). *J. Exp. Zool.* **5,** 245.
Witters, L. A., Friedman, S. A., and Bacon, G. W. (1981). *Proc. Natl. Acad. Sci. U.S.A.* **78,** 3639.
Wolosewick, J. J., and Porter, K. R. (1979). *J. Cell Biol.* **82,** 114.
Yamada, K. M., and Olden, K. (1978). *Nature (London)* **275,** 179.
Yamashiro, D. J., Fluss, S. R., and Maxfield, F. R. (1983). *J. Cell Biol.* **97,** 929.
Yarden, Y., Gabbay, M., and Schlessinger, J. (1981). *Biochim. Biophys. Acta* **674,** 188.
Yotsuyanagi, Y. (1955). *Nature (London)* **176,** 1208.
Zampighi, G., Corless, J. M., and Robertson, J. D. (1980). *J. Cell Biol.* **86,** 190.
Zimmerman, R., Hennig, B., and Neupert, W. (1981). *Eur. J. Biochem.* **116,** 455.
Zwizinski, C., Schleyer, M., and Neupert, W. (1983). *J. Biol. Chem.* **258,** 4071.

## Chapter 4

# THE INTERPHASE NUCLEUS

## I. GENERAL BACKGROUND[1]

DNA is located in the cell nucleus. From this organizational center are derived the orders and messages to build up appropriate proteins in the surrounding cytoplasm. The nucleus receives its food and energy supplies from the cytoplasm. It is not surprising that, as we shall soon see, the wall that separates the nucleus from the cytoplasm (the nuclear membrane or envelope) is so porous that it allows almost free communication between the two major compartments of the cell.

Figures 1 and 2 show the appearance of the cell nucleus under the phase-contrast and electron microscopes. It is surrounded by a double nuclear membrane, which can easily be observed, for instance, in growing oocytes or amebas. Most of the nuclear content consists of chromatin, which may have a loose (euchromatin) or more compact (heterochromatin) organization; condensed chromatin is often attached to the nuclear membrane, as shown in Fig. 2. In the center of the nucleus, refringent and dense nucleoli can easily be seen with the light or phase-contrast microscopes. These structures are bathed in an amorphous nuclear sap, which becomes very abundant in the large nuclei (germinal vesicles) of growing oocytes. As mentioned in Chapter 1, nuclei are generally almost spherical, but may occasionally display a very irregular shape (as in the silk glands of silkworms; Fig. 3). The highly condensed nuclei of the spermatozoa are another example of morphological nuclear differentiation.

There are many methods (mentioned in Chapter 2) for the isolation of whole nuclei (treatment of the cells with citric acid or detergents, differential centrifugation of homogenates made in isotonic or hypertonic sucrose, etc.). Isolation in nonaqueous media (Behrens, 1938; Allfrey *et al.,* 1952) has the advantage that soluble substances (ions, amino acids, nucleotides) are retained in the nuclei; although this technique is tedious and its yield low, one should not forget that nuclei isolated from nonaqueous media lose both RNA and proteins when they are treated with sucrose or citrate (Ficq and Errera, 1955; Kay *et al.,* 1956).

In ''Biochemical Cytology,'' Chapter 4 was entitled ''The Nucleus of the Resting Cell''; hence, we spoke of the resting nucleus. At that time, a resting cell was a nondividing cell. However, work done during the past 25 years has clearly shown that the nucleus of a nondividing cell (thus, of an interphase cell) is not

[1]Reviewed by Hancock and Boulikas (1982).

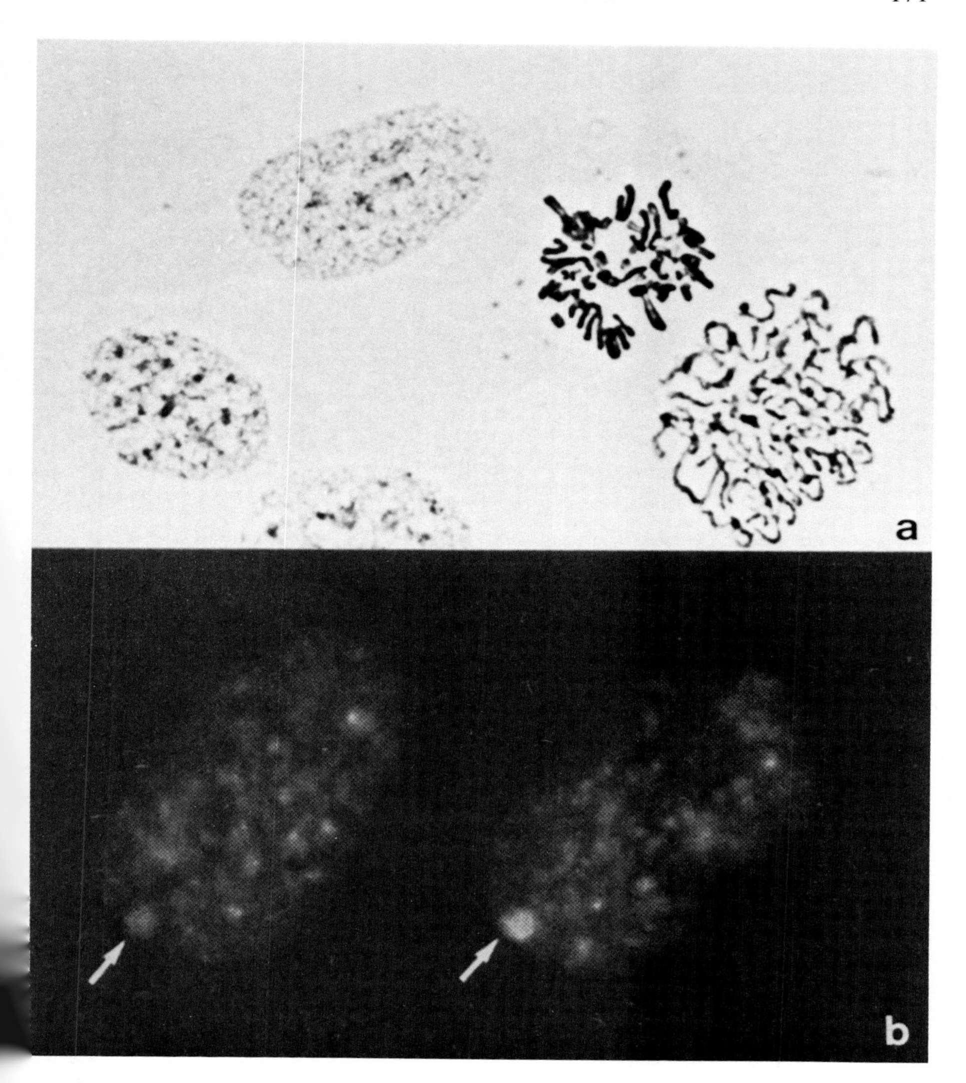


FIG. 1. (a) Cultured human diploid fibroblasts stained with orcein; left: interphase nuclei; right: prophase and metaphase. (b) Interphase nuclei stained with the fluorescent dye atebrin; arrows show the condensed, inactive X chromosome called the "Barr body." [(a) Courtesy of D. Scott. (b) Jonge *et al.* (1982). Reprinted by permission from *Nature* **295,** p. 625. Copyright © 1982 Macmillan Journals Limited.]

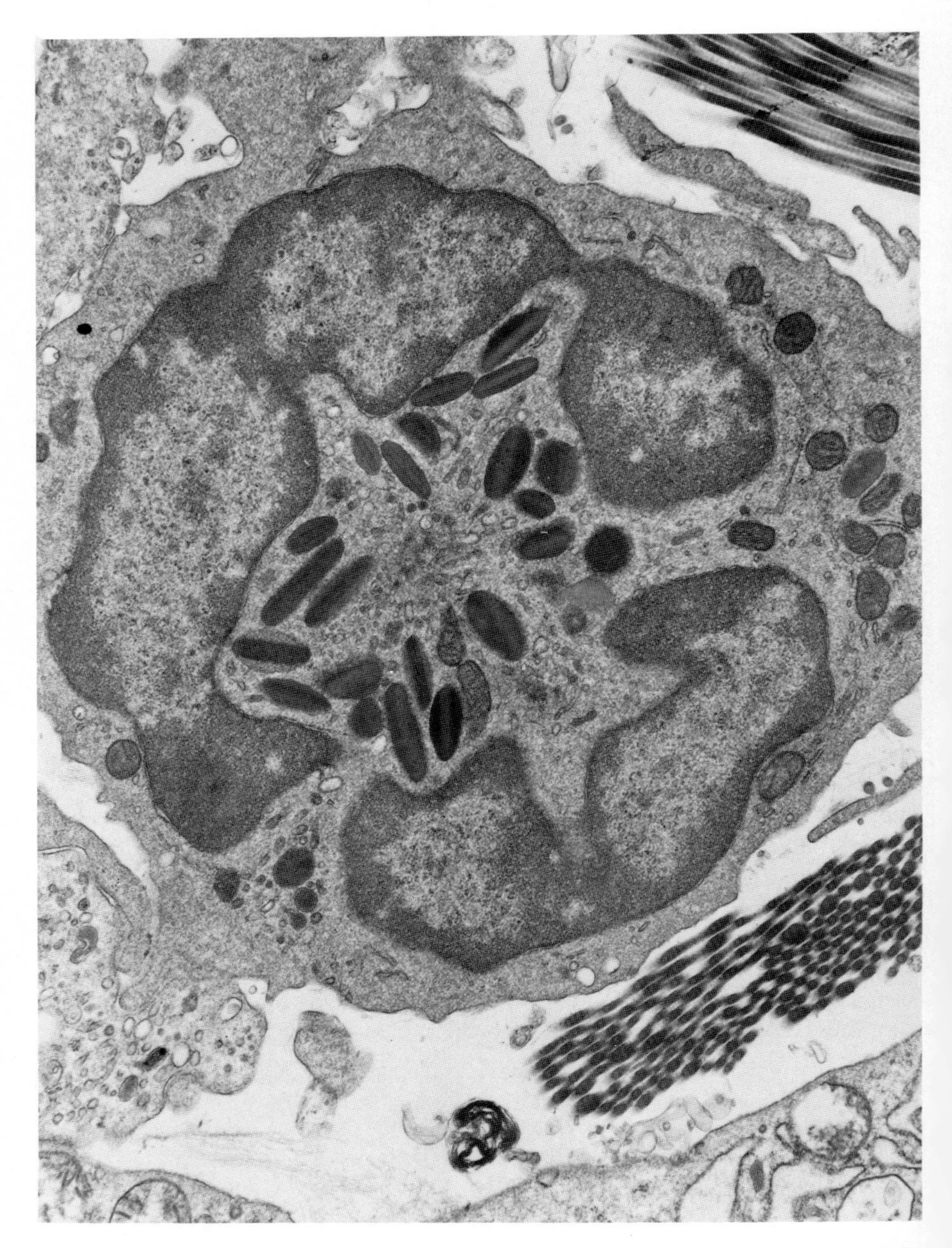

FIG. 2. Electron micrograph of an eosinophil leukocyte; lobulated nucleus with condensed chromatin close to the nuclear membrane. The large granules in the cytoplasm are responsible for eosin staining. Collagen fibers, at the right, surround the leukocyte. [Courtesy of N. Van Lerberghe.]

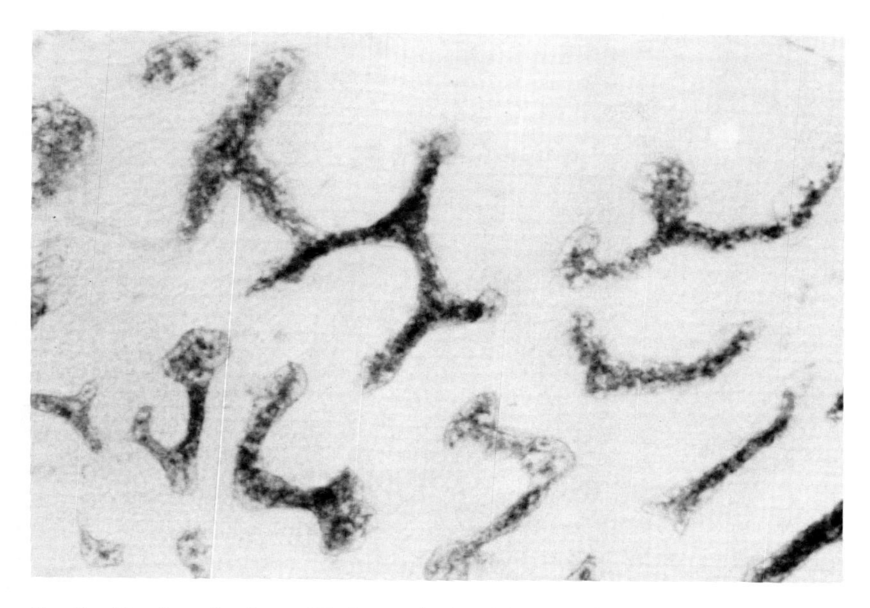

FIG. 3. Very irregular shape of Feulgen-stained nuclei in the silk glands of silkworms. [Brachet (1957).]

resting at all. In general, it is engaged in RNA synthesis (transcription) and, if the cell is preparing for division, in DNA replication. Transcription of chromosomal and nucleolar (ribosomal) genes will be discussed in this chapter; DNA replication, which is closely linked with cell division, will be discussed in Chapter 5.

The main chemical components of isolated nuclei are DNA (mainly in chromatin), RNA (mainly in the nucleolus), basic proteins [histones and high-mobility group (HMG) proteins associated with DNA in chromatin], and a host of nonhistone proteins. Mirsky (1947) called them "residual proteins" because they cannot be extracted, like the nucleohistones, by 1 *M* NaCl. Indeed, when nucleohistones are extracted from isolated nuclei using this method, " ghosts" of the nuclei remain; this was shown first by Jeener in 1946.

## II. THE NUCLEAR MATRIX CONCEPT[2]

Jeener's nuclear ghosts were forgotten for almost 30 years. Interest in them revived when Berezney and Coffey (1974, 1975) showed that isolated liver nuclei, after extraction with salts, nucleases, and detergents, retain a nuclear matrix or scaffold (Fig. 4). It is a fibrillar granular intranuclear network that is

[2]Reviewed by Agutter and Richardson (1980), Hancock (1982), and Bouteille *et al.* (1983).

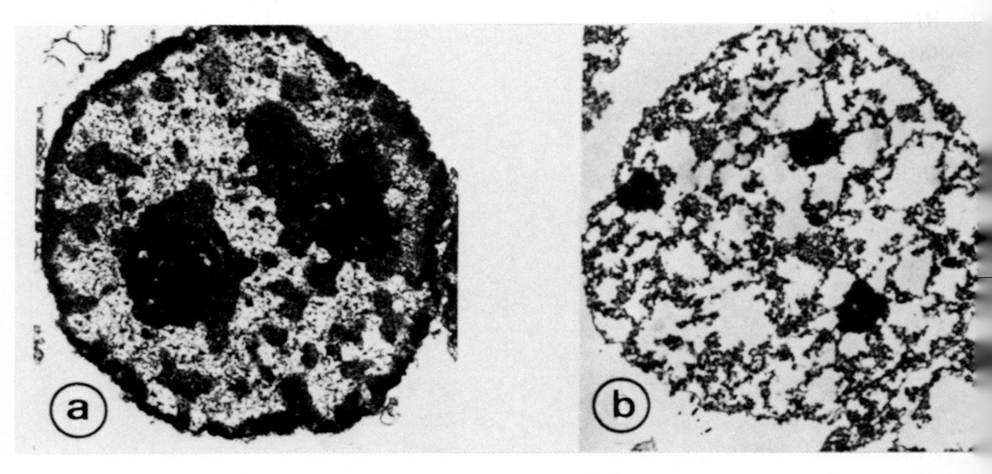


FIG. 4. The nuclear protein matrix. (a) Isolated nucleus seen in the electron microscope. (b) The same nucleus after removal of chromatin, RNA, and phospholipids. [Berezney and Coffey (1974).]

closely linked to the nuclear membrane; the electron microscope shows that the extracted nuclei contain, in addition to remnants of the nuclear membrane, residues of the nucleoli (Berezney and Coffey, 1974; Kaufmann *et al.*, 1981). At first, many cytologists believed that the nuclear matrix is just a preparation artifact due to harsh treatments (with salts, enzymes, or detergents); today the arguments for its reality are overwhelming. As we shall see, it probably plays a major role in DNA replication and transcription, as well as in the maintenance of nuclear morphology. It may even be endowed with contractility, since actin is one of its components (Capco *et al.*, 1982; Nakayasu and Ueda, 1983). However, the presence of myosin in the nuclear matrix has not been reported so far.

The main constituent of the nuclear matrix is a proteinaceous complex, originally called ''matricin'' by Berezney (1980); about 8% of the DNA remains associated with matricin, as well as ribonucleoprotein complexes (RNPs). Three (Berezney, 1980) or four (Peters and Comings, 1980) major proteins can be separated from the matrix protein fibrils; according to a more recent study by Capco *et al.* (1982), the proteins of the nuclear matrix are very different from those of chromatin and of the cytoplasmic cytoskeleton. There is good evidence that DNA is attached in loops, between 20 and 100 kb long, to the nuclear matrix (Berezney and Smith, 1981; Basler *et al.*, 1981). According to Bodnar *et al.* (1983), two proteins are tightly bound to DNA at its nuclear matrix attachment sites. Their role is to anchor DNA sequences to the matrix. It seems that specific DNA sequences are associated with the nuclear matrix anchorage sites (Small *et al.*, 1982; Goldberg *et al.*, 1983). There is also good evidence for a role of the nuclear matrix in DNA transcription. Newly synthesized heterogeneous nuclear

RNA (hnRNA, to be discussed later in this chapter) is hooked by two proteins to the nuclear matrix (van Eckelen and van Verrooÿ, 1981). But the strongest evidence comes from studies on hormone-induced gene transcription. As will be discussed later in more detail (Volume 2, Chapter 2), administration of estrogens to hens induces the synthesis of egg white proteins (ovalbumin, conalbumin, etc.) by the oviduct; this synthesis of a small number of proteins occurs on a large scale and is due to the increased transcription of the corresponding genes in the target (oviduct) cells.

Interestingly, it has been reported that the nuclear matrix in the nuclei of the target tissues possesses specific receptors for steroid hormones (Barrack and Coffey, 1980; Buttyan *et al.*, 1983). It has also been reported that the ovalbumin gene sequences are linked to the nuclear matrix in the hen oviduct cells, but not in brain cells that do not produce egg white proteins (Robinson *et al.*, 1982). The primary transcription products (pre-mRNAs) of two of these genes (encoding ovalbumin and ovomucoid, respectively) are also associated with the nuclear matrix of hen oviduct nuclei. This provides good evidence for the view that the egg protein genes are transcribed on the nuclear matrix of target cell nuclei; the transformation of the mRNA precursors into mature mRNAs (the so-called RNA processing, to be discussed later in this chapter) also takes place on the nuclear matrix (Ciejek *et al.*, 1982, 1983). Finally, Robinson *et al.* (1983) have studied the effects of hormone withdrawal on the preferential association of the ovalbumin genes with the nuclear matrix in hen oviduct cells. Estrogen withdrawal results in the selective dissociation of the ovalbumin gene sequences from the nuclear matrix; these sequences reassociate with the matrix after hormonal restimulation. Similar results have been reported by Jost and Seldran (1984), who studied the chicken vitellogenin II gene. The gene is associated with the nuclear matrix in chicken liver, which responds to estrogen stimulation by vitellogenin synthesis, but not with the nuclear matrix in the oviduct, which does not respond. This association of the gene with the nuclear matrix of the liver cells precedes vitellogenin mRNA synthesis in hormone-stimulated animals. The gene is no longer associated with the nuclear matrix when the hormonal stimulation period ceases. Mirkovitch *et al.* (1984) have recently produced evidence for the presence of specific attachment sites of the DNA loops on the nuclear scaffold or matrix. This would be the case for the genes coding the histones and for the genes encoding proteins that are synthesized when cells are subjected to a short heat treatment (heat shock genes) (Small *et al.*, 1985). In both cases, the attachment sites of the DNA molecule to the nuclear matrix could be identified. All these experiments provide strong evidence for the view that the nuclear matrix plays an important role in both DNA replication and transcription, a conclusion drawn again by Jackson and Cook (1985). In addition, this matrix provides the nucleus with a cytoskeleton distinct from the cytoplasmic cytoskeleton described in the preceding chapter. The concept that an intranuclear network links together nucleoli, neosynthesized DNA and RNA, and the innermost layer of the nucleolar membrane (the so-called pore–lamina complex, which we shall now examine) is quite satisfying and now rests on solid evidential ground.

## III. THE NUCLEAR MEMBRANE (NUCLEAR ENVELOPE)

The first observations on the ultrastructure of the nuclear membrane were made by Callan and Tomlin (1950) on dissected and ruptured germinal vesicles of amphibian oocytes. They found that the nuclear membrane was composed of two sheets: an outer and an inner layer separated by an intermembrane space. They also discovered the presence in the membrane of numerous pores or annuli (Fig. 5). More recent work (reviewed by Franke, 1977; Harris, 1978; Maul, 1977; Green 1982) has confirmed this general picture and has added many important details. It is now generally accepted that, as first proposed by Gay (1955), the nuclear membrane is derived from the endoplasmic reticulum. When the nuclear membrane re-forms at the end of cell division, the telophase chromosomes are surrounded by pieces of endoplasmic membranes, which fuse together to build up a continuous nuclear membrane. Electron microscopy has demonstrated the presence of ribosomes on the cytoplasmic side of the outer nuclear membrane, which presumably possesses the ribophorins needed to bind ribosomes to the endoplasmic reticulum membrane (see Chapter 3, Section III,B). According to Sikstrom *et al.* (1976), nuclear membranes from liver nuclei contain all the classic marker proteins of the endoplasmic reticulum: glucose-6-phosphatase, cytochromes P450 and $b_5$, and 5′-nucleotidase. However, the generally accepted morphological continuity between the endoplasmic reticulum and the nuclear membrane has been questioned by Richardson and Maddy (1980b).

The pores are not simple holes boring the nuclear membrane; on the contrary, they are very complex structures. As shown in Fig. 6 (Roberts and Northcote, 1970; Franke, 1977; Unwin and Milligan, 1982), they are octagonal and surrounded by a circular annulus; pores and annuli together form the "nuclear pore complexes." The diameter of the nuclear pores varies between 50 and 80 nm, according to the cell type. They are partially closed by a central granule that may correspond, as suggested by Franke (1977), to ribonucleoprotein particles moving out of the nucleus into the cytoplasm. However, this interpretation remains more a likely hypothesis than an actual fact. The annuli that reinforce the pores are made up of eight granules (about 15 nm) present on both the nuclear and cytoplasmic faces of the nuclear membrane. This description of the nuclear pore complexes is probably a gross oversimplification. A reinvestigation of the nuclear membrane of amphibian oocytes (the material first used by Callan and Tomlin in 1950) has shown that the pores are often clustered in triplets and that the pore complex is formed from 40 subunits measuring 30 nm (Schatten and Thoman, 1978; Unwin and Milligan 1982). The number of pores varies greatly from cell to cell and is related to the intensity of RNA synthesis in the nucleus.

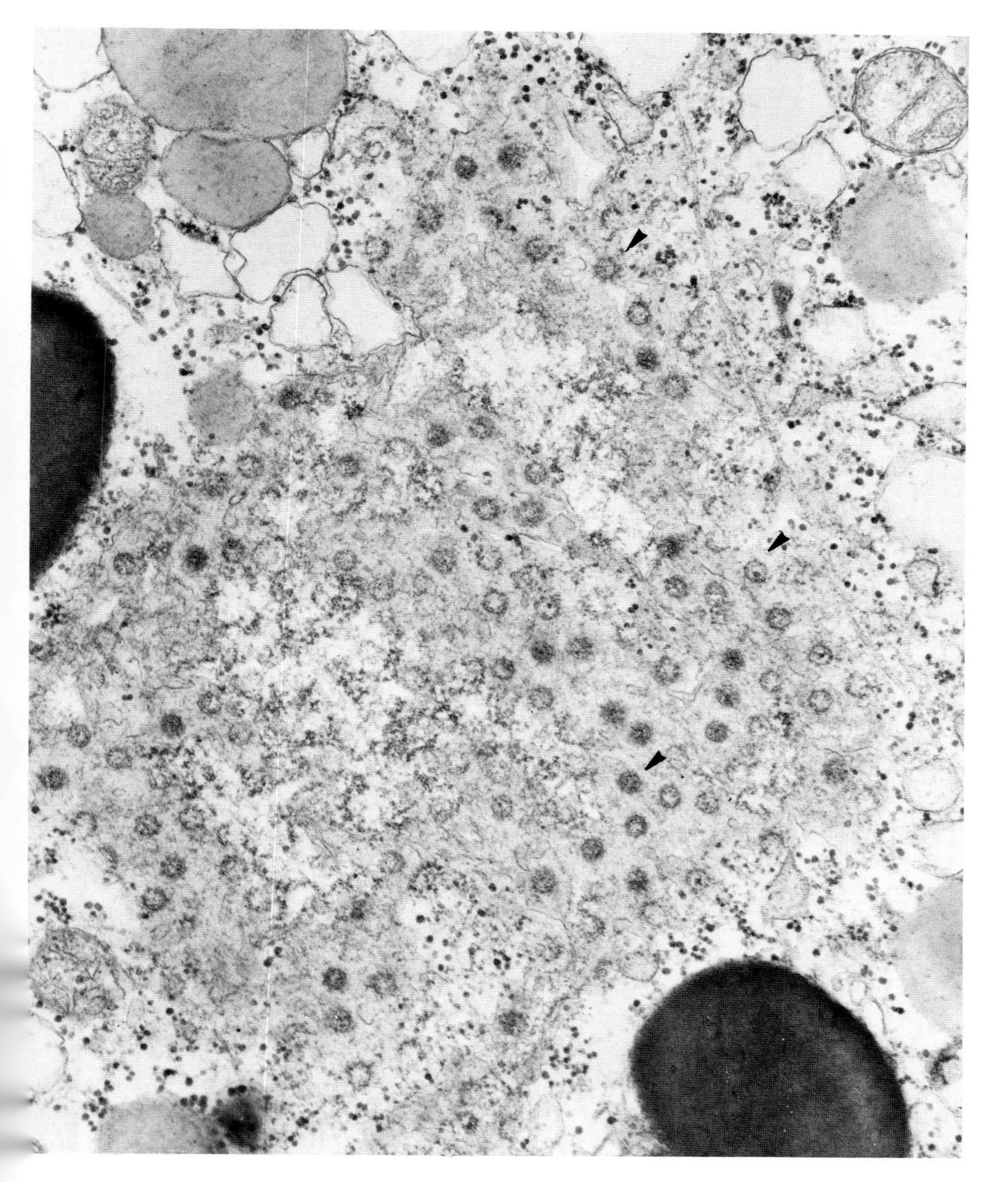

FIG. 5. Tangential section of the nuclear envelope in a *Pleurodeles* cleaving egg. Arrowheads show the annuli. [Courtesy of M. Geuskens.]

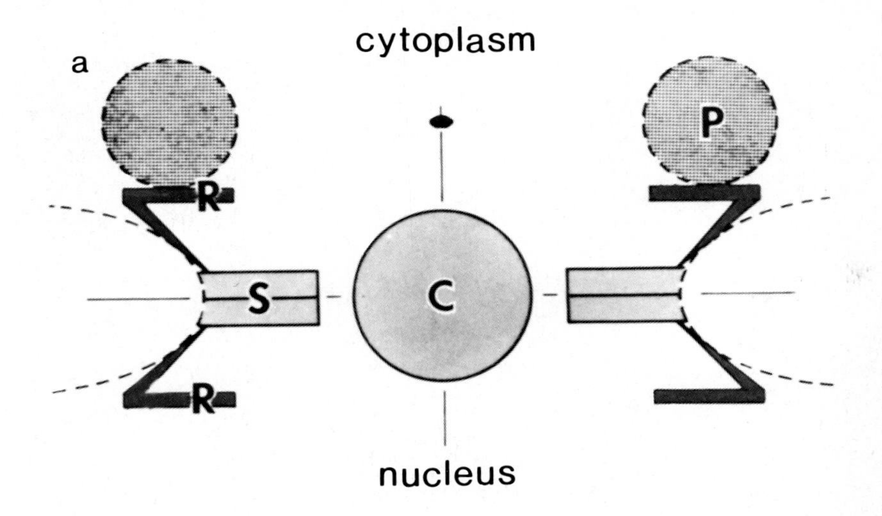


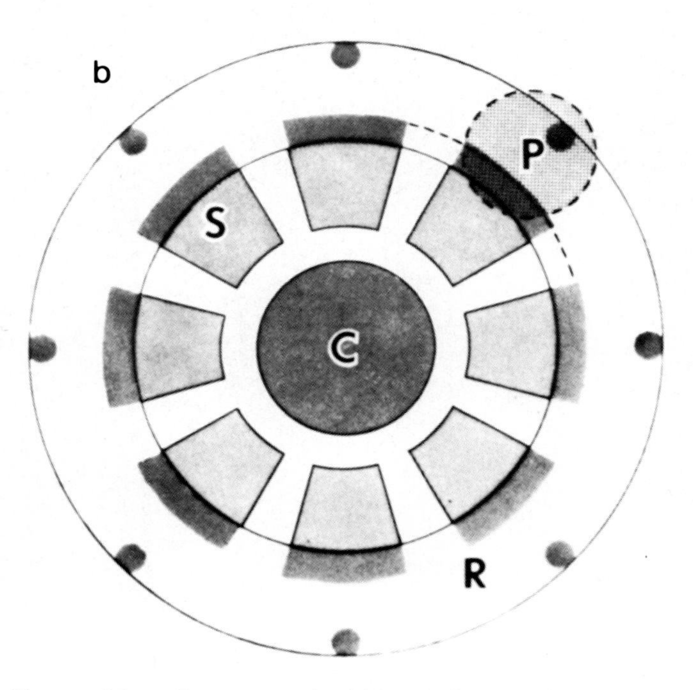


FIG. 6. Diagram of the nuclear pore complex (a) in central cross section and (b) in projection down the octad axis. C, central granule (plug); S, spokes; R, rings. P, octagonally arranged particles resembling ribosomes. [Unwin and Milligan (1982). Reproduced from *The Journal of Cell Biology,* 1982, **93,** p. 75, by copyright permission of The Rockefeller University Press.]

For instance, according to Scheer (1973), the total number of pores in *Xenopus* oocytes, where (as we shall see in Volume 2, Chapter 2) RNA synthesis is particularly active, reaches 40 million per nucleus by the end of oogenesis (about 60 pores/$\mu m^2$). In contrast, the nuclei of red blood cells of the same species, in which RNA synthesis is negligible, have only 150–300 pores (about 3 pores/$\mu m^2$). The nuclear membrane of spermatozoa, which do not synthesize RNA, also has very few nuclear pores.

Amphibian oocytes have also played a part in the discovery of another component of the nuclear envelope, the lamina. If germinal vesicles are isolated in the presence of a detergent, the electron microscope reveals a "fibrillar reticulum" linked to a glycoprotein lamella, the lamina (Scheer *et al.*, 1976) (Fig. 7). In association with the interior (nuclear) face of the nuclear membrane, it forms a 30-nm-thick fibrous cortex (Schatten and Thoman, 1978). In the germinal vesicle of *Spisula* (clam) oocytes, this fibrous lamina holds together the nuclear pores (60 $nm^2$), according to Maul and Avdalović (1980).

These nuclear pore–lamina complexes are not specific for oocyte nuclei, but are probably present in all cells. This ubiquity has allowed progress in their molecular characterization. As early as 1975, Aaronson and Blobel isolated nuclear pores and showed that they were composed of protein. According to Krohne *et al.* (1978), the nuclear membrane of *Xenopus* oocytes is composed of 10 major and 15 minor proteins, but only 2 proteins remain in the pore complexes after treatment with the detergent Triton X-100. In a more recent paper, Krohne *et al.* (1981) concluded that there are two major proteins in the lamina of *Xenopus* red blood cells, three in liver cells, and only one ($M_r$ 62,000) in oocyte nuclei. According to a definition proposed by Gerace and Blobel (1980) in a paper on the subject, the nuclear envelope lamina is a supramolecular assembly of proteins associated with the internal face of the nuclear membrane, which together form the pore–lamina complex; three proteins, the A, B, and C lamins, are its main constituents. Interestingly, during mitosis, the A and C lamins become soluble, while lamin B remains attached to dispersed fragments of the nuclear membrane. These authors also suggest that depolymerization of the nuclear membrane lamina at prophase results from the phosphorylation of the lamins. In agreement with this suggestion, Lam and Kasper (1979) showed that the nuclear membrane possesses an endogenous protein kinase that is inhibited by $Ca^{2+}$ and activated by $Mg^{2+}$ and that phosphorylates one of the proteins present in the nuclear membrane pores. Thus, there are reasons to believe that nuclear membrane breakdown at mitosis and meiosis is controlled by a protein factor that is inactivated by $Ca^{2+}$ and activated by $Mg^{2+}$ [the so-called maturation promoting factor (MPF), which will be discussed in Chapter 5 and Volume 2, Chapter 2]. That lamins A, B, and C are the predominating proteins in the nuclear lamina fraction has been confirmed by Shelton *et al.* (1980). However, it is still too early for all workers to agree fully, since each has used different study

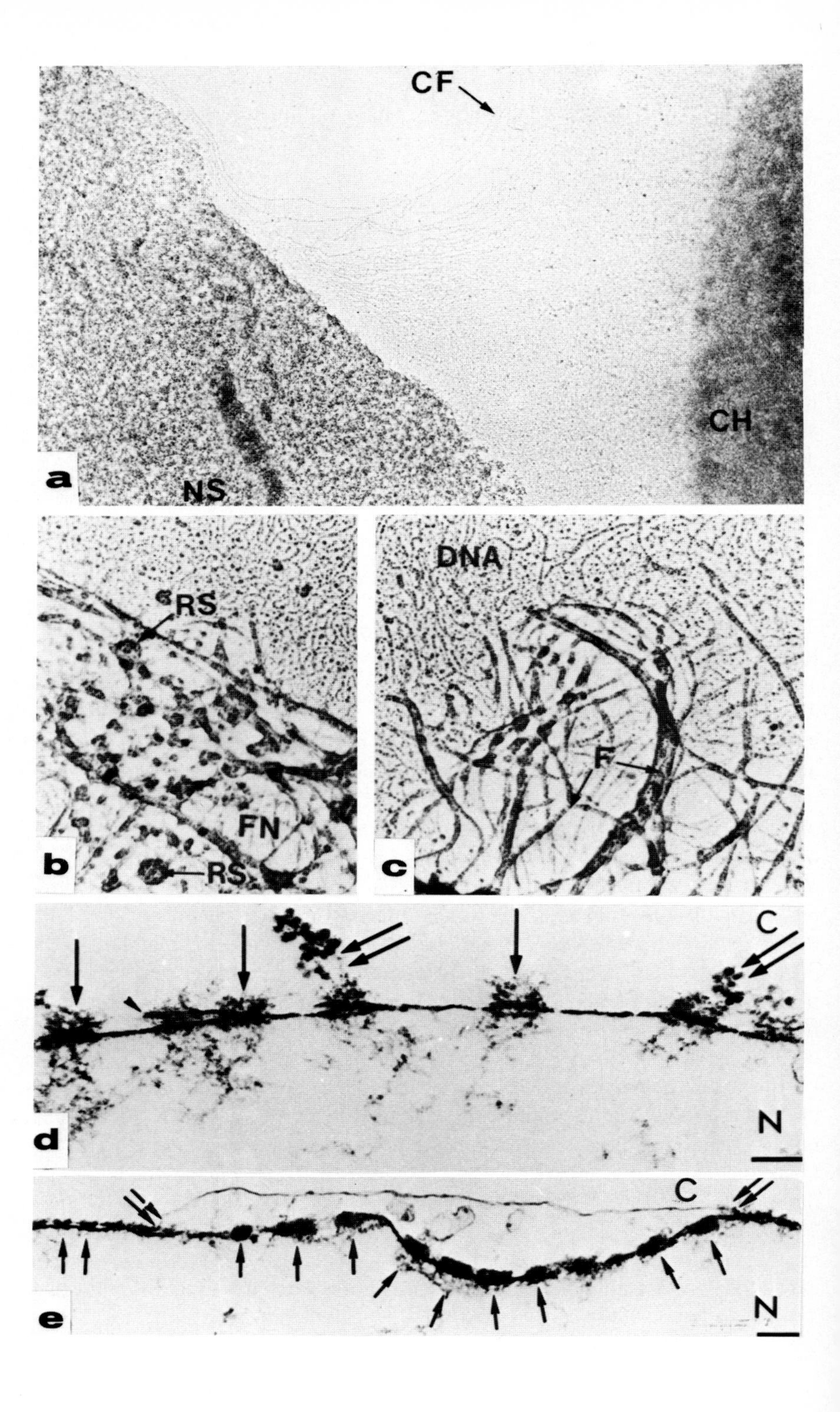
CF
CH
a
NS
RS
FN
b
RS
DNA
F
c
C
N
d
C
e
N

materials. According to Richardson and Maddy (1980a), the pore–lamina complex is composed of 93.6% protein, 6% RNA (of high complexity, according to Clawson and Smuckler, 1980), and only 0.4% lipids. They detected 2 major proteins and 10 minor ones in the nuclear pore complex and only 1 in the fibrous lamina present on the inner phase of the nuclear membrane. In germinal vesicles of *Spisula,* Maul and Avdalović (1980) found eight proteins (three of them phosphorylated) in the fibrous lamina–pore complex and only one in the nuclear membrane itself. In the same material, fluorescent antibodies directed against lamin B stained mainly the nuclear membrane; but the nuclear matrix was also stained, reinforcing the idea that there is a morphological and molecular continuity between the two (Baglia and Maul, 1983). Immunofluorescence has shown that, in the nuclear envelope, only the nuclear lamina (thus, the nucleoplasmic side of the membrane) contains lamins A, B, and C (which have common antigenic determinants); the nuclear pores remain unstained. In dividing cells, after nuclear membrane breakdown, all one can see is a dispersed labeling of the cytoplasm (Burke *et al.*, 1983). There is growing evidence for the view that a glycoprotein (Gerace *et al.*, 1982) and lamin B (Lebel and Raymond, 1984) anchor the pore–lamina complex to the inner nuclear envelope membrane; both are integral proteins of the inner nuclear membrane, and they bind it to the peripheral lamina.

It has been suggested that the fibrous lamina material is closely associated with the outermost layer of peripheral chromatin, which would constitute the nuclear shell (Bouvier *et al.*, 1980; Hubert *et al.*, 1981). The evidence is based largely on ultrastructural studies, but isolation of a fraction containing 1% of the total DNA associated with the fibrous lamina has been described by Hubert *et al.* (1981). This finding again raises the old—and still unanswered—question of the possible relationships between the nuclear membrane and DNA replication. In this area, there are many contradictory reports in the literature. O'Brien *et al.* (1972) claim that DNA synthesis in HeLa cells occurs in contact with the nuclear membrane. Wise and Prescott (1973), feel that there is no relationship between the two. There are also contradictions regarding the localization of DNA polymerase α, an enzyme directly involved in DNA replication. Herzberg *et al.* (1981) are of the opinion that it is associated with the nuclear membrane, while according to Brown *et al.* (1981), who worked with nucleate and anucleate

---

FIG. 7. (a) Nuclear "shell" (cortical chromatin associated with the pore–lamina complex). The dissociated nuclear shell is attached to its internal chromatin by numerous chromatin fibrils (CF). (b) Ring-like structures (RS) with a central dot, interconnected by a fibrous network (FN). (c) Long flexuous fibers (F) associated with naked DNA strands (Bouvier *et al.*, 1980). (d,e) Ultrathin sections of nuclear envelopes isolated from newt oocytes. The outer nuclear membrane has been partially removed, especially in the interpore regions [resistant outer membrane portions are shown by short arrowheads in (d)]. It may be detached over larger distances (between the double arrows in (e) from the inner membrane and the pore complex material. The arrows show pore complexes, and the double arrows in (d) show cytoplasmic polyribosomes). [Part (c) from Bouvier *et al.* (1980); part (d) from Scheer *et al.* (1976).]

fragments of cells, the major part of DNA polymerase α is in the cytoplasm surrounding the nucleus; thus, the nucleus itself would contain only small amounts of the enzyme. Perhaps the aforementioned studies on the nuclear matrix (of which the pore–lamina complex is a part) will provide an explanation of a phenomenon that is still poorly understood, but is important for our understanding of the cellular mechanisms of DNA replication.

We have seen that the nuclear membrane contains, besides the enzymes characteristic of the endoplasmic reticulum, an endogenous protein kinase which could be involved in the breakdown of the nuclear membrane during cell division. It possesses another enzyme that plays an important role in nucleocytoplasmic exchanges, in particular RNA transport into the cytoplasm (review by Agutter, 1984). This enzyme is a nucleoside triphosphatase (NTPase), for which ATP is one of the substrates (Agutter *et al.*, 1976; Clawson *et al.*, 1980). The efflux of ribonucleoprotein particles from isolated nuclei is strongly stimulated by addition of ATP; this ATP-dependent efflux is regulated by cytoplasmic proteins that exert either a positive or negative control on the process. Translocation of ribonucleoprotein particles, including ribosomes, through the nuclear pores requires hydrolysis of NTPs by the nuclear envelope NTPase. However, according to Clawson *et al.* (1980), NTPase is abundant in the nuclear membrane of liver nuclei, but not in the pore complexes. More recently, Clawson *et al.* (1984) have confirmed that the enzyme is associated with the nucleocytoplasmic face of the nuclear membrane and the adjacent heterochromatin but is absent from the nuclear pores. It is unlikely that the nuclear membrane possesses its own system for energy production. Jarasch and Franke (1974) demonstrated that the cytochrome oxidase activity which seemed to be associated with the nuclear membrane is due to mitochondrial contamination. However, this conclusion has been denied by Zbarsky (1978) who has reviewed the enzymatic activities of the nuclear envelope.

The permeability of the nuclear membrane will be discussed here and again in more detail in Volume 2, Chapter 2, because most of our knowledge of this important aspect of nucleocytoplasmic interactions stems from studies on amphibian oocytes. The existence of an intensive two-way traffic between the nucleus and the cytoplasm—presumably through the nuclear membrane–pore complexes—was already well established when Gurdon (1970) reviewed the subject of nuclear membrane permeability. From the experiments of Gurdon, Feldherr, Bonner, De Robertis, and others on amphibian oocytes, the following general picture emerged. The RNAs synthesized in the germinal vesicle move unidirectionally from the nucleus to the cytoplasm, provided that their nuclear precursors have been properly processed (see the next section). According to Scheer (1973) this traffic is intensive, since 2.6 molecules of rRNA (with a molecular weight higher than 1 million) move out of the amphibian germinal vesicle every minute through each of its 40 million pores. This process can be visualized only if one assumes that the giant rRNA molecules—which are associated with proteins—undergo conformational changes (unfolding) during their passage through the pores (Franke, 1977; Clawson and Smuckler, 1982). At the same time, proteins synthesized in the cytoplasm move into the nucleus; this has best been shown by the experiments of Gurdon (1970), Bonner (1975), Feldherr (1975), De Robertis *et al.* (1978), and others, who injected labeled proteins into the cytoplasm of *Xenopus* oocytes, followed by autoradiography. The general outcome of these experiments was that the intake of proteins by the nucleus did not depend on size or electrical charge; some proteins, but not others, accumulate in the nucleus. This is particularly true of the proteins that, under normal conditions, are accumulated in the nucleus. These karyophilic proteins (which may have a molecular weight as high as 120,000) might bear a signal that allows their penetration and accumulation in the nucleus. However, this selective uptake of proteins by the germinal vesicle is not due to the permeability of the nuclear membrane, since it takes place even when a hole has been bored in this membrane with a fine needle (Feldherr and Ogburn, 1980). Similar experiments on *Amoeba proteus* have shown that, as in amphibian oocytes, the nuclear membrane is freely permeable to proteins in both directions; more than 50% of the proteins injected in the cytoplasm accumulate in the nucleus (Goldstein and Ko, 1981). These results cannot be explained unless one assumes the existence of specific high-affinity binding sites for proteins in the nucleus. Discrimination in the uptake of soluble proteins is displayed even by isolated HeLa nuclei, where accumulation of nuclear proteins does not require energy (Cox, 1982).

However, it is not certain that the results obtained with oocytes and amebas can be generalized to the somatic cells of higher organisms. Reynolds and Tedeschi (1984) injected a number of substances into the nuclei of living mammalian cells and found that the nuclear membrane is permeable to molecules that have radii between 2.4 and 2.8 nm; larger molecules are retained in the injected nuclei. The conclusion of these experiments is that the pore radius lies between 3.4 and 6.5 nm.

## IV. CHROMATIN

The regulation of gene activity and expression begins with chromosomal DNA and associated proteins (histones and non-histone chromosomal proteins) and ends in the cytoplasm, where proteins coded by the structural DNA genes are synthesized on polyribosomes. The subject of chromatin is so vast that it has to be treated here in only a superficial way; the finer details are left to textbooks and review papers on molecular biology. A complete book on DNA chromatin and chromosomes has been published by Bradbury *et al.* (1981). Following an overview of chromatin localization in the interphase nucleus, we shall deal with DNA, with basic chromosomal proteins (histones in particular), with the molecular organization of nucleohistones, and with non-histone chromosomal proteins. Our discussion will end with DNA transcription and the processing of the

RNA transcripts. DNA replication, although it takes place in the interphase nucleus, is so closely linked to cell division that it must be left for the next chapter.

## A. Microscopic Structure of Chromatin

Long ago, cytologists noticed that chromatin consists of two distinct parts: euchromatin and heterochromatin (Heitz, 1934). Euchromatin, which is believed to contain the major genes, has a less compact structure than the strongly condensed heterochromatin. Figure 2 shows that at the EM level, chromatin is more condensed in some parts of the nucleus (in general, immediately under the nuclear membrane and in the vicinity of the nucleoli) than in others, where a much looser structure is observed. The heterochromatin regions of resting nuclei, which are the strongly Feulgen-positive regions of the chromatin "network," are often called "chromocenters." The base composition of DNA in such strongly heterochromatic regions usually differs from that of euchromatin. These regions are often composed of tandem repetitions of nearly identical stretches of DNA, while the base sequence of euchromatin DNA presents a much greater sequence heterogeneity. Typical heterochromatic structures are the nucleolar organizers composed of highly repeated genes coding for the ribosomal RNAs; the kinetochores (centromeres) of mitotic chromosomes; the chromocenters, which link together the giant chromosomes of *Drosophila* salivary gland nuclei (see Chapter 5); the Y chromosomes; and the so-called Barr bodies (Fig. 1b) present in the nuclei of females from many mammalian species, including man. These strongly condensed Barr bodies correspond to the genetically inactive X chromosome, while the other member of the XX pair is genetically active (Lyons phenomenon) and is not visible in Feulgen-stained interphase nuclei.

Except for the ribosomal genes present in the nucleolar organizers, the heterochromatic regions of the nucleus are genetically inactive. Autoradiography at the EM level shows that they are rarely the sites of RNA synthesis; the latter is, of course, absolutely required for gene expression. Genetic analysis has shown that heterochromatic segments of chromosomes contain few or no major genes; however, translocation of an active, euchromatic gene close to heterochromatin may strongly affect the expression of this gene (position effect). It is thus possible that heterochromatin plays an important role in the complex regulation of gene activity. This vitally important phenomenon, which remains poorly understood in eukaryotes and which will be discussed in Volume 2, Chapter 2, has been reviewed by D. D. Brown (1981). Another general characteristic of heterochromatin is that its DNA replicates late in those cells preparing for division. Thus, heterochromatin DNA is less active, compared to euchromatin DNA, in both transcription and replication. Variable methods for separating loose and condensed chromatins have shown that fractions enriched in the former are also enriched in DNA-coding sequences and non-histone proteins (Murphy *et al.*, 1973; Warnecke *et al.*, 1974; Gosden and Mitchell, 1975; Georgieva *et al.*, 1981). Despite all of the differences between euchromatin and heterochromatin, their molecular structure is the same. Both are composed of DNA–histone complexes, called "nucleosomes," which will be discussed later in this chapter. For this reason, the interest in the classic distinction between "facultative" and "constitutive" heterochromatins has greatly diminished. Facultative heterochromatin is simply euchromatin in a more condensed form because the conformation of the chromatin fibers has become more compact. Constitutive heterochromatin is composed of highly repeated DNA sequences (ribosomal DNA, "satellite" DNA); such repeated sequences are relatively easy to isolate, and their localization in heterochromatin can be easily demonstrated by *in situ* hybridization (Gall *et al.*, 1971; Barsacchi and Gall, 1972; Hennig, 1972).

When chromatin of isolated nuclei is spread under the electron microscope, it is dispersed in a network of 25- to 30-nm thick fibrils (Fig. 8). These fibrils form many loops that attach to the nuclear matrix (or cage). According to Lepault *et al.* (1980), the major part of chromatin is associated with the inner nuclear membrane, where it forms a nuclear cortex. As we have seen, attachment of DNA loops to the nuclear cage, which remains intact after extraction of cells with detergents and consists essentially of RNA and proteins, seems to be required for transcription. In other words, genes become active when their DNA sequences bind to the cage.

The ultrastructure of chromatin in conventional EM thin sections will be briefly examined. At higher magnifications than those in Fig. 2, the fibrous structure of chromatin is very apparent (Fig. 9); the average diameter of the fibers (about 8–9 nm) is similar in sectioned material and in spread chromatin. However, there are marked differences in thickness among different fibers. Work done in the laboratory of W. Bernhard has shown that electron-dense granules and fibrils can be seen around and between the chromatin threads. These are called the "perichromatin fibrils," the "perichromatin granules," and the "interchromatin fibrils" (Fig. 10; reviewed by Fakan and Puvion, 1980; Puvion-Dutilleul, 1983). Cytochemical studies at the EM level have shown that these fibrils are not composed of DNA, but of ribonucleoproteins. The existing data are interpreted in the following way by Puvion and his colleagues. Perichromatin fibrils contain the first products of RNA synthesis, the so-called heterogeneous nuclear RNA (hnRNA), associated with proteins; perichromatin granules are believed to be storage forms of both rRNA precursors and nascent hnRNA and result from the winding of nascent ribonucleoproteins. Finally, the perichromatin granules are transformed into interchromatin fibrils, which migrate into the interchromatin space when chromatin undergoes decondensation. This attempt to visualize under the electron microscope the complex process of DNA transcription is certainly interesting, but will remain hypothetical as long as

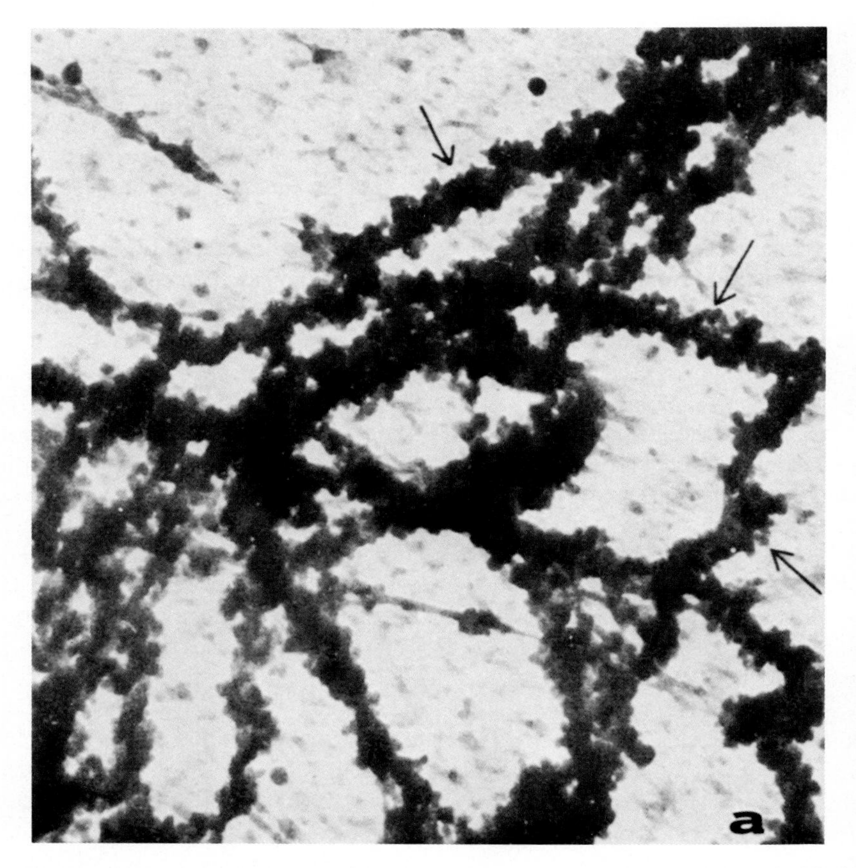


FIG. 8. Spread chromatin seen under the electron microscope. (a) Network observed by Ris 25 years ago (in Brachet, 1957, ''Biochemical Cytology''). (b) A spread metaphase chromosome shows thin chromatin fibers emerging from a coarser, partially spread network. [Courtesy of D. De Kegel.]

the interpretation of the images seen on the micrographs is not based on more direct experimentation.

## B. CHEMICAL COMPOSITION

### *1. DNA*

When the author was writing ''Biochemical Cytology'' in 1957, it was believed that the DNA content of all cells from the same species was exactly the same (except during DNA replication) and that DNA was metabolically stable. The author never had great sympathy for these views. Therefore, he concluded (Brachet, 1957, pp. 88, 89) that DNA shows a tendency to remain constant per

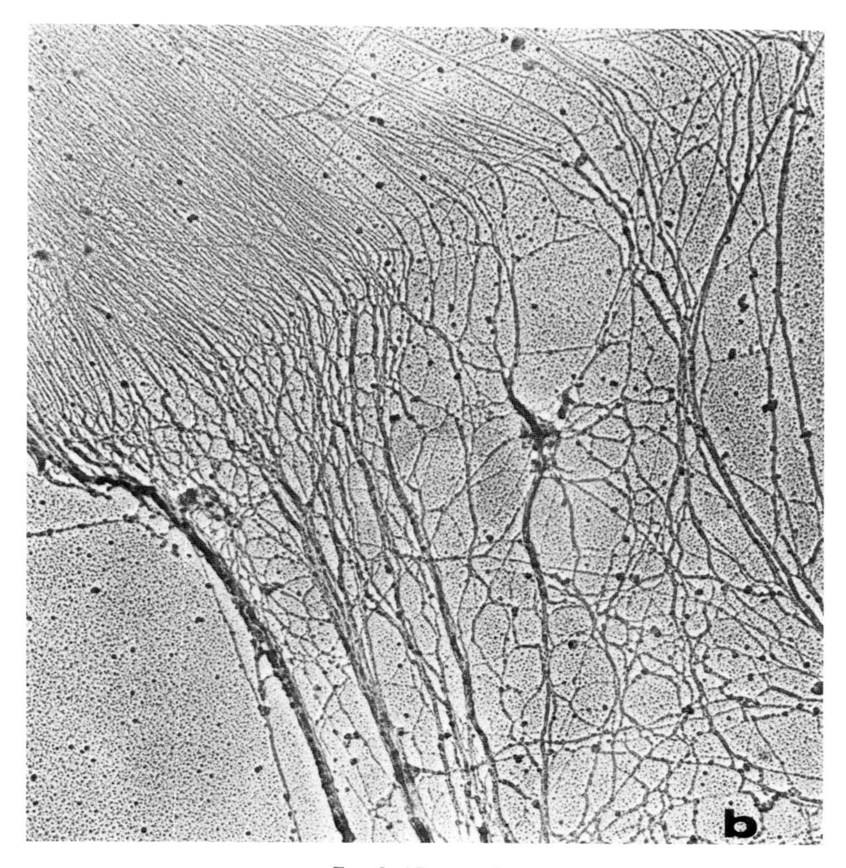


FIG. 8. (*Continued*)

nucleus, but its constancy is not absolute; metabolically, DNA is a very stable substance, but again, stability is not absolute. These conclusions were somewhat reinforced by the finding that the DNA content per nucleus (which could be determined by Feulgen cytophotometry or chemical analysis of isolated nuclei) varies greatly from one species to another. For instance, among the amphibians, newts and salamanders have almost 10 times more DNA per nucleus than frogs (this is the so-called C-paradox, C being the DNA complement of the species). Since urodeles are unlikely to have 10 times more genes than anurans (and indeed, they do not), it is clear that a very large amount of nuclear DNA is not made of coding genes. In fact, the major part of DNA is made of what has been called "selfish" (Orgel and Crick, 1980) or even "junk" DNA because it plays no genetic role. If this is so, why should total DNA remain absolutely constant

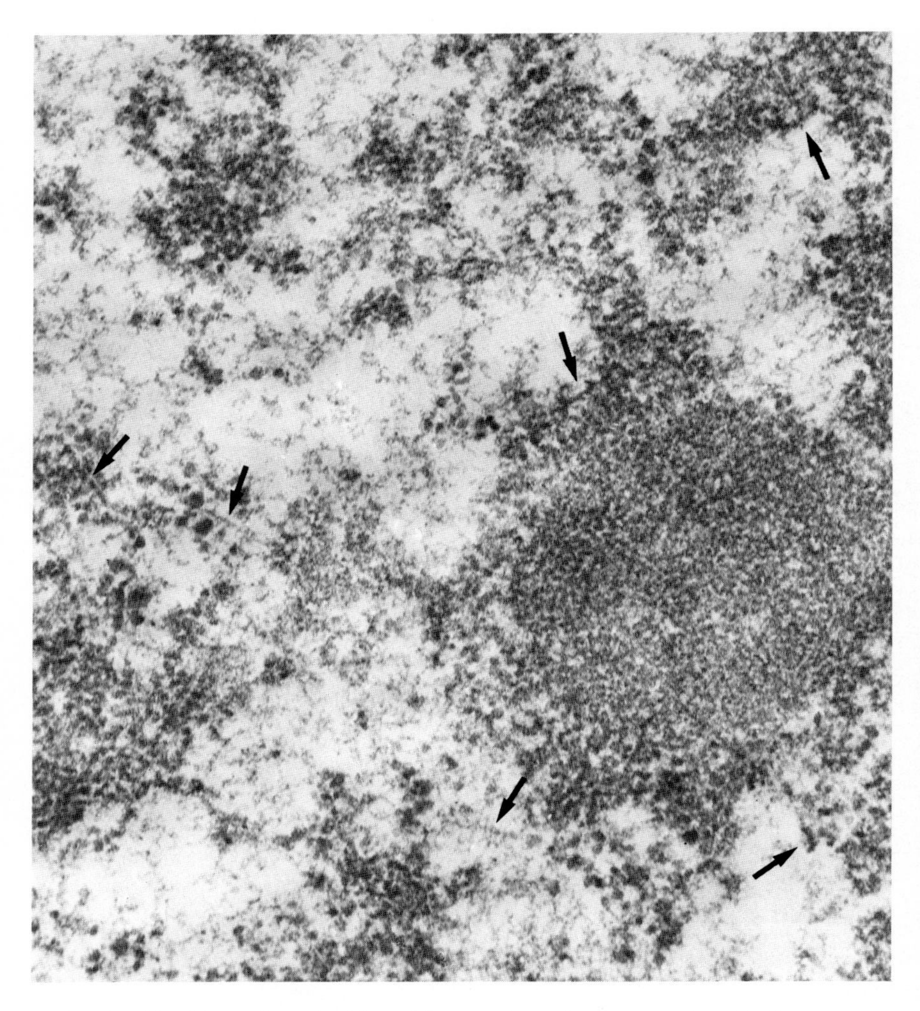

FIG. 9. Ultrathin section of a *Pleurodeles* gastrula nucleus; arrows show fibrillar structures. [Courtesy of M. Geuskens.]

per nucleus and metabolically completely stable? That such absolute stability was unlikely was shown by the studies of Ficq and Pavan (1957). Using autoradiography, they localized portions of a chromosome that could be the site of DNA replication, although the bulk of this chromosome is apparently made of stable DNA. Methodology currently available allows the isolation of pure genes and precise analysis of their chemical composition. These methods have demonstrated that DNA is far more diverse than it was believed to be a few years ago.

As we shall see, DNA not only undergoes point mutations, but other changes as well (deletions, insertions, transpositions, duplications, gene conversion, amplification or loss of genes, etc.). These investigations into the finer details of DNA have greatly increased our interest in this most fascinating and important molecule.

The double helix structure of DNA proposed by Watson and Crick in 1953 is too well known to be discussed here; this famous arrangement (Fig. 11) has the

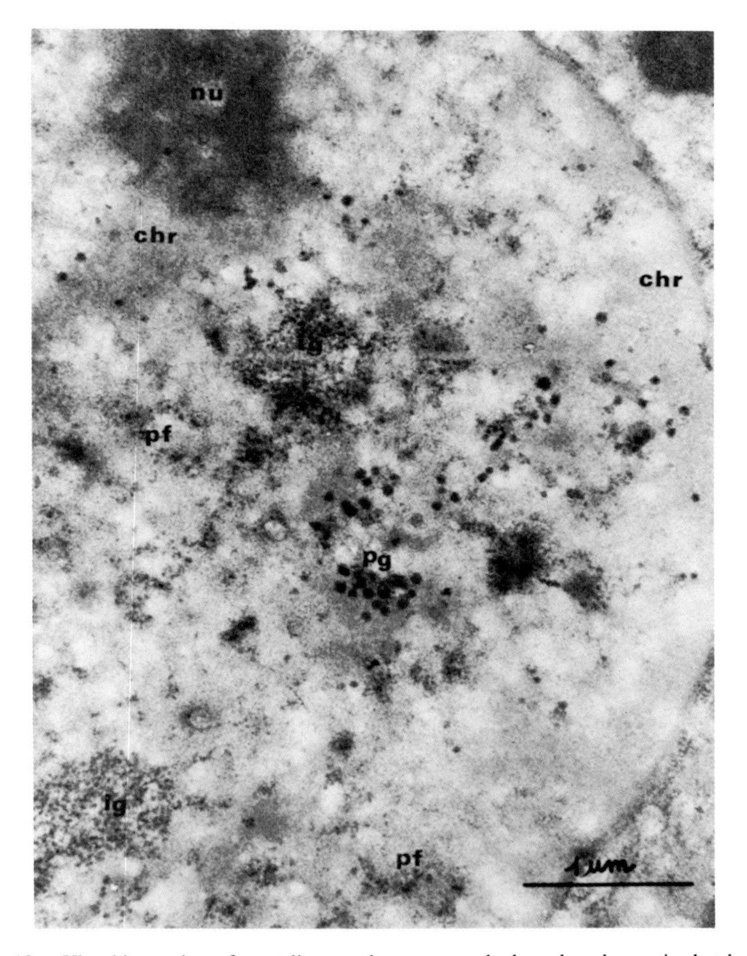


FIG. 10. Ultrathin section of a rat liver nucleus. nu, nucleolus; chr, chromatin that has been bleached by the technique used; ig, interchromatin granules; pg, perichromatin granules; pf, perichromatin fibers. [Puvion and Bernhard (1975). Reproduced from The Journal of Cell Biology, 1975, **67,** p. 204, by copyright permission of The Rockefeller University Press.]

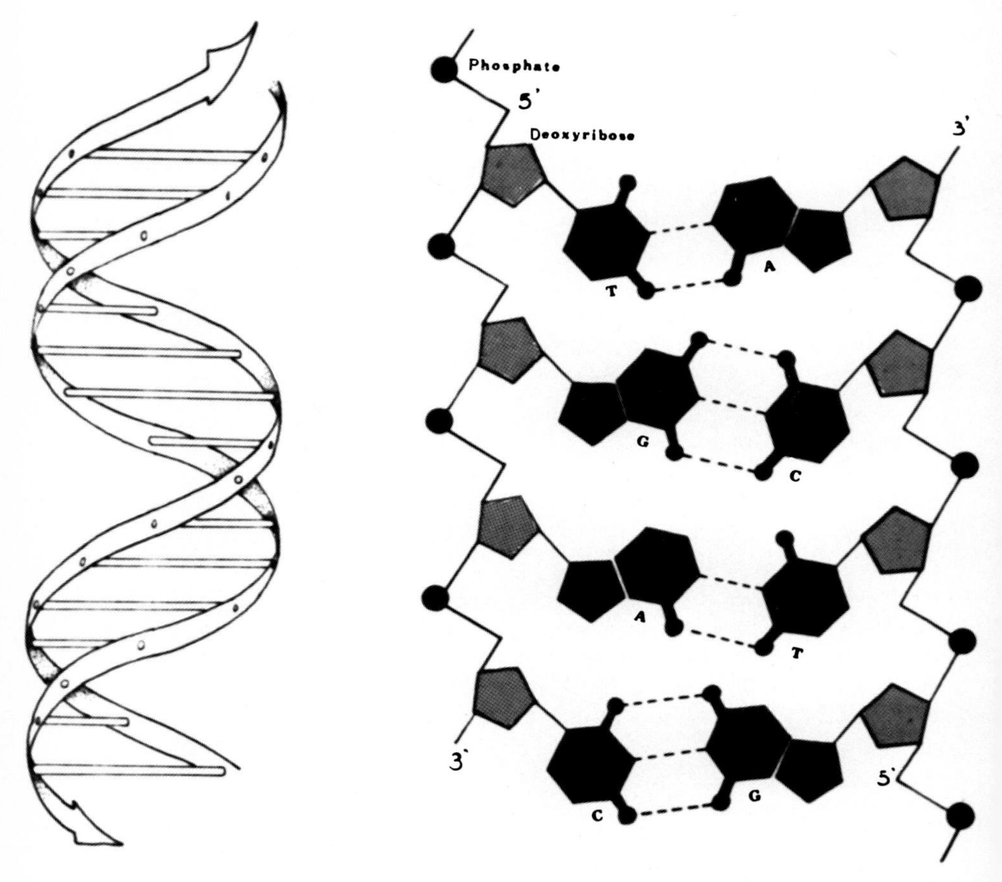


FIG. 11. Left: the DNA double helix, as proposed by Watson and Crick in 1953. Right: diagram of a segment of the double helix; the bases adenine (A) and thymine (T), on the one hand, and guanine (G) and cytosine (C), on the other, are linked by hydrogen bonds.

admirable simplicity of genius. The complementary structure of the two DNA chains beautifully explains DNA replication if one assumes that if the two chains unwind and separate, each serves as a template for the formation of its complement. The model also explains the possibility of transcription of one of the DNA strands into complementary RNA. Any error, such as the introduction of an abnormal base, will lead to mismatching and mutation. Last, but certainly not least, the Watson-Crick model has an exceptional merit: all of its predictions have been fulfilled by experimentation.

The now classic molecular structure of DNA is beyond the scope of this book. We must restrict ourselves to some questions, related to this structure, that are important for cell biologists. The very structure of the DNA molecule immediately raises a problem: the double helix is only 2 nm thick, but it is very long; 1 pg of DNA corresponds to 31 cm of DNA, and it is easy to calculate from cytophotometry data that one of our diploid cells contains about 2 m of DNA. This raises the major problem of how a huge molecule is packed in a limited volume; this problem is even more acute for mitotic chromosomes, in which compaction of DNA is about 7000-fold. It is relatively easy to imagine a supercoiling of the DNA molecules in order to form the 200-Å fibers seen under the microscope, but this is still insufficient to accommodate all of the DNA in chromatin. We shall come back to this problem when we discuss the molecular and supramolecular organization of chromatin.

The classic Watson-Crick structure described in Fig. 11 applies to the so-called B-form of DNA, which is typical of DNA molecules under physiological conditions. But, as pointed out by A. Rich and his colleagues, stretches of the right-handed B-DNA double helix may locally unwind and assume the left-handed configuration typical of the so-called Z-DNA (zigzag DNA). The B → Z-DNA transition is favored by local arrangements of the nucleotide sequences; poly(dT–dG) sequences, made up of about 50 alternating dT and dG sequences, are potential Z-DNA–forming sequences. These sequences are highly repeated in the human genome; it contains more than 10,000 stretches of $(dT–dG)_n$ sequences that can adopt the Z-configuration (Hamada and Kakunega, 1982; Hamada *et al.*, 1982). Similarly, Jeang and Hayward (1983) have reported that the complementary $(dC–dA)_n$ tracts potentially capable of forming Z-DNA punctuate the mammalian genome with a frequency of more than $10^5$ per cell. Another factor that favors the B-DNA → Z-DNA transition is methylation of cytosine residues, which destabilizes B-DNA and stabilizes Z-DNA (Fradin *et al.*, 1982; Fujii *et al.*, 1982). Current interest in Z-DNA, CACA sequences and DNA methylation stems from the fact that they might play a role in the still enigmatic mechanisms that control gene expression; we shall often return to this important question in this book.

Work done in 1960–1965 has yielded important information on the general organization of the genome. Isopycnic centrifugation in cesium chloride gradients has shown that it is often possible to separate satellite from bulk DNA because they have different equilibrium densities. Some are heavy and rich in G + C sequences, such as *Xenopus* ribosomal DNA (rDNA), while others are light and have a high content of A + T sequences (as in mouse satellite DNA). Satellites are made up of highly repetitive, simple DNA sequences (repeated up to $10^7$ times in mouse satellite DNA) and are, as already mentioned, located in heterochromatin.

Another very useful technique has been the study of renaturation kinetics in sheared denatured preparations of DNA. Heating separates the two DNA helices from each other in a process called "denaturation" (Thomas, 1954) or melting.

When denatured DNA is allowed to cool slowly, the separated strands will reanneal, the speed depending on sequence homology; repetitive sequences will find their partners (i.e., the complementary homologous sequences) faster than the unique (or single copy) sequences that, 10 years ago, were believed to correspond to individual genes. Using this hybridization technology, Davidson *et al.* (1973, 1975; reviewed by Davidson and Britten, 1973) found that the genome of almost all metazoans is built on the same general pattern: one-half of the DNA is composed of repetitive sequences that are 300 nucleotides long interspersed with nonrepetitive (single-copy) sequences that are 800 nucleotides long. The other one-half of the genome is composed of unique sequences separated by stretches of repetitive sequences longer than 4000 nucleotides.

This general pattern of DNA sequence organization in the genome is further complicated by the existence of palindromes (Wilson and Thomas, 1974), also called ''inverted repeats'' or ''snapback sequences.'' They are made up of DNA sequences that read the same way in both directions (like the name of the once too famous general, Lon Nol); as the result of base pairing, palindromes form hairpin structures (Fig. 12). After melting, palindromes reassociate almost immediately, in view of the perfect sequence homology of the two inverted repeats that form them. According to Dott *et al.* (1976), no less than $2 \times 10^6$ palindromes, with an average size of 190 nucleotides and representing 6% of the total genome, are present in the human haploid genome. A more recent report by Biezunski (1981a,b), based on EM observations, concluded that there are about 30,000 palindromes (inverted repeat pairs) in *Drosophila* and between 224,000 and 320,000 in the mouse genome. In both cases, the palindromes are often seen in clusters. We can only hypothesize about their function: they might be initiating sites for DNA replication, play a role in the folding of the chromosomes, act

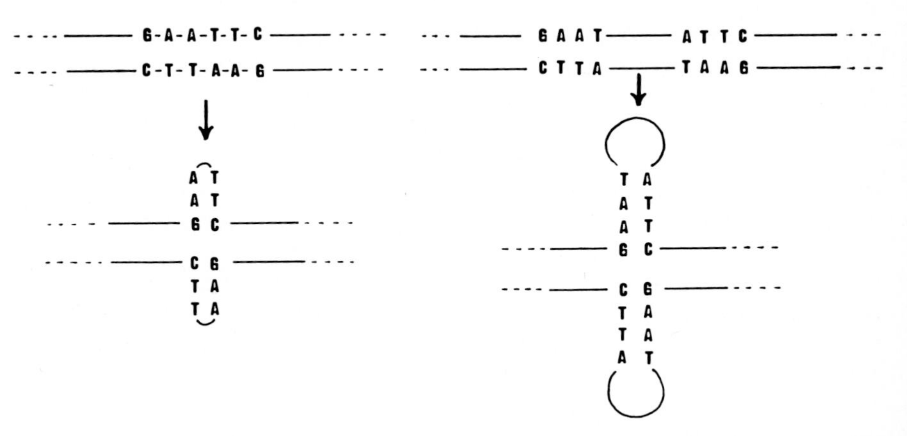


FIG. 12. Schematic representation of palindromes. Left: a hairpin palindrome; right: two successive palindromes give rise to a looped hairpin structure.

as binding sites for regulatory proteins, etc. They might also act as signals or punctuation marks on the huge DNA double helix; the existence of such a punctuation mechanism is also suggested by the fact that A + T-rich regions are distributed according to a definite pattern in human DNA.

It was long believed that repetitive DNA sequences are not transcribed and that all single-copy sequences are transcribed in mRNAs. This oversimplified view is no longer held. The tandemly repeated ribosomal (see Section V) and histone genes are very actively transcribed; in sea urchins, the histone genes are reiterated at least 300 times in the genome (Kedes *et al.*, 1971; Birnstiel, *et al.*, 1974; Kedes *et al.*, 1975; Weinberg *et al.*, 1975; reviewed by Kedes, 1979). Ribosomal and histone genes are not organized like the monotonous mouse DNA satellite, but are tandemly repeated many times and are separated from each other by noncoding stretches of DNA called "spacers" (Fig. 23). Repetition of histone genes also varies greatly from one species to another (from several hundreds in sea urchins to only 10–20 in mammals, according to Jacob *et al.*, 1976, and Stephenson *et al.*, 1981). Reiteration of histone and ribosomal genes clearly serves a useful purpose: it provides the cell with large amounts of their gene products, histone mRNAs, and rRNAs, respectively.

In contrast to the ribosomal genes present in the nucleolar organizers, the true satellite DNAs (present in heterochromatic regions as part of the chromosomes) are probably never transcribed. The classic example is the A + T-rich mouse satellite DNA, which is composed of highly repeated ($10^7$ times), very short sequences (shorter than 20 bp). One should make a distinction between satellite DNAs and dispersed repeated sequences (reviewed by Singer, 1982a), exemplified by the so-called Alu family[3] in humans ("Alu-like" in other mammals). These short repeated sequences (500 bp), in contrast to those of the satellite DNAs, are very widely dispersed in the genome, where they are scattered in 600,000 copies (reviewed by Schmid and Jelinek, 1982). Other families of dispersed repeated sequences are known to exist in the human genome—for instance, the Hinf (Shimizu *et al.*, 1983) and KpnI (Di Giovanni *et al.*, 1983) families. There is evidence that these scattered, short sequences are transcribed: in polyribosomes or in nonpolysomalcytoplasm, cells contain small mRNA sequences that have been copied on the Alu sequences (Calabretta *et al.*, 1981; Haynes and Jelinek, 1981; Krayev *et al.*, 1982), but their role remains unknown. Similar observations have been reported for *Xenopus* by Spohr *et al.* (1981) and for the chicken by Stumph *et al.* (1981). The mystery is further intensified by the fact that Alu sequences have been found in small, circular DNA molecules that can be recovered from an increasing number of isolated nuclei (Krolewski *et al.*, 1982; Calabretta *et al.*, 1982; Bertelsen *et al.*, 1982; McGrath and Emerson, 1982; Stanfield and Helinski, 1984). However, their significance is still un-

[3]Alu-1 is a restriction enzyme.

known. They might be intermediaries in the chromosomal rearrangements or gene amplification processes (which will be discussed later); alternatively, they could be products of DNA degradation. Their very existence, however, reinforces the view that the old dogma of DNA absolute stability is incorrect.

Finally, it should be pointed out that, according to Rosbach *et al.* (1975), more than 70% of the so-called unique or single-copy sequences do not code for proteins. Many or perhaps all of them are transcribed. As we shall see later, the great majority of these transcripts are destroyed in the nucleus itself, never reaching the cytoplasmic ribosomes. Thus, the situation we are currently facing is far more complex than the one that was accepted only a few years ago.

Recombinant DNA methodology (see the special issue of "Science," Nov. 19, 1980, for details) has completely modified our view of the eukaryotic genome as a static structure. Dynamic changes in the DNA molecules can lead to genome instability, which is in sharp contrast to the ideas that prevailed in the 1950s. We can now isolate, in amounts sufficient for detailed analysis, almost any gene of interest. Genomic DNA can be cut at specific sites with restriction endonucleases and the fragments inserted into plasmids (small circular DNA molecules) or phages that are present in bacteria. Plasmids or phages containing the recombinant eukaryotic DNA are used to infect other strains of bacteria; their multiplication leads to the formation of bacterial clones that can be screened by hybridizing their DNA with specific radioactive probes—for instance, the radioactive mRNA of the gene of interest or its cDNA (complementary DNA). A cDNA is a radioactive DNA copy of an mRNA; it is obtained with the viral enzyme reverse transcriptase, which copies RNA into DNA. If the mRNA coding for the protein of interest is not available in sufficient amounts to synthesize its cDNA, and if part of the amino acid sequence of this protein is known, there is another way of isolating the gene: constructing a synthetic oligonucleotide corresponding to the partial amino acid sequence. These unique-sequence synthetic DNA probes will bind specifically to the gene and allow its isolation. The final goal of this type of work is to establish the base sequence of the entire gene and its 5′- and 3′-flanking regions. Today, it is often easier to deduce the amino acid sequence of a protein by sequencing the corresponding gene (or the cDNA copy of its mRNA) and to use the genetic code than to directly establish the amino acid sequence of this protein (Maxam and Gilbert, 1977). The subject has been admirably summarized by Gilbert (1981a,b) and by Sanger (1981a,b) in their Nobel Prize lectures.

One of the first big surprises to result from recombinant DNA work was the discovery of "intervening sequences" or introns (reviewed by Doel, 1980; Lewin, 1981a), which interrupt the vast majority of eukaryotic genes (Fig. 13). Introns are intragenic noncoding sequences that are not represented in cytoplasmic mRNAs. As in the spacers that separate genes, their base composition varies more from species to species (and sometimes even from one indi-

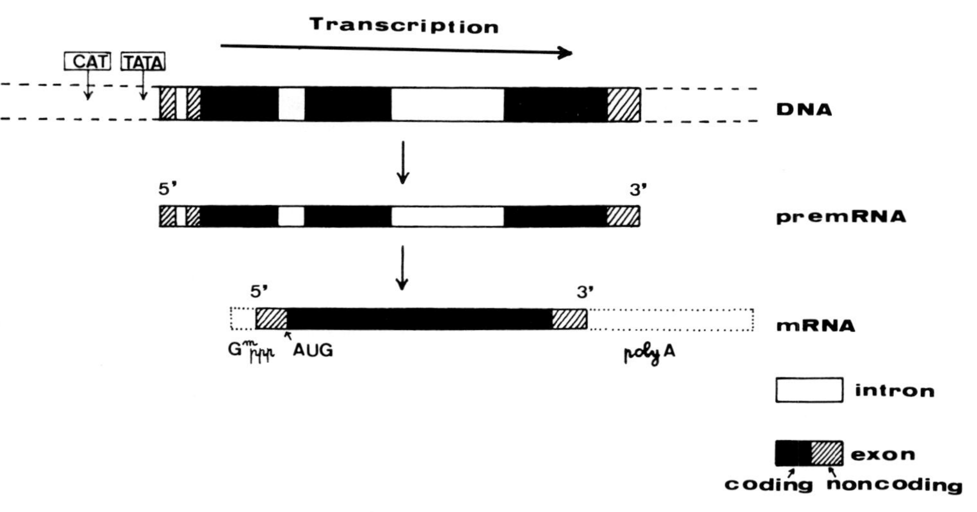


FIG. 13. Schematic representation of the transcription of a eukaryotic gene. The gene is interrupted by intervening noncoding sequences (introns), which are transcribed and then removed during the maturation of the pre-mRNA into mRNA. In the mature mRNA, the cap (Gppp) has been added at the 5′-end and the poly(A) sequence at the 3′-end; noncoding sequences are present at the ends of the molecule. AUG is the initiation codon for translation in the polyribosomes. The CAAT and TATA boxes are promoters of transcription by RNA polymerase II. [Modified from Coutelle (1981).]

vidual to another within the same species) than that of the sequences that code for a specific protein. The role played by the introns still remains a mystery. Reports indicate that the genes coding for $\alpha_2$ type I collagen of the chicken has more than 50 introns (which must be exactly removed after their transcription, as will be seen later), while the $\alpha_2$ interferon gene (like the histone genes) has no introns at all (Ohkubo *et al.*, 1980; Vogeli *et al.*, 1981; Lawn *et al.*, 1981). Yet, both genes are functional and code for the corresponding RNAs and proteins. It might be that, as proposed by Gilbert (1978), the presence of introns once presented selective advantages during evolution (unfortunately, we were not there to witness that event). Interestingly, the base composition of the 5′-flanking regions of the aforementioned collagen and interferon genes present remarkable homologies. Both contain sequences (the so-called TATA and CAAT boxes) that are required for enzymatic transcription of many eukaryotic genes. It has been suggested (Brown, 1981) that removal of intron sequences (splicing) during the maturation of the nuclear mRNA precursors into the smaller mature cytoplasmic mRNAs might be necessary for the passage of mRNAs from the nucleus to the cytoplasm. However, according to Gruss *et al.* (1981), deletion of a unique intron does not suppress (although it reduces) mRNA translation; mRNA splicing can thus be bypassed without harming the cell. On the other hand, Harbers *et al.*

(1984) have recently made an unexpected discovery: introduction of a mouse leukemia virus into the first intron of the $\alpha_1$ (I) collagen gene of a fertilized mouse egg was followed by the death of the embryo after 12 days. This lethal mutation was due to the lack of collagen production. It seems that this particular intron plays a crucial role in tissue-specific collagen expression.

It has been recently proposed that the intron–exon junctions (also called "splice junctions") correspond to definite locations on the surface of the encoded protein. For several proteins, proteolysis cleavage points correspond to splice junctions (Craik *et al.*, 1982). Ny *et al.* (1984) have analyzed the gene encoding a proteolytic enzyme, plasminogen activator. It has 14 exons that code for, respectively, the signal peptide, the propeptide, and various domains of the heavy chain. The structural domains of the protein correlate well with the exon–intron pattern of the gene. However, histones are composed of several domains, although they are encoded by intronless genes (reviewed by Blake, 1985).

Introns that interrupt genes and spacers between genes, since they have undergone considerable change in nucleotide sequences during evolution, obviously introduce some complexity into the rigid organization of the genome. However, other changes are required to introduce (in addition to classic mutations) real instability (reviewed by Khesin, 1980; Brown, 1981) into the genome. One of these changes is the loss of genetic material due to chromosome elimination. This takes place, on a large scale, in *Ascaris* and in some insects (*Sciara*, cecidyomids) during early cleavage of fertilized eggs (Fig. 14). Part of the genomic DNA in such cases is germ-line limited. Genes necessary for germ cell formation would be selectively eliminated and destroyed in the cytoplasm, as exemplified by Boveri's classic chromatin diminution in *Ascaris*. As pointed out by Brown (1981), one cannot exclude the possibility that this process might be more frequent than is generally believed; it might take place in a discrete way during the cleavage of eggs from many species. However, it should be pointed out that, according to the latest report on chromatin diminution in *Ascaris* (Roth and Moritz, 1981), germ line–limited DNA is composed of two families of satellite DNAs (tandemly repeated sequences of, respectively, 125 and 131 bp) and that there is no good evidence for the elimination of structural genes in *Ascaris*. Furthermore, elimination of satellite DNA during chromatin diminution in *Ascaris* is incomplete. Nevertheless, it is clear that the rule of DNA constancy per nucleus cannot hold true when the DNA contents of somatic and germ cells are compared in *Ascaris* and in some insects. The opposite of gene loss, i.e., gene amplification, [reviewed by Schimke (1984), in which molecular models are presented and discussed] has long been known to occur in *Xenopus*, in which oocytes possess 1000 times more ribosomal genes than somatic cells (see Volume 2, Chapter 2). This is by no means the only case of gene amplification known today. In *Drosophila*, the genes coding for chorion proteins are amplified, prior to their expression in the follicle cells that surround the oocytes

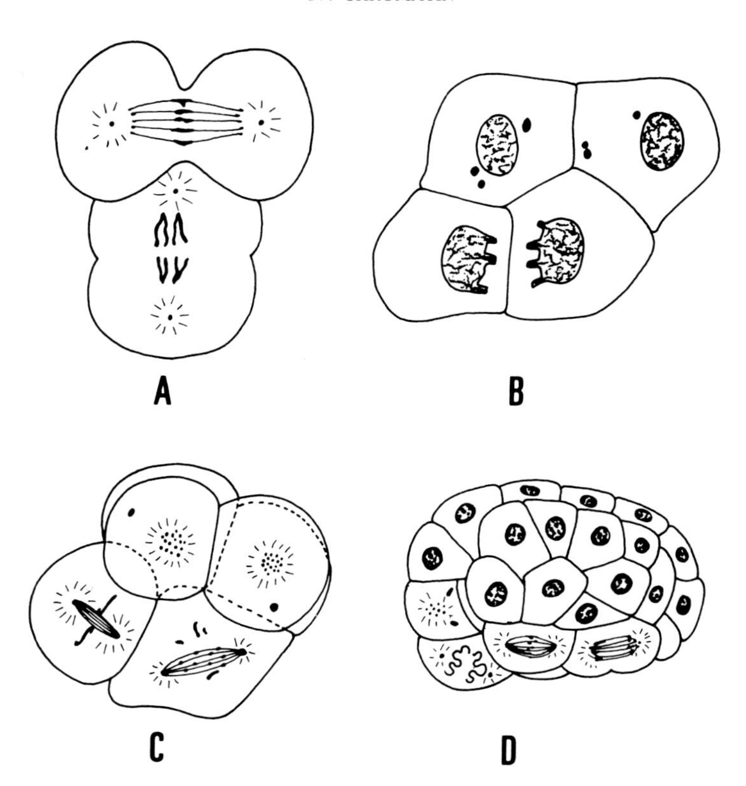


FIG. 14. Chromatin diminution in *Ascaris* after Boveri (1887). The cells that have lost chromosome segments will become somatic cells. The cell that, in the lower left corner of (D), has retained the full chromosome complement will give rise to the germ cells.

(Spradling and Mahowald, 1980). When cells resistant to the cancer drug methotrexate (an inhibitor of dihydrofolate reductase, an enzyme involved in DNA synthesis) are selected, large quantities of the reductase and its mRNA are found. It was also unexpectedly found that the number of dihydrofolate reductase genes had also greatly increased. Thus, there is a selective amplification of the dihydrofolate reductase and its neighboring genes; between 700 and 1000 copies of the dihydrofolate gene are dispersed on three chromosomal sites, including a long, homogeneously staining, DNA-containing segment that is absent in control cells (Fig. 15) (Nunberg *et al.*, 1978; Alt *et al.*, 1978; Millbrandt *et al.*, 1981; Berenson *et al.*, 1981). *In situ* molecular hybridization has clearly shown that the amplified dihydrofolate reductase genes are located in the long, homogeneously staining chromosomal segment, and that they are also present in pairs of very small chromosomes, called "double minutes." They appear when the genes are amplified and replicated in an autonomous fashion (Haber and Schimke, 1981).

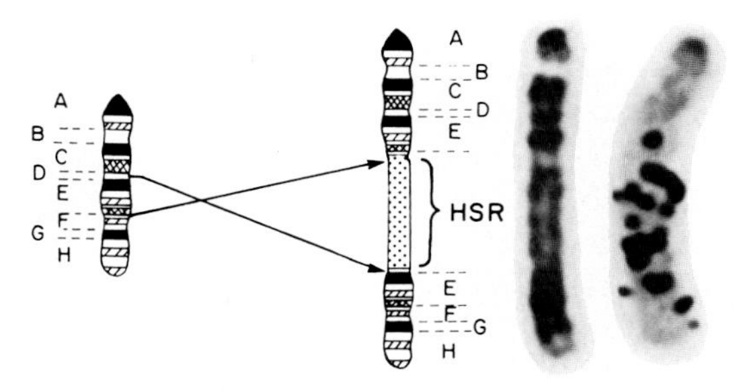


FIG. 15. Left: ideograms of a normal mouse chromosome and of a chromosome with a homogeneously staining region (HSR) corresponding to amplified dihydrofolate reductase genes in methotrexate-resistant cells. This is demonstrated, at the extreme right, by *in situ* hybridization using [$^3$H]cDNA (radioactive DNA copy of dihydrofolate reductase mRNA). [Berenson *et al.* (1981).]

In addition to the homogeneously staining, long chromosomal segment and the double minutes, circular chromosomes were found in the nuclei of methotrexate-treated cells by Bostock and Tyler-Smith (1981). These data show that there are extensive DNA rearrangements during gene amplification. The double minutes, which contain three to four copies of the dihydrofolate reductase gene, disappear together with the amplified genes when the cells are cultured in a methotrexate-free medium (P. C. Brown *et al.*, 1981). They are unequally distributed at mitosis, which leads to an unequal distribution of the dihydrofolate reductase genes (Kaufmann *et al.*, 1981).

Similar observations have been made for cells resistant to inhibitors of aspartotranscarbamoylase (Wahl *et al.*, 1979; Padgett *et al.*, 1982; de Saint Vincent *et al.*, 1981) and for the gene coding for metallothionein I, a sulfur-containing protein that binds heavy metals. In cells selected for resistance to cadmium, the metallothionein I genes are amplified three to five times, with a concomitant appearance of three small surnumerary chromosomes (Beach and Palmiter, 1981). Thus, in the presence of enzyme inhibitors, somatic cells are almost selectively able to amplify the gene coding for the enzyme, which is the target for the inhibitor.

Comparable observations have been made on trypanosomes by Pays *et al.*, (1981a,b; reviewed by Bernards, 1985; Borst and Cross, 1982; Borst *et al.*, 1983). When pathogenic trypanosomes multiply in the bloodstream of humans or other mammals, the infected host produces antibodies against the glycoprotein present in the surface coat of the invading parasite. Curiously, when the host's immune response becomes powerful enough to reduce the number of trypanosomes present in the bloodstream, the parasite covers its surface with a new, chemically different surface antigen. The number of such variable surface antigens is quite large (100 or more), and all are coded by different genes, which are expressed when new surface antigens are synthesized and secreted by the trypanosomes in the bloodstream. Analysis of the DNA present in the trypanosomes has clearly shown that the gene that is expressed when a specific antigen appears on the cell surface has formed an extra copy, and that it is this extra copy that is expressed. Other mechanisms (gene conversion) also operate to switch one surface antigen to another.

There is currently a good deal of interest in multigene families. In addition to the well-known tandem multigene families of histone and ribosomal genes, it has been found that proteins such as actin and tubulin are encoded by more than one gene per haploid genome (briefly reviewed by Firtel, 1981). For instance, there are 17 actin genes in the slime mold *Dictyostelium,* 6 in *Drosophila,* more than 11 in the sea urchin, and around 10 in humans, in whom actin-coding sequences are distributed in many different regions of the genome (Humphries *et al.*, 1981). In humans, different genes code for the cytoskeletal β- and γ-actins, but only one for the α-actins of skeletal and cardiac muscle (Engel *et al.*, 1982; Ponte *et al.*, 1983). In chickens the α- and β-tubulin genes also form a dispersed multigene family. Four genes are located on four different chromosomes, and the β genes are on at least two different chromosomes (Cleveland *et al.*, 1981). In addition, analysis of the genome by base sequencing has disclosed the unexpected existence of pseudogenes (reviewed by Little, 1982), which are incomplete (but otherwise normal) copies of a bona fide gene (human β-tubulin, for instance, according to a study by Cowan *et al.*, 1981) and do not code for any protein. Their role, if any, remains unknown, particularly since the discovery that some pseudogenes have the typical characteristics of processing: splicing of the introns and addition of a poly(A) tail (as in the processing of mRNA precursors into mature mRNAs). These processed or intronless pseudogenes are probably numerous, since several examples are already known in the globin, tubulin, immunoglobulin metallothionein, I, and dihydrofolate reductase gene families. A likely explanation for the existence of processed pseudogenes is that (Wilde *et al.*, 1982a,b) they are DNA copies of mRNAs, like the cDNAs prepared *in vitro* by treatment of an mRNA with viral reverse transcriptase. These DNA copies of RNAs are then inserted into the genome by a mechanism similar or identical to that of the insertion of RNA viruses (retroviruses, see Volume 2, Chapter 3) into host DNA. If so, genetic information could go back to the genome through an RNA intermediate which seems to contradict the central dogma of molecular biology (reviewed by Baltimore, 1985). However, as far as is known, pseudogenes are not expressed and this new genetic information gained via a RNA intermediate is useless for the cell. This conclusion is reinforced by the finding that a dihydrofolate reductase pseudogene is not amplified, in contrast to the true gene, when cells are treated with methotrexate (Masters *et al.*, 1983). As expected, the number of pseudogenes discovered since the review by Little in 1982 is steadily increasing, [see the more recent review by Vanin (1984)] and the progress made in this field substantiates the suggestions made by Wilde *et al.*

(1982a,b) when they studied two human β-tubulin pseudogenes and concluded that they have different origins. One of them has an intron, but there is a deletion in the coding region of the gene; the other, like the mRNA, has no intron, but possesses a polyadenylation signal (an AATAAA sequence) and an oligoadenylate tract at its 3′ end. This pseudogene is flanked by a short direct repeat. It probably results from reverse transcription of an mRNA followed by reintegration of the resulting cDNA in the genome. A similar immunoglobulin (Ig) "processed" pseudogene was found by Battey *et al.* (1982) on human chromosome 9 (the *Ig* locus is on chromosome 14); it has lost three introns and gained a poly(A) tail, and it is flanked by an 11-bp sequence that is directly repeated 150 bp upstream of the first coding domain. This strongly suggests that DNA sequences can move about the genome via RNA intermediates. Numerous pseudogenes have been found for the snRNA (small nuclear RNA, to be discussed later in this chapter) gene family. Again two possible origins have been proposed: either DNA duplication or deletion, or RNA-mediated insertion (Denison and Weiner, 1982; Bernstein *et al.*, 1983). Pseudogenes have also been found in the cytochrome *c* multigene family, and it has been proposed that they result from the insertion in the genome of DNA copies of cytochrome *c* mRNA (Scarpulla and Wu, 1983). It has been pointed out by Denison and Weiner (1982) and by Sharp (1983) that there is a striking similarity between this subclass of processed pseudogenes and the Alu family of short repetitive sequences. Both would result from the integration into the genome of DNA copies of RNA transcripts. We shall return to this point when we discuss transposable elements.

The "great dynamic turnover" of the genome has been well exemplified in a discussion of the evolution of the β-globin gene clusters by Lewin (1981c). This evolution did not occur (as was believed) by point mutations alone; more drastic changes (insertions, deletions) have allowed more rapid evolutionary changes. Figure 16 (Lewin, 1981c) shows the arrangement of the β-globin gene family. It is known that hemoglobin is made up of heme and the two α- and β-globin chains; the latter are coded for by genes localized on different chromosomes. Both α- and β-genes are, in fact, clusters of α- or β-like genes that code for different globins synthesized during various periods of life. As shown in Fig. 16, there are five β-like genes in the cluster that are utilized successively at the beginning of embryonic life and during very early and very late fetal life. They are disposed, in the gene cluster, in the order in which they will be expressed during life (starting from ϵ and ending with β; this is also true of the α-globin genes). It is believed that the α- and β-globin genes arose from the duplication of an ancestral gene some 500 million years ago. One of the reasons for believing this is that the introns of both genes are in homologous positions; the lengths of the introns have not changed much, but their base composition (under the impact of mutations) has changed considerably during evolution. The coding sequences of the globin genes, in contrast, have remained surprisingly stable. There is a

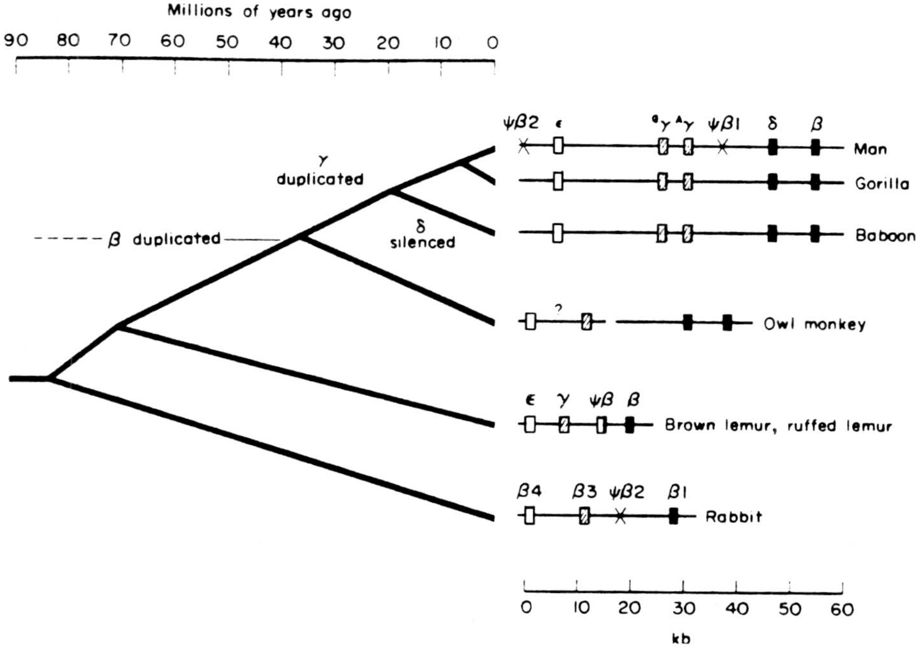


FIG. 16. Phylogeny of the β-globin genes clusters. Their maps show the sequence in which certain genes changed during evolution. The stability of the cluster in the Old World monkeys, apes, and humans is indicated by the constant size and organization of their gene clusters. The positions of pseudogenes are shown by crosses. [Lewin (1981b). Copyright 1981 by the American Association for the Advancement of Science.]

marked degree of homology between our hemoglobin genes and those coding for *leg*-hemoglobin present in the root nodules of leguminous plants. Both the α- and β-globin gene clusters include two pseudogenes. Their base composition has remained as stable during evolution as that of the corresponding genes, suggesting that they play some role other than coding for a protein. Curiously, in the α-locus, one of the two pseudogenes is not in the cluster and has been transposed to another region of the genome.

This genetic accident can occur even in an apparently well-linked tandem multigene family. Childs *et al.* (1981) have coined the term ''orphons'' for genetic elements that have lost their families (the sea urchin histone genes family, in the present case). Elements that leave the gene family live a dispersed, solitary life. In the sea urchin genome, there are more than 50 histone gene orphons. They display great individualism (polymorphism), since they vary in number from one sea urchin to another.

Orphons are merely a category of ''mobile dispersed genes'' or ''transposable elements'' (reviewed by Calos and Miller, 1980; Spradling and Rubin, 1981;

Finnagen, 1985). These transposable elements are middle repetitive sequences, of identical or very similar composition, that are dispersed throughout the genome. The aforementioned *Alu* sequences of our genome probably originated in the same way as the transposable elements that are now being extensively studied in yeast and in *Drosophila.* In this insect, there are several types of transposable elements called *copia*, 412,297 mdg1, mdg3, B104, P elements, foldback elements, gypsy, F-factors, etc. One of them is 4.5 kb long and is flanked on both sides by a 276-nucleotide sequence arranged in the same direction (direct repeat), as shown in Fig. 17. The gene can be transcribed. There are about 50 copies of *copia* mRNA in the cytoplasm of the *Drosophila* cell (Young, 1979; Falkenthal and Lengyel, 1980; Flavell *et al.*, 1980). Like *copia,* the mobile genetic elements called mdg3 by the Russian workers who discovered and studied them (Ilyin *et al.*, 1980; Bayev *et al.*, 1980; Tchurikov *et al.*, 1980) have a highly variable chromosomal localization, as shown by *in situ* hybridization: they can occupy 15 or 17 sites dispersed on all four *Drosophila* chromosomes (Fig. 18). There are about 15 copies of mdg3 in *Drosophila* embryos, but their number increases 13 times in cultured cells. Like *copia,* mdg3 is made up of 5.5-kbp flanked by two repeated sequences of 300–500 bases and is transcribed in a 26 S mRNA. Still another class of transposable elements in *Drosophila* is FB (foldback), which differs from *copia* by a greater variability. Again, there are marked differences in chromosomal localization among strains (Truett *et al.*, 1981; Potter, 1982). The relatively recently discovered F-family of transposable elements (Di Nocera *et al.*, 1983) is present in *Drosophila* in 50 copies localized at 25 sites in both euchromatin and heterochromatin; these F-elements end with 30 adenylate residues, which are preceded by the polyadenylation signal AATAAA. They bear an obvious resemblance to both the processed pseudogenes and the human *Alu* sequences. On the whole, transposable elements represent a sizable proportion (about 5%) of the total *Drosophila* genome.

Since the days of T. H. Morgan and his school, *Drosophila* has been extremely useful to geneticists. It is likely that this fly will attract more and more interest, since we now know that integration of "jumping" transposable se-

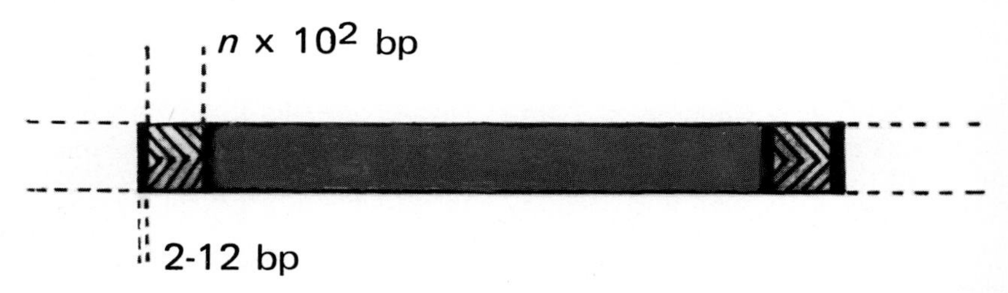


FIG. 17. Schematic organization of an inserted mobile genetic element (or a provirus). The gene (dark) is flanked on both sides by direct DNA sequence repeats (n.$10^2$ bp) bounded by short inverted repeats (2–12 bp). [Modified from Flavell (1981).]

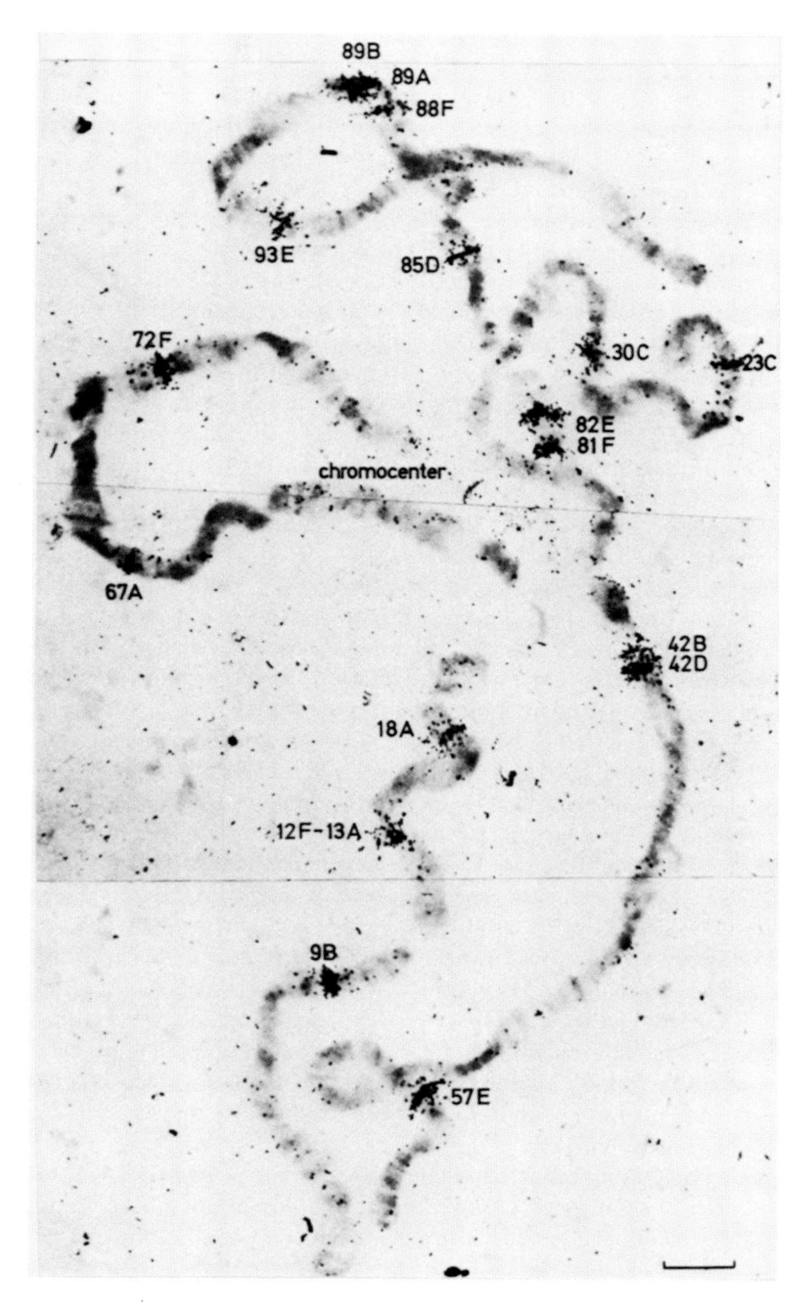


FIG. 18. Detection of the 17 insertion sites of the mdg3 transposable genetic elements by *in situ* hybridization on the four *Drosophila* salivary gland chromosomes. [Ilyin *et al.* (1980).]

quences in many sites of the genome may induce mutations as a consequence of gene rearrangements. For instance, insertion of *copia* into one of the introns (Pirrotta and Bröckl, 1984) of the *white* genetic locus induces the *apricot* eye color as the result of genome rearrangement (Goldberg *et al.*, 1982). Large transposable elements discovered by Paro *et al.* (1983) bear the *white, apricot,* and *roughest* genes; they can be transposed to about 100 sites scattered throughout the *Drosophila* genome. This should make it possible to study the regulation of these genes when they are integrated into new surroundings. The *white crimson* ($w^c$) mutation ($w^c$ is an unstable allele of the *white* locus) is due to the insertion of a member of the foldback (FB) transposable elements family; loss of the FB element, which is capable of precise excision at high frequencies, leads to phenotypic reversion (Collins and Rubin, 1983). The *hairy* mutation, which affects the formation of segments in the larva, is due to insertion of a *gypsy* element (Holmgren, 1984). Particularly interesting are the results obtained by Spradling and Rubin (1981) and Rubin and Spradling (1982). Introducing a P-factor bearing a wild-type copy of the *rosy* gene in *rosy* mutants of *Drosophila,* they succeeded in restoring the normal phenotype; for the first time, the old dream of "gene therapy" has become a reality. In other experiments, which will be presented in Volume 2, Chapter 2, the same authors succeeded in incorporating cloned $rosy^+$ genes in the germ line of $rosy^-$ embryos. Hereditary genetic transformation via transposable element vectors is thus now possible in *Drosophila.* In yeast (reviewed by Williamson, 1983), the transposable element *Ty1,* which is very similar to *copia,* has been extensively studied by Gafner and Philippsen (1980) and Kingsman *et al.* (1981). Finally, Hoffmann-Liebermann *et al.* (1985) have reported that a pseudogene of one of the sea urchin histones (histone H2B) contains a large (32-kb-long) transposable element that has been named *TU-1*. It is a member of a heterogeneous family of transposable elements and has similarities to the FB family of *Drosophila*. There is no doubt that future work will disclose the presence of transposable elements in a growing number of animal and plant species.

If all dispersed middle repetitive sequences (including those of the *Alu* family in humans) originated from mobile dispersed genes of the *copia* type, one might wonder about the origin of the transposable elements. As pointed out by Flavell (1981), there is a striking similarity between transposable genetic elements and RNA tumor viruses (retroviruses). Because of the viral enzyme reverse transcriptase, which copies the viral RNA genome into complementary DNA, the resulting provirus can be inserted at any point in the host genome (see Volume 2, Chapter 3 for additional details). The general structure of all of the transposable elements currently known is the same as that of the inserted retroviruses (Flavell, 1981). As shown in Fig. 18, DNA sequence repeats a few hundred nucleotide pairs long are found at each end of both transposable elements and integrated proviruses; in all cases, these long terminal repeats are bounded by short (2–12 nucleotide pairs) inverted repeats. The similarity between *copia* and retroviruses is further emphasized by the fact that cultured *Drosophila* cells (which possess a much larger number of transposable genetic elements than embryonic cells) contain 1–10 circular copies of *copia* elements that look similar to the circular proviral forms of retroviruses (Flavell and Ish-Horowicz, 1981). These circular copies of *copia* replicate extrachromosomally by the same semiconservative mechanisms (to be discussed in the next chapter) as chromosomal DNA (Sinclair *et al.*, 1983). In addition, Shiba and Saigo (1983) have found that cultured *Drosophila* cells contain retrovirus-like particles; their RNA is homologous to that of *copia*. Ilyin *et al.* (1984) also found extrachromosomal circular copies of the *Mdg1, Mdg3,* and *copia* mobile genetic elements in cultured *Drosophila* cells. These cells also contain DNA–RNA complexes between Mdg1 and Mdg3 DNAs and poly(A)$^+$ RNAs (Arkhipova *et al.*, 1984). These findings reinforce the current belief that the mobile genetic elements of *Drosophila* result from the reverse transcription of cellular RNAs. Analogies to retroviruses can also be found for the dispersed middle repetitive sequences of the *Alu* family. In both humans and hamsters, the *Alu* sequences are flanked by direct repeats, as in *copia* and retroviruses inserted in the host genome (Haynes *et al.*, 1981; Rogers, 1983). Further work will certainly lead to increased understanding of this fascinating problem.

That similar biochemical mechanisms operate for the insertion of proviruses (DNA copies of a retrovirus), transposable elements, pseudogenes, and dispersed middle repetitive sequences is strongly suggested, as we have just seen, by the analysis of the flanking DNA base sequences. However, one point still worries biochemists: retroviruses possess a gene coding for reverse transcriptase, the enzyme that copies viral RNA into proviral DNA; reverse transcriptase activity is absolutely required for integration of the virus into host DNA. But it is not yet clear whether reverse transcriptase activity is present in normal, uninfected cells. Mondal and Hofschneider (1983) have reinvestigated the question; they found that reverse transcriptase can be detected in the nuclei and cytoplasm of normal quail embryos. Of the nuclear enzyme, 60% is in a free, easily extractable form, but 85% of the cytoplasmic enzyme is bound to particles that might be endogenous retroviruses. Have endogenous or exogenous retroviruses been instrumental in the generation and insertion of our *Alu* repeated sequences and pseudogenes during evolution? Do all the so-called normal cells harbor retroviral particles? More biochemical work on reverse transcriptase activities, including the isolation and characterization of the enzyme, is needed before we can answer these intriguing questions.

In contrast to the random, undirected gene rearrangements due to insertion of transposable genetic elements is directed gene rearrangements. The already mentioned production of an additional gene copy when the variable surface antigens in trypanosomes switch from one to another provides an example of such rearrangements. The best-studied case so far, however, is that of the immunoglobulin genes (briefly reviewed by Marx, 1981b, and Gough, 1981). The problems raised by the amazing specificity and variety of immune responses to a host of different antigens are so complex that immunology is now a science in its own right. This complexity extends to the organization of the genes coding for immunoglobulins. As was first shown by Tonegawa, the organization of the genes coding for the variable and constant parts of the antibodies is not the same in embryos and in differentiated antibody-producing cells. These genes are separated in the embryo and linked together in plasmatocytes as the result of deletions and insertions of DNA segments (Hozumi and Tonegawa, 1976; Brack and Tonegawa, 1977; Max *et al.*, 1979)—somatic recombination, in addition to somatic mutations, thus takes place after embryogenesis (Tonegawa, 1983). It is probable that this process is an exception and does not take place during differentiation of cells other than lymphocytes, since it does not occur when differentiating cells begin to produce hemoglobin, ovalbumin, or silk fibroin.

Another mechanism for increasing genetic diversity is arousing considerable interest today. It is gene conversion, a phenomenon discovered many years ago by Lindegren. Analyzing spore segregation in yeast, he occasionally found a 3 : 1 (instead of the expected 2 : 2) segregation ratio for given genetic markers; he concluded that a gene had been replaced by another one and called this rare event "gene conversion." The subject has been reviewed by Gough (1982) and Baltimore (1981). Gene conversion is a nonreciprocal transfer of information from one DNA duplex to another. It can make new genes by combining elements of previous ones and leads to DNA rearrangements. It takes place between various members of a multigene family or the same gene family. As pointed out by Baltimore (1981), it may occur between genes located anywhere in the genome. It has been suggested that the recent evolution of the immunoglobulin (*Ig*) variable genes results from divergence by gene conversion (Bentley and Rabbitts, 1983); this would be due to a deletion shortening the distance between two *Ig* loci (Ollo and Rougeon, 1983). Intergenic exchanges (gene conversion) of a short internal DNA segment from one gene to the corresponding sequence of another gene can also explain the generation of polymorphism among the H-2 genes (the histocompatibility locus, responsible for graft acceptance or rejection) (Weiss *et al.*, 1983). Molecular models of gene conversion have been proposed by Gough (1982), Baltimore (1981), and Szostak *et al.* (1983). They have tried to explain how gene conversion, contrary to the unequal crossing-overs that take place at meiosis or mitosis, may either increase or decrease the number of genes in a gene family. Only genes that display great homology exchange their DNA, and only one of the two genes undergoes conversion.

A totally different way to modify the genome is DNA methylation (reviewed by Razin and Riggs, 1980; Wigler, 1981; Razin and Cedar, 1984; Razin and Szyf, 1984), a process to which we shall return when we discuss embryonic development and cell differentiation. The biochemical mechanisms of DNA methylation and demethylation have been recently discussed by Razin and Szyf (1984). DNA can be methylated on some of its cytosine residues by DNA methyltransferases; 5-methylcytosine usually represents between 2 and 7% of the total DNA cytosines. Analysis with restriction endonucleases that recognize methylcytosines in the DNA molecule has shown that 90% of the methylated cytosines are in CpG sequences. No demethylation enzymes have been isolated and purified, but their existence remains a good possibility. However, demethylation could take place as a result of DNA replication where the pattern of the two strands is conserved (Fig. 19). This might explain, as pointed out by Brown (1981), why a certain number of rounds of DNA replication are generally needed before embryonic cells become committed to differentiation (see Volume 2, Chapters 2 and 3). The interest in the problem of DNA methylation stems from the fact that, in general, undermethylated or demethylated genes are more active in transcription than their methylated counterparts. For instance, Mandel and Chambon (1979) showed that methylation of the ovalbumin gene is lower in the oviduct, where the gene is expressed, than in other tissues of the hen. Somatic cell DNA is less methylated than that of the transcriptionally inactive spermatozoa in sea urchins (Bird and Taggart, 1980). The DNA of the locus of the already discussed β-, γ-, and δ-globin genes is fully methylated in spermatozoa but only sparsely methylated in the cells that express the genes (van der Ploeg and Flavell, 1980); during the switching of fetal to embryonic and adult α-globins, there is a striking correlation between undermethylation and gene activity (Weintraub *et al.*, 1981). According to a more recent paper by Mavilio *et al.* (1983), the switch from embryonic to fetal and from fetal to adult hemoglobin

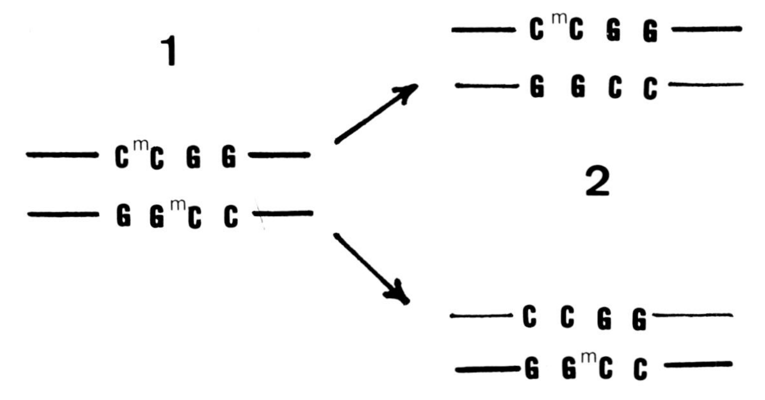


FIG. 19. Schematic representation of DNA undermethylation as a result of DNA replication. The hemimethylated DNA molecules shown in part 2 are good substrates for DNA methylases, but repeated DNA replication may lead to undermethylation. $^{m}$C, 5-methylcytosine. [Modified from Taylor and Jones (1979).]

is accompanied by undermethylation of the sequences flanking the three genes involved in globin synthesis (successively, the $\epsilon$-, $\gamma$-, and $\beta$-globin genes). Work has shown that uv irradiation activates silent metallothionein-I genes in cadmium-resistant cells; simultaneously, one of the metallothionein-I alleles undergoes complete demethylation (Lieberman *et al.*, 1983). It has long been known that the transcriptionally inactive mouse satellites and heterochromatic kinetochores are heavily methylated.

However, the rule that undermethylation and active transcription are associated has a growing number of exceptions. In *Xenopus* oocytes, the amplified ribosomal genes are strongly undermethylated while the somatic ribosomal genes are heavily methylated, yet both are very actively transcribed (Brown and Dawid, 1968; Dawid *et al.*, 1970). There is no change in the methylation pattern of the histone genes during sea urchin egg development, although their transcription increases considerably after cleavage (Bird *et al.*, 1979). This is not a peculiarity of eggs and embryos. In adult *Xenopus,* somatic 5 S DNA is as strongly methylated in cells that express the gene (liver cells) as in those that do not express it (red blood cells) (Sims *et al.*, 1983). Two genes involved in DNA synthesis (adenosylphosphoribosyltransferase and dihydrofolate reductase genes) are equally methylated in all mammalian tissues (Stein *et al.*, 1983). Vedel *et al.* (1983) have compared DNA sequences from fetal and adult hepatocytes and kidney and hepatoma cells in the rat. Fetal liver and hepatoma (a liver tumor) synthesize $\alpha$-fetoprotein, while adult liver synthesizes albumin. The two proteins are closely related, but are encoded by different genes; kidney does not synthesize the two proteins to a detectable extent. The analysis of the methylation pattern of the two genes in liver hepatoma, and kidney leads to the general conclusion that changes in DNA methylation are not responsible for the changes in gene activity that occur during the late phases of rat development. Highly methylated genes can be transcribed, and active genes are not always undermethylated. These negative conclusions do not exclude the possibility that DNA demethylation can be one of the factors involved in cell differentiation. As we shall see in more detail in later chapters, agents that inhibit DNA methylation and therefore lead to its undermethylation (5-azacytidine in particular) may induce the differentiation of cultured fibroblasts into muscle cells, chondrocytes, and adipocytes (Taylor and Jones, 1979; Jones and Taylor, 1980). 5-Azacytidine also induces the expression of the mouse metallothionein-I gene, according to Compere and Palmiter (1981); as previously noted, Lieberman *et al.* (1983) have obtained similar results in uv-irradiated cells. In conclusion, there is no absolute rule regarding the undermethylation–gene activity relationship. The results reported so far vary greatly from one biological system to another (as one might expect in view of the great diversity displayed by all living organisms). A similar conclusion has been drawn by Bird (1984) in his recent critical review on the importance of DNA methylation in gene control.

Since, in this chapter, we are considering primarily normal cells, we will not discuss all of the accidents to which DNA moleules may be vulnerable, such as mismatching of pair bases, intercalation of chemicals between the two DNA strands distorting the double helix, substitution of a normal by an abnormal base such as 5-azacytidine or bromodeoxyuridine, binding of carcinogenic agents, single- or double-strand breaks (by radiation or toxic chemicals), etc. However, one should bear in mind that these are frequent occurrences.

Fortunately, the cell has mechanisms that allow either complete or imperfect repair of DNA damage. One should also keep in mind that DNA is always associated with DNA-binding proteins (reviewed by Champoux, 1978; Duguet, 1981), which play a very important role in modifying the conformation and flexibility of DNA molecules, as well as playing a role in DNA transcription and replication.

### 2. *Histones and Other Acid-Soluble Chromosomal Proteins*

Histones are small ($M_r$ 11,000–20,000) basic proteins that can easily be isolated by acid extraction of isolated nuclei. They are ubiquitous among eukaryotes and, with one exception (H1), they have been remarkably conserved during evolution. Five main histone fractions are found in all cells: H1, H2A, H2B, H3, and H4. They owe their general properties to their high basic amino acid content. H1 is a lysine-rich histone, H2A and H2B are moderately rich in lysine, and H3 and H4 are arginine-rich (see Bradbury, 1975, for a competent discussion of the five histone types). In bird erythrocytes, a histone called H5 progressively replaces histone H1 during erythropoietic differentiation; this leads to a very strong condensation of chromatin, with concomitant reduction of gene activity (Lasters *et al.*, 1981). During spermatogenesis, there is also a progressive replacement of the classic histones by arginine-rich "sperm histones." The highly condensed chromatin of ripe spermatozoa is inactive in transcription, and it is likely that intensive chromatin condensation is important in spermatozoa for the protection of paternal genes against damage. In the extreme case of fish sperm, histones are replaced by protamines, polypeptides that are exceedingly rich in arginine. Cells that have a low rate of proliferation accumulate a histone called $H_1^\circ$ (Pieler *et al.*, 1981); this basic protein shares immunological determinants with both bird histone H5 and histone H1 (Pehrson and Cole, 1981; Mura and Stollar, 1981).

It was long believed that histones play a major role in the regulation of gene activity, but it is now established that they are essentially structural constituents of chromatin. In particular, histones H2A, H2B, H3, and H4 spontaneously form octamers around which a DNA molecule can wind; such a histone tetramer–DNA complex constitutes a nucleosome, which will be discussed later in this chapter.

Refined methods for histone separation have shown that these proteins are heterogeneous and may be subdivided into several subclasses. For instance,

HeLa cells possess two distinct H1 histones, called H1A and H1B, with an $M_r$ of 21,000 and 22,000, respectively (Ajiro *et al.*, 1981). The main reason for histone heterogeneity is that they easily undergo modifications, which are catalyzed by enzymes present in the cell nucleus; histone phosphorylation (by histone kinases) and acetylation (by histone acetylases) are important processes because they affect the structure of chromatin in such a way that DNA transcription is facilitated or impeded. Histone modifications can thus play a modulatory role in the control of gene activity.

Bradbury *et al.* (1973) and Lake (1973) have found that histone H1 phosphorylation plays a very important role in chromosome condensation during mitosis. Similarly, the compact configuration of satellite DNA in DNA heterochromatic regions is believed to be due to the fixation of phosphorylated histone H1 to DNA (Blumenfeld *et al.*, 1978); it has, indeed, been shown that histones H1 and H3 are superphosphorylated in condensed chromatin (Gurley *et al.*, 1978).

While histone phosphorylation reduces transcriptional activity in chromatin by inducing its condensation, histone acetylation has the opposite effect. This was found in studies on the effects of butyrate on cells in culture. It was discovered that butyrate inhibits the deacetylation of acetylated histones both *in vivo* and *in vitro* (Vidali *et al.*, 1978; Candido *et al.*, 1978; Sealy and Chalkley, 1978). As a result, butyrate treatment leads to an accumulation of hyperacetylated histones (in particular, H3 and H4) and the arrest of cell division. When chromatin undergoes condensation, as in the highly condensed metaphase chromosomes, histone acetylation is decreased (Gómez-Lira and Bode, 1981). Closely related to the histones is a chromosomal protein called A24 protein. It is composed of histone H2A conjugated covalently to another basic protein called ubiquitin (or HMG 20, to indicate that it belongs to the "high-mobility-group" proteins). This 8000-dalton A24 protein is freed from the histone conjugate during mitosis (Matsui *et al.*, 1979; Wu *et al.*, 1981). According to these authors, metaphase chromosomes possess all classes of histones, but they are no longer "ubiquitinated." The subject has been reviewed by Levinger and Varshavsky (1982), who concluded that the A24 protein is mainly associated with actively transcribed genes and that it modifies the interactions between adjacent nucleosomes. Many other proteins are enzymatically bound to ubiquitin; ubiquitin adenylate is an intermediary in the reaction (Haas *et al.*, 1983).

There is no need to discuss in detail here the organization and transcription of histone genes, since they have been the subject of several detailed and excellent reviews (Kedes, 1979; Hohmann, 1981; Hentschel and Birnstiel, 1981). As we have already seen, families of histone genes may comprise several hundred genes (as in *Drosophila* and sea urchins). The exact sequence of these genes in the histone "operon" is known in many cases. However, the word "operon" is misleading, since there is absolutely no evidence that the transcription product of the histone gene cluster is polycistronic. Each gene, which is separated from its neighbor by a spacer (which varies from one sea urchin to another, according to Kohn and Kedes, 1979, who also showed that sea urchin histone genes have no introns), is individually transcribed. In a remarkable paper, Grosschedl and Birnstiel (1980) identified the putative regulatory and promoter sequences of the sea urchin H2A histone gene. However, one should recall that the sea urchin histone gene organization is far from universal. In humans, mice, chicken, and toads, the histone genes are scattered and separated by long stretches of nonhistone DNA, and there are only a few tens of histone gene copies per genome.

Histones are synthesized at a low level in nondividing cells; the newly synthesized histones immediately move into the nucleus, where they are quickly incorporated into chromatin in the form of nucleosomes (Wu *et al.*, 1983). This small amount of histone synthesis, which takes place in cells where DNA replication is absent, probably compensates for the degradation of preexisting molecules. Histones are unstable molecules that turn over in nonproliferating organs such as liver and kidney; each histone turns over individually, according to Djondjurov *et al.* (1983).

Finally, we should briefly discuss the high-mobility group (HMG) proteins discovered by Goodwin *et al.* (1973). As their name implies, these acid-soluble proteins are characterized by rapid migration during electrophoresis. The main representatives of this group are the HMG-1 and HMG-2 (with variants 2A and 2B) and the HMG-14 and -17 proteins, the former of which have a much higher molecular weight than the latter. There is good evidence that HMG-14 and -17 proteins bind selectively to regions of chromatin that are active in transcription (Weisbrod *et al.*, 1980; Gazit and Cedar, 1980; Weisbrod and Weintraub, 1981; Weisbrod, 1982b); it is probable that their binding favors the access of the transcribing enzymes (RNA polymerases) to DNA by modifying the molecular structure of chromatin. According to Isackson and Reeck (1981), HMG-14 and -17 proteins bind preferentially to single-stranded DNA, and their content decreases in metaphase chromosomes. All we know about the role of HMG-1 and -2 proteins in the living cell is that they are lost when erythroleukemic Friend cells are committed to differentiate (Seyedin *et al.*, 1981). Thus, the presence of HMG proteins, like histone modification by phosphorylation or acetylation, might control DNA transcription by locally modifying the molecular organization of chromatin. This conclusion is reinforced by recent findings: undermethylated sequences (which are potentially active in transcription) are associated with HMGs, while DNA sequences rich in methylcytosine are preferentially associated with histone H1 (which is involved in chromatin condensation and inactivation) (Ball *et al.*, 1983). Injection into fibroblasts of antibodies raised against histones and HMG-17 inhibits transcription, while anti–HMG-1 and anti–HMG-2 antibodies have no effect; the antibodies, after binding to histones and HMG-17, probably prevent the movement of the RNA-synthesizing enzymes

(RNA polymerases) along chromatin. However, Seale *et al.* (1983) failed to find a correlation between HMG-14 and -17 content and transcriptional activity in chromatin. Westermann and Grossbach (1984) recently reported that only regions of the giant *Drosophila* salivary glands chromosomes (see Chapter 5) that are transcriptionally active contain HMG-14. Since this protein is absent from these regions (puffs) prior to gene activation, a role of HMG-14 in gene activation is clearly suggested.

### C. Molecular Organization of Chromatin: Nucleosomes[4]

As has occurred more than once in the history of molecular biology, important supramolecular structures were described by electron microscopists before their biological significance became established through the use of chemical and physical techniques. The pioneering work of Olins and Olins (1974) showed that spread chromatin has a beaded structure. As shown in Fig. 20, the beads (which are around 10 nm in diameter) are equidistant on a thread of DNA that links them together. The beads, called "$\nu$-bodies" by Olins and Olins (1974), are now called "nucleosomes." The use of physicochemical techniques has provided us with a great deal of information on their chemical nature and their organization at the molecular level (reviewed by Kornberg, 1977).

The molecular structure of chromatin has been analyzed by ultracentrifugation in a sucrose gradient followed by gel electrophoresis of the fragments obtained after progressive digestion of isolated chromatin by deoxyribonucleases (DNases). DNA is not evenly digested, but is cut into fragments about 200 bases long (or multiples of 200 bases). Figure 21 shows the presence of monomeres, dimeres, trimeres, etc., in the digest. A considerable amount of work, done simultaneously in a number of laboratories and summarized by Kornberg (1977), who played a major part in these studies, has led to the following conclusions. The nucleosomes ($\nu$-bodies) are composed of a histone octamer "core" formed by the association of two molecules of each of the four histones—H2A, H2B, H3, and H4. In addition, a DNA thread, coiled around the histone core (Oudet *et al.*, 1975), links adjacent nucleosomes (Fig. 22). The distance between two nucleosomes, called the "repeat length," is about 200 bp of DNA. The histone octamer core is not a sphere but a disk of $5.5 \times 11 \times 11$ nm. The DNA that is wound around the histone core is 146 bp long and makes a $1\frac{3}{4}$ turn around the core, which has been confirmed by the x-ray crystallography analysis of Finch *et al.* (1981). The structure of the nucleosome core and the number of base pairs of DNA wound around it are remarkably constant; nucleosome organization seems to be identical in euchromatin, heterochromatin, and metaphase chromosomes (Bostock *et al.,* 1976; Bokhon'ko and Reeder, 1976). The repeat length of nucleosomes (about 200 bp) may vary from one tissue to another due to the fact that the length of the spacer DNA, which links adjacent nucleosomes, is not

[4]Reviewed by Cartwright *et al.* (1982) and Butler (1983).

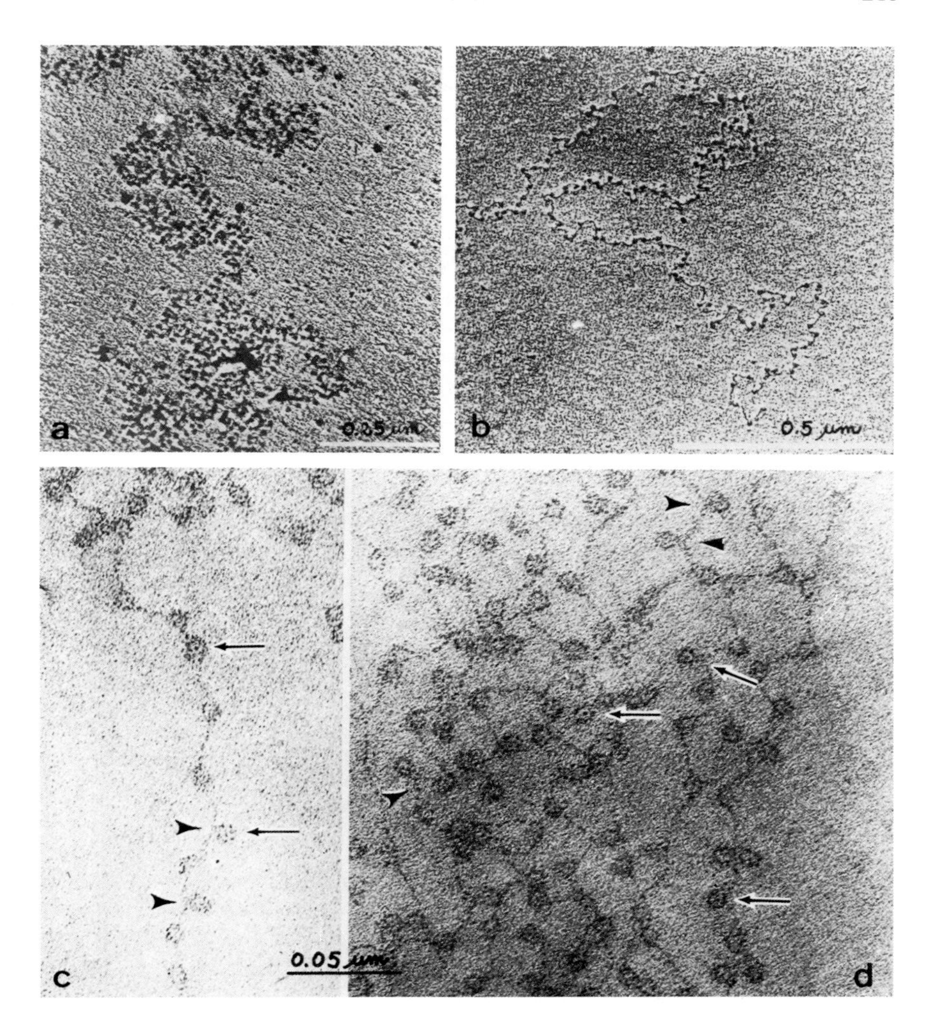


FIG. 20. Nucleosomes. (a) Chicken liver chromatin spilling out of nuclei lysed directly on a grid (shadow casting). (b) Same as (a), but after trypsin digestion to remove histone H1. (c,d) Electron micrographs of uranyl-stained spreads of chromatin fibers from isolated chicken erythrocyte nuclei. Thin arrows denote nucleosomes showing a central spot of stain; thick arrowheads denote nucleosomes with lateral association with the connecting strand. [(a) Oudet *et al.* (1975), copyright by M.I.T.(c,d) Olins and Olins (1976). Reproduced from *The Journal of Cell Biology,* 1976, **68,** p. 789, by copyright permission from The Rockefeller University Press.]

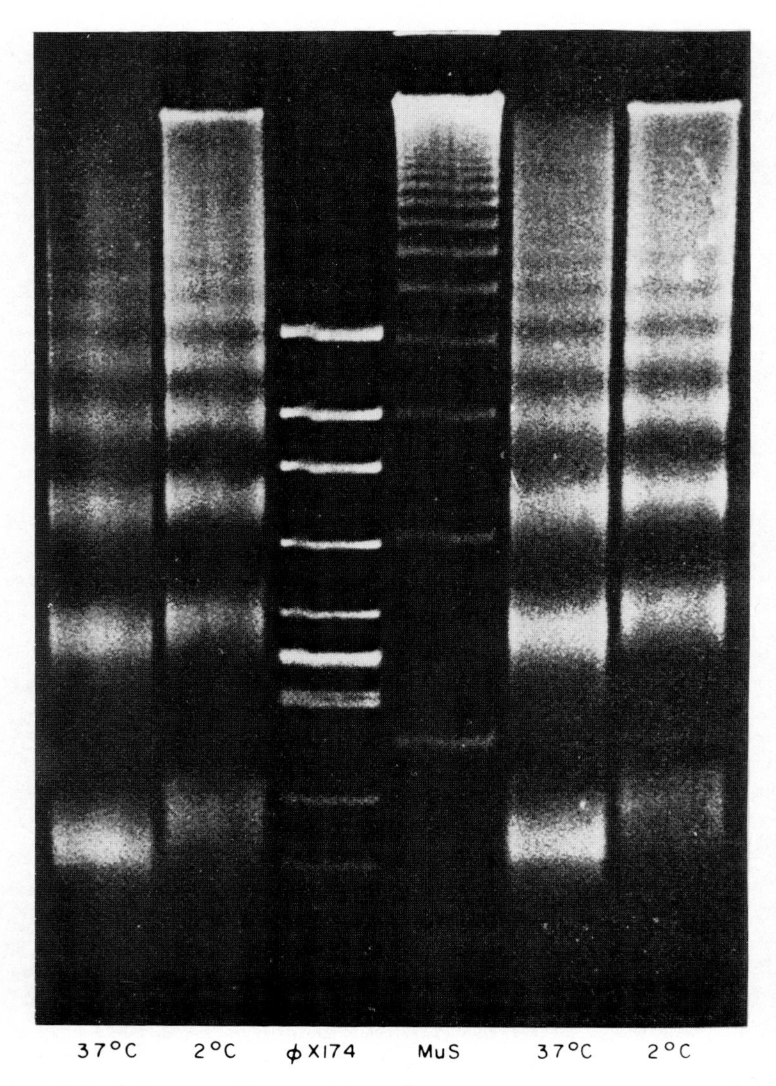


FIG. 21. Polyacrylamide gel electrophoresis of micrococcal nuclease digests of chromatin at 37° and 2° and of restriction enzyme fragments of bacteriophage $\phi \times 174$ and mouse satellite DNA (MuS). The DNA has been extracted after digestion of isolated nuclei. MuS was used for calibration. The beads correspond to multiples of 200 bases in the digest (monomers, dimers, trimers, etc.). [Noll and Kornberg (1977). With permission from *J. Mol. Biol.* Copyright: Academic Press Inc. (London) Ltd.]

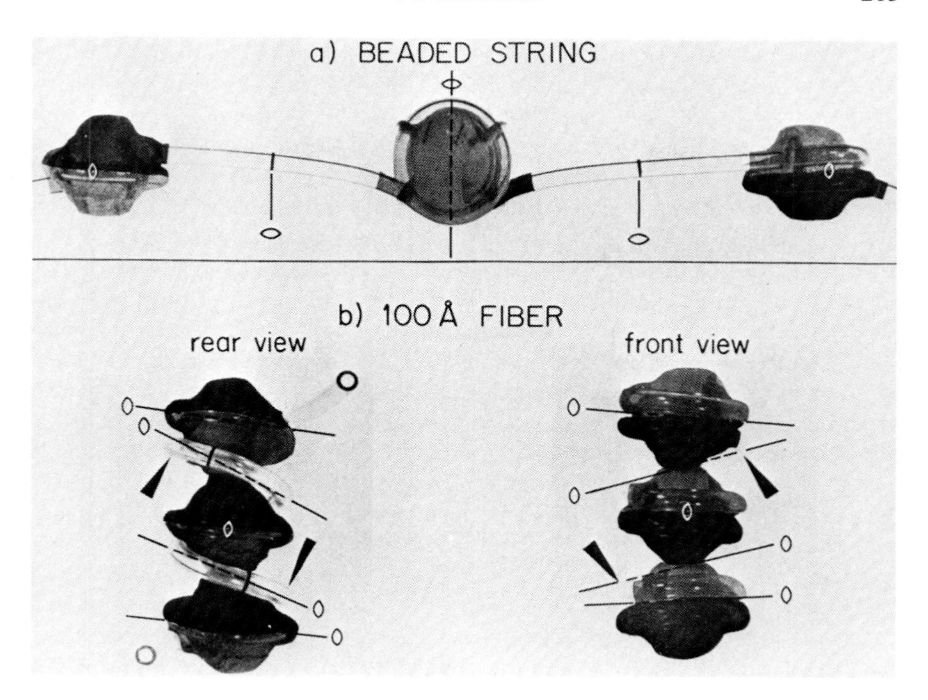


FIG. 22. Schematic representation of the chromatin organization in nucleosomes. In this model, the nucleosome core is represented by tennis balls and DNA by plastic tubing. (a) The beads-on-a-string configuration results from unwinding of part of the DNA and loss of histone H1 during sample preparation. (b) In the 10-nm fiber, the nucleosomes are in close contact. [Worcel (1977).]

necessarily the same in all cells. For instance, the length of the repeating unit is greater in sea urchin sperm (260 bp) than in sea urchin embryo nuclei (220 bp) (Spadafora *et al.*, 1976; Keichline and Wassarman, 1977). According to Lohr *et al.* (1977), the length of the internucleosomal spacer or linker may vary from 20 to 50 bp. A more recent estimation by McGhee *et al.* (1980) gives a range of 0–80 bp for its length. According to a report by Prunell and Kornberg (1982), the length of linker DNA between nucleosomes is highly variable within a hepatic cell. Histone H1 and, in bird erthrocytes, histone H5 (Frado *et al.*, 1983) are bound to the linker DNA; however, linker DNA is more susceptible to mild DNase digestion than are the nucleosomes, which is why chromatin, as we have seen in Fig. 21, is cut by the enzyme into mono-, di-, and trinucleosomes. Transcriptionally active genes have an increased nucleosomal spacing; their long linkers increase their susceptibility to DNase digestion (Smith, *et al.*, 1983).

The DNase susceptibility of chromatin has become a very important subject since Weintraub and Groudine (1976) and Garel and Axel (1976) discovered that genes that are expressed are much more sensitive to DNase digestion than others (reviewed by Elgin, 1981, and Weisbrod, 1982a). For example, globin and ovalbumin genes display greater susceptibility to DNase in the chromatin of the organ in which they are expressed (hematopoietic cells and oviduct, respectively,) than in chromatin from tissues that do not synthesize these proteins. For instance, Bellard *et al.* (1980) have shown that pancreatic DNase preferentially digests the ovalbumin gene in oviduct chromatin and the β-globin gene in chicken red blood cells. The high DNase suceptibility of a DNA segment [for instance, the gene extra copy in the aforementioned case of the variable antigens studied by Pays *et al.* (1981a) in trypanosomes] is taken as a strong argument for the view that this segment corresponds to an active gene. Gene activity requires gene transcription by an enzyme, namely, an RNA polymerase. It thus follows that gene activity is correlated with a particular conformation of DNA in chromatin, a conformation that allows free access of RNA polymerase molecules to the DNA segments that will be transcribed. How these conformation changes take place in chromatin is not yet fully understood, but we have already uncovered some of the factors involved. We have seen that histone H1 phosphorylation is somehow associated with the condensation of chromatin; dephosphorylation of histone H1 by a protein phosphatase, as well as core histone acetylation, should locally loosen the chromatin and favor the access of RNA polymerases to the DNA template. We have also seen that the HMG-14 and -17 proteins seem to be required for the DNase sensitivity of regions in the DNA molecule (Gazit and Cedar, 1980; Levy-Wilson, 1981); however, this has been questioned by Nicolas *et al.* (1983). There is also a frequent correlation between susceptibility to this enzyme and undermethylation of active gene sequences. In this respect, it may be significant that, in some cells at least, the major portion of DNA methyltransferase activity is associated with the internucleosomal linkers (Creusot and Christman, 1981). It should be added that DNase sensitivity is not, according to one report, limited to the active gene, but may extend for 250–300 bp upstream of the 5′-end of the gene itself. This is the case for the preproinsulin II gene in the chromatin of cells that synthesize insulin, but not in other cells (Wu and Gilbert, 1981). According to Lawson *et al.* (1982), the DNase-sensitive region of chicken oviduct nuclei extends over 100 kb, comprising the ovalbumin gene, the neighboring X and Y genes, and the adjacent sequences. However, this DNA region is insensitive to DNase in chicken red blood cells, and liver and sperm nuclei. Interestingly, Bellard *et al.* (1982) found that a 2500-bp region at the 5′-end of the ovalbumin gene is not organized in nucleosomes in oviduct nuclei, while it has the classic nucleosomal structure in red blood cell nuclei. A similar finding has been reported for the mouse ribosomal genes: they are not arranged in nucleosomes in cells very active in ribosome production (Davies *et al.*, 1983). Widmer *et al.* (1984) recently found that when the gene that produces the salivary proteins in *Drosophila* larvae (the *BR* locus, to be discussed in Chapter 5) is hyperactivated by pilocarpin, it looses its nucleosomal organization. Nucleosomes can easily be detected in the inactive *BR* gene of *Drosophila* cultured cells.

Another factor that seems to play a role in DNase sensitivity is the conformation of the DNA molecules themselves. Immunological studies on *Drosophila* giant chromosomes (see chapter 5) have suggested that regions believed to be less active in transcription are overmethylated and contain a larger proportion of Z-DNA (left-handed, zigzag DNA) (Nordheim *et al.*, 1981). However a recent reinvestigation of this question by Hill *et al.* (1984) has shown that isolated *Drosophila* chromosomes do not bind antibodies against Z-DNA unless they have been fixed with acetic acid or ethanol. Artifacts can never be ruled out under these conditions.

Curiously, the DNase-hypersensitive sites do not always have the same localization in interphase nuclei and in the highly condensed metaphase chromosomes. Chromatin condensation seems to alter the nuclease-hypersensitive cleavage sites (Kuo *et al.*, 1982). This agrees with the conclusion drawn by Nicolas *et al.* (1983) that DNase hypersensitivity depends on the "supranucleosomal" organization of the chromatin, which will soon be discussed.

The reader who wishes to learn more about our present knowledge of active chromatin is advised to read the excellent reviews by Weisbrod (1982a) and Reeves (1984); nuclease sensitivity, the role of HMG-14 and -17, histone acetylation, DNA undermethylation, and the formation of Z-DNA are thoroughly discussed. Thus, the complex regulation mechanisms at the level of DNA transcription are finally being uncovered, but one cannot escape the feeling that other exciting breakthroughs lie ahead.

Compact chromatin (as in sperm heads and avian red blood cells) is inactive or poorly active in contrast to loose chromatin. This raises the question of whether actively transcribed chromatin is always organized in nucleosomes. As we shall see later in more detail, gene transcription can be visualized under the electron microscope using a technique devised by Miller and Beatty in 1969. As shown in Fig. 23, the transcription units present, in spread chromatin, a characteristic "Christmas tree" configuration. Longer lateral fibrils, corresponding to ribonucleoproteins containing the growing RNA chains, grow from their point of origin on the DNA fiber; they are made up of newly synthesized RNA molecules associated with proteins. In the nucleolar chromatin shown in Fig. 23, these transcription units are separated by nontranscribed spacers. On the DNA fiber, granules are seen at regular intervals. It is generally accepted that they correspond to molecules of the transcribing enzyme, an RNA polymerase. Their presence might obscure the typical "beads-on-a-string" pattern characteristic of the nucleosomes. For this reason, there are still discrepancies in the interpretation of electron micrographs obtained using spreading techniques that are not

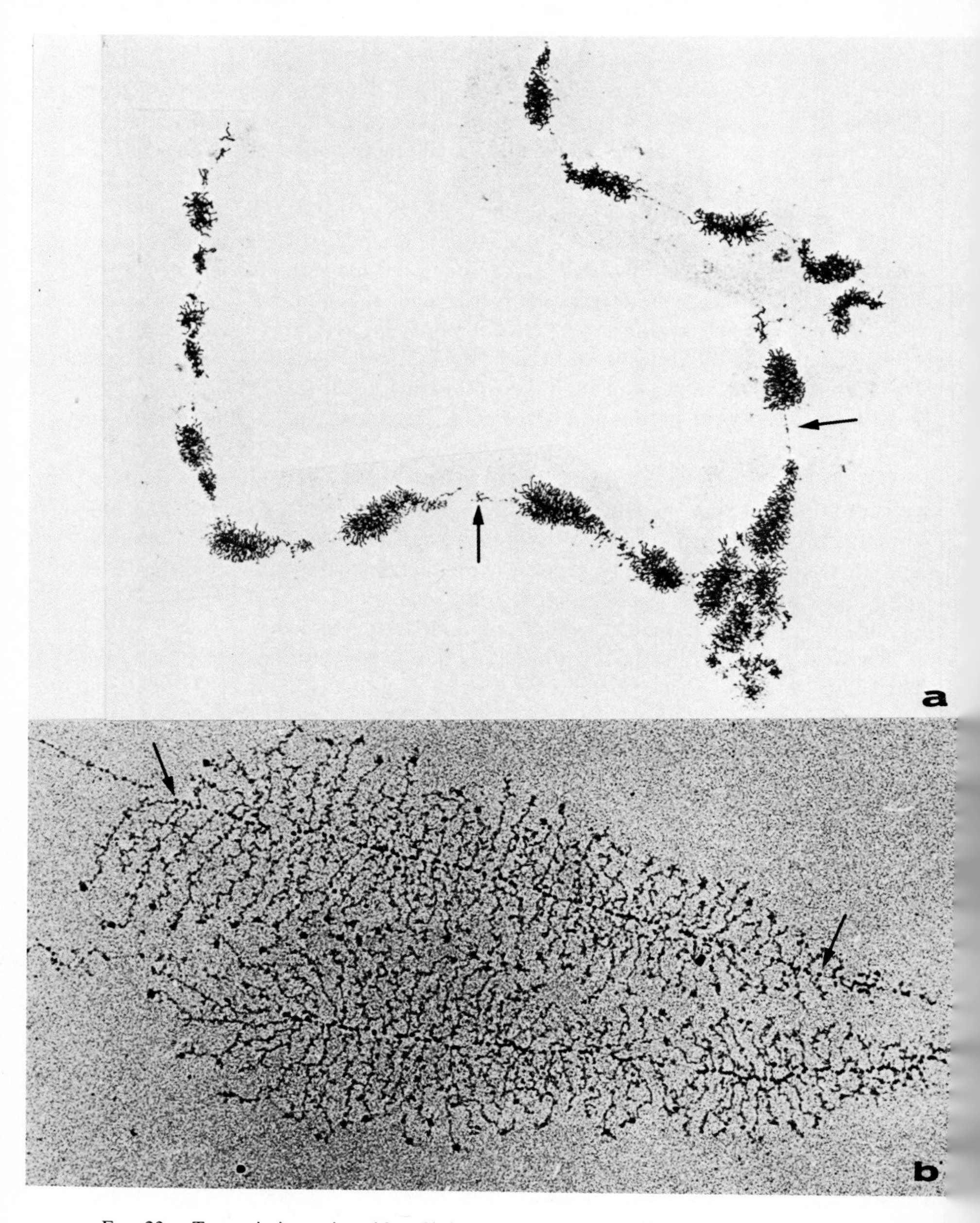


FIG. 23. Transcription units with a Christmas tree configuration in nucleolar ribosomal genes. The lateral fibrils correspond to ribonucleoproteins containing rRNA chains growing from their point of origin on the DNA fiber. (a) Arrows show ''spacers'', i.e., untranscribed regions of the DNA fiber. (b) Arrows correspond to RNA polymerase I molecules. [Courtesy of Dr. O. L. Miller, Jr.]

always perfectly reproducible. For Franke *et al.* (1980), DNA is extended in transcriptionally active chromatin and is packed into nucleosomes when it is inactive; however, there are no nucleosomes in the transcriptionally inactive spacers between the individual ribosomal genes (transcription units). On the other hand, Martin *et al.* (1980) reported that nucleosomes can be observed within active transcription units in regions where RNA polymerase molecules do not obscure the ultrastructure of chromatin. A more recent paper by Scheer *et al.* (1981) provides evidence that, in the myxomycete *Physarum,* many transcription units and even their adjacent nontranscribed regions are not organized in nucleosomes. It has also been reported that the chromatin of dinoflagellates does not display the beads-on-a-string pattern. The conclusion that can be drawn from presently available evidence is that the organization of chromatin in nucleosomes is the general rule, but that it would be a mistake to believe that this organization is a stable, static one. Conformational and biochemical changes certainly occur when transcription takes place, and one should keep in mind the possibility that some of the nucleosomes—if not all—are transient structures that can open up and then re-form. As we have seen, the biochemical work of Bellard *et al.* (1982), Davis *et al.* (1983), Karpov *et al.* (1984), and Widmer *et al.* (1984) supports the view expressed by Franke and his colleagues that active regions of chromatin are not organized in nucleosomes.

The winding of DNA around the nucleosome cores allows a five- to sevenfold packing of the double-helix molecule. As already pointed out, folding of the 10-nm nucleosome chain in the 20- to 30-nm fibers seen in spread chromatin can be explained by a solenoidal mechanism. X-ray crystallography suggests that such a solenoidal, supranucleosomal structure would comprise about six nucleosomes per turn (Finch and Klug, 1976). For Worcel *et al.* (1981), the supranucleosomal structure (a 25-nm fiber) is a zigzag helical ribbon that may be twisted. Other models, based largely on the reconstitution of nucleosomes (i.e., mixing of DNA and the various histones under different conditions), have been proposed by Weintraub (1980), McGhee *et al.* (1980, 1983), and Butler and Thomas (1980). They all conclude that the supranucleosomal structure results from solenoid formation (Fig. 24), but a rope-like model has been proposed by Cavazza *et al.* (1983). This conclusion is shared by Butler (1984) in a more recent study, while Woodcock *et al.* (1984) have proposed a slightly different model: chromatin folding would result from the formation of a zigzag ribbon, which is then coiled in a double helical structure (helical ribbon model).

Another kind of supranucleosomal arrangement has been proposed by Franke and his colleagues (1980). The formation of globular superbeads (18–32 nm in diameter) in transcriptionally inactive chromatin. Some biochemical evidence in favor of such a model has been recently brought forward by Weintraub (1984). According to him, inactive genes are located in supranucleosomal particles containing 20–40 kb of DNA maintained by H1 and H5 histones; active genes are

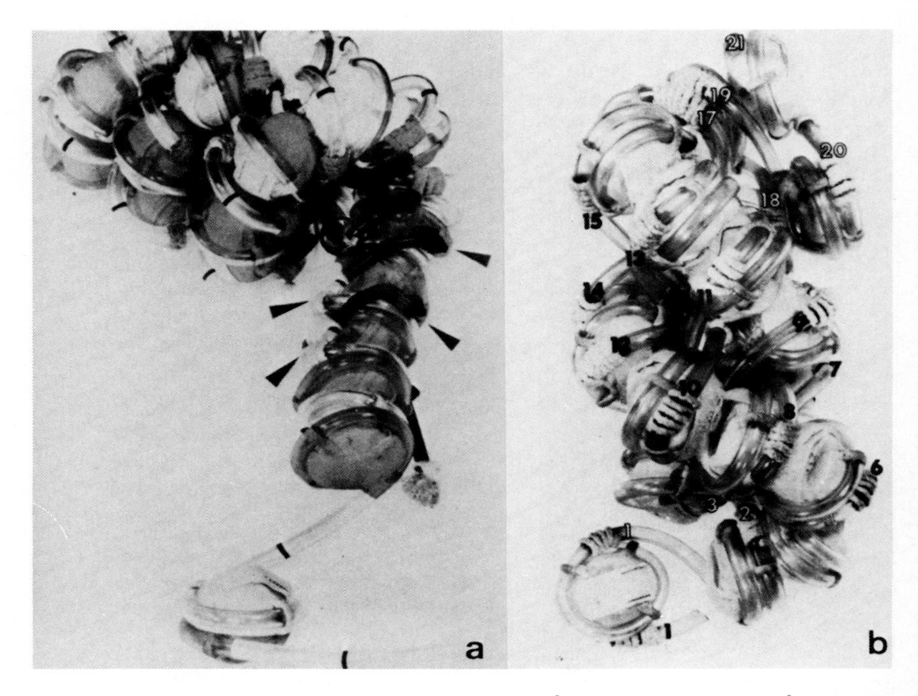


FIG. 24. Supranucleosomal structure. Coiling of a 100-Å fiber into a 200- to 300-Å chromatin fiber. (a) Worcel model (1977), in which the arrows point to the Hl termini interactions with the 100-Å fiber. (b) More recent model of Worcel *et al.* (1981) in which the 250-Å chromatin fiber is seen as a twisted ribbon. The spacer DNA remains relaxed as it moves from one nucleosome to the next, and there are 1.75 DNA turns per nucleosome.

contained in similar particles, but the polynucleosomes are not held together after DNase cleavage of the linker DNA. The evidence obtained in Franke's laboratory is based on electron microscopy that shows that condensed chromatin contains chains of closely juxtaposed chromatin granules (Fig. 25). Interestingly, if a purified circular DNA is injected into the nucleus of a *Xenopus* oocyte, it is packed into globular superbeads about 30 nm diameter (Fig. 26) (Trendelenburg and Gurdon, 1978; Scheer *et al.*, 1980). Zentgraf *et al.* (1980) have further shown that the chromatin of isolated nuclei can be transformed into globular supranucleosomal units by modifications of the pH or ionic strength of the medium. It should be added that whether one accepts the solenoid or the superbead model, the degree of compaction of the DNA molecule would be around 30-fold, which is still a long way from the 5000 to 10,000-fold packing of DNA in metaphase chromosomes. We do not know how to fill this enormous gap, but it is very likely that the nuclear matrix (and the chromosomal scaffold to be examined later) play an important role in the folding of the chromatin fibers at the cellular level. Chromatin loops almost certainly hook and wrap around a proteinaceous matrix.

The fact that circular DNA, after injection into the nucleus of a *Xenopus* oocyte, undergoes condensation has led to interesting attempts to identify the chemical factors that convert a mixture of DNA and histones into nucleosomes. One of these factors is nucleoplasmin, a major acidic, phosphorylated nuclear protein that was first isolated from *Xenopus* oocyte nuclei (Krohne and Franke, 1980a,b; Laskey *et al.*, 1978). This protein is present in all nuclei, except in the highly condensed nuclei of spermatozoa and chicken erythrocytes (Krohne and Franke, 1980a,b). Nucleoplasmin, which is very abundant in the nuclear sap, is a pentamer that binds to histones, but not to DNA (Earnshaw *et al.*, 1980). Other factors probably involved in the folding and unfolding of chromatin fibers are the DNA topoisomerases (reviewed by Champoux, 1978). An ATP-dependent DNA topoisomerase I (called the "nicking–closing enzyme," indicating that it can open and seal circular DNA molecules), present in *Drosophila* embryos, converts circular DNA into a huge network of catenated rings, which is similar to the organization of kDNA (the mitochondrial DNA of trypanosomes, Chapter 3,

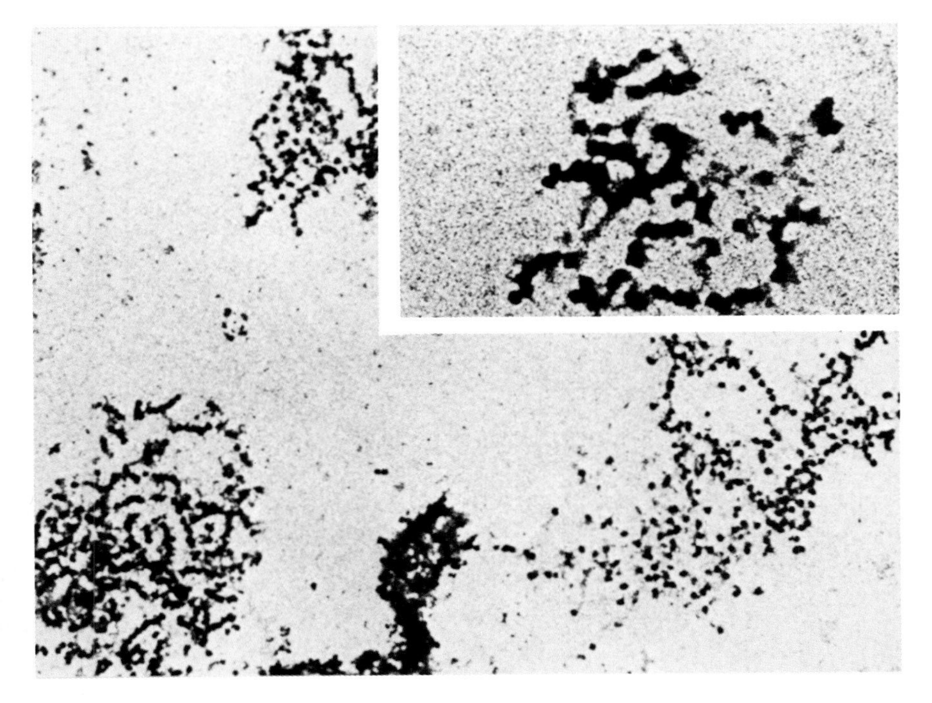

FIG. 25. Large (18- to 32-nm) chromatin granules representing supranucleosomal orders of chromatin packing in spread chromatin. Arrays of "superbeads." [Franke *et al.* (1980).]

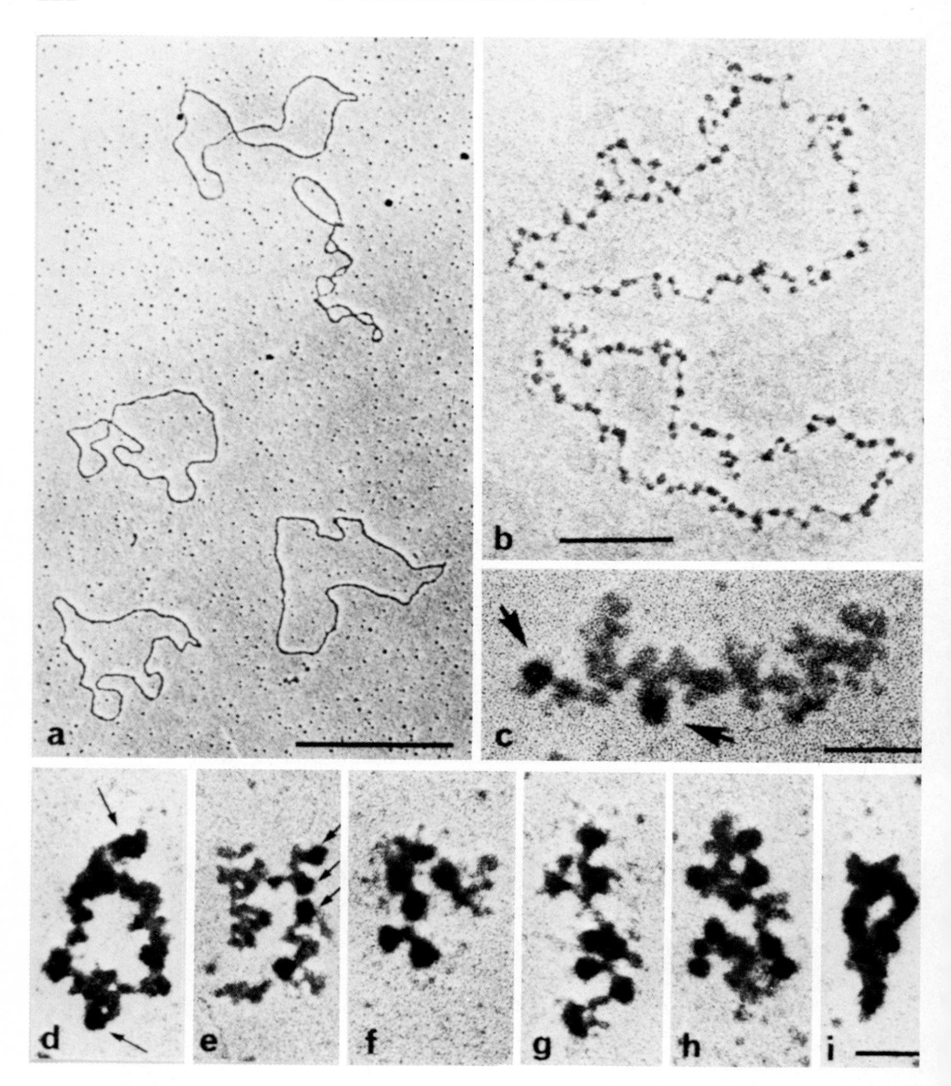


FIG. 26. Conversion of plasmid DNA (a) into nucleosomal and higher-order chromatin structures after injection into the nuclei of *Pleurodeles* oocytes, as seen in spread preparations made 3 hr after injection. The plasmids are recovered in the form of open beaded chromatin circles (b) or dense aggregates (c–i). [Scheer *et al.* (1980).]

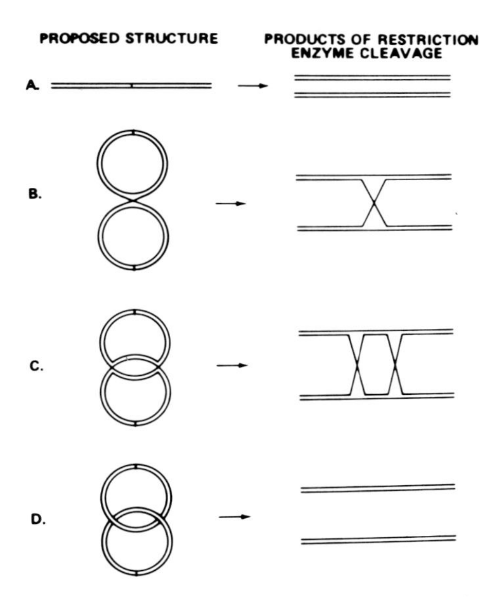


FIG. 27. Models explaining the action of *Drosophila* ATP-dependent DNA topoisomerase. Models shown involve (A) covalently linked linear tandem arrays, (B) a single exchange of DNA strands between two duplexes (also covalently linked), (C) two-strand exchanges between duplexes, and (D) catenated or interlocked rings. [Hsieh and Brutlag (1980). Copyright by M.I.T.]

shown in Fig. 27). Other DNA topoisomerases relax closed duplex DNA molecules, as shown in Fig. 27 (Hsieh and Brutlag, 1980). DNA topoisomerases I introduce single-strand breaks in the DNA duplex molecule, while DNA topoisomerases II introduce double-strand breaks. The biological function of these enzymes is not known for certain; it is possible that, besides playing a role in chromatin condensation, they are instrumental in inserting into the genome the mobile genetic elements discussed earlier. Such a role would fit with their nicking–resealing abilities. How nucleoplasmin and DNA topoisomerases (which bind, respectively, to histones and DNA) might be involved in nucleosome assembly has been discussed by Laskey and Earnshaw (1980).

### D. NON-HISTONE CHROMOSOMAL PROTEINS

These proteins can be isolated together with DNA from purified chromatin (reviewed by Elgin and Weintraub, 1975); the topoisomerases and many other non-histone DNA-binding proteins conform to this definition. A broader definition was used in ‘‘Biochemical Cytology’’ (Brachet, 1957), where we spoke of ‘‘residual nuclear proteins’’ for all of the proteins that remain after nucleohistone has been selectively extracted. As pointed out by Elgin and Weintraub (1975), non-histone chromosomal proteins overlap with, but are not identical to, the numerous acidic proteins and phosphoproteins that can be extracted from isolated nuclei. In fact, it is not always easy to decide whether a given protein is present in chromatin or in the nuclear sap, or in both. Furthermore, as already pointed out, loss of proteins through the leaky nuclear envelope or nonspecific adsorption of proteins on the nuclei during the isolation procedure is difficult to rule out in the absence of adequate controls.

The main trouble with non-histone chromosomal proteins compared to histones is their great number—a number that has steadily increased as methodology to isolate them has improved. In their 1975 review, Elgin and Weintraub spoke of 15–20 major proteins. However, a more recent report by Peterson and McConkey (1976) stated that, in HeLa cells, there are about 450 non-histone proteins associated with chromatin and 300 other proteins in the nuclear sap; the method used detected about 500 proteins in the cytoplasm. According to Stein (1978), 200–500 different proteins difficult proteins are present in nuclei. A group of 60 proteins, including actin, is found in all three cell compartments. Unpolymerized tubulin is also associated with chromatin (according to Menko and Tan, 1980, as well as actomyosin (Comings and Harris, 1975).

Interest in non-histone chromosomal proteins stems from the fact that, in contrast to histones, they display a marked tissue specificity (Bekhor *et al.*, 1974). Furthermore, there is considerable evidence for the view that active euchromatin contains more non-histone proteins than condensed heterochromatin (see, for instance, the work of Comings *et al.*, 1977, on *Drosophila*). This has led to the general view that non-histone chromosomal proteins must play an important role in the regulation of gene activity, a subject that was reviewed by Stein *et al.*, in 1974. There is no reason to doubt that non-histone chromosomal proteins might play such a role, as well as affecting the molecular organization of chromatin. However, it should be borne in mind that, at present, we know next to nothing about this problem.

Among the non-histone chromosomal proteins are many enzymes involved in the metabolism of DNA and its associated proteins. The most important of these enzymes are the DNA and RNA polymerases. In addition, there are endonucleases that can introduce single-strand nicks or double-strand cuts into the DNA molecules (Machray and Bonner, 1981), DNA ligase, and poly(ADP-ribose) polymerase, which are involved in DNA repair, DNA methyltransferases, histone acetyl- and methyltransferases, histone kinases (all the enzymes that are involved in DNA and histone modification); DNA topoisomerases I and II; a specific histone protease, protein kinases, which can phosphorylate endogenous non-histone proteins; phosphatases, etc. Thus, as expected, all of the major enzymes required for DNA and RNA synthesis and breakdown, as well as for modification of DNA and its associated proteins, are present in preparations of isolated chromatin.

Of particular importance are the DNA and RNA polymerases, which are directly involved in DNA replication and DNA transcription, respectively. There are three major species of DNA polymerases called α, β, and γ (reviewed by Weissbach, 1977). It is widely believed, but not yet proved, that the α-enzyme (which is specifically inhibited by aphidicolin) plays a major role in DNA replication and the β-enzyme in DNA repair (these two processes will be examined in Chapter 5). DNA polymerase-γ is the only DNA polymerase present in mitochondria and is certainly involved in mitochondrial DNA replication (see, for instance, Hübschler *et al.*, 1979). The function of the three DNA polymerases and their general mechanism of action are the same: they copy a single-strand DNA segment added as a template in a reaction mixture, which comprises the enzyme and the four deoxyribonucleoside triphosphates (dATP, dGTP, dCTP, and dTTP) needed as precursors for DNA synthesis (Fig. 28). Although DNA polymerases-α and -β must necessarily be attached to chromatin DNA in order to function, this does not imply that the enzymes are always located in chromatin or even in the nucleus. In fact, DNA polymerase activity is almost entirely cytoplasmic in unfertilized sea urchin or frog eggs. As we shall see in Volume 2,

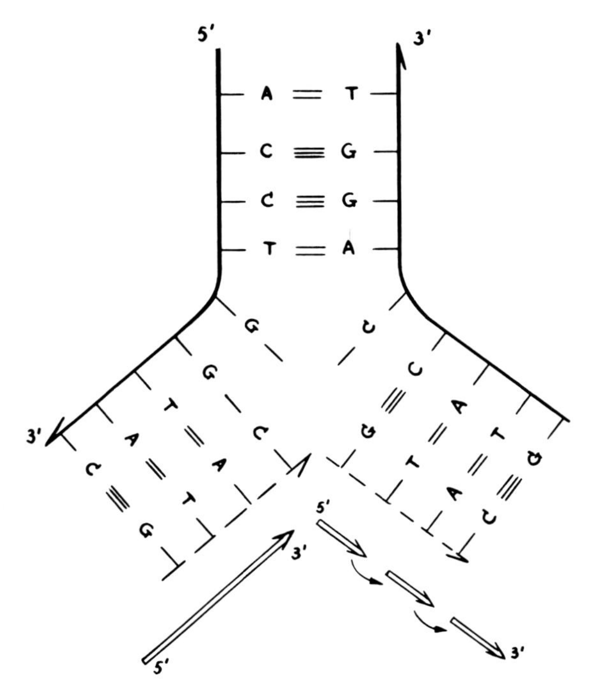


FIG. 28. Schematic representation of DNA replication by a DNA polymerase. For details, see Chapter 5. [Courtesy of P. Van Gansen.]

Chapter 2, during the rapid cleavages that follow fertilization, enzymatic activity progressively moves from the cytoplasm to the nuclei.

There are also three major species of RNA polymerases (I, II, and III) in all eukaryotic cells specializing in the synthesis of the major types of RNAs. RNA polymerase I, which binds to the nucleolar DNA, is responsible for the synthesis of the 28 S and 18 S ribosomal RNAs; RNA polymerase II directs the synthesis of the mRNAs, and can be selectively inhibited by α-amanitin; finally, RNA polymerase III specializes in the synthesis of the low molecular weight RNAs (5 S RNA and tRNAs, in particular). In addition, mitochondria have their own RNA polymerase, which is a smaller and simpler molecule than the nuclear enzymes. As in bacteria, mitochondrial RNA polymerase is specifically inhibited by rifampicin. The overall mechanism of action of all of the RNA polymerases is the same: they transcribe a DNA template into an RNA copy (Fig. 13). A mixture of all four ribonucleoside triphosphates (ATP, GTP, CTP, UTP) serves as a precursor.

Like the DNA polymerases, the RNA polymerases are not always bound to chromosomal DNA. Thus, it is possible to separate template-bound (or engaged) RNA polymerases from free enzymes. The ratio between the two constitutes one of the control mechanisms of transcription, for instance, in amphibian oocytes (Thomas *et al.*, 1980).

As will be seen, the first products of the three RNA polymerases are not the rRNAs, mRNAs, 5 S RNAs, or tRNAs as they are found in the cytoplasm. Maturation (processing) of larger precursors into the final products is one of the most important activities of the cell nucleus.

## E. Nuclear RNAs: Chromosomal DNA Transcription[5]

The transcription of the ribosomal genes, coding for the 28 and 18 S rRNA, takes place on the DNA of nucleolar organizers. Since this is a typically nucleolar function, it will be discussed in the next section.

The RNA population present in the nucleus, which results from the transcription of chromosomal DNA associated with chromatin fibrils, is very heterogeneous. For this reason, it is known as "heterogeneous nuclear RNA (hnRNA)." hnRNA becomes associated with proteins as soon as it has been transcribed on its DNA template; it is, therefore, in the form of nascent ribonucleoprotein particles (hnRNP). These hnRNP were called "informofers" by Lukanidin *et al.* (1972), who concluded that they were typically nuclear particles and were thus absent from the cytoplasm. Their possible role in the control of genetic activity was discussed by Georgiev in 1972 and by Pederson in 1983. About 20 proteins are present in the 30 S hnRNP (informofers), where they

[5]Reviewed by Breathnach and Chambon (1981), Paul (1982), Lewin (1981), and Knowler (1983).

represent 80% of the particles (Kulguskin *et al.*, 1980). However, according to Economidis and Pederson (1983), hnRNA is complexed *in vivo* with only six proteins to form hnRNPs; the assembly of hnRNA and proteins is a very fast process that takes place as soon as the hnRNAs are transcribed, even before the addition of a polyadenylate tract to the mRNA precursors. According to Pederson (1983), only the transcripts of intron-containing genes would be assembled in hnRNPs that play an important role in mRNA processing.

It was quickly discovered that hnRNA comprises very large RNA molecules (4–9 μm long, according to Holmes and Bonner, 1978) and that it differs, among other things, from polysomal mRNA by the presence of highly repetitive sequences. There are numerous "snapback" RNA sequences that easily produce hairpins and loops that resist digestion by pancreatic RNase (Holmes and Bonner, 1978; Ryskov *et al.*, 1973; Jelinek *et al.*, 1974).

Improvements in nucleic acid technology—in particular, the introduction of the powerful molecular hybridization techniques—have led to a better characterization of the hnRNAs. Brandhorst and McConkey (1974) found that hnRNA molecules are unstable and have an average half-life of only 23 min; their rate of synthesis is rapid ($5.4 \times 10^{-2}$ μg/cell/min), and they correspond to as much as 7% of total cellular RNA. Surprisingly, only 2% of the hnRNAs synthesized in the nucleus are found in cytoplasmic polyribosomes. In agreement with this last conclusion, Hough *et al.* (1975) and Levy *et al.* (1976) found that the complexity of polyribosomal mRNA is 10 times less than that of hnRNA. According to Ryffel (1976), about 10,000 RNA species never leave the nucleus.

It is thus clear that only a minor part of hnRNA can serve as a precursor for cytoplasmic mRNA; the role—if any—of the bulk of hnRNA remains a mystery. Davidson *et al.* (1977) suggested that hnRNA plays a structural role rather than that of an mRNA precursor. More recently, Davidson and Britten (1979) proposed an interesting model in which hnRNA, particularly the transcribed middle repetitive sequences, is involved in the regulation of gene activity.

To be expressed, a gene must be transcribed in an mRNA molecule that is transferred to the cytoplasm, where it is translated into the protein coded by this particular gene. A good deal of work has been done on the complexity of the mRNA species present in the cytoplasm with the hope of establishing the total number of active genes present in an organism. Unfortunately, this number is not known. It has been estimated for the mouse as a maximum of 40,000–50,000 genes by Affara *et al.* (1977). In general, around 10,000 different mRNA sequences have been found in association with the polyribosomes, but it has been reported that the human brain possesses as many as 100,000 distinct mRNAs. Further, according to Ordahl and Caplan (1978), the more modest myoblasts (undifferentiated presumptive muscle cells) contain as many 124,000 different mRNA species.

It is sad to have to admit our ignorance of the number of active genes in humans and mice at a time when such rapid progress has been made in our understanding of the molecular mechanisms of gene transcription (reviewed by the editors of *Science* 209, 1980; Marx, 1981a; Lewin, 1981b; Coutelle, 1981; Baralle, 1983). A lengthy discussion of the numerous and complex steps that lead from the gene to cytoplasmic mRNA is outside the scope of the present book. They are summarized in Fig. 13.

In Fig. 13, the gene (DNA) supposedly has one intron separating two exons (coding sequences that will be expressed). Flanking the 5′-end (thus upstream) are sequences that are important for the initiation of *in vitro* transcription of RNA polymerase II. One of these sequences is the "TATA box" (Goldberg–Hogness box), which is similar to the promoter region found in bacteria. In 60 eukaryotic genes, this sequence has been found at about 30 nucleotides to the left of the 5′-extremity of the gene itself. It is believed to play an important role in the initiation of transcription, since replacement, by *in vitro* engineering, of the TATA sequence by a TAGA sequence almost suppresses the *in vitro* transcription of the conalbumin gene (Wasylyk *et al.*, 1980a). However, a study by Zarucki-Schultz *et al.* (1982) has shown that replacement of the TATA sequence by a TGTA sequence (by *in vitro* mutagenesis) does not impede the correct initiation of ovalbumin gene transcription, but *in vitro* expression of this gene is greatly decreased. Farther to the left (at about 20 bases upstream from the first base of the coding gene) is the "CCAAT box," which seems to be required for the efficient initiation of gene transcription. However, different results are obtained when one studies *in vivo* transcription, as Grosschedl and Birnstiel (1980) did. The sea urchin histone H2A gene was injected into *Xenopus* oocytes after *in vitro* modification of the 5′-flanking sequences. It was found that the presence of the TATA box is not absolutely indispensable for *in vivo* transcription, that deletion of the CCAAT box leads to an unexpected 20-fold increase in gene transcription, and that a sequence beginning 120 nucleotides upstream from the gene is required for transcription. Thus, its deletion greatly decreases the *in vivo* transcription of the injected sea urchin H2A histone gene, while its initiation requires the cooperation of three control regions [called modulator, selector, and initiator sequences (Fig. 29)]. Unexpectedly, promotion of sea urchin histone gene transcription after injection into *Xenopus* oocytes depends on an AT-rich sequence present in the spacer, which separates the histone genes from each other. *In vivo* transcription, after its introduction into mouse cells of the rabbit β-globin gene, also requires the integrity of three regions located in the 109-nucleotide stretch 5′ of the initiation codon: they are the TATA box region, the CCAAT region, and the −100 region (located 100 nucleotides upstream from the 5′-end of the gene). Point mutations in any of these three regions decrease *in vivo* transcription by a factor of 2 (Dierks *et al.*, 1983). It is believed that the role of the 5′-flanking regions is to modify the already discussed nucleosomal organization of chromatin in such a way that the initiation site is exposed to RNA

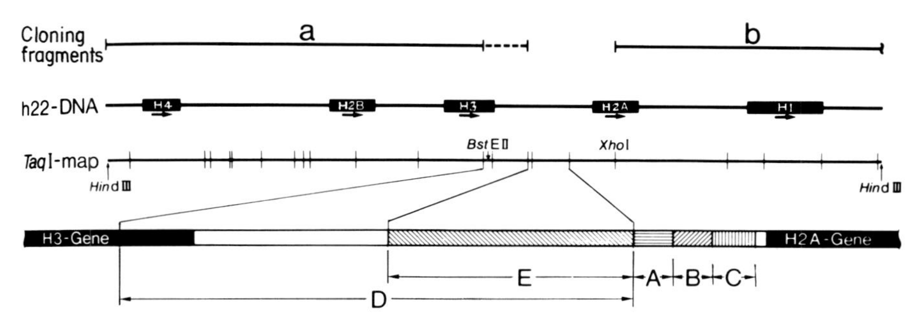


FIG. 29. Modulators (segments A and E), selector (segment B), and initiator (segment C, which includes the TATA box) are present in the spacer between the H3 and H2A histone genes. They are essential for the promotion of H2A histone gene transcription after injection into a *Xenopus* oocyte nucleus. These conclusions result from experiments in which DNA had been manipulated *in vitro* prior to its injection. [Grosschedl and Birnstiel (1980).]

polymerase II. We shall see later that a different mechanism for the control of gene transcription is at work when 5 S ribosomal RNA is synthesized by RNA polymerase III in *Xenopus* oocytes. Control of tRNA and 5 S RNA genes involves the binding of protein factors to regulatory sequences in the genes themselves (reviewed by Lassar *et al.*, 1983).

Figure 13 also shows the processing of the mRNA precursors. The first transcript, which is present in hnRNA, is a full copy of the gene, including its introns. In addition, other noncoding sequences are also transcribed at both the 5′- and 3′-ends of the gene. Furthermore, a methylated "cap" is added at the 5′-end of the primary transcript, and in many (but not all) cases, a stretch of polyadenylic acid sequences [the poly(A) tail] is added at the 3′-end of the mRNA precursor (reviewed by Baralle, 1983). The consensus AAUAAA sequence, common to all mRNAs, is the polyadenylation signal commanding the addition of adenylate residues at the 3′-end of the mRNA molecule. A single U to G transversion in the polyadenylation signal has no effect on polyadenylation, but it prevents the cleavage step that immediately precedes it (Montell *et al.*, 1983). A mutation in the polyadenylation signal (AATAAG instead of AATAAA) of the human α-globin gene is responsible for a form of α-thalassemia (production by patients of hemoglobin deficient in its α chains) (Higgs *et al.*, 1983). Histone mRNAs are an example of messengers that are devoid of the 3′-poly(A) sequence and of a polyadenylation signal; such poly(A)$^-$mRNAs form a class distinct from the more frequent poly(A)$^+$-mRNAs. However, Wells and Kedes (1985) recently discovered a histone variant (H3.3) gene which has introns and codes for a poly(A)$^+$ mRNA. The cap structure at the 5′-end is a short nucleotide sequence ending in methylguanine (Adams and Cory, 1975; Perry *et al.*, 1975; reviewed by Perry, 1976; Shatkin,

1976; Banerjee, 1980); its probable function is binding mRNA to ribosomes. Its necessity for proper functioning of an mRNA is shown by the fact that a structural analog of the cap, 7-methylguanosine monophosphate, inhibits the *in vitro* translation of HeLa cell mRNAs (Weber *et al.*, 1976). The 3′-poly(A) sequence plays an important role in the stability of mRNAs, as shown by experiments in which normal poly(A)$^+$ or artificially produced poly(A)$^-$ globin mRNAs were injected into *Xenopus* oocytes (Marbaix *et al.*, 1975) or HeLa cells (Huez *et al.*, 1981). Enzymatically deadenylated globin mRNA had a much lower stability. Stability is, of course, as important for the fate of individual mRNAs as their synthesis for the establishment of a steady-state mRNA population in the cytoplasm. However, this function of the poly(A) tail in mRNA stability does not seem to be valid for all mRNAs; it might also play a direct role in protein synthesis (Jacobson and Favreau, 1983). That the pre-mRNA molecules contain long, untranslated sequences (more than 1000 nucleotides long) has been shown by Shapiro and Schimke (1975) for the ovalbumin gene and, more recently, by Hofer and Darnell (1981) for the β-globin gene. In the latter case, deletion by endonuclease treatment of a 1000-nucleotide sequence beyond the poly(A) addition site must occur before polyadenylation can take place.

An important event in the processing of the mRNA precursors is the correct removal of the transcribed introns and the accurate joining of the split fragments (splicing). How this occurs is not known for certain, but a possible explanation was presented by Lerner *et al.* (1980) when they discovered that a small nuclear RNA (snRNA) called $U_1$ had a 5′-nucleotide sequence that was complementary to the splice junctions (i.e., intron–exon borders). Base pairing of $U_1$ with the splice junctions might permit the correct alignment of the intron–exon borders. More recently, Calvet *et al.* (1982) have shown that another snRNA ($U_2$ snRNA) is also base-paired to hnRNA and is therefore probably involved in its processing. Padgett *et al.* (1983) have presented a strong argument in favor of the view that $U_1$ RNA is actually involved in the splicing of mRNA precursors. They prepared monoclonal antibodies against $U_1$ ribonucleoproteins ($U_1$ RNP) and found that they inhibit the *in vitro* splicing of the mRNA precursor of an adenovirus. $U_1$ RNP binds selectively to the 5′-splice junction of the mouse β-globin gene, reinforcing the role played by this snRNA and associated proteins in mRNA processing (Mount *et al.*, 1983).

There is currently a great deal of interest in a possible role of intron RNA is splicing [reviewed by Davies (1984)] and in the fate of the excised intron RNAs. It was recently found that *in vitro* splicing of mRNA precursors occurs through lariat intermediates. The excised RNA intron has the shape of a circle with a branch, which might be the signal for its degradation (Padgett *et al.*, 1984; Ruskin *et al.*, 1984). RNA sequences in the intron itself are involved in splicing, according to Ruskin *et al.* (1984).

There is good evidence that in mitochondrial DNA of yeast cells, an intron flanking the cytochrome *b* gene codes for an ''mRNA maturase'' responsible for the splicing and maturation of cytochrome *b* mRNA. This maturase would destroy the RNA sequence that codes for its own synthesis, thus creating a negative feedback mechanism (Lazowska *et al.*, 1980). It is not known whether a similar mechanism applies to nuclear DNA transcripts. But work by Zaug *et al.* (1983) on the splicing of an intron in the nuclear but extrachromosomal rRNA gene of the flagellate *Tetrahymena* has led to curious and interesting results. This intron is excised as a linear RNA molecule that undergoes cyclization *in vitro* in the absence of proteins. This autocatalytic cyclization, like splicing, is a cleavage-ligation reaction; its molecular mechanism has been elucidated by Zaug *et al.* As pointed out by Cech *et al.* (1983), this splicing autocatalytic reaction requires neither energy not protein. Unexpectedly, the excised intron has sequences homologous to fungal (yeast) mitochondrial DNA. This may indicate a common origin of some nuclear and mitochondrial introns and the existence of common splicing mechanisms in the nucleus and the mitochondria.

The RNA splicing mechanisms have been discussed by Cech (1983, 1985). There are three different classes of splicing mechanisms, but all of them are two-step reactions; cleavage and ligation follow each other. The three classes correspond to tRNAs, mRNAs encoded by nuclear genes, and mitochondrial mRNAs and rRNAs, respectively. Details about the molecular mechanisms of the three splicing mechanisms can be found in Cech's reviews.

An interesting consequence of splicing was discovered simultaneously by Gilbert, Darnell, and Doolittle in 1978: different mRNAs can be found for the same gene as the result of different splicings. Another way to explain the existence of two different mRNAs for a single gene is that they derive from multiple initiation sites (Zehner and Paterson, 1983). More recently, Masters *et al.* (1983) reported that three different dihydrofolate reductase mRNAs derive from the same gene by different transcription or RNA processing events; alternative processing of the gene transcript allows a single *Drosophila* gene to encode two different polypeptides. This is possible because an intron of this gene contains a polyadenylation signal. In recent years, cases where a unique gene generates two or even three different mRNAs by alternative RNA splicing have accumulated.

Going back to the control of transcription, we should mention the current interest in enhancer DNA sequences (reviewed by Hogan, 1983; Perry, 1984; Reeder, 1984). It has been reported by Conrad and Botchan (1982) that the human genome contains multiple copies of nucleotide sequences homologous to the regulatory sequences of a DNA-containing oncogenic virus (SV40); these sequences activate the neighboring genes. These enhancers have a "core" sequence of eight DNA nucleotides, which might act on neighboring regions as long as several kilobases, where they increase the transcription of promoter sequences. Immunoglobin genes have enhancer sequences located upstream of the TATA box; they propagate a signal upstream of the region where initiation of transcription is possible. The highly conserved ATTGGGAT core sequence lies 60–80 bases upstream of the initiation site (Perry, 1984). Other enhancer-like sequences turn on the genes coding for insulin and chymotrypsin in the pancreas. Ribosomal genes have also enhancer sequences. They are located in the spacers that separate the tandem repeated genes; elements of the spacers exert a strong enhancing effect on transcription of the ribosomal genes by RNA polymerase I (Reeder, 1984).

It is important to stress the importance of the role played by certain proteins in the control of transcription. This is obvious for the DNA polymerases, but there is no doubt that the cells additionally contain a number of transcription protein factors, which, by binding selectively to regions adjacent to the gene or to the gene itself, affect transcription by the RNA polymerases in a positive or negative way. It is probable that the enhancer sequences are controled by a small group of proteins that bind to the DNase-hypersensitive regions. A specific protein would recognize a particular enhancer; this could change the structure of the gene and allow its transcription by a RNA polymerase. The first case in which the necessity of a transcription factor has been recognized is that of the 5 S genes of the *Xenopus* oocytes, which will be discussed later. According to Kaye *et al.* (1984), specific proteins bind to the 5′ flanking region of the ovalbumin gene and make it more accessible to nuclease attack. Similar results have been reported by Wu (1984), and there is little doubt that more and more cases of protein control of gene activity will be found in the near future. These positive and negative controls of gene activity by proteins encoded by other genes are called *trans-regulation.* Specific genes are recognized by proteins and are turned on or off; these regulatory proteins might bypass the control of gene activity by DNA regulatory sequences. We all know that proteins are made by DNA via RNA intermediates, but it is important to realize that proteins can activate or inactivate the genes.

Finally, a specific control of gene transcription might be due to the endogenous production of anti-sense (nonsense) RNAs. Izant and Weintraub (1984) have excised the protein coding sequence of a cloned thymidine kinase (*TK*) gene and reinserted it in reverse orientation. This anti-sense *TK* gene has been co-injected with the normal gene in $TK^-$ fibroblasts. This decreased the number of viable cells by fourfold and diminished greatly thymidine incorporation into DNA. Inhibition was probably due to formation of a duplex between *TK* mRNA and anti-sense *TK* RNA. It is not known whether such duplex RNA molecules form in intact cells, but in the future, anti-sense RNAs should provide an elegant method for inhibiting specifically the expression of a given gene. Injection of anti-sense RNAs in eggs should lead to interesting results regarding the role played by specific genes in morphogenesis and embryonic differentiation (reviewed by Weintraub *et al.*, 1985).

This overview of the mechanisms involved in gene transcription shows their complexity. It seem almost miraculous that genes are correctly expressed; this is, of course, due to the existence of a whole array of very precise control mechanisms. Among the control mechanisms that have been just discussed, one should recall that cells contain protein factors that activate the RNA polymerases. As will be discussed in Volume 2, Chapter 2, 5 S rRNA transcription requires a "transcription factor" that binds to the gene itself. Similarly, initiation of transcription of tRNA$^{arg}$ genes after their introduction into a *Xenopus* oocyte nucleus requires the sequestration of two RNA polymerase III transcription factors; again, we are dealing with an intragenic control (Sharp *et al.*, 1983). In *Drosophila,* RNA polymerase II requires two transcription factors, called A and B. Factor B binds specifically to a 65-bp segment of DNA that includes the TATA box, the starting point of transcription, and a portion of the leader region (Parker and Topol, 1984). In addition, nonprotein factors may also be required for efficient gene expression. Recently, Birchmeier *et al.* (1984) found that a small (60-nucleotide) RNA extracted from sea urchin embryo nuclei is required for the correct processing of sea urchin H3 histone pre-mRNA in *Xenopus* oocyte nuclei. Understanding the detailed mechanisms of gene expression remains an important task for the future. We shall not understand cell differentiation unless we know them.

We should conclude with a few problems that are of direct concern to cell biologists. One of them has already been mentioned: the ultrastructure of recently transcribed hnRNA sequences. We have seen the interpretation given by Fakan and Puvion (1980) of the various structures (perichromatin fibrils, perichromatin granules, interchromatin fibrils) seen under the electron microscope, namely, that they represent various stages of the intranuclear evolution of hnRNP and preribosomal particles. Another question deals with whether or not the proteins that are associated with nuclear hnRNA and cytoplasmic mRNA are the same. A paper by Greenberg and Setyono (1981) has clearly shown that the seven proteins that are associated with cytoplasmic mRNA are completely different from the more numerous proteins associated with hnRNA in the nucleus. It seems that when mRNA crosses the nuclear membrane, it eliminates the accompanying nuclear proteins and associates in the cytoplasm with already existing proteins (those belonging to a free pool). Unfortunately, we know nothing about the role played by nuclear proteins in hnRNP and by cytoplasmic proteins in mRNP except that they can protect the RNAs against degradation by ribonucleases. The fact that one of the mRNP proteins is specifically linked to the poly(A) tail might possibly lead to a better understanding of the mechanisms that control mRNA stability.

It is, of course, important for cell biologists to have at their disposal specific inhibitors of RNA synthesis. The classic inhibitor of DNA transcription is actinomycin D, which binds to G + C-rich regions and intercalates between the two DNA helices. At low concentrations, it preferentially inhibits rRNA synthesis, at least in cultured cells. α-*Amanitin,* a specific inhibitor of RNA polymerase II, is very useful for *in vitro* studies, but its utilization for *in vivo* experiments is often restricted by its low permeability. A derivative of benzimidazole (abbreviated DRB) specifically inhibits hnRNA synthesis, according to Tamm *et al.* (1976). Cordycepin triphosphate (an adenine nucleotide) is widely used to inhibit polyadenylation of mRNA precursors (Rose *et al.*, 1977).

Finally, associated with the cell nucleus are six small nuclear RNAs (snRNAs) (reviewed by Knowler and Wilks, 1980), which are about 100 nucleotides long, and are called, respectively, $U_1$, $U_2$, etc. $U_3$ is accumulated in the nucleolus, while $U_6$ is believed to be present in the nuclear perichromatin granules, where it might conceivably play a role in the processing and transport of hnRNA (Epstein *et al.*, 1980). We have just seen that a similar role in mRNA precursor processing (splicing) has been proposed by Lerner *et al.* (1980) for $U_1$ snRNA. Work has shown that the $U_1$ coding gene is part of a multigene family composed of 6–10 copies per haploid genome. Like the much larger mRNA precursors, $U_1$ snRNA is synthesized by RNA polymerase II; whether its flanking sequence contains the usual TATA box is not clear (Roop *et al.*, 1981; Monstein *et al.*, 1982). An unexpected observation is that the snRNAs $U_1$, $U_2$, and $U_3$ are coded by 24 loci in humans, but most of these loci are untranscribed pseudogenes (Denison *et al.*, 1981). Clearly, snRNAs are "mischievous little devils" that enjoy playing tricks on the investigators. However, one thing is certain: they warrant further investigation since they are not, as was long believed, meaningless fragments resulting from hnRNA processing.

## V. THE NUCLEOLUS

### A. General Background

The nucleoli are easily visible, under the light microscope, as refringent, almost spherical, highly basophilic bodies. They often display ameboid movements, may contain vacuoles, and are easily displaced by mild centrifugation of oocytes (where they are particularly large and conspicuous) to the centrifugal pole of the nuclear membrane. In the past, extrusion of entire nucleoli during oogenesis was frequently reported, but improvement in fixation techniques and the advent of electron microscopy have shown that these observations were artefictual.

The nucleolus is a typical nuclear structure that, like chromatin, is composed of DNA, RNA, and proteins. Its DNA/RNA ratio is much lower than that in chromatin, and the diversity of the DNA sequences present in the "nucleolar organizers" is also much smaller. As a result, the functions of the nucleolus are more limited than those of chromatin. It does not produce a large variety of messengers, but only the numerous ribosomes that carry out the orders of the messengers sent out by chromosomal DNA to the cytoplasm. Continuous production of ribosomes is necessary for life. In *Xenopus,* for example, a mutation (called "anucleolate") whereby nucleolar DNA is deleted occurs; this heterozygous condition (*nu*-o/+), in which the cells have only one nucleolus instead of two, is not lethal. However, crosses between two such heterozygotes will produce in their progeny 25% of anucleolate *nu*-o (absence of both nucleoli) individuals. Such mutants die after about 1 week, since they are unable to synthesize 28 and 18 S rRNAs (Brown and Gurdon, 1965). Their relatively limited survival time is due to the presence in the egg cytoplasm of a huge store of ribosomes of maternal origin. Interestingly, the *nu*-o mutants synthesize 5 S RNA at a normal rate, which showed for the first time that the 5 S RNA genes are not located in the nucleoli (Wallace and Birnstiel, 1966). Indeed, Pardue *et al.* (1973) showed by *in situ* hybridization that the 5 S RNA genes are dispersed on many chromosomes and are absent from the nucleoli in *Xenopus*.

In *Drosophila,* RNA–DNA hybridization experiments have demonstrated that the amount of rRNA hybridized to DNA is proportional to the number of heterochromosomes (Fig. 30). These experiments proved that the *Drosophila* nucleolar organizers are localized in the X and Y chromosomes (Ritossa and Spiegelman, 1965). In addition, it was found that the so-called *bobbed* (*bb*) mutants of

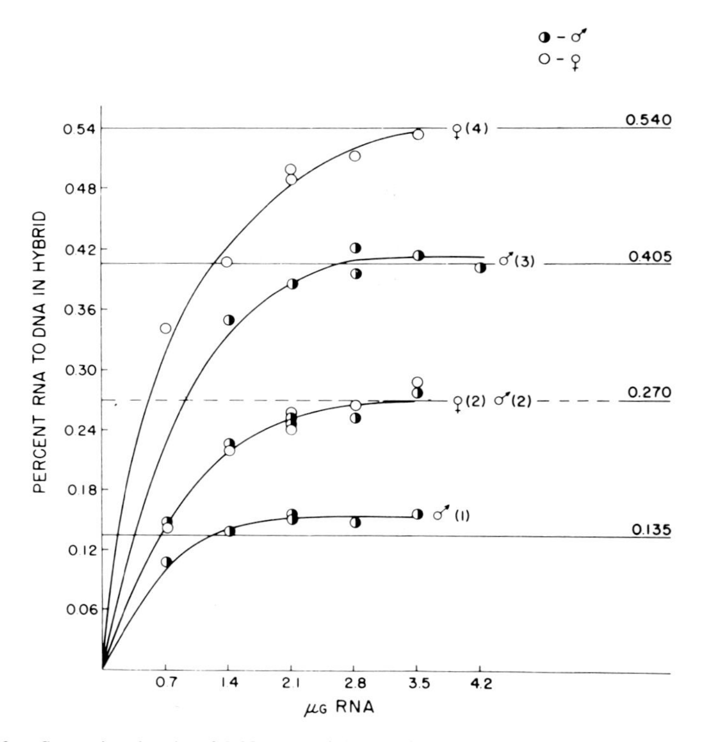


FIG. 30. Saturation levels of DNA containing various dosages of nucleolar organizer (NO) region after DNA–ribosomal RNA hybridization. The dosage of NO is indicated by the numbers in parentheses. Numerical values of the plateaus are given to the right. The amount of rRNA hybridized is proportional to the number of NOs. [Ritossa and Spiegelman (1965).]

*Drosophila* are deficient in ribosomal genes; extreme mutants of this category are lethal if they have less than 30 (instead of 130 in the wild type) copies of the ribosomal genes in their nucleolar organizers.

That the nucleolus plays an important role in cell economy is further shown by the fact that its destruction by irradiation impedes the transfer of all of the RNAs from the nucleus to the cytoplasm (Ďeak, 1973).

## B. ULTRASTRUCTURE[6]

The nucleolus has no limiting membrane. Thus, nucleoli come into direct contact with the surrounding chromatin and nuclear sap. As shown in Fig. 31, a typical nucleolus is composed of a granular cortex and a fibrillar core. In close contact with the latter is nucleolar DNA, i.e., the nucleolar organizer (NOR), which is responsible for all ribosomal RNA (28 and 18 S) synthesis (reviewed by Hernandez-Verdun, 1983). Cytochemical methods have shown that the nucleolus (in particular, its granular cortex) is rich in RNA (Caspersson and Schultz, 1939; Brachet, 1940). In cells that do not synthesize appreciable amounts of ribosomes, for example, very young oocytes or cleaving eggs, only the fibrillar core is present. Autoradiography has shown that synthesis of the rRNA precursor begins in close proximity to the nucleolar organizer; its maturation to the 28 and 18 S rRNA species takes place in the nucleus. The granules present in the nucleolar cortex are believed to correspond to precursors of the cytoplasmic ribosomes.

Suppression of rRNA synthesis by a small amount of actinomycin D exerts "dramatic"[7] effects on nucleolar morphology; the granular and fibrillar regions are segregated (as shown in Fig. 32). However, this alteration of nucleolar topography is not specific for inhibitors of RNA synthesis and is frequently found when cells are subjected to stress or injury. Treatment of cells with various chemicals can lead to the appearance of granules in the area of the nucleoli. These nucleolar perichromatin granules probably correspond to degradation forms of abnormally synthesized or processed pre-rRNP nucleolar particles. In contrast to the already mentioned perichromatin granules found elsewhere in the nucleus, their formation is not inhibited by treatment with DRB (which arrests hnRNA synthesis) (Puvion *et al.*, 1981).

Finally, a report by Franke *et al.* (1981) shows that, at least in *Xenopus* oocytes, the nucleoli have a nucleolar skeleton that remains intact after extraction with salt or nuclease. It is a meshwork of 4-nm-thick filaments made of a major

[6]Reviewed by Goessens (1984).

[7]This word is being used more and more frequently in scientific papers. We think that it should be reserved for more tragic events than increases in enzyme activities or morphological changes that do not lead to death.

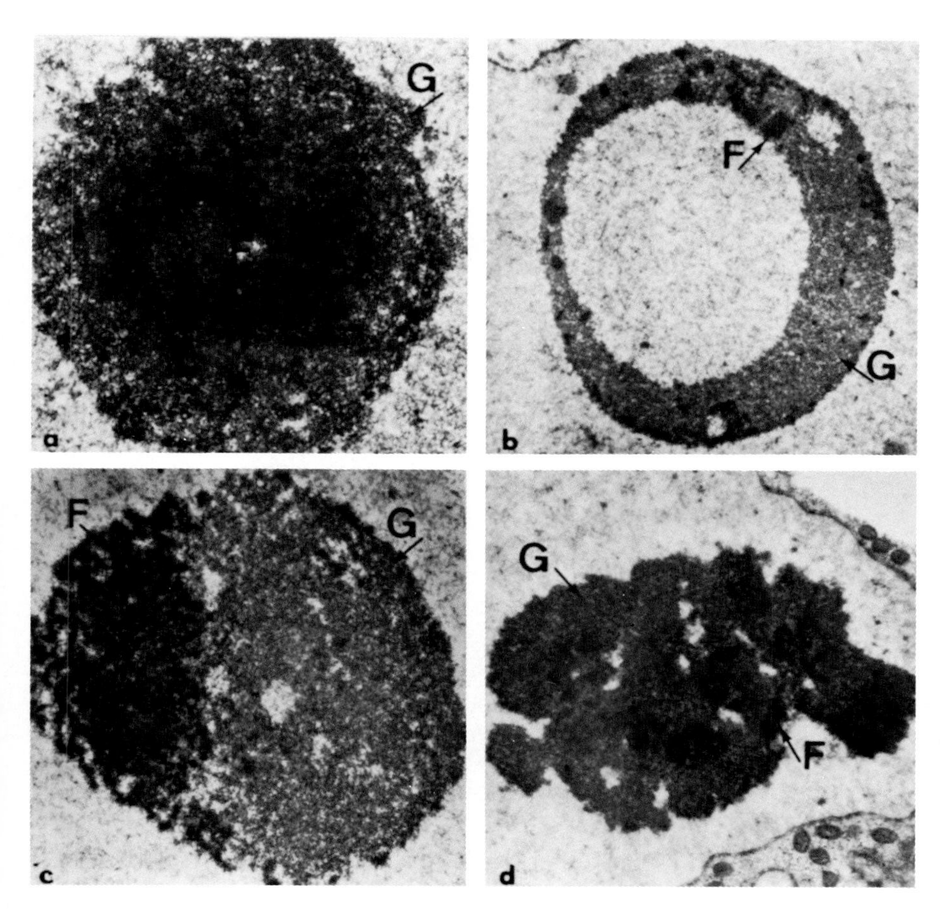


FIG. 31. Ultrastructure of the nucleoli in oocyte nuclei of four different amphibian species: (a) *Xenopus*, (b) *Triturus*, (c) *Bufo*, (d) *Pleurodeles*. G, granular constituent. F, fibrillar constituent. [Courtesy of P. Van Gansen.]

protein ($M_r$ 140,000) (Fig. 33). The nucleolar skeleton is independent of the other extraction-resistant structures (nuclear matrix and lamina).

In the following sections, we shall deal successively with nucleolar DNA, RNA, and proteins. In the last group, two major phosphorylated proteins called B23 and C23 are of particular interest to cytochemists because they stain with silver nitrate and thus allow the detection of the nucleolar organizers (Jones *et al.*, 1981; Lischwe *et al.*, 1981). Experiments with inhibitors of RNA and protein synthesis have suggested that silver staining by the so called Ag-AS-

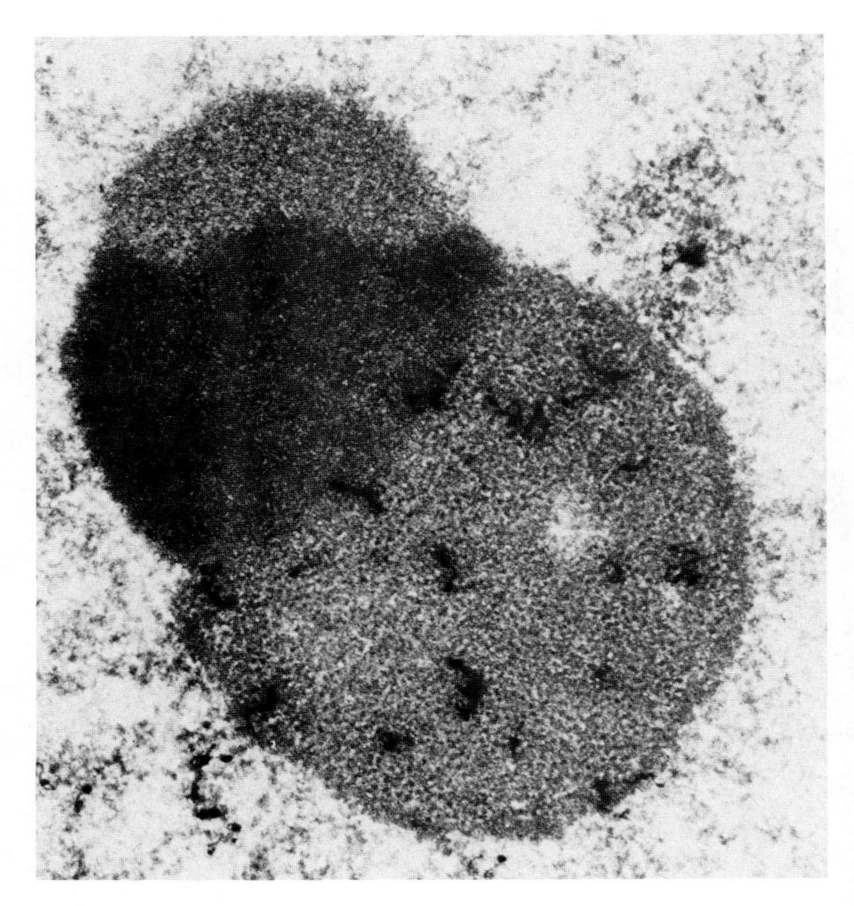

FIG. 32. Segregation between the granular and fibrillar parts of a nucleolus in cultured simian kidney cells treated with actinomycin D (1 μg/ml over 7 hr). The cells had received a short (5-min) [$^3$H]uridine pulse before the actinomycin D treatment; the ultrathin sections have been submitted to autoradiography. [Courtesy of M. Geuskens.]

NOR reaction is positive only if the nucleolar organizers are actively involved in rRNA synthesis (Hubbell *et al.*, 1980). However, Boloukhère (1984) found a positive silver staining of the fibillar component of the nucleoli (Fig. 34) at all stages of *Xenopus* oogenesis, although the rRNA-synthesizing activity of the nuclear organizers of small oocytes is barely detectable. Utilization of this reaction at the ultrastructural level has shown that silver staining is limited to the fibrillar part of the nucleoli (Hernandez-Verdun *et al.*, 1980; Dimova *et al.*, 1982; Boloukhère, 1984) whether nucleolar chromatin is transcribed or not.

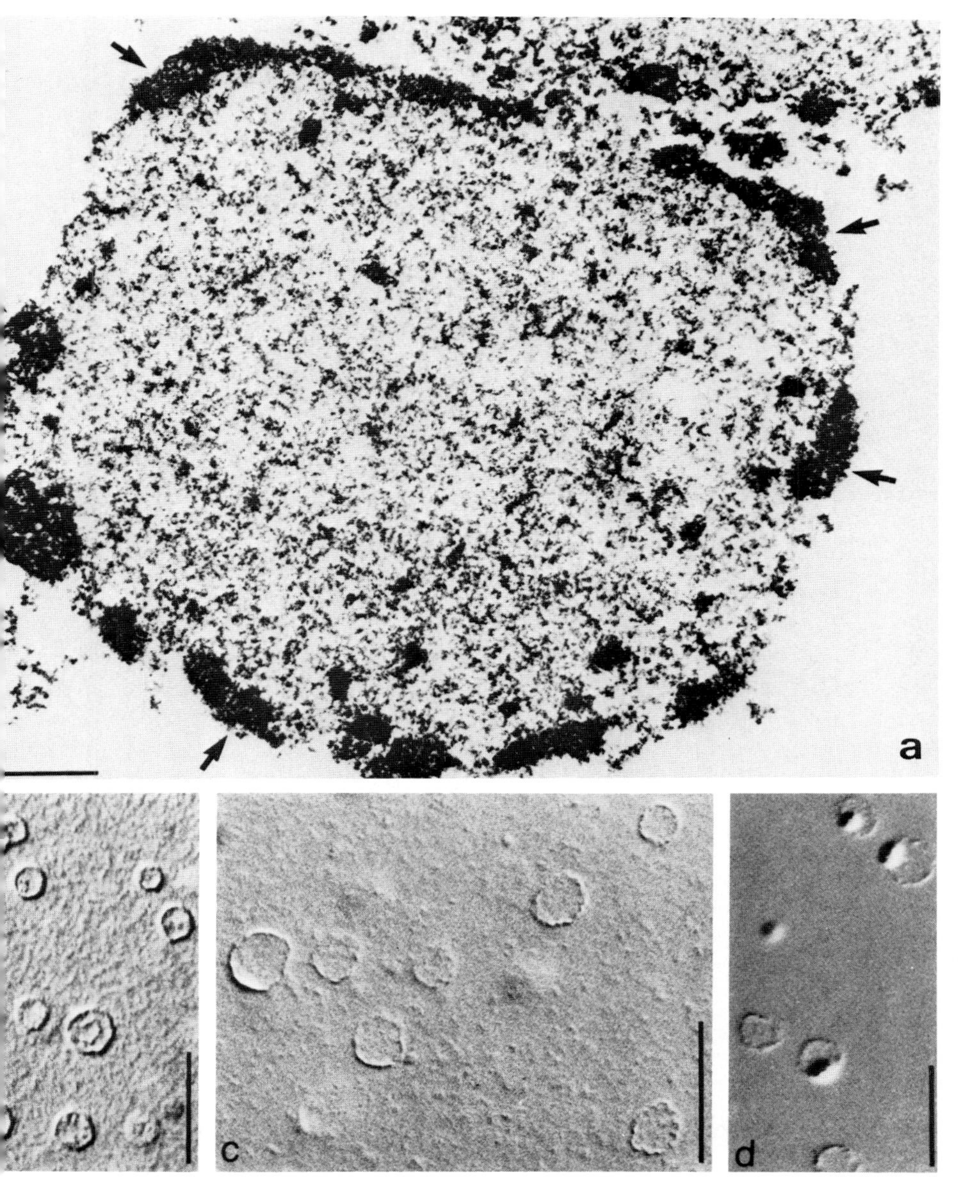


FIG. 33. Nucleolar skeleton of *Xenopus* oocytes. (a) After treatment of isolated nucleoli with buffer containing 1 m*M* KC1 and 1% Triton X-100. Arrows show peripheral densely stained aggregates (22,000 X). (b) After treatment of isolated nucleoli with DNase and RNase (1700 X). (c) After treatment with DNase and repeated washes in low-salt buffer, the remaining skeletal compound appears as a ring (1700 X). (d) After incubation in 1 m*M* Tris-HC1 buffer (pH 7.21), a ring-like cortical structure surrounding a dense aggregate can be seen (750 X). [Franke *et al.* (1981). Reproduced from *The Journal of Cell Biology,* 1981, **90,** p. 292, by copyright permission of The Rockefeller University Press.]

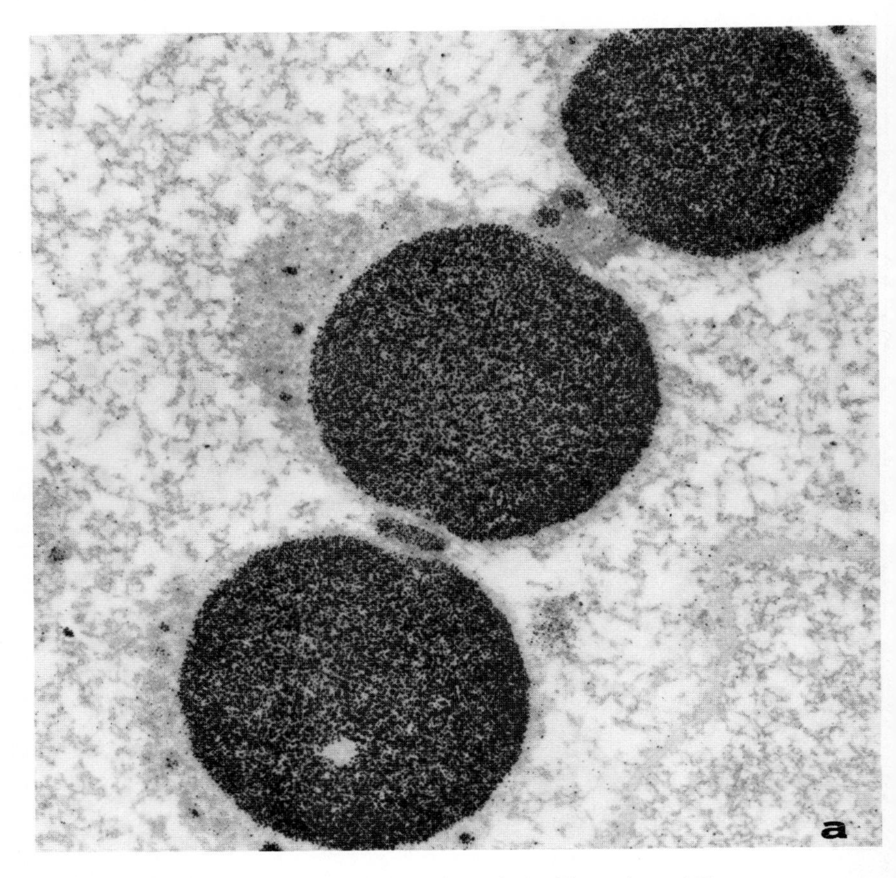

FIG. 34. Silver staining (Ag-AS-NOR reaction) of ultrathin sections of *Xenopus* oocytes nucleoli. (a) Intense staining in previtellogenic oocytes, which do not synthesize rRNA (see Chapter 7). (b) There are very few silver grains (arrowheads) in the nucleoli of vitellogenic oocytes, which synthesize large amounts of rRNA. [Courtesy of M. Boloukhère.]

## C. RIBOSOMAL DNA (rDNA) AND ITS TRANSCRIPTION

Brown and Dawid (1968) succeeded in separating the ribosomal DNA (rDNA) from bulk DNA in young *Xenopus* oocytes (Volume 2, Chapter 2). This achievement was made possible by two favorable circumstances: the fact that rDNA has a higher G + C content (and, hence, a higher buoyant density) than bulk DNA, from which it can be separated by ultracentrifugation and the existence of a 1000-fold amplification of the ribosomal genes at the onset of *Xenopus* oogenesis. Because of the occurrence of selective synthesis of ribosomal genes, the oocyte contains about $2 \times 10^6$ copies of the ribosomal cistrons as compared to 1000 in

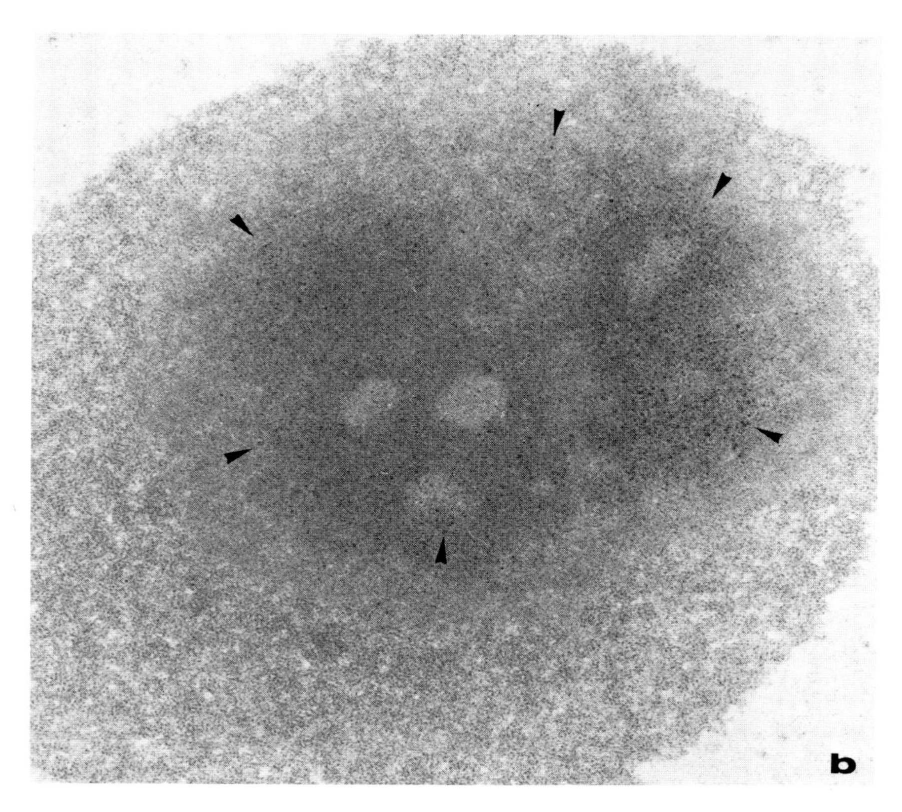


FIG. 34. (*Continued*)

somatic cells (see Volume 2, Chapter 2). Thus, in the *Xenopus* oocyte, nature has done what genetic engineering invented by human intelligence is doing now: to provide molecular biologists with sufficient material to allow a detailed study of the "anatomy" of a single highly purified gene.

The *Xenopus* oocyte is so remarkable that it will be discussed in detail in Volume 2, Chapter 2, and will only be briefly summarized here. The organization of *Xenopus* oocyte ribosomal cistrons is unique. Like the histone genes in sea urchin eggs, the ribosomal genes in *Xenopus* are composed of tandem repeated sequences separated by spacers. However, as for the histone genes, this type of organization is far from universal. This diversity does not make our task easier. The interested reader will find a detailed treatment of ribosomal gene organization and processing of its primary transcripts in the reviews by Hadjiolov (1980) on mammalian cells and by Long and Dawid (1980) on *Drosophila*.

Figure 35 presents a simplified diagram of the organization, transcription, and

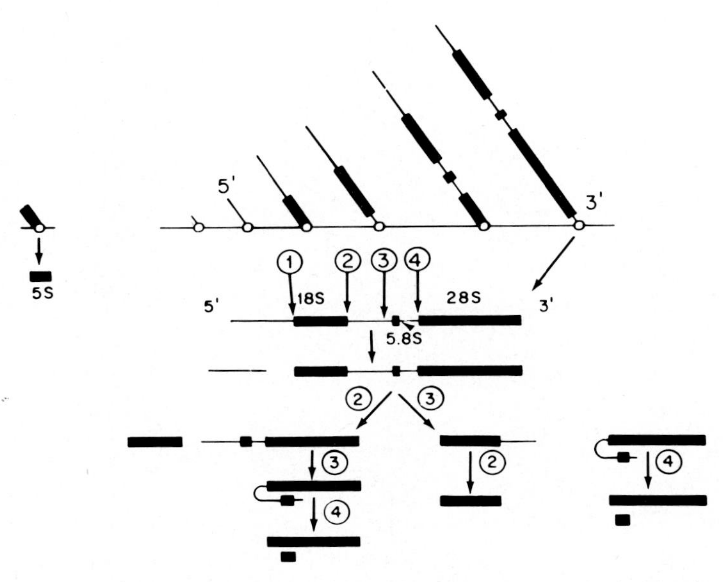


FIG. 35. Production of a transcription unit by nucleolar ribosomal genes (above) and processing of the rRNA precursor (below). The initial transcript or pre-rRNA (40–45 S, according to the species) is growing from the 5′ to the 3′ end. Its sequences are complementary to the following gene regions: external transcribed spacer, 18 S rRNA internal transcribed spacer containing a covalently bound 5,8 S sequence and 28 S rRNA. Sites 1 to 4 are the positions of cleavage. Two temporal orders of cleavage at sites 2 and 3 are given: 2 → 3 (*Xenopus*) and 3 → 2 (HeLa cell). [Perry (1976).]

processing of the *Xenopus* ribosomal genes; Fig. 23 has already shown the appearance of nucleolar organizers after being spread and photographed under the electron microscope. The work done by Dawid *et al.* (1970), Wensink and Brown (1971), Brown and Sugimoto (1973), and Dawid and Wellauer (1976) has firmly established the structure of the 28 and 18 S genes. Wellauer and Dawid (1973, 1974) have determined how the initial transcript (pre-rRNA) is processed into mature 28 and 18 S rRNAs. *Xenopus* ribosomal genes are composed of the tandem repetition of $8.7 \times 10^6$-dalton units; there are about 450 of these units in a nucleolar organizer. Between the adjacent 28 and 18 S cistrons are spacers made up of a transcribed DNA piece (600,000 daltons) and a nontranscribed fragment of variable length and base composition. While the rDNA sequences are identical in two *Xenopus* species (*X. laevis* and *X. borealis*), their spacers are very different (Wellauer and Reeder, 1975). The base composition of these spacers varies even from one *X. laevis* to another (Reeder *et al.*, 1976). Transcription of the gene and processing of the primary transcript (a 40 S molecule in *Xenopus*) occur in the nucleolus itself (Saiga and Higashinakagawa, 1979); the primary transcript (i.e., the transcription unit shown in Fig. 35) includes the transcribed spacer, which is destroyed during processing. It should be added that the transcription units seen under the electron microscope often display abnormalities. Correct transcription of the large ribosomal genes is not easy, and frequent failure is thus not surprising (Franke *et al.*, 1976). While transcription is carried on by nucleolar RNA polymerase I, it is likely that pre-rRNA processing results from endonucleolytic cleavage catalyzed by nucleolar ribonucleases (Kelley and Perry, 1971; Mirault and Scherrer, 1972; Winicov and Perry, 1974; Eichler and Eales, 1983). Finally, one should mention a remarkable achievement by Salim and Maden (1981), who were able to deduce the sequence of *Xenopus* 18 S rRNA from the base sequence (which is 1825 nucleotides long) of the corresponding gene. The region that shows a high degree of homology with yeast 18 S rRNA is peculiar in that it contains almost all of the RNA methyl groups.

*Drosophila* (reviewed by Long and Dawid, 1980) is exceptional in that about 50% of the ribosomal genes localized in the heterochromosomes are interrupted by introns. The so-called type I insertions have 1.0 or 0.5 kb, and their presence inactivates the gene. Analysis of their base sequence shows that they are flanked by duplications and that they arose from insertion of *copia*-like transposable elements into rDNA (Dawid and Rebbert, 1981). According to Beckingham (1981), 5% of the rDNA genes have introns in the fly *Calliphora,* while almost 100% are found in *D. virilis,* 50% are found in *D. melanogaster,* and none are found in vertebrates. This is another unexplained mystery of nature.

Mammalian cells have 150–700 rDNA cistrons per haploid genome. In HeLa cells, the first preribosomal transcript is larger than that found in *Xenopus* (45 S, 14,000 nucleotides). As in *Xenopus,* the tandem-repeated genes are separated by spacers; while some of these spacers are part of the tandem repeats, others are interspersed (as in *Drosophila*) in the human genome (Higuchi *et al.*, 1981). The classic pathway of the 45 S pre-rRNA processing is as follows:

$$45\ S \rightarrow 41\ S \begin{array}{l} \nearrow\ 32\ S \rightarrow 28\ S \\ \searrow\ 20\ S \rightarrow 18\ S \end{array}$$

A paper by Bowman *et al.* (1981) indicates that the situation is probably even more complex. In mouse cells, there are at least two additional pre-rRNA cleavage pathways involving 34, 37, and 26 S previously unknown intermediates. In mice, all tissues, including sperm, possess methylated and unmethylated rDNA fractions, their proportion varying according to the strain of mouse. Only the unmethylated fractions are sensitive to DNase digestion, which suggests that hypermethylated genes are transcriptionally inactive (Bird *et al.*, 1981).

In contrast to the redundancy of the rRNA genes, the number of gene copies coding for the cytoplasmic *ribosomal proteins* is low. There is no amplification of these genes in *Xenopus* oocytes, their number being only one to two in the amphibian genomes. The number of genes coding for ribosomal proteins is between 7 and 20 both in humans and the mouse (Monk *et al.*, 1981). However, in the mouse, these genes are dispersed on different chromosomes, and there is no indication of a clustering, as is the case for mammalian rRNA genes (D'Eustachio *et al.*, 1981).

## D. Nucleolar Proteins

Work done in H. Busch's laboratory has shown that isolated nucleoli contain both histones (Orrick *et al.*, 1973) and non-histone proteins that include the two silver-staining phosphoproteins B23 and C23 (Olson *et al.*, 1975; Ochs *et al.*, 1983). Phosphorylation of nucleolar proteins results from the action of a protein kinase, which is localized in the nucleolus itself (Grummt, 1974). Among the non-histone nuclear proteins, RNA polymerase I and RNases are, of course, particularly important from the biochemical viewpoint. Recent work by Scheer and Rose (1984) has demonstrated that RNA polymerese I is a strictly nuclear enzyme, closely associated with the ribosomal genes in the fibriller portion of the nucleoli and the nucleolar organizers of metaphase chromosomes (which do not synthesize rRNA).

It is possible to isolate preribosomal particles from purified preparations of nucleoli (reviewed by Hadjiolov, 1980). As Pederson and Kumar (1971) have shown, there are two kinds of such particles, with sedimentation constants of 80 and 55 S, respectively. The dimensions of the largest ones are 23 × 18 nm, which should allow free passage through the nuclear pore complex (Simard *et al.*, 1973). The 80 S particles contain the totality of the 45 S preribosomal RNA, while the 55 S particles possess smaller rRNAs (25–40 S and 10–25 S), which are intermediates in the processing of the 45 S precursor. It now seems clear that the large 80 S particles are the precursors of the 55 S particles. The first would be involved in the transcription of the ribosomal genes, while the second would function in the maturation of the 45 S first transcription product (Stevens *et al.*, 1980). Both 80 and 55 S particles also contain the small 5 S rRNA that, as we have seen, is synthesized elsewhere in chromatin.

All of the proteins present in the preribosomal particles are necessarily of cytoplasmic origin. The nucleolus should thus be visualized as a center where nucleic acids (28, 18, 5 S rRNAs) and proteins are assembled. Some of these nucleolar proteins are also found in cytoplasmic ribosomes, but others are non-ribosomal (Prestayko *et al.*, 1974; Kumar and Subramaniam, 1975; Kuter and Rodgers, 1976; Lastick, 1980). The 80 S preribosomes already contain typical ribosomal proteins; many others are added afterward and associate with RNA during the processing of 45 S pre-rRNA. Therefore, the 55 S particles (which are the precursors of the cytoplasmic 80 S ribosomes) possess many ribosomal and some nonribosomal proteins. It is believed that the missing ribosomal proteins are added in the cytoplasm. Todorov *et al.* (1983) have studied 80 and 41 S nucleolar preribosomes. The former contain the 45 S pre-rRNA, and the second possess the 21 S pre-rRNA. These preribosomes contain 17 proteins characteristic of the cytoplasmic ribosomes. Two proteins are added to the 40 S preribosomes in the nucleolus; four others are added later, probably outside the nucleolus. There is thus a sequential addition of proteins during the formation of the cytoplasmic ribosome's small subunits.

The function of the nucleolus has long been a mystery. In "Biochemical Cytology," we concluded that although the function of the nucleolus still remains enigmatic, there is no doubt that it is very important in the metabolism of RNA and proteins. The idea that the nucleolus was a reserve material or a byproduct of nuclear metabolism can confidently be discarded. We now know that the nucleolus indeed plays a very important role in protein synthesis, as was suggested by T. Caspersson in 1941, but this role is an indirect one. The nucleolus is a factory for the production of ribosomes that are exported to the cytoplasm and are used for protein synthesis. The "boss" of the factory is ribosomal DNA, and its "office" is in the nuclear organizer located at the very heart of the cell.

## VI. THE NUCLEAR SAP (NUCLEOPLASM)

Very little is known about the chemical composiion of nuclear sap in somatic cells. Its existence can be demonstrated by the fact that, in ultracentrifuged liver cells, chromatin and nucleoli can be sedimented in the intact nucleus, away from the nuclear sap (Claude, 1943). More is known about the nuclear sap or larger cells, in particular, *Xenopus* oocytes; this aspect of the problem will be dealt with in Volume 2, Chapter 2.

The composition of the nuclear sap is extremely complex, since so many cell constituents travel back and forth from the nucleus to the cytoplasm. Preribosomal particles (which are said to be 30 times more abundant in the nucleus than in the cytoplasm), mRNAs associated with proteins that differ from those present in cytoplasmic mRNP, breakdown products of the processing of the various kinds of RNA molecules, snRNAs, and other products are present in the nuclear sap. In addition, the nuclear sap contains a host of proteins (at least 300 non-histone proteins, according to Peterson and McConkey, 1976). One of the major nuclear sap proteins is nucleoplasmin, which is involved in nucleosome assembly. Its localization in the nuclear sap is indicated by the fact that it is missing from sperm and bird erythrocyte nuclei, which have no or very little nucleoplasm (Franke *et al.*, 1980). It is also known that the non-histone HMG proteins accumulate in the nucleus when they are injected into the cytoplasm of HeLa cells (Rechsteiner and Kuehl, 1979). It is likely that histones display a similar behavior, judging from what we know about the *Xenopus* oocyte nucleus (Volume 2, Chapter 2).

One of the proteins identified in the cell sap of HeLa cells by Peterson and

McConkey (1976) is actin. It is worth repeating here that if mammalian cells are treated with high concentrations (7.5–20%) of dimethyl sulfoxide or with the bivalent ion ionophore A23187 in the presence of $Mg^{2+}$, paracrystals of actin, tropomyosin, and α-actinin form in the nucleus (Osborn and Weber, 1980).

## VII. BIOCHEMICAL ACTIVITIES OF ISOLATED NUCLEI

### A. Energy Production

Whether isolated nuclei actually respire and produce ATP remains one of the vexing questions facing cell biologists. Their respiration, if it exists at all, is very low compared to that of the cytoplasm, and the possibility of contamination of the nuclei by cytoplasmic organelles (mitochondria, in particular) is difficult to rule out. We have already seen that there is a disagreement concerning the presence or absence of cytochrome oxidase in the nuclear membrane. Jarasch and Franke (1974) believe that the activities found are due to contamination with mitochondria, Zbarsky (1978) is of the opposite opinion.

Even though one cannot exclude the possibility that the cell nucleus has a very low respiration and possibly some glycolytic activity, the conclusion that energy production is derived from the cytoplasm seems unavoidable. Synthesis of both DNA and RNA in the nucleus obviously requires ATP production by the mitochondria, which ceases when phosphorylative oxidations are inhibited.

### B. Protein Synthesis

Is there a nuclear protein synthesis? Are all the nuclear proteins of cytoplasmic origin? These are other vexing questions.

In ''Biochemical Cytology'' (Brachet, 1957, pp. 101–105), we discussed at length the experiments of Allfrey *et al.* (1957) on isolated thymocyte nuclei. These isolated nuclei incorporated large amounts of labeled amino acids into their proteins; thus, the conclusion was drawn that nuclear protein synthesis could amount to as much as 30% of the total. Thymocytes, however, are peculiar cells because their nucleus is surrounded by only a thin layer of cytoplasm; for that reason, it is probable that contamination of the nuclei with intact cells occurred. We concluded in 1957 that it is likely that interphase (thymus) nuclei are capable of incorporating large amounts of amino acid into protein. These experiments on thymus nuclei should be repeated using more modern methods. In addition, one should study the effects of the many inhibitors of protein synthesis that are now available. The autoradiography studies of Ficq and Errera (1955), which suggested that uncontaminated thymus nuclei incorporate labeled amino acids, should also be repeated using high-resolution autoradiography techniques.

More recent work, done on cells other than thymocytes, shows that very little, if any, nuclear protein synthesis occurs: less than 0.2% of the total protein synthesis in nuclei isolated from sea urchin blastulae (Allen and Wilt, 1975), and 0.4–0.8% in HeLa cell nuclei (Chatterjee *et al.*, 1977).

More recently, unexpected results have been reported by Umansky *et al.* (1980). They claim that isolated nuclei and even chromatin incorporate labeled amino acids in the absence of added ATP. The incorporation is inhibited by chloramphenicol (70%) and puromycin (40%), but not by cycloheximide; it is blocked by DNase, but not by RNase. The $M_r$ of the synthesized polypeptides is low (about 6000 daltons). Since addition of $NAD^+$ stimulates incorporation, the authors suggest that poly(ADP-ribose) (whose precursor is $NAD^+$) is the source of energy. Clearly, it is doubtful that isolated nuclei actually synthesize proteins, and the biochemical mechanisms by which this could occur require further investigation.

### C. Nucleic Acid Synthesis

Isolated nuclei incorporate thymidine into DNA in the presence of all four deoxyribonucleoside triphosphates. However, DNA synthesis is limited to the elongation of already existing DNA chains. Isolated nuclei can only terminate DNA synthesis already initiated *in vivo* (Lynch *et al.*, 1970; Arms, 1971). So far, we have not been able to induce the appearance of new initiation sites for DNA synthesis in isolated nuclei.

On the other hand, isolated nuclei are very active in RNA synthesis, since they have a high content of the three kinds of RNA polymerases. They are a rich source of these enzymes for *in vitro* RNA synthesis. The presence of all four ribonucleoside triphosphates, as well as that of an energy-producing system, is, of course, required for prolonged RNA synthesis by isolated nuclei. Integrity of a nuclear DNA template is, of course, also necessary. Treatment with DNase, actinomycin D, α-amanitin, and histones inhibit RNA synthesis in isolated nuclei. It has been claimed that the addition of certain RNA fractions. (Frenster, 1965) and of non-histone proteins enhances RNA production in these *in vitro* systems, but this still requires confirmation. RNA labeled in such *in vitro* experiments always leaks out of the nuclei. Since this process is accelerated by the addition of ATP, it has been suggested that the nucleoside triphosphatase present in the nuclear envelope is responsible for the transfer of RNA from the nucleus to the cytoplasm. We shall not discuss further the results obtained with systems that have an obvious interest for biochemists but remain artificial for cell biologists; more important are the studies done *in vivo* on large unicellular organisms (Volume 2, Chapter 1) and on *Xenopus* oocytes (Volume 2, Chapter 2).

## REFERENCES

Aaronson, R. P., and Blobel, G. (1975). *Proc. Natl. Acad. Sci. U.S.A.* **72,** 1007.
Adams, J. M., and Cory, S. (1975). *Nature (London)* **255,** 28.
Adams, R. L. P., and Burdon, R. H. (1982). *CRC Crit. Rev. Biochem.* **13,** 349.

Affara, N. A., Jacquet, M., Jakob, H., Jacob, F., and Gros, F. (1977). *Cell (Cambridge, Mass.)* **12,** 509.
Agutter, P. S. (1984). *Subcell Biochem.* **10,** 281.
Agutter, P. S., and Richardson, J. C. W. (1980). *J. Cell Sci.* **44,** 395.
Agutter, P. S., MacArdle, H. J., and McCaldin, B. (1976). *Nature (London)* **263,** 165.
Ajiro, K., Borun, T. W., and Cohen, L. H. (1981). *Biochemistry* **20,** 1445.
Allen, W. R., and Wilt, F. H. (1975). *Exp. Cell Res.* **97,** 151.
Allfrey, V. G., Stern, H., Mirsky, A. E., and Saetren, H. (1952). *J. Gen. Physiol.* **35,** 529.
Allfrey, V. G., Mirsky, A. E., and Osawa, S. (1957). *J. Gen. Physiol.* **40,** 451.
Allfrey, V. G., Littau, V. C., and Mirsky, A. E. (1965). *J. Cell Biol.* **21,** 213.
Alt, F. W., Kellems, R. E., Bertino, J. R., and Schimke, R. T. (1978). *J. Biol. Chem.* **253,** 1357.
Arkhipova, I. R., Gorelova, T. V., Ilyin, Y. V., and Schuppe, N. G. (1984). *Nucl. Ac. Res.* **12,** 7533.
Arms, K. (1971). *Dev. Biol.* **26,** 497.
Baglia, F. A., and Maul, G. G. (1983). *Proc. Natl. Acad. Sci. U.S.A.* **80,** 2285.
Ball, D. J., Gross, D. S., and Garrard, W. T. (1983). *Proc. Natl. Acad. Sci. U.S.A.* **80,** 5430.
Baltimore, D. (1981). *Cell (Cambridge, Mass.)* **24,** 592.
Baltimore, D. (1985). *Cell (Cambridge, Mass.)* **40,** 481.
Banerjee, A. K. (1980). *Microbiol. Rev.* **44,** 175.
Baralle, F. E. (1983). *Int. Rev. Cytol.* **81,** 71.
Barrack, E. R., and Coffey, D. S. (1980). *J. Biol. Chem.* **255,** 7265.
Barsacchi, G., and Gall, J. G. (1972). *J. Cell Biol.* **54,** 580.
Basler, J., Hastie, N. D., Pietras, D., Matsui, S. I., Sandberg, A. A., and Berezney, R. (1981). *Biochemistry* **20,** 6921.
Battey, J., Max, E. E., McBride, W. O., Swan, D., and Leder, P. (1982). *Proc. Natl. Acad. Sci. U.S.A.* **79,** 5956.
Bayev, A. A., Jr., Krayev, A. S., Lyubomirskaya, N. V., Ilyin, Y. V., Skryabin, K. G., and Georgiev, V. (1980). *Nucleic Acids Res.* **8,** 3263.
Beach, L. R., and Palmiter, R. D. (1981). *Proc. Natl. Acad. Sci. U.S.A.* **78,** 2110.
Beckingham, K. (1981). *J. Mol. Biol.* **149,** 141.
Behrens, M. (1938). *Hoppe-Seyler's Z. Physiol. Chem.* **253,** 185.
Bekhor, I., Lapeyre, J. N., and Kim, J. (1974). *Arch. Biochem. Biophys.* **161,** 1.
Bellard, M., Kuo, M. T., Dretzen, G., and Chambon, P. (1980). *Nucleic Acids Res.* **8,** 2737.
Bellard, M., Dretzen, G., Bellard, F., Oudet, P., and Chambon, P. (1982). *EMBO J.* **1,** 223.
Bentley, D. L., and Rabbitts, T. H. (1983). *Cell (Cambridge, Mass.)* **32,** 181.
Berenson, R. J., Francke, U., Dolnick, C. K., and Bertino, J. R. (1981). *Cytogenet. Cell Genet.* **29,** 143.
Berezney, R. (1980). *J. Cell Biol.* **85,** 641.
Berezney, R., and Buchholtz, L. A. (1981). *Biochemistry* **20,** 4995.
Berezney, R., and Coffey, D. S. (1974). *Biochem. Biophys. Res. Commun.* **60,** 1410.
Berezney, R., and Coffey, D. S. (1975). *Science* **189,** 291.
Berezney, R., and Smith, H. C. (1981). *J. Cell Biol.* **91,** 67a.
Bernards, A. (1985). *Biochim. Biophys. Acta* **284,** 1.
Bernstein, L. B., Mount, S. M., and Weiner, A. M. (1983). *Cell (Cambridge, Mass.)* **32,** 461.
Bertelsen, A. H., Humayum, M. Z., Karfopoulos, S. G., and Rush, M. G. (1982). *Biochemistry* **21,** 2076.
Biezunski, N. (1981a). *Crhomosoma* **84,** 87.
Biezunski, N. (1981b). *Chromosoma* **84,** 111.
Birchmeier, C., Schümperli, D., Sconzo, G., and Birnstiel, M. L. (1984). *Proc. Natl. Acad. Sci. U.S.A.* **81,** 1057.

Bird, A. P. (1984). *Nature (London)* **307,** 503.
Bird, A. P., and Taggart, M. (1980). *Nucleic Acids Res.* **8,** 1485.
Bird, A. P., Taggart, M. H., and Smith, B. A. (1979). *Cell (Cambridge, Mass.)* **17,** 889.
Bird, A. P., Taggart, M. H., and Gehring, C. A. (1981). *J. Mol. Biol.* **152,** 1.
Birnstiel, M., Telford, J., Weinberg, E., and Stafford, D. (1974). *Proc. Natl. Acad. Sci. U.S.A.* **71,** 2900.
Blake, C. C. F. (1985). *Int. Rev. Cytol.* **93,** 149.
Blumenfeld, M., Orf, J. W., Sina, B. J., Kreber, R. A., Callahan, M. A., Mullins, J., and Snyder, L. A. (1978). *Proc. Natl. Acad. U.S.A.* **75,** 866.
Bodnar, J. W., Jones, C. J., Coombs, D. H., Pearson, G. D., and Ward, D. C. (1983). *Mol. Cell. Biol.* **3,** 1567.
Bokhon'ko, A., and Reeder, R. H. (1976). *Biochem. Biophys. Res. Commun.* **70,** 146.
Bouloukhère, M. (1984). *J. Cell Sci.* **65,** 73.
Bonner, W. M. (1975). *J. Cell Biol.* **64,** 421.
Borst, P., and Cross, G. A. M. (1982). *Cell (Cambridge, Mass.)* **29,** 291.
Borst, P., Bernards, A., Van der Ploeg, L. H. T., Michels, P. A. M., Lieu, A. Y. C., De Lange, T., and Kooter, J. M. (1983). *Eur. J. Biochem.* **137,** 383.
Bostock, C. J., and Tyler-Smith, C. (1981). *J. Mol. Biol.* **153,** 219.
Bostock, C. J., Christie, S., and Hatch, F. T. (1976). *Nature (London)* **262,** 516.
Bouteille, M., Bouvier, D., and Seve, A. P. (1983). *Int. Rev. Cytol.* **83,** 135.
Bouvier, D., Hubert, J., and Bouteille, M. (1980). *J. Ultrastruct. Res.* **73,** 288.
Bowman, L. H., Rabin, B., and Schlessinger, D. (1981). *Nucleic Acids Res.* **9,** 4951.
Brachet, J. (1940). *Arch. Biol.* **51,** 151.
Brachet, J. (1957). ''Biochemical Cytology.'' Academic Press, New York.
Brack, C., and Tonegawa, S. (1977). *Proc. Natl. Acad. Sci. U.S.A.* **74,** 5652.
Bradbury, E. M. (1975). *Curr. Top. Dev. Biol.* **9,** 1.
Bradbury, E. M., Capenter, B. G., and Rattle, H. W. E. (1973). *Nature (London)* **241,** 123.
Bradbury, E. M., Inglis, R. J., and Matthews, H. R. (1974). *Nature (London)* **247,** 257.
Bradbury, E. M., Maclean, N., and Matthews, H. R. (1981). ''DNA, Chromatin and Chromosomes.'' Blackwell, Oxford.
Brandhorst, B. P., and McConkey, E. H. (1974). *J. Mol. Biol.* **85,** 451.
Breathnach, R., and Chambon, P. (1981). *Annu. Rev. Biochem.* **50,** 349.
Brinkley, B. R., Fistel, S. H., Marcum, J. M., and Pardue, R. L. (1980). *Int. Rev. Cytol.* **63,** 59.
Brinkley, B. R., Cox, S. M., Pepper, D. A., Wible, M., Brenner, S. L., and Pardue, R. L. (1981). *J. Cell Biol.* **90,** 554.
Brown, D. D. (1981). *Science* **211,** 667.
Brown, D. D., and Dawid, I. B. (1968). *Science* **160,** 272.
Brown, D. D., and Gurdon, J. B. (1965). *Proc. Natl. Acad. Sci. U.S.A.* **51,** 139.
Brown, D. D., and Sugimoto, K. (1973). *Cold Spring Harbor Symp. Quant. Biol.* **38,** 501.
Brown, D. D., Wensink, P. C., and Jordan, E. (1971). *Proc. Natl. Acad. Sci. U.S.A.* **68,** 3175.
Brown, M., Bollum, F. J., and Chang, L. M. S. (1981). *Proc. Natl. Acad. Sci. U.S.A.* **78,** 3049.
Brown, P. C., Beverley, S. M., and Schimke, R. T. (1981). *Mol. Cell. Biol.* **1,** 1077.
Burke, B., Tooze, J., and Warren, G. (1983). *EMBO J.* **2,** 361.
Busslinger, M., Rusconi, S., and Birnstiel, M. L. (1982). *EMBO J.* **1,** 27.
Butler, P. J., and Thomas, J. O. (1980). *J. Mol. Biol.* **140,** 505.
Butler, P. J. G. (1983). *CRC Crit. Rev. Biochem.* **15,** 57.
Butler, P. J. G. (1984). *EMBO J.* **3,** 2599.
Buttyan, R., Olsson, C. A., Sheard, B., and Kallos, J. (1983). *J. Biol. Chem.* **258,** 14366.
Calabretta, B., Robberson, D. L., Maizel, A. L., and Saunders, G. F. (1981). *Proc. Natl. Acad. Sci. U.S.A.* **78,** 6003.

Calabretta, B., Robberson, D. L., Barrera-Saldãna, H. A., Lambrou, T. P., and Saunders, G. F. (1982). *Nature (London)* **296,** 219.
Callan, H. G., and Tomlin, S. G. (1950). *Proc. R. Soc. London, Ser. B* **137,** 367.
Calos, M. P., and Miller, J. H. (1980). *Cell (Cambridge, Mass.)* **20,** 579.
Calvet, J. D., Meyer, L. M., and Pederson, T. (1982). *Science* **217,** 456.
Candido, E. P., Reeves, R., and Davie, J. R. (1978). *Cell (Cambridge, Mass.)* **14,** 105.
Capco, D. G., Wan, K. M., and Penman, S. (1982). *Cell (Cambridge, Mass.)* **29,** 847.
Cartwright, I. L., Abmayr, S. M., Fleischmann, G., Lowenhaupt, K., Elgin, S. C. R., Keene, M. A., and Howard, G. C. (1982). *CRC Crit. Rev. Biochem.* **13,** 1.
Caspersson, T., and Schultz, J. (1939). *Nature (London)* **143,** 602.
Cavazza, B., Trefilletti, V., Pioli, F., Ricci, E., and Patrone, E. (1983). *J. Cell Sci.* **62,** 81.
Cech, T. R. (1983). *Cell (Cambridge, Mass.)* **34,** 713.
Cech, T. R. (1985). *Int. Rev. Cytol* **93,** 3.
Cech, T. R., Tanner, N. K., Tinoco, I., Jr., Weir, B. R., Zuker, M., and Perlman, P. S. (1983). *Proc. Natl. Acad. Sci. U.S.A.* **80,** 3903.
Champoux, J. J. (1978). *Annu. Rev. Biochem.* **47,** 449.
Chatterjee, N. K., Dickerman, H. W., and Beach, R. A. (1977). *Arch. Biochem. Biophys.* **183,** 228.
Childs, G., Maxson, R., Cohn, R. H., and Kedes, L. (1981). *Cell (Cambridge, Mass.)* **23,** 651.
Ciejek, E. M., Nordstrom, J. L., Tsai, M. J., and O'Malley, B. W. (1982). *Biochemistry* **21,** 4945.
Ciejek, E. M., Tsai, M. J., and O'Malley, B. W. O. (1983). *Nature (London)* **306,** 607.
Claude, A. (1943). *Biol. Symp.* **10,** 111.
Clawson, G. A., and Smuckler, E. A. (1980). *Biochem. Biophys. Res. Commun.* **96,** 812.
Clawson, G. A., and Smuckler, E. A. (1982). *J. Theor. Biol.* **65,** 607.
Clawson, G. A., James, J., Woo, C. H., Friend, D. S., Moody, D., and Smuckler, E. A. (1980). *Biochemistry* **19,** 2748.
Clawson, G. A., Friend, D. S., and Smuckler, E. A. (1984). *Exp. Cell Res.* **155,** 310.
Clerc, R. G., Bucher, P., Strub, K., and Birnstiel, M. L. (1983). *Nucleic Acids Res.* **11,** 8641.
Cleveland, D. W., Hughes, S. H., Stubblefield, E., Kirschner, M. W., and Varmus, H. E. (1981). *J. Biol. Chem.* **256,** 3130.
Cohn, R. H., and Kedes, L. H. (1979). *Cell (Cambridge, Mass.)* **18,** 855.
Collins, M., and Rubin, G. M. (1983). *Nature (London)* **303,** 259.
Comings, D. E., and Harris, D. C. (1975). *Exp. Cell Res.* **96,** 161.
Comings, D. E., Harris, D. C., Okada, T. A., and Homlquist, G. (1977). *Exp. Cell Res.* **105,** 349.
Compere, S. J., and Palmiter, R. D. (1981). *Cell (Cambridge, Mass.)* **25,** 233.
Compton, J. L., Bellard, M., and Chambon, P. (1976). *Proc. Natl. Acad. Sci. U.S.A.* **73,** 43.
Conrad, S. E., and Botchan, M. R. (1982). *Mol. Cell. Biol.* **2,** 949.
Cook, P. R., and Brazell, I. A. (1980). *Nucleic Acids Res.* **8,** 2895.
Coutelle, C. (1981). *Biochem. J.* **197,** 1.
Cowan, N. J., Wilde, D., Chow, L. T., and Wefald, F. C. (1981). *Proc. Natl. Acad. Sci. U.S.A.* **78,** 4877.
Cox, G. S. (1982). *J. Cell Sci.* **58,** 363.
Craik, C. S., Sprang, S., Fletterick, R., and Rutter, W. J. (1982). *Nature (London)* **299,** 180.
Creusot, F., and Christman, J. K. (1981). *Nucleic Acids Res.* **9,** 5359.
Darnell, J. E., Jr. (1978). *Science* **202,** 1257.
Davidson, E. H., and Britten, R. W. (1973). *Q. Rev. Biol.* **48,** 565.
Davidson, E. H., and Britten, R. J. (1979). *Science* **204,** 1052.
Davidson, E. H., Hough, B. R., Amenson, C. S., and Britten, R. J. (1973). *J. Mol. Biol.* **77,** 1.
Davidson, E. H., Galau, G. A., Angerer, R. C., and Britten, R. J. (1975). *Chromosoma* **51,** 25.
Davidson, E. H., Klein, W. H., and Britten, R. J. (1977). *Dev. Biol.* **55,** 69.
Davies, R. W. (1984). *BioSci. Rep.* **4,** 707.

Davis, A. H., Reudelhuber, T. L., and Garrard, W. T. (1983). *J. Mol. Biol.* **167,** 133.
Dawid, I. B., and Rebbert, M. L. (1981). *Nucleic Acids Res.* **9,** 5011.
Dawid, I. B., and Wellauer, P. K. (1976). *Cell (Cambridge, Mass.)* **8,** 443.
Dawid, I. B., Brown, D. D., and Reeder, R. H. (1970). *J. Mol. Biol.* **51,** 341.
Déak, I. I. (1973). *J. Cell Sci.* **13,** 395.
Denison, R. A., and Weiner, A. M. (1982). *Mol. Cell. Biol.* **2,** 815.
Denison, R. A., Van Arsdell, S. W., Bernstein, L. B., and Weiner, A. M. (1981). *Proc. Natl. Acad. Sci. U.S.A.* **78,** 810.
De Robertis, E. M., Longthorne, R. F., and Gurdon, J. B. (1978). *Nature (London)* **272,** 254.
de Saint Vincent, B. R., Delbrück, S., Eckhart, W., Meinkoth, J., Vitto, L., and Wahl, G. (1981). *Cell (Cambridge, Mass.)* **27,** 267.
D'Eustachio, P., Meyuhas, D., Ruddle, F., and Perry, R. P. (1981). *Cell (Cambridge, Mass.)* **24,** 307.
Dierks, P., Van Ooyen, A., Cochran, M. D., Dobkin, C., Reiser, J., and Weissmann, C. (1983). *Cell (Cambridge, Mass.)* **32,** 695.
Di Giovanni, L., Haynes, S. R., Misra, R., and Jelinek, W. R. (1983). *Proc. Natl. Acad. Sci. U.S.A.* **80,** 6533.
Dimova, R. N., Markov, A. V., Gajdardjieva, K. C., Dabeva, M. D., and Hadjiolov, A. A. (1982). *Eur. J. Cell Biol.* **28,** 272.
Di Nocera, P. P., Digan, M. E., and Dawid, I. B. (1983). *J. Mol. Biol.* **168,** 715.
Djondjurov, L. P., Yancheva, N. Y., and Ivanova, E. C. (1983). *Biochemistry* **22,** 4095.
Doel, M. T. (1980). *Cell Biol. Int. Rep.* **4,** 433.
Doenecke, D., and Gallwitz, H. (1982). *Mol. Cell. Biochem.* **44,** 113.
Doerfler, W. (1983). *Annu. Rev. Biochem.* **52,** 93.
Doolittle, W. F. (1978). *Nature (London)* **272,** 581.
Dott, P. J., Chuang, C. R., and Saunders, G. F. (1976). *Biochemistry* **15,** 4120.
Duguet, M. (1981). *Biochimie* **63,** 649.
Dunn, A. R., and Gough, N. (1984). *Trends Biochem. Sci.* **9,** 81.
Earnshaw, W. C., Honda, B. M., Laskey, R. A., and Thomas, J. O. (1980). *Cell (Cambridge, Mass.)* **21,** 373.
Economidis, I. V., and Pederson, T. (1983). *Proc. Natl. Acad. Sci. U.S.A.* **80,** 1599.
Ehrlich, M., Gama-Sosa, M. A., Huang, L. H., Midgett, R. M., Kuo, K. C., McCune, R. A., and Gehrke, C. (1982). *Nucleic Acids Res.* **10,** 2709.
Eichler, D. C., and Eales, S. J. (1983). *J. Biol. Chem.* **258,** 10049.
Einck, L., and Bustin, M. (1983). *Proc. Natl. Acad. Sci. U.S.A.* **80,** 6735.
Elgin, S. C. R. (1981). *Cell (Cambridge, Mass.)* **27,** 413.
Elgin, S. C. R., and Weintraub, H. (1975). *Annu. Rev. Biochem.* **44,** 737.
Emerson, B. M., and Felsenfeld, G. (1984). *Proc. Natl. Acad. Sci. U.S.A.* **81,** 95.
Engel, J., Gunning, P., and Kedes, L. (1982). *Mol. Cell. Biol.* **2,** 674.
Epstein, P., Reddy, R., Henning, D., and Busch, H. (1980). *J. Biol. Chem.* **255,** 8901.
Fakan, S., and Puvion, E. (1980). *Int. Rev. Cytol.* **65,** 255.
Falkenthal, S., and Lengyel, J. A. (1980). *Biochemistry* **19,** 5842.
Feldherr, C. M. (1975). *Exp. Cell Res.* **93,** 411.
Feldherr, C. M., and Ogburn, J. A. (1980). *J. Cell Biol.* **87,** 589.
Ficq, A., and Errera, M. (1955). *Biochim. Biophys. Acta* **16,** 45.
Ficq, A., and Pavan, C. (1957). *Nature (London)* **180,** 983.
Finch, J. T., and Klug, A. (1976). *Proc. Natl. Acad. Sci. U.S.A.* **73,** 1897.
Finch, J. T., Brown, R. S., Rhodes, D., Richmond, T., Rushton, B., Lutter, L. C., and Klug, A. (1981). *J. Mol. Biol.* **145,** 757.
Finnagan, D. J. (1985). *Int. Rev. Cytol.* **93,** 281.

Firtel, R. A. (1981). *Cell (Cambridge, Mass.)* **24,** 6.

Flavell, A. (1981). *Nature (London)* **289,** 10.

Flavell, A. J., and Ish-Horowicz, D. (1981). *Nature (London)* **292,** 591.

Flavell, A. J., Ruby, S. W., Toole, J. J., Roberts, B. E., and Rubin, G. M. (1980). *Proc. Natl. Acad. Sci. U.S.A.* **77,** 7107.

Fradin, A., Manley, J. L., and Prives, C. L. (1982). *Proc. Natl. Acad. Sci. U.S.A.* **79,** 5142.

Frado, L. L., Mura, C. V., Stollar, B. D., and Woodcock, C. L. F. (1983). *J. Biol. Chem.* **258,** 11984.

Francke, C., Edström, J. E., McDowall, A. W., and Miller, O. L., Jr. (1982). *EMBO J.* **1,** 59.

Franke, W. W. (1977). *Biochem. Soc. Symp.* **42,** 125.

Franke, W. W., Scheer, U., Spring, H., Trendelenburg, M. F., and Krohne, G. (1976). *Exp. Cell Res.* **100,** 233.

Franke, W. W., Scheer, U., Zentgraf, H., Trendelenburg, M. F., Müller, U., Krohne, G., and Spring, H. (1980). *Results Probl. Cell Differ.* **11,** 15.

Franke, W. W., Kleinschmidt, J. A., Spring, H., Krohne, G., Grund, C., Trendelenburg, M. F., Stoekhr, M., and Scheer, U. (1981). *J. Cell. Biol.* **90,**289.

Frenster, J. H. (1965). *Nature (London)* **206,** 680.

Fujii, S., Wang, A. H. J., Van der Marel, G., Van Boom, J. H., and Rich, A. (1982). *Nucleic Acids Res.* **10,** 7879.

Gafner, J., and Philippsen, F. (1980). *Nature (London)* **286,** 444.

Gall, J. G., Cohen, E. H., and Polan, M. L. (1971). *Chromsoma* **33,** 319.

Garel, A., and Axel, R. (1976). *Proc. Natl. Acad. Sci. U.S.A.* **73,** 3966.

Gay, H. (1955). *Proc. Natl. Acad. Sci. U.S.A.* **41,** 370.

Gazit, B., and Cedar, H. (1980). *Nucleic Acids Res.* **8,** 5143.

Georgiev, G. P. (1972). *Curr. Top. Dev. Biol.* **7,** 1.

Georgieva, E. I., Pashev, I. G., and Tsanev, R. G. (1981). *Biochim. Biophys. Acta* **652,** 240.

Gerace, L., and Blobel, G. (1980). *Cell (Cambridge, Mass.)* **19,** 277.

Gerace, L., Ottaviano, Y., and Kondor-Koch, C. (1982). *J. Cell Biol.* **95,** 826.

Gilbert, W. (1978). *Nature (London)* **271,** 501.

Gilbert, W. (1981a). *Science* **214,** 1305.

Gilbert, W. (1981b). *Biosci. Rep.* **1,** 353.

Goessens, G. (1984). *Int. Rev. Cytol.* **87,** 107.

Goldberg, G. I., Collier, I., and Cassel, A. (1983). *Proc. Natl. Acad. Sci. U.S.A.* **80,** 6887.

Goldberg, M. L., Paro, R., and Gehring, W. J. (1982). *EMBO J.* **1,** 93.

Goldstein, L., and Ko, C. (1981). *J. Cell Biol.* **88,** 516.

Gomez-Lira, M. M., and Bode, J. (1981). *FEBS Lett.* **127,** 228.

Goodwin, G. M., Sanders, C., and Johns, E. W. (1973). *Eur. J. Biochem.* **38,** 14.

Gosden, J. R., and Mitchell, A. R. (1975). *Exp. Cell Res.* **92,** 131.

Gough, N. (1981). *Trends Biochem. Sci.* **6,** 203.

Gough, N. (1982). *Trends Biochem. Sci.* **7,** 307.

Green, N. M. (1982). *Nature (London)* **297,** 287.

Greenberg, J. R., and Setyono, B. (1981). *Biol. Cell.* **41,** 67.

Gross, K., Schaffner, W., Telford, J., and Birnstiel, M. (1976). *Cell (Cambridge, Mass.)* **8,** 479.

Grosschedl, R., and Birnstiel, M. L. (1979). *Proc. Natl. Acad. Sci. U.S.A.* **77,** 1432.

Grosschedl, R., and Birnstiel, M. L. (1980). *Proc. Natl. Acad. Sci. U.S.A.* **77,** 7102.

Grosschedl, R., and Birnstiel, M. L. (1982). *Proc. Natl. Acad. Sci. U.S.A.* **79,** 297.

Groudine, M., Kohwi-Shigematsu, T., Gelinas, R., Stamatoyannopoulos, G., and Papayannopoulou, T. (1983). *Proc. Natl. Acad. Sci. U.S.A.* **80,** 7551.

Grummt, I. (1974). *FEBS Lett.* **39,** 125.

Gruss, P., Efstratiadis, A., Karathanasis, S., König, M., and Khoury, G. (1981). *Proc. Natl. Acad. Sci. U.S.A.* **78,** 6091.

Gurdon, J. B. (1970). *Proc. R. Soc. London, Ser. B* **176,** 303.
Gurley, L. R., Walters, R. A., Barham, S. S., and Deaven, L. L. (1978). *Exp. Cell Res.* **111,** 373.
Haas, A. L., Warms, J. V. B., and Rose, I. A. (1983). *Biochemistry* **22,** 4388.
Haber, D. A., and Schimke, R. T. (1981). *Cell (Cambridge, Mass.)* **26,** 355.
Hadjiolov, A. A. (1980). *Subcell. Biochem.* **7,** 1.
Hamada, H., and Kakunaga, T. (1982). *Nature (London)* **298,** 396.
Hamada, H., Petrino, M. G., and Kakunaga, T. (1982). *Proc. Natl. Acad. Sci. U.S.A.* **79,** 6465.
Hancock, R. (1982). *Biol. Cell.* **46,** 105.
Hancock, R., and Boulikas, T. (1982). *Int. Rev. Cytol.* **79,** 165.
Hancock, R., and Hughes, M. E. (1982). *Biol. Cell.* **44,** 201.
Harbers, K., Kuehn, M., Delius, H., and Jaenisch, R. (1984). *Proc. Natl. Acad. Sci. U.S.A.* **81,** 1504.
Harris, J. R. (1978). *Biochim. Biophys. Acta* **515,** 55.
Haynes, S. R., and Jelinek, W. R. (1981). *Proc. Natl. Acad. Sci. U.S.A.* **78,** 6130.
Haynes, S. R., Toomey, T. P., Leinwand, L., and Jelinek, W. R. (1981). *Mol. Cell. Biol.* **1,** 573.
Heitz, E. (1934). *Biol. Zentralbl.* **54,** 588.
Henikoff, S., Sloan, J. S., and Kelly, J. D. (1983). *Cell (Cambridge, Mass.)* **34,** 405.
Hennig, W. (1972). *J. Mol. Biol.* **71,** 407.
Hentschel, C. C., and Birnstiel, M. L. (1981). *Cell (Cambridge, Mass.)* **25,** 301.
Hernandez-Verdun, D. (1983). *Biol. Cell.* **49,** 191.
Hernandez-Verdun, D., Hubert, J., Bourgeois, C. A., and Bouteille, M. (1980). *Chromosoma* **79,** 349.
Herzberg, M., Bibor-Hardy, V., and Simard, R. (1981). *Biochem. Biophys. Res. Commun.* **100,** 644.
Herzog, M., and Soyer, M. O. (1981). *Eur. J. Cell Biol.* **23,** 295.
Higgs, D. R., Goodbourn, S. E. Y., Lamb, J., Clegg, J. B., Weatherall, D. J., and Proudfoot, N. J. (1983). *Nature (London)* **306,** 398.
Higuchi, R., Stang, H. D., Browne, J. K., Martin, M. O., Huot, M., Lipeles, J., and Salser, W. (1981). *Gene* **15,** 177.
Hill, R. J., Watt, F., and Stollar, B. D. (1984). *Exp. Cell Res.* **153,** 469.
Hoffer, E., and Darnell, J. E., Jr. (1981). *Cell (Cambridge, Mass.)* **23,** 585.
Hoffmann-Liebermann, B., Liebermann, D., Kedes, L. H., and Cohen, S. N. (1985). *Molec. Cellul. Biol.* **5,** 991.
Hogan, B. (1983). *Nature (London)* **306,** 313.
Hohmann, P. (1981). *Int. Rev. Cytol.* **71,** 41.
Hollis, G. F., Hieter, P. A., McBride, O. W., Swan, D., and Leder, P. (1982). *Nature (London)* **296,** 321.
Holmes, D. S., and Bonner, J. (1978). *Biochemistry* **12,** 2330.
Holmgren, R. (1984). *EMBO J.* **3,** 569.
Hough, B. R., Smith, M. J., Britten, R. J., and Davidson, E. H. (1975). *Cell (Cambridge, Mass.)* **5,** 291.
Hozumi, N., and Tonegawa, S. (1976). *Proc. Natl. Acad. Sci. U.S.A.* **73,** 3628.
Hsieh, T. S., and Brutlag, D. (1980). *Cell (Cambridge, Mass.)* **21,** 115.
Hubbell, H. R., Lau, Y. F., Brown, R. L., and Hsu, T. C. (1980). *Exp. Cell Res.* **129,** 139.
Hubert, J., Bouvier, D., Arnoult, J., and Bouteille, M. (1981). *Exp. Cell Res.* **131,** 446.
Hübscher, U., Kuenzle, C. C., and Spadari, S. (1979). *Proc. Natl. Acad. Sci. U.S.A.* **76,** 2316.
Huez, G., Bruck, C., and Cleuter, Y. (1981). *Proc. Natl. Acad. Sci. U.S.A.* **78,** 908.
Humphries, S. E., Whithall, R., Minty, A., Buckingham, M., and Williamson, R. (1981). *Nucleic Acids Res.* **9,** 4895.
Ilyin, Y. V., Chmeliauskaite, V. G., Ananiev, E. V., and Georgiev, G. P. (1980). *Chromosoma* **81,** 27.

Ilyin, Y. V., Schuppe, N. G., Lyubomirskaya, N. V., Gorelova, T. V., and Arkhipova, I. R. (1984). *Nucl. Ac. Res.* **12,** 7517.
Isackson, P. J., and Reeck, G. R. (1981). *Nucleic Acids Res.* **9,** 3779.
Ish-Horowicz, D. (1982). *Nature (London)* **296,** 806.
Izant, J. G., and Weintraub, H. (1984). *Cell* **34,** 1007.
Jackson, D. A., and Cook, P. R. (1985). *EMBO J.* **4,** 919.
Jackson, D. A., McCready, S. J., and Cook, P. R. (1981). *Nature (London)* **292,** 522.
Jacob, E. (1976). *Eur. J. Biochem.* **65,** 275.
Jacob, E., Malacinski, G., and Birnstiel, M. L. (1976). *Eur. J. Biochem.* **69,** 45.
Jacobson, A., and Favreau, M. (1983). *Nucleic Acids Res.* **11,** 6353.
Jagadeeswaran, P., Forget, B. G., and Weissman, S. M. (1981). *Cell (Cambridge, Mass.)* **26,** 141.
Jarasch, E. D., and Franke, W. W. (1974). *J. Biol. Chem.* **249,** 7245.
Jeang, K. T., and Hayward, G. S. (1983). *Mol. Cell. Biol.* **3,** 1389.
Jeener, R. (1946). *C.R. Séances Soc. Biol.* **140,** 1139.
Jelinek, W., Molloy, G., Fernandez-Munoz, R., Salditt, M., and Darnell, J. E. (1974). *J. Mol. Biol.* **82,** 361.
Jones, C. E., Busch, H., and Olson, M. O. J. (1981). *Biochim. Biophys. Acta* **667,** 209.
Jones, P. A., and Taylor, S. M. (1980). *Cell (Cambridge, Mass.)* **20,** 85.
Jonge, A. J. R. de, Abrahams, P. J., Westerbeld, A., and Bootsma, D. (1982). *Nature (London)* **295,** 625.
Jost, J. P., and Seldron, M. (1984). *EMBO J.* **3,** 2005.
Karpov, V. L., Preobrazhenskaya, O. V., and Mirzabekov, A. D. (1984). *Cell* **36,** 423.
Kaufman, R. J., Brown, P. C., and Schimke, R. T. (1981). *Mol. Cell. Biol.* **1,** 1084.
Kaufmann, S. H., Coffey, D. S., and Shaper, J. H. (1981). *Exp. Cell Res.* **132,** 105.
Kay, E. R., Smellie, R. M. S., Humphrey, G. F., and Davidson, J. N. (1956). *Biochem. J.* **62,** 160.
Kaye, J. S., Bellard, M., Dretzen, G., Bellard, F., and Chambon, P. (1984). *EMBO J.* **3,** 1137.
Kedes, L. H. (1979). *Annu. Rev. Biochem.* **48,** 837.
Kedes, L. H., Birnstiel, M. L., Purdom, I. F., and Williamson, R. (1971). *Nature (London), New Biol.* **230,** 165.
Kedes, L. H., Chang, A. C. Y., Houseman, D., and Cohen, S. N. (1975). *Nature (London)* **255,** 533.
Keichline, L. D., and Wassarman, P. M. (1977). *Biochim. Biophys. Acta* **475,** 139.
Kelley, D. E., and Perry, R. P. (1971). *Biochim. Biophys. Acta* **238,** 357.
Khesin, R. B. (1980). *Mol. Biol.* **14,** 949.
King, C. R., and Piatigorsky, J. (1984). *J. Biol. Chem.* **259,** 1822.
Kingsman, A. J., Gimlich, R., Clarke, L., Chinault, A. C., and Carbon, J. (1981). *J. Mol. Biol.* **145,** 619.
Knowler, J. T. (1983). *Int. Rev. Cytol.* **84,** 103.
Knowler, J. T., and Wilks, A. F. (1980). *Trends Biochem. Sci.* **5,** 268.
Kohn, R. H., and Kedes, L. H. (1979). *Cell (Cambridge, Mass.)* **18,** 855.
Kornberg, R. D. (1977). *Annu. Rev. Biochem.* **46,** 931.
Krayev, A. S., Markusheva, T. V., Kramerov, D. A., Ryskov, A. P., Skryabin, K. G., Bayev, A. A., and Georgiev, G. P. (1982). *Nucleic Acids Res.* **23,** 7461.
Krohne, G., and Franke, W. W. (1980a). *Exp. Cell Res.* **129,** 167.
Krohne, G., and Franke, W. W. (1980b). *Proc. Natl. Acad. Sci. U.S.A.* **77,** 1034.
Krohne, G., Franke, W. W., and Scheer, U. (1978). *Exp. Cell Res.* **116,** 85.
Krohne, G., Dabauvalle, M. C., and Franke, W. W. (1981). *J. Mol. Biol.* **151,** 121.
Krolewski, J. J., Bertelsen, A. H., Humayum, M. Z., and Rush, M. G. (1982). *J. Mol. Biol.* **154,** 399.

Kulguskin, V. V., Krichevskaya, A. A., Lukanidin, E. M., and Georgiev, G. P. (1980). *Biochim. Biophys. Acta* **609,** 410.
Kumar, A., and Subramanian, A. R. (1975). *J. Mol. Biol.* **94,** 409.
Kuo, M. T., Iyer, B., and Schwarz, R. J. (1982). *Nucleic Acids Res.* **10,** 4565.
Kuter, D. J., and Rodgers, A. (1976). *Exp. Cell Res.* **102,** 205.
Lake, R. S. (1973). *Nature (London)* **242,** 145.
Lam, K., and Kasper, C. (1979). *Biochemistry* **18,** 307.
Laskey, R. A., and Earnshaw, W. C. (1980). *Nature (London)* **286,** 763.
Laskey, R. A., Honda, B. M., Mills, A. D., and Finch, J. T. (1978). *Nature (London)* **275,** 416.
Lassar, A. B., Martin, P. L., and Roeder, R. G. (1983). *Science* **222,** 740.
Lasters, U., Muyldermans, S., Wyns, L., and Hamers, R. (1981). *Biochemistry* **20,** 1104.
Lastick, S. M. (1980). *Eur. J. Biochem.* **113,** 175.
Lawn, R. M., Gross, M., Houck, C. M., Franke, A. E., Gray, P. V., and Goeddel, D. V. (1981). *Proc. Natl. Acad. Sci. U.S.A.* **78,** 5435.
Lawson, G. M., Knoll, B. J., March, C. J., Woo, S. L. C., Tsai, M. J., and O'Malley, B. W. (1982). *J. Biol. Chem.* **257,** 1501.
Lazowska, J., Jacq, C., and Slonimski, P. P. (1980). *Cell (Cambridge, Mass.)* **22,** 333.
Lebel, S., and Raymond, Y. (1984). *J. Biol. Chem.* **259,** 2693.
Lepault, J., Bram, S., Escaig, J., and Wray, W. (1980). *Nucleic Acids Res.* **8,** 265.
Lerner, M. R., Boyle, J. A., Mount, S. M., Wolin, S. L., and Steitz, J. A. (1980). *Nature (London)* **283,** 220.
Levinger, L., and Varshavsky, A. (1982). *Cell (Cambridge, Mass.)* **28,** 375.
Levy, W. B., Johnson, C. B., and McCarthy, B. J. (1976). *Nucleic Acids Res.* **3,** 1777.
Levy-Wilson, B. (1981). *Proc. Natl. Acad. Sci. U.S.A.* **78,** 2189.
Lewin, R. (1981a). *Science* **212,** 28.
Lewin, R. (1981b). *Science* **214,** 426.
Lieberman, M. W., Beach, L. R., and Palmiter, R. D. (1983). *Cell (Cambridge, Mass.)* **35,** 207.
Liebermann, D., Hoffman-Liebermann, B., Weinthal, J., Childs, G., Maxson, R., Mauron, A., Cohen, S. N., and Kedes, L. (1983). *Nature (London)* **306,** 342.
Lischwe, M. A., Richards, R. L., Busch, R. K., and Busch, H. (1981). *Exp. Cell Res.* **136,** 101.
Littauer, U. Z., and Soreq, H. (1982). *Prog. Nucleic Acid Res. Mol. Biol.* **27,** 53.
Little, P. F. R. (1982). *Cell (Cambridge, Mass.)* **28,** 683.
Lohr, D., Corden, J., Tatchell, K., Kovacic, T., and Van Holde, K. E. (1977). *Proc. Natl. Acad. Sci. U.S.A.* **74,** 79.
Long, E. D., and Dawid, I. B. (1980). *Annu. Rev. Biochem.* **49,** 727.
Lukanidin, E. M., Olsnes, S., and Pihl, A. (1972). *Nature (London), New Biol.* **240,** 90.
Lynch, W. E., Brown, R. F., Umeda, T., Langreth, S. G., and Liebermann, I. (1970). *J. Biol. Chem.* **245,** 3911.
McGhee, J. D., Rau, D. C., Charney, E., and Felsenfeld, G. (1980). *Cell (Cambridge, Mass.)* **22,** 87.
McGhee, J. D., Nickol, J. M., Felsenfeld, G., and Rau, D. C. (1983). *Cell (Cambridge, Mass.)* **33,** 831.
McGrath, J. P., and Evenson, D. P. (1982). *Gamete Res.* **5,** 379,
Machray, G. C., and Bonner, J. (1981). *Biochemistry* **20,** 5466.
Mandel, J. L., and Chambon, P. (1979). *Nucleic Acids Res.* **7,** 2081.
Marashi, F., Baumbach, L., Rickles, R., Sierra, F., Stein, J. L., and Stein, G. S. (1981). *Science* **215,** 683.
Marbaix, G., Huez, G., Burny, A., Cleuter, Y., Hubert, E., Leclercq, M., Chantrenne, H., Soreq, H., Nudel, U., and Littauer, U. Z. (1975). *Proc. Natl. Acad. Sci. U.S.A.* **72,** 3065.

Martin, K., Osheim, Y. N., Beyer, A. L., and Miller, O. L., Jr. (1980). *Results Probl. Cell Differ.* **11,** 37.
Marx, J. L. (1981a). *Science* **212,** 653.
Marx, J. L. (1981b). *Science* **212,** 1015.
Masters, J. N., Keeley, B., Gay, H., and Attardi, G. (1982). *Mol. Cell. Biol.* **2,** 498.
Masters, J. N., Yang, J. K., Cellini, A., and Attardi, G. (1983). *J. Mol. Biol.* **167,** 23.
Matsui, S. I., Seon, B. K., and Sandberg, A. A. (1979). *Proc. Natl. Acad. Sci. U.S.A.* **76,** 6386.
Maul, G. G. (1977). *Int. Rev. Cytol., Suppl.* **6,** 71.
Maul, G. G., and Avdalović, N. (1980). *Exp. Cell Res.* **130,** 229.
Mavilio, F., Giampaolo, A., Carè, A., Migliaccio, G., Calandrini, M., Russo, G., Pagliardi, G. L., MastroBerardino, G. Marinucci, M., and Peschle, C. (1983). *Proc. Natl. Acad. Sci. U.S.A.* **80,** 6907.
Max, E. E., Seidman, J. G., and Leder, P. (1979). *Proc. Natl. Acad. Sci. U.S.A.* **76,** 3450.
Maxam, A. M., and Gilbert, W. (1977). *Proc. Natl. Acad. Sci. U.S.A.* **74,** 560.
Menko, A. S., and Tan, K. B. (1980). *Biochim. Biophys. Acta* **629,** 359.
Milbrandt, J. D., Heintz, N. H., White, W. C., Rothman, S. M., and Hamlin, J. L. (1981). *Proc. Natl. Acad. Sci. U.S.A.* **78,** 6043.
Miller, O. L., Jr., and Beatty, B. R. (1969). *Science* **164,** 955.
Mills, A. D., Laskey, R. A., Black, P., and De Robertis, E. M. (1980). *J. Mol. Biol.* **139,** 561.
Mirault, M. E., and Scherrer, K. (1972). *FEBS Lett.* **20,** 233.
Mirkovitch, J., Mirault, M. E., and Laemmli, U. K. (1984). *Cell* **39,** 223.
Mirsky, A. E. (1947). *Cold Spring Harbor Symp. Quant. Biol.* **12,** 143.
Mondal, H., and Hoschneider, P. H. (1983). *Biochem. Biophys. Res. Commun.* **116,** 303.
Monk, R. J., Meyuhas, O., and Perry, R. P. (1981). *Cell (Cambridge, Mass.)* **24,** 301.
Monstein, H. J., Westin, G., Philipson, L., and Petterson, U. (1982). *EMBO J.* **1,** 133.
Montell, C., Fisher, E. F., Caruthers, M. H., and Berk, A. J. (1983). *Nature (London)* **305,** 600.
Moreau, J., Matyash-Smirniaguina, L., and Scherrer, K. (1981). *Proc. Natl. Acad. Sci. U.S.A.* **78,** 1341.
Mount, S. M., Pettersson, I., Hinterberger, M., Karmas, A., and Steitz, J. A. (1983). *Cell* **33,** 509.
Mura, C. V., and Stollar, B. D. (1981). *J. Biol. Chem.* **256,** 9767.
Murphy, E. C., Hall, S. H., Shepherd, J. H., and Weiser, R. S. (1973). *Biochimie* **12,** 3843.
Nabeshima, Y. I., Fujii-Kuriyama, Y., Muramatsu, M., and Ogata, K. (1984). *Nature (London)* **308,** 333.
Nakayasu, H., and Ueda, K. (1983). *Exp. Cell Res.* **143,** 55.
Nicolas, R. H., Wrigt, C. A., Cockerill, P. N., Wyke, J. A., and Goodwin, G. H. (1983). *Nucleic Acids Res.* **11,** 753.
Nishizuka, Y. (1984). *Nature* **308,** 693.
Noll, M., and Kornberg, R. D. (1977). *J. Mol. Biol.* **109,** 395.
Nordheim, A., Pardue, M. L., Lafer, E. M., Möller, A., Stollar, B. D., and Rich, A. (1981). *Nature (London)* **294,** 417.
Nunberg, J. H., Kaufman, R. J., Schimke, R. T., Urlaub, G., and Chasin, L. A. (1978). *Proc. Natl. Acad. Sci. U.S.A.* **75,** 5553.
Ny, T., Elgh, F., and Lund, B. (1984). *Proc. Natl. Acad. Sci. U.S.A.* **81,** 5355.
O'Brien, R. L., Sanyal, A. B., and Stanton, R. H. (1972). *Exp. Cell Res.* **70,** 106.
Ochs, R., Lischwe, M., O'Leary, P., and Busch, H. (1983). *Exp. Cell Res.* **146,** 139.
Ohkubo, H., Vogeli, Y., Mudryj, M., Avvedimento, E., Sullivan, M., Pastan, I., and de Crombrughe, B. (1980). *Proc. Natl. Acad. Sci. U.S.A.* **77,** 7059.
Olins, A. L., and Olins, D. E. (1974). *Science* **183,** 330.
Olins, A. L., and Olins, D. E. (1976). *J. Cell Biol.* **68,** 789.
Olio, R., and Rougeon, F. (1983). *Cell (Cambridge, Mass.)* **32,** 515.

Olson, M. O. J., Ezrailson, E. G., Guetzow, K., and Busch, H. (1975). *J. Mol. Biol.* **97,** 611.
Ordahl, C. P., and Caplan, A. I. (1978). *J. Biol. Chem.* **253,** 7683.
Orgel, L. E., and Crick, F. H. C. (1980). *Nature (London)* **284,** 604.
Orrick, L. R., Olson, M. O. J., and Busch, H. (1973). *Proc. Natl. Acad. Sci. U.S.A.* **70,** 13.
Osborn, M., and Weber, K. (1980). *Exp. Cell Res.* **129,** 103.
Oudet, P., Gross-Bellard, M., and Chambon, P. (1975). *Cell* **4,** 281.
Padgett, R. A., Wahl, G. M., and Stark, G. R. (1982). *Mol. Cell. Biol.* **2,** 293.
Padgett, R. A., Mount, S. M., Steitz, J. A., and Sharp, P. A. (1983). *Cell (Cambridge, Mass.)* **35,** 101.
Padgett, R. A., Konarska, M. M., Grabowski, P. J., Hardy, S. F., and Sharp, P. A. (1984). *Science* **225,** 898.
Pardue, M. L., Brown, D. D., and Birnstiel, M. L. (1973). *Chromosoma* **42,** 191.
Parker, C. S., and Topol, J. (1984). *Cell (Cambridge, Mass.)* **36,** 357.
Paro, L., Goldberg, M. L., and Gehring, W. J. (1983). *EMBO J.* **2,** 853.
Paul, J. (1982). *Biosci. Rep.* **2,** 63.
Pays, E., Lheureux, M., and Steinert, M. (1981a). *Nature (London)* **292,** 265.
Pays, E., Van Meirvenne, N., LeRay, D., and Steinert, M. (1981b). *Proc. Natl. Acad. Sci. U.S.A.* **78,** 2673.
Pederson, T. (1983). *J. Cell Biol.* **97,** 1321.
Pederson, T., and Kumar, A. (1971). *J. Mol. Biol.* **61,** 655.
Pehrson, J. R., and Cole, R. D. (1981). *Biochemistry* **20,** 2298.
Perry, R. P. (1960). *Exp. Cell Res.* **20,** 216.
Perry, R. P. (1976). *Annu. Rev. Biochem.* **45,** 605.
Perry, R. P. (1984). *Nature* **310,** 14.
Perry, R. P., Kelley, D. E., Friederici, K., and Rottman, F. (1975). *Cell (Cambridge, Mass.)* **4,** 387.
Peters, K. E., and Comings, D. E. (1980). *J. Cell Biol.* **86,** 135.
Peterson, J. L., and McConkey, E. H. (1976). *J. Biol. Chem.* **251,** 548.
Pieler, C., Adolf, G. R., and Swetly, P. (1981). *Eur. J. Biochem.* **115,** 329.
Pirrotta, V., and Bröckl, C. (1984). *EMBO J.* **3,** 563.
Poirier, G. G., De Murcia, G., Jongstra-Bilen, J., Niedergang, C., and Mandel, P. (1982). *Proc. Natl. Acad. Sci. U.S.A.* **79,** 3423.
Ponte, P., Gunning, P., Blau, H., and Kedes, L. (1983). *Mol. Cell. Biol.* **3,** 1783.
Potter, S. S. (1982). *Nature (London)* **297,** 201.
Prestayko, A. W., Klomp, G. R., Schmoll, D. J., and Busch, H. (1974). *Biochemistry* **13,** 1945.
Prunell, A., and Kornberg, R. D. (1982). *J. Mol. Biol.* **154,** 515.
Puvion, E., and Bernhard, (1975). *J. Cell Biol.* **67,** 204.
Puvion, E., Puvion-Dutilleul, F., and Leduc, E. H. (1981). *J. Ultrastruct. Res.* **76,** 181.
Puvion-Dutilleul, F. (1983). *Int. Rev. Cytol.* **84,** 57.
Razin, A., and Cedar, H. (1984). *Intern. Rev. Cytol.* **92,** 159.
Razin, A., and Riggs, A. D. (1980). *Science* **210,** 604.
Razin, A., and Szyf, M. (1984). *Biochim. Biophys. Acta* **782,** 331.
Rechsteiner, M., and Kuehl, L. (1979). *Cell (Cambridge, Mass.)* **16,** 901.
Reeder, R. H. (1984). *Cell* **38,** 349.
Reeder, R. H., Brown, D. D., Wellauer, P. K., and Dawid, I. B. (1976). *J. Mol. Biol.* **105,** 507.
Reeves, R. (1984). *Biochim. Biophys. Acta* **782,** 343.
Reynolds, C. R., and Tedeschi, H. (1984). *J. Cell Sci.* **70,** 197.
Richardson, J. C. W., and Maddy, A. H. (1980a). *J. Cell Sci.* **43,** 253.
Richardson, J. C. W., and Maddy, A. H. (1980b). *J. Cell Sci.* **43,** 269.
Ritossa, F. M., and Spiegelman, S. (1965). *Proc. Natl. Acad. Sci. U.S.A.* **53,** 737.

Roberts, K., and Northcote, D. H. (1970). *Nature (London)* **228,** 385.
Robinson, S. I., Nelkin, B. D., and Vogelstein, B. (1982). *Cell (Cambridge, Mass.)* **28,** 99.
Robinson, S. I., Small, D., Idzerda, R., McKnight, G. S., and Vogelstein, B. (1983). *Nucleic Acids Res.* **11,** 5113.
Rogers, J. (1983). *Nature (London)* **305,** 101.
Roop, D. R., Kristo, P., Stumph, W. E., Tsai, M. J., and O'Malley, B. W. (1981). *Cell (Cambridge, Mass.)* **23,** 671.
Rosbach, M., Campo, M. S., and Gummerson, K. S. (1975). *Nature (London)* **258,** 682.
Rose, K. M., Bell, L. E., and Jacob, S. T. (1977). *Biochim. Biophys. Acta* **475,** 548.
Roth, G. E., and Moritz, K. B. (1981). *Chromosoma* **83,** 169.
Rubin, G. M., and Spradling, A. C. (1982). *Science* **218,** 348.
Rungger, D., and Crippa, M. (1977). *Prog. Biophys. Mol. Biol.* **31,** 247.
Ruskin, B., Krainer, A. R., Maniatis, T., and Green, M. R. (1984). *Cell* **38,** 317.
Ryffel, G. U. (1976). *Eur. J. Biochem.* **62,** 417.
Ryskov, A. P., Saunders, G. F., Farashyan, V. R., and Georgiev, J. P. (1903). *Biochim. Biophys. Acta* **312,** 152.
Saiga, H., and Higashinakagawa (1979). *Nucleic Acids Res.* **6,** 1929.
Salim, M., and Maden, E. H. (1981). *Nature (London)* **291,** 205.
Sanders, M. M. (1981). *J. Cell Biol.* **91,** 579.
Sanger, F. (1981a). *Science* **214,** 1205.
Sanger, F. (1981b). *Biosci. Rep.* **1,** 3.
Sano, H., and Sager, R. (1982). *Proc. Natl. Acad. Sci. U.S.A.* **79,** 3584.
Sass, H. (1982). *Cell (Cambridge, Mass.)* **28,** 269.
Scarpulla, R. C., and Wu, R. (1983). *Cell. (Cambridge, Mass.)* **32,** 473.
Schatten, G., and Thoman, M. (1978). *J. Cell Biol.* **77,** 517.
Scheer, U. (1973). *Dev. Biol.* **30,** 13.
Scheer, U., and Rose, K. M. (1984). *Proc. Natl. Acad. Sci. U.S.A. 81,* 1431.
Scheer, U., Kartenbeck, D., Trendelenburg, M. F., Stadler, J., and Franke, W. W. (1976). *J. Cell Biol.* **69,** 1.
Scheer, U., Sommerville, J., and Müller, U. (1980). *Exp. Cell Res.* **129,** 115.
Scheer, U., Zentgraf, H., and Sauer, H. W. (1981). *Chromosoma* **84,** 279.
Schimke, R. T. (1984). *Cell* **37,** 705.
Schmid, C., and Jelinek, W. R. (1982). *Science* **216,** 1065.
Schon, E., Evans, T., Welsh, J., and Efstratiadis, A. (1983). *Cell (Cambridge, Mass.)* **35,** 837.
Schwartz, H. E., Lockett, T. J., and Young, M. W. (1982). *J. Mol. Biol.* **157,** 49.
Seale, R. L., Annunziato, A. T., and Smith, R. D. (1983). *Biochemistry* **22,** 5008.
Sealy, L., and Chalkley, R. (1978). *Cell (Cambridge, Mass.)* **14,** 115.
Seyedin, S. M., Pehrson, J. R., and Cole, R. D. (1981). *Proc. Natl. Acad. Sci. U.S.A.* **78,** 5988.
Shapiro, D. J., and Schimke, R. T. (1975). *J. Biol. Chem.* **250,** 1759.
Shapiro, J. A., and Cordell, B. (1982). *Biol. Cell.* **43,** 31.
Sharp, P. A. (1983). *Nature (London)* **301,** 471.
Sharp, S., Dingermann, T., Schaack, J., Sharp, J. A., Burke, D. J., and De Robertis, E. M. (1983). *Nucleic Acids Res.* **11,** 8677.
Shatkin, A. J. (1976). *Cell (Cambridge, Mass.)* **9,** 645.
Sheffery, M., Marks, P. A., and Rifkind, R. A. (1984). *J. Mol. Biol.* **172,** 417.
Shelton, K. R., Higgins, L. L., Cochran, D. L., Ruffolo, J. J., and Egle, P. M. (1980). *J. Biol. Chem.* **255,** 10978.
Shiba, T., and Saigo, K. (1983). *Nature (London)* **302,** 119.
Shimizu, Y., Yoshida, K., Ren, C. S., Fujinaga, K., Rajagopdlan, S., and Chinnadurai, G. (1983). *Nature (London)* **302,** 587.

Sikstrom, R., Lanoix, J., and Bergeron, J. J. M. (1976). *Biochim. Biophys. Acta* **448,** 88.
Simard, R., Sakr, F., and Bachellerie, J. P. (1973). *Exp. Cell Res.* **81,** 1.
Sims, M. A., Doering, J. L., and Hoyle, H. D. (1983). *Nucleic Acids Res.* **11,** 277.
Sinclair, J. H., Sang, J. H., Burke, J. F., and Ish-Horowicz, D. (1983). *Nature (London)* **306,** 198.
Singer, M. F. (1982a). *Int. Rev. Cytol.* **76,** 67.
Singer, M. F. (1982b). *Cell (Cambridge, Mass.)* **28,** 433.
Sinibaldi, R. M., and Morris, P. W. (1981). *J. Biol. Chem.* **256,** 10735.
Small, D., Nelkin, B., and Vogelstein, B. (1982). *Proc. Natl. Acad. Sci. U.S.A.* **79,** 5911.
Small, D., Nelkin, B., and Vogelstein, B. (1985). *Nuc. Acids Res.* **13,** 2413.
Smith, H. C., and Berezney, R. (1980). *Biochem. Biophys. Res. Commun.* **97,** 1541.
Smith, R. D., Seale, R. L., and Yu, J. (1983). *Proc. Natl. Acad. Sci. U.S.A.* **80,** 5505.
Sommerville, J., and Hill, R. J. (1973). *Nature (London), New Biol.* **245,** 104.
Spadafora, C., Noviello, L., and Geraci, G. (1976). *Cell Differ.* **5,** 225.
Spohr, G., Reith, W., and Sures, I. (1981). *J. Mol. Biol.* **151,** 573.
Spradling, A. C., and Mahowald, A. P. (1980). *Proc. Natl. Acad. Sci. U.S.A.* **77,** 1096.
Spradling, A. C., and Rubin, G. M. (1981). *Annu. Rev. Genet.* **15,** 219.
Stanfield, S. W., and Helinski, D. R. (1984). *Mol. Cell. Biol.* **4,** 173.
Stein, G. H. (1978). *Methods Cell Biol.* **17,** 271.
Stein, G. S., Spelsberg, T. C., and Kleinsmith, L. J. (1974). *Science* **183,** 817.
Stein, R., Razin, A., and Cedar, H. (1982). *Proc. Natl. Acad. Sci. U.S.A.* **79,** 3418.
Stein, R., Sciaky-Gallili, N., Razin, A., and Cedar, H. (1983). *Proc. Natl. Acad. Sci. U.S.A.* **80,** 2422.
Stephenson, E. C., Erba, H. P., and Gall, J. G. (1981). *Nucleic Acids Res.* **9,** 2281.
Stevens, B. J., Nicoloso, M., and Amalric, F. (1980). *Biol. Cell.* **39,** 171.
Stumph, W. E., Kristo, P., Tsai, M. J., and O'Malley, B. W. (1981). *Nucleic Acids Res.* **9,** 5383.
Szostak, J. W., Orr-Weaver, T. L., Rothstein, R. J., and Stahl, F. W. (1983). *Cell (Cambridge, Mass.)* **33,** 25.
Tamm, I., Hand, R., and Caliguiri, L. A. (1976). *J. Cell Biol.* **69,** 229.
Taylor, S. M., and Jones, P. A. (1979). *Cell (Cambridge, Mass.)* **17,** 771.
Tchurikov, N. A., Zelentsova, E. S., and Georgiev, G. P. (1980). *Nucleic Acids Res.* **8,** 1243.
Thomas, R. (1954). *Biochim. Biophys. Acta* **14,** 231.
Thomas, R., Heilporn-Pohl, V., Hanocq, F., Pays, E., and Bolouk̀hère, M. (1980). *Exp. Cell Res.* **127,** 63.
Todorov, I. T., Noll, F., and Hadjiolov, A. A. (1983). *Eur. J. Biochem.* **131,** 271.
Tonegawa, S. (1983). *Nature (London)* **302,** 575.
Trendelenburg, M. F., and Gurdon, J. B. (1978). *Nature (London)* **276,** 292.
Truett, M. A., Jones, R. S., and Potter, S. S. (1981). *Cell (Cambridge, Mass.)* **24,** 753.
Umansky, S. R., Matinyan, K. S., and Tokarskaya, V. I. (1980). *Eur. J. Biochem.* **105,** 117.
Unwin, P. N. T., and Milligan, R. A. (1982). *J. Cell Biol.* **93,** 63.
van der Ploeg, L. H. T., and Flavell, R. A. (1980). *Cell (Cambridge, Mass.)* **19,** 947.
Vanin, E. F. (1984). *Biochim. Biophys. Acta* **782,** 231.
van Eckelen, C. A. G., and van Venrooij, W. J. (1981). *J. Cell Biol.* **88,** 554.
Vedel, M., Gomez-Garcia, M., Sala, M., and Sala-Trepat, J. (1983). *Nucleic Acids Res.* **11,** 4335.
Vidali, G., Boffa, L. C., Bradbury, E. M., and Allfrey, V. G. (1978). *Proc. Natl. Acad. Sci. U.S.A.* **75,** 2239.
Vogeli, G., Okhubo, H., Sobel, M. E., Yamada, Y., Pastan, I., and de Crombrugghe, B. (1981). *Proc. Natl. Acad. Sci. U.S.A.* **78,** 5334.
Vogelstein, B., Pardoll, D. M., and Coffey, D. S. (1980). *Cell (Cambridge, Mass.)* **22,** 79.
Wahl, G. M., Padgett, R. A., and Stark, G. R. (1979). *J. Biol. Chem.* **254,** 8679.
Wallace, H., and Birnstiel, M. L. (1966). *Biochim. Biophys. Acta* **114,** 296.

Warnecke, P., Kruse, K., and Harbers, E. (1974). *Biochim. Biophys. Acta* **331,** 295.
Wasylyk, B., Derbyshire, R., Guy, A., Molko, D., Roget, A., Teoule, R., and Chambon, P. (1980a). *Proc. Natl. Acad. Sci. U.S.A.* **77,** 7024.
Wasylyk, B., Kedinger, C., Corden, J., Brison, O., and Chambon, P. (1980b). *Nature (London)* **285,** 367.
Watson, J. D., and Crick, F. H. C. (1953). *Nature (London)* **171,** 737.
Weber, L. A., Feman, E. R., Hickey, E. D., Williams, M. C., and Baglioni, C. (1976). *J. Biol. Chem.* **251,** 5657.
Weinberg, E. S., Overton, G. C., Shutt, R. H., and Reeder, R. H. (1975). *Proc. Natl. Acad. Sci. U.S.A.* **72,** 4815.
Weintraub, H. (1980). *Nucleic Acids Res.* **8,** 4745.
Weintraub, H. (1984). *Cell* **38,** 17.
Weintraub, H., and Groudine, M. (1976). *Science* **193,** 848.
Weintraub, H., Larsen, A., and Groudine, M. (1981). *Cell (Cambridge, Mass.)* **24,** 333.
Weintraub, H. Izant, J. C., and Harland, R. M. (1985). *Trends Genet.* **1,** 22.
Weisbrod, S. T. (1982a). *Nature (London)* **297,** 289.
Weisbrod, S. T. (1982b). *Nucleic Acids Res.* **10,** 2017.
Weisbrod, S. T., and Weintraub, H. (1981). *Cell (Cambridge, Mass.)* **23,** 391.
Weisbrod, S. T., Groudine, M., and Weintraub, H. (1980). *Cell (Cambridge, Mass.)* **19,** 289.
Weiss, E. H., Mellor, A., Golden, L., Fahrner, K., Simpson, E., Hurst, J., and Flavell, R. A. (1983). *Nature (London)* **301,** 671.
Weissbach, A. (1977). *Annu. Rev. Biochem.* **46,** 25.
Wellauer, P. K., and Dawid, I. B. (1973). *Proc. Natl. Acad. Sci. U.S.A.* **70,** 2827.
Wellauer, P. K., and Dawid, I. B. (1974). *J. Mol. Biol.* **89,** 379.
Wellauer, P. K., and Reeder, R. H. (1975). *J. Mol. Biol.* **94,** 151.
Wells, D., and Kedes, L. (1985). *Proc. Natl. Acad. Sci. U.S.A.* **82,** 2834.
Wensink, P. C., and Brown, D. D. (1971). *J. Mol. Biol.* **60,** 235.
Westermann, R., and Grossbach, V. (1984). *Chromosoma* **90,** 355.
Widmer, R. M., Lucchini, R., Lezzi, M., Meyer, B., Sogo, J. M., Edström, J. E., and Koller, Th. (1984). *EMBO J.* **3,** 1635.
Wiegand, R. C., and Brutlag, D. L. (1981). *J. Mol. Chem.* **256,** 4578.
Wigler, M. H. (1981). *Cell (Cambridge, Mass.)* **24,** 285.
Wilde, C. D., Crowther, C. E., Cripe, T. P., Lee, M. G. S., and Cowan, N. J. (1982a). *Nature (London)* **297,** 83.
Wilde, C. D., Crowther, C. E., and Cowan, N. J. (1982b). *Science* **217,** 549.
Williamson, V. M. (1983). *Int. Rev. Cytol.* **83,** 1.
Wilson, D. A., and Thomas, C. A., Jr. (1974). *J. Mol. Biol.* **84,** 115.
Winicov, I., and Perry, R. P. (1974). *Biochemistry* **13,** 2908.
Wise, G. E., and Prescott, D. M. (1973). *Proc. Natl. Acad. Sci. U.S.A.* **70,** 714.
Woodcock, C. L. F., Frado, L. L. Y., and Rattner, J. B. (1984). *J. Cell Biol.* **99,** 42.
Worcel, A. (1977). *Cold Spring Harbor Symp. Quant. Biol.* **42,** 313.
Worcel, A., Strogatz, S., and Riley, D. (1981). *Proc. Natl. Acad. Sci. U.S.A.* **78,** 1461.
Wu, C. (1984). *Nature* **309,** 229.
Wu, C., and Gilbert, W. (1981). *Proc. Natl. Acad. Sci. U.S.A.* **78,** 1577.
Wu, R. S., Kohn, K. W., and Bonner, W. M. (1981). *J. Biol. Chem.* **256,** 5916.
Wu, R. S., Perry, L. J., and Bonner, W. M. (1983). *FEBS Lett.* **162,** 161.
Young, M. W. (1979). *Proc. Natl. Acad. Sci., U.S.A.* **76,** 6274.
Zamb, T. J., and Petes, T. D. (1982). *Cell (Cambridge, Mass.)* **28,** 355.

Zarucki-Schulz, T., Tsai, S. Y., Itakura, K., Soberon, X., Wallace, R. B., Tsai, M. J., Woo, S. L. C., and O'Malley, B. W. (1982). *J. Biol. Chem.* **257,** 11070.
Zaug, A. J., Grabowski, P. J., and Cech, T. R. (1983). *Nature (London)* **301,** 578.
Zbarsky, I. B. (1978). *Int. Rev. Cytol.* **54,** 295.
Zehner, Z. E., and Paterson, B. M. (1983). *Nucl. Acids Res.* **11,** 8317.
Zentgraf, H., Müller, U., and Franke, W. W. (1980). *Eur. J. Cell Biol.* **23,** 171.
Zhang, X. Y., and Hörz, W. (1982). *Nucleic Acids Res.* **10,** 1481.

CHAPTER 5

# CELL DIVISION

## I. GENERAL BACKGROUND

There are two characteristics of cell division that sharply distinguish living organisms from nonliving machines: reproduction and heredity. When cells divide, the two daughter cells have the same heredity because they have received identical DNA molecules (both in amount and in sequence organization). In culture, they will give rise to identical strains until mutations and DNA rearrangements induce the inevitable diversification that characterizes all living organisms. Finally, cell division remains the only method of reproduction for all asexual organisms. Sexual reproduction will be touched upon when we discuss oocytes and eggs in Volume 2, Chapter 2.

In ''Biochemical Cytology,'' Chapter 5 was entitled ''Mitosis,'' not ''Cell Division.'' This change is due to the fact that, in 1957, little was known of the existence of a cell cycle. Although this idea had been proposed as early as 1953 by Howard and Pelc, it had little immediate impact on cytological research and thinking. The progress in this field in the years that followed was the subject of a book by Prescott (1976); a review by Hochhauser *et al.* (1981) includes about 1200 references on the subject. Today we know that the cell prepares for mitotic division by a series of complex events that ultimately lead to a revolution: mitosis, which produces two (in theory, at least) genetically identical daughter cells. We may call this a revolution because it is accompanied by a complete rearrangement of almost all of the cell constituents that have been considered so far.

The preparatory events that culminate in mitosis are dominated by DNA replication. This phase of DNA synthesis (Fig. 1) is called the ''S phase'' of the cell cycle. Prior to DNA synthesis is the ''$G_1$ phase,'' which lasts for a variable length of time (it is even absent during the cleavage of fertilized eggs in most animal species). DNA replication is followed by a ''$G_2$ phase,'' in which biochemical changes prepare for nuclear membrane breakdown, condensation of chromatin into individual chromosomes, disappearance of the nucleoli, and rearrangements of the cell surface and cytoskeleton. Cells that permanently or temporarily lie outside the cell cycle are said to be in the ''$G_0$'' phase. Under the influence of appropriate stimuli, they may eventually enter into $G_1$ and then follow a normal cell cycle.

Revolutions are seldom bloodless. Thus, cell death during the mitotic revolution is a frequent occurrence. If sea urchin or frog eggs are kept under adverse

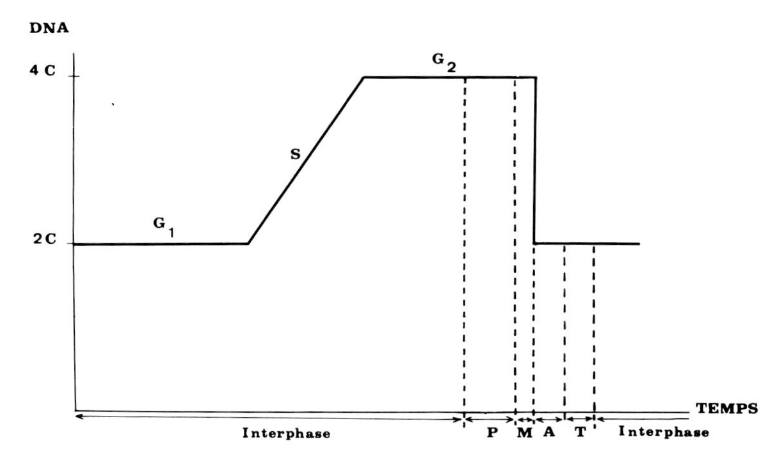


FIG. 1. DNA replication during the cell cycle. Cells that are preparing for cell division have a diploid DNA complement (2C) and are in the $G_1$ period of interphase. During the S phase, nuclear DNA is replicated and reaches 4C. After a lag ($G_2$ phase), the DNA content falls back to the $G_1$ value during mitotic anaphase. P, prophase; M, metaphase; A, anaphase; T, telophase. [Original drawing by P. Van Gansen.]

conditions for a few hours, many cells die; almost all of them were in mitosis at the time of their death. If the adjective "dramatic" is to be applied to biological processes, it is certainly apropos for the changes that occur between interphase and mitotic cell division.

All students of biology or medicine have been exposed to several descriptions of mitosis during their studies. The entire subject was treated masterfully by D. Mazia in 1977; a very brief description of mitotic cell division will thus be sufficient there.

As shown in Fig. 2a, several important changes occur during prophase. The nucleus swells and the chromosomes, which are usually not visible in normal interphase nuclei, become recognizable as long threads; the nucleoli disappear simultaneously and small asters surround the centrioles, forming the so-called centrosomes or poles. The centrosomes, which result from the separation of the two centriolar elements, which constitute a diplosome (see Chapter 3), and the astral rays move to opposite sides of the nucleus. The nuclear membrane then breaks down, and the nuclear sap mixes with the surrounding cytoplasm. The prometaphase (Fig. 2b) stage is reached when the asters and spindle (i.e., the so-called mitotic or achromatic apparatus) have formed; the chromosomes (which may already have divided longitudinally into two sister chromatids) still lie at random in the middle of a short spindle.

At metaphase (Fig. 2c), all of the chromosomes are located at the equatorial plate in the middle of the spindle, which has increased in size; they are attached

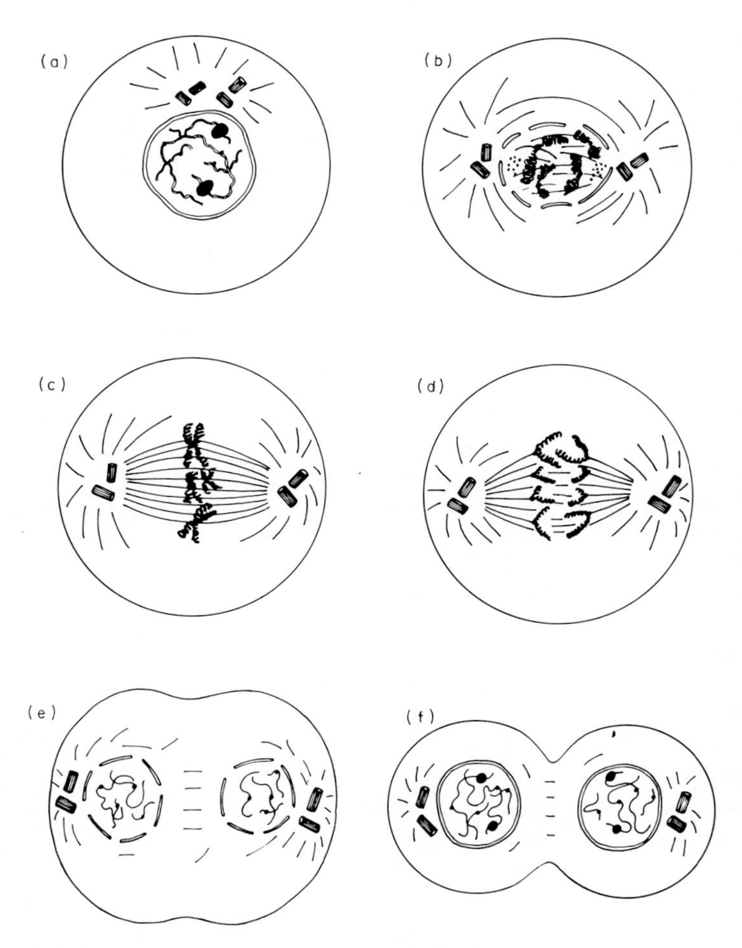


FIG. 2. Schematic representation of mitosis in an animal cell. (a) Interphase: nucleus with nuclear membrane, nucleoli, chromatin; the centrioles have already replicated. (b) Prophase: breakdown of the nuclear membrane, condensation of chromatin into distinct chromosomes, migration of the centrioles toward the poles. (c) Metaphase: the chromosomes are attached to the spindle fibers by their kinetochores (centromeres); asters have developed around the centrioles. (d) Anaphase: the daughter chromosomes are moving toward the poles. (e) Early telophase: the nuclear membranes are re-forming and the asters are regressing. (f) Late telophase: the daughter nuclei are now complete, with nucleoli, chromatin, and a nuclear membrane; an equatorial furrow progresses at the equator until it divides the cytoplasm into two halves (cytokinesis). [Original drawing by P. Van Gansen.]

to spindle fibers by so-called chromosomal fibers attached to a specialized chromosomal region, the kinetochore or centromere. Other spindle fibers, running throughout the whole spindle, directly link one pole with the other. Equatorial plates are important for cytological work, since it is at this stage that chromosome counting is easiest; for this purpose, agents such as colchicine, which arrest mitosis in metaphase, are extremely useful. The fission between daughter chromatids is usually visible in metaphase chromosomes, especially when squash methods are used. Metaphase chromosomes are shorter and thicker than they are at earlier stages.

The separation of the daughter chromosomes begins at the kinetochores. At anaphase (Fig. 2d), the chromosomes move toward the opposite poles, forming a V shape; centromeres are visible between the two arms of the V. The asters, especially in cleaving eggs, acquire maximal dimensions and come close to the cortical layer.

By telophase (Fig. 2e), the chromosomes have migrated very close to the poles. In eggs, they form irregular vesicles that soon fuse together, so that individual chromosomes cease to be distinguishable. When the asters have reached their maximal size, the furrow that will separate the cytoplasm into two daughter cells first appears; this furrow progresses rapidly, while the mitotic apparatus regresses. During this period, the nuclear membrane becomes more visible and the nucleoli reappear. The small region where the two daughter cells are still held together is called the "midbody" (Fig. 2f). Its disappearance signals an independent existence for the two daughter cells.

Needless to say, mitosis shows very different patterns in various cells and species. In many species of protozoa, for instance, the spindle is entirely intranuclear. Plant cells have no centrioles or asters when they divide, and the formation of new cell walls is not due to constriction by a furrow. The outgrowth of the central part of the spindle and the intervention of numerous Golgi bodies form the phragmoplast.

This chapter will begin with an investigation of the various phases of the cell cycle ($G_0$, $G_1$, S, $G_2$, and M). We shall then examine the structure of metaphase chromosomes and of chromosomes that are visible in intact nuclei (giant polytene and lampbrush chromosomes). The molecular organization of the mitotic apparatus and a study of furrow formation (cytokinesis) will follow. After a brief presentation of the cyclic biochemical changes that take place during mitosis, the main mitotic abnormalities will be described. We shall then consider the effects of mitotic poisons and discuss the factors that control (positively or negatively) mitotic activity in normal cells. Finally, a brief comparison will be made between mitosis and meiosis. The latter, like amplification of the ribosomal genes at the onset of oogenesis in *Xenopus* and endoreduplication in giant salivary gland chromosomes, can be considered as a nonmitotic cell cycle.

## II. The Cell Cycle[1]

### A. Cell Synchronization

Cell synchronization is almost a prerequisite for analysis of the cell cycle. In general, synchronization is obtained by treating the cells with agents that inhibit DNA synthesis, arresting them at the $G_1 \rightarrow S$ transition of the cell cycle; the synchronized cells can be released from the block by washing away the inhibitor. The most commonly used inhibitors of DNA synthesis are excess thymidine; hydroxyurea (HU), which inhibits ribonucleotide reductase (the enzyme that converts ribonucleotides into the deoxyribonucleotides required for DNA synthesis); and fluorodeoxyuridine (Frd Urd), which blocks the activity of thymidylate synthetase (the enzyme that converts dUMP into TMP). More recently, Pedrali-Noy *et al.* (1980) have stressed the advantage of using aphidicolin, a nontoxic, reversible inhibitor of DNA polymerase-$\alpha$, for reversibly arresting cells at the $G_1 \rightarrow S$ transition.

In other methods of synchronization, the cells are arrested in metaphase by treatment with inhibitors of microtubule (MT) assembly such as colchicine or colcemid; under proper conditions, the metaphase-arrested cells will re-form a mitotic apparatus and undergo normal cell division after removal of the drug. A more elegant way to isolate cells undergoing mitosis is to shake the cell culture gently. Cells undergoing mitosis round up and detach more easily from the substratum than the more strongly adhering interphase cells. Mitotic cells are thus obtained under almost physiological conditions, but the yield is relatively small. By far the best method for obtaining naturally synchronized cells is to use cleaving eggs.

Synchronization of the cell population is obviously a great advantage in the analysis of the cell cycle. Its progress can be followed by the combination of two techniques: (1) cytophotometry measurements of the DNA content after staining with the Feulgen reaction or with fluorescent dyes, which bind selectively or preferentially to DNA (quinacrine, acridine orange, ethidium bromide, DAPI, etc.), and (2) autoradiography after thymidine incorporation (only the cells that are in the S phase of DNA synthesis will be labeled, unless DNA has been damaged and is being repaired). Fusion of interphase- and metaphase-arrested cells also provides a way to identify the position of a cell in its cycle. As we shall see, such fusion experiments lead to ''premature chromosome condensation'' in the interphase cell if it has entered the cell cycle. Finally, the use of a fluorescence-activated cell sorter (FRCS) allows the separation of large amounts of cells that are in the $G_1$ and $G_2$ phases, respectively (after they have been stained with a DNA-specific fluorochrome; since these cells differ in their DNA content, the

[1]Reviewed by Pardee *et al.* (1978), Yanishevsky and Stein (1981), Gelfant (1981), and Hochhauser *et al.* (1981).

two populations can be readily separated from each other). For an application of this efficient but very expensive technique, see Moser *et al.* (1981).

### B. $G_0$ and $G_1$ Phases of the Cell Cycle

$G_0$ cells are, by definition, not cycling cells; this is the case for fully, terminally differentiated cells (such as our neurons, which never divide after birth). Overcrowded, confluent cells in cell cultures also stop dividing; however, dilution in fresh serum or addition of growth factors allows them to reenter $G_1$, and thus to again become cycling cells. Very old fibroblasts (after about 50 passages for human diploid fibroblasts) grow and stop dividing, but fusion with a young cell can, in some cases at least, induce DNA synthesis in the nucleus of the old cell. However, the reverse result (inhibition of DNA synthesis in the young nucleus) seems to be a more frequent outcome of fusion experiments between young and old cells, as we shall see in Volume 2, Chapter 3. Cycling cancer cells can no longer return to a noncycling condition.

These few examples show that it is not always easy to distinguish $G_0$ and $G_1$ cells. The same is true of the potential of the various territories in an egg or in young embryos. We know what these territories will become in the whole embryo. However, if these territories are explanted and cultivated out of the embryo, their potentiality (capacity to differentiate into a given organ) may be increased by embryonic regulation or decreased by lack of contact with an inducing tissue. Therefore, we are never sure that we have found the ideal conditions for the maximal differentiation of an embryonic explant. In the same way, we are never sure that we know the ideal conditions for inducing a differentiated $G_0$ cell to reenter the cell cycle.

The molecular factors that prevent the cells from cycling and that induce them to remain in $G_0$ are poorly known. It has been recently reported by Iida and Yahara (1984) that cells entering in $G_0$ synthesize a set of "stress" proteins. These high-molecular-weight proteins are similar to the "heat shock proteins" (hsp), which are synthesized when cells are submitted to heat or hypotony treatments, etc. It is possible that these *hsp* maintain the cells in $G_0$, but it is also possible that they are simply indicative of stress conditions in cells permanently maintained in $G_0$.

There is still a good deal of discussion about the significance of the $G_1$ phase for the following events of the cell cycle. The only factor on which there is universal agreement is its variable length of time. As already mentioned, there is no $G_1$ period in the cleaving eggs of most animal species; in cell cultures, the duration of $G_1$ varies from one cell type to another. This has led to the development of a probabilistic instead of a deterministic model by Smith and Martin (1973). The $G_1$ period is thought to be divided into two phases: A and B. The first would last for a variable length of time, in contrast to a constant time for the second. We do not know what signal induces the passage from the A to the B phase of $G_1$. Darzynkiewicz *et al.* (1980), using a cell sorter, separated $G_1$ cells that had been stained with an RNA-specific fluorochrome. Their experiments led to the conclusion that cells in $G_1A$ have a lower RNA content than cells in $G_1B$. For this reason, $G_1A$ cells would be unable to replicate their DNA immediately and thus to enter into the S phase. According to Campisi *et al.* (1982), accumulation of a protein with a half-life of 2.5 hr is needed to overcome the "restriction point" in $G_1$. It should be added that the transition probability model of Smith and Martin (1973) is not universally accepted. A new model involving simple random transitions has been proposed by van Zoelen *et al.* (1981). Rønning and Lindmo (1981) have pointed out that there are at least two transition points during the cell cycle, since the duration of the $S+G_2+M$ period is also not constant. The interested reader should consult the review of Edmunds and Adams (1981), who propose that, although its length is variable, the cell cycle is a biological clock.

It is generally assumed that $G_1$ is a period in which the cell accumulates the machinery required for DNA synthesis. Elegant experiments in which cells at various stages of the cell cycle have been fused together led Rao *et al.* (1978) to conclude that the synthesis of inducers of DNA replication, which are known to be present in S-phase cells (see below for a more detailed presentation of cell cycle analysis by somatic hybridization experiments), are accumulated during $G_1$. More recently, Stancel *et al.* (1981) have shown that the length of the $G_1$ phase can be reduced to as little as 1 hr if DNA replication is slowed down by low concentrations of HU. For Stancel *et al.*, the $G_1$ phase would not serve to prepare for the following S phase; it would merely be an interruption between the completion of one chromosome cycle and the next.

We have only a fragmentary view of the biochemical changes that take place during the $G_1$ phase and are probably responsible for the induction of DNA synthesis. Most of the work has been done on fibroblasts in culture, and we know little or nothing about the biochemical events that induce skin or intestinal cells to divide *in situ*.

Almost all investigators have focused on fibroblasts arrested in $G_0$–$G_1$ by saturation density or starvation in nutrients; dilution of the cultures and addition of fresh serum lead to rapid proliferation. But serum is a very complex mixture. Its favorable effects on cell proliferation are largely due to the presence in serum of growth factors (reviewed by Das, 1982, Rozengurt, 1983b, and Schlessinger *et al.*, 1983 see Section I,4 for details). Among these mitogens, epidermal growth factor (EGF) and platelet-derived growth factor (PDGF) have been particularly studied. Both are relatively small polypeptides ($\hat{M}r$ = 6000 for EGF and 32,000 and 35,000 for PDGF I and II, respectively); both bind to cell surface–specific receptors and are internalized by receptor-mediated endocytosis (see Chapter 3). The response to mitogens is usually monitored by following the increased incorporation of [$^3$H]thymidine into DNA that characterizes the onset of the S phase;

less is known about the effects of mitogens on $G_0$–$G_1$ cells. These effects are complex because the cells receive a multiplicity of signals before they begin to replicate DNA. Among the pleiotropic effects exerted by both EGF and PDGF are activation of the cell surface ($Na^+$,$K^+$)–ATPase; increased amino acid, glucose, and uridine transport, activation of tryosine-specific kinases; increase in $Na^+$ uptake, often with a concomitant increase in the cell's internal pH; synthesis of rRNAs and mRNAs, followed by increased protein synthesis; and finally, DNA replication. But the exact sequence of these events is not known and may, of course, vary greatly from one cell type to another.

Given these general remarks, let us consider some particulars. There is no doubt that ions play an important role in the control of cell proliferation. It is well established that a low $Ca^{2+}$ concentration in the medium decreases mitotic activity in normal (but not malignant) cells (Boynton *et al.*, 1983). A threshold concentration of potassium ion is needed for the $G_1 \rightarrow S$ transition. EGF and other mitogens increase the $K^+$ concentration in early $G_1$, which may be one of the factors leading later to increased protein synthesis (Lopez-Rivas *et al.*, 1982). It is also clear that an $Na^+$ influx is required to stimulate DNA synthesis (Owen and Villereal, 1983a); however, according to the same authors (Owen and Villereal, 1983b), addition of growth factors leads not only to an $Na^+$ intake but also to a to a $Ca^{2+}$ loss. The $Na^+$ influx might thus be mediated by changes in the intracellular free $Ca^{2+}$ concentration, but it might also result in EGF- or PDGF-treated cells, either from an activation of the ($Na^+/K^+$) pump (Moolenaar *et al.*, 1982) or from an $Na^+/H^+$ exchange (Schuldiner and Rozengurt, 1982; Rothenberg *et al.*, 1982, 1983; Cassel *et al.*, 1983; Moolenaar *et al.*, 1983). Such an $Na^+/H^+$ exchange, which is inhibited by the diuretic amiloride, was first discovered in fertilized sea urchin eggs, as we shall see in Volume 2, Chapter 2; it results in an increase in the internal pH ($pH_i$) in both eggs and mitogen-treated cells. This increase in $pH_i$, controlled by an $Na^+/H^+$ exchange, functions as a transmembrane signal transducer in the action of growth factors (Burns and Rozengurt, 1983). It is probably an early effect of these factors and precedes by many hours the initiation of DNA synthesis. According to Chefouleas *et al.* (1984), changes in calmodulin play an important role in the reentry of quiescent $G_0$ cells into the cell cycle. The regulatory role of $Ca^{2+}$ and calmodulin in cell growth has been recently reviewed by Veigl (1984).

As we have seen, cells have another mechanism for increasing their concentration of free calcium ions. Release of $Ca^{2+}$ from intracellular stores follows the breakdown of phosphoinositides with the production of inositol triphosphate ($IP_3$) and diacylglycerol. $IP_3$ mobilizes $Ca^{2+}$, whereas diacylglycerol activates proteinkinase C, which is probably involved in the increases in $Na^+/H^+$ exchange (stimulation of the $Na^+/H^+$ antiport) and intracellular pH. This cascade of events is increased by addition to the cells of the growth factors PDGF (Berridge, 1984; Berridge *et al.*, 1984) and EGF or serum (Moolenaar *et al.*,

1984a,b). $Ca^{2+}$ mobilization takes place within a few seconds in the stimulated cells. The effects of the growth factors can be mimicked by addition of a synthetic diacylglycerol to the fibroblasts (Rozengurt *et al.*, 1984). In the treated cells, a 80-kilodalton protein is phosphorylated by proteinkinase C and DNA synthesis is increased. It was concluded that generation of diacylglycerols at the cell membrane is a mitogenic signal.

There is increasing evidence for the view that the amiloride-sensitive $Na^+/H^+$ exchange is responsible for the increase in internal pH that follows addition of mitogens or serum to $G_1$-arrested cells (Lopez-Rovas *et al.*, 1984; L'Allemain *et al.*, 1984). Particularly convincing is the finding by Pouysségur *et al.* (1984) that a mutation which abolishes the $Na^+/K^+$ exchange prevents the growth of fibroblasts at a neutral or acidic pH. An increase in the internal pH is certainly a prerequisite for the initiation of DNA synthesis in many cells; how the two are linked together is unfortunately not yet known.

Molenaar *et al.* (1984a,b) have recently shown that a synthetic diacylglycerol and a phorbol ester that stimulates cell proliferation (see Chapter 3, Volume 2) activate the $Na^+/H^+$ exchange and increase the internal pH of the treated cells. The exchange is set in motion by protein kinase C, but an early rise in free calcium ions is not required for the activation of the $Na^+/H^+$ exchanges. It is likely that protein kinase C phosphorylates the exchanges at the internal face of the plasma membrane. This mechanism is probably valid for all the agents that stimulate cell division. Recent evidence shows that the increase in free $Ca^{2+}$ precedes the rise in $pH_i$ when $G_1$ cells are stimulated to divide (Hersketh *et al.*, 1985).

Other important changes may take place at the cell surface level. Lectins (ConA, phytohemagglutinins) that bind to carbohydrate residues of the cell surface glycoproteins induce DNA replication in many cells. That these superficial sugar residues are important for receiving the mitogenetic signals is shown by the fact that tunicamycin, an inhibitor of glycoprotein synthesis, prevents the initiation of DNA replication (Jimenez de Asua *et al.*, 1983).

How the mitogenic signals received at the cell surface are transmitted to the nucleus is not known for certain; the little we know will be discussed later in this chapter. It is probable that, as proposed by Rozengurt (1983a,b), cyclic nucleotides and ion fluxes mediate the mitogenic signals, at least in fibroblasts. Agents that increase the cAMP content (cholera toxin, prostaglandin $E_1$, methylxanthines) stimulate DNA synthesis (Rozengurt *et al.*, 1983a). The mitogen PDGF also increases the cAMP content of fibroblasts, according to Rozengurt *et al.* (1983b). In contrast to what was generally believed, this increase might be a mitogenic signal for quiescent, confluent fibroblasts.

Stimulation of fibroblasts by growth factors results in changes in protein synthesis that follow the early membrane events. For instance, O'Farrell (1983) has reported that several proteins increase and others decrease in amount when cell proliferation is stimulated by growth factors; particularly conspicuous is an increase in the synthesis of two nuclear proteins. PDGF induces the synthesis of several new mRNAs before density-arrested fibroblasts enter into the S phase (Hendrickson and Scher, 1983). At the beginning of the $G_0$–$G_1$ period, PDGF also induces the synthesis of a 29,000-dalton protein that has been named PI; it is localized in the fibroblast nuclei and might be the PDGF mediator, according to Olashaw and Pledger (1983). Another protein, called p53, is arousing a good deal of interest today because it is particularly abundant in cancer cells; it is also present, at a lower level, in normal proliferating cells. Toward the end of $G_1$, p53 mRNA increases in serum-stimulated cells (Mercer *et al.*, 1982); injection into fibroblasts of an anti-p53 serum prevents DNA synthesis (Winclester, 1983). If cancer cells are induced to differentiate and therefore become arrested in $G_1$, their content in p21 mRNA and in p21 itself markedly decreases (Oren and Levine, 1983; Shen *et al.*, 1983).

One would expect quiescent cells to contain factors that prevent them from entering into mitosis. Negative as well as positive controls presumably operate during the $G_0$–$G_1$ period (Adlakha *et al.*, 1983, 1984; Polunovsky *et al.*, 1983). It has indeed been found that density-dependent inhibition of growth seems to be due to the release of inhibitory factors in the medium (McMahon *et al.*, 1984); one of these factors is a protein ($M_r$ = 40,000) that inhibits DNA synthesis (Harel *et al.*, 1983). Another protein seems to be involved in the arrest of cells in $G_0$–$G_1$. It is histone $H_1^\circ$, which accumulates in the nuclei of cells arrested in $G_1$ at saturation density (Chabanas *et al.*, 1983). If cells are treated with butyrate, initiation of DNA synthesis does not occur at the $G_1$–S transition; in such cells there is an increase in $H_1^\circ$ histone in the nucei (D'Anna *et al.*, 1982) and DNA replication does not take place (Littlefield *et al.*, 1982).

We have tried to reconstruct the chain of events that proceed from the binding of a mitogen to its cell surface receptor to the initiation of nuclear DNA synthesis. Changes in ion movements, $pH_i$, cyclic nucleotides, etc. would finally lead to the synthesis of proteins required for DNA replication. However, according to Rønning and Lindmo (1983), the $G_1 \rightarrow S$ transition is not prevented by cycloheximide, the classic inhibitor of cytoplasmic protein synthesis. Clearly, much more work is needed before we have a clear understanding of the role played by the $G_1$ phase in the initiation of the S phase of the cell cycle.

## C. The S Phase

### *1. DNA Replication*

In 1957 we concluded, from an analysis of the then available data, that in most cases DNA synthesis is a characteristic event of interphase. At present, it can be stated that in all cases this is true.

Autoradiography and cytophotometry studies have clearly shown that one should distinguish between early and late DNA replication. This became apparent when Bostock and Prescott (1971) discovered that replication of satellite

DNA in mouse cells is a late event in the S phase. In general, replication of heterochromatin follows that of euchromatin. Recently, Goldman *et al.* (1984) came to the conclusion that only the early replicating genes are transcribed: tissue-specific genes are replicated early in cells where these genes are expressed, late in the other cells. The genes required for vital cell functions (the so-called housekeeping genes) and therefore for cell survival are among the early-replicated ones.

There has been, and still is, a considerable amount of interest in the intranuclear localization of the initiation sites of DNA synthesis. In 1973, Franke *et al.* concluded, in contradiction to previous work, that DNA synthesis is not initiated at the nuclear membrane. There is now growing evidence for the view that DNA synthesis begins on the nuclear matrix. For instance, Pardoll *et al.* (1980) found that the nuclear matrix contains only 15% of the total DNA, but 90% of the DNA synthesized during a 30-sec [$^3$H]thymidine pulse is attached to the matrix. Similar results have been obtained by Vogelstein *et al.* (1980) and Berezney and Buchholtz (1981). Using a preparation in which only 1–2% of the total DNA remained attached to the proteinaceous nucleoskeletal matrix, Berezney and Buchholtz (1981) found that 50% of this DNA is replicated after a 1-min thymidine pulse. According to their calculations, 125 loops of 80 kb (about 25 μm of DNA) would be matrix attached. A DNA fraction enriched in replicating DNA molecules can be obtained when isolated HeLa cell nuclei are digested with endonuclease *Eco*RI (Valenzuela *et al.*, 1983). It has been hypothesized that replicating DNA loops are motile with respect to their membrane matrix anchorage sites (Vogelstein *et al.*, 1980; Cook and Lang, 1984). The possibility that the enzymes needed for DNA replication are closely associated with the nuclear matrix deserves serious consideration. According to Reddy and Pardee (1980), six enzymes involved in DNA replication constitute a multienzyme complex (called ''replitase'' by these authors). Experiments on cells separated into karyoplasts and cytoplasts have shown that replitase is located in the cytoplasm during the $G_1$ phase. During the S phase it moves into the nucleus, where it might well be associated with the nuclear matrix. The main enzyme involved in DNA synthesis, DNA polymerase α, moves from the cytoplasm into the nucleus, where it becomes tightly associated with the nuclear envelope before the onset of DNA synthesis (Smith and Berezney, 1983; Matsukage *et al.*, 1983).

### 2. *Molecular Mechanisms of DNA Synthesis*

We previously pointed out (Brachet, 1957, p. 181) that there was some evidence that, during liver regeneration (another process in which $G_0$ cells enter the cell cycle), ''one-half of the cellular DNA is synthesized, the other half being retained'' (Fusijawa and Sibatani, 1954; Kihara *et al.*, 1956; Daoust *et al.*, 1956). We also pointed out that these results and conclusions ''would fit nicely'' with the (then recent) Watson–Crick (1953) model of DNA molecular structure. As pointed out by Watson and Crick (1953), the double-stranded molecular structure proposed for DNA easily explains gene and chromosome reduplication. Since the structure of one of the polynucleotide chains determines the structure of the other, one need only postulate that the two strands become separated from each other at the time of cell division; each of these strands might then, by a template mechanism, be able to reproduce its counterpart. Two identical molecules of DNA will then be formed. In both of them, one-half comes from the preexisting strand; the other is newly synthesized (Fig. 28, Chapter 4).

This semiconservative DNA replication mechanism has been established beyond any possible doubt. Definite proof came from the famous experiments of Meselson and Stahl (1958) on bacteria in which the replication of $^{15}$N-labeled DNA was followed during two cell generations after transfer to $^{14}$N-containing medium. As shown in Fig. 3b, after the first replication cycle, all of the DNA has a density intermediate between that of heavy ($^{15}$N) and light ($^{14}$N) DNA. After the next replication cycle, one-half of the DNA has the intermediate ($^{15}N - {}^{14}N$) density and the other half is made up of light $^{14}$N DNA.

Taylor *et al.* (1957) demonstrated that semiconservative DNA replication is also the rule in eukaryotes and that this rule is valid for chromosomes. This was shown by autoradiography after a [$^3$H]thymidine pulse. As shown in Fig. 3a, only one of the sister chromatids is radioactive, indicating that its DNA had been newly synthesized during the S phase; the other chromatid is unlabeled, demonstrating that it was conserved during DNA replication.

Around 1973 (reviewed by Taylor, 1974), it was discovered that DNA synthesis is discontinuous and that it requires the presence of an RNA primer (Fig. 4). After very short (1 min or less) pulses with labeled thymidine, only small DNA pieces (200–1000 nucleotides long) are synthesized. These small DNA pieces are called the "Okazaki fragments," in honor of the Japanese molecular biologist who first discovered that DNA replication is discontinuous in bacteria. In both eukaryotes and prokaryotes Okazaki fragments are associated with a small RNA molecule (Wagar and Huberman, 1973). According to Taylor *et al.* (1973, 1975), Okazaki fragments measure about 0.5 μm, are single-stranded, and are quickly integrated into larger DNA molecules by ligation of the ends. The RNA primer molecules that are attached covalently to small, newly synthesized DNA fragments are short (only 30 bases long) and are short-lived (Hunter and Francke, 1974). Fox *et al.* (1973) estimated a length of 50–100 nucleotides for the RNA primer, which is associated with nascent, single-stranded DNA, but according to a more recent paper by Tseng and Goulian (1977), the small RNA initiator of DNA replication has only 8–11 nucleotides. Its sequence varies from one initiation site of DNA replication to another, and it might provide a signal for the initiation of DNA synthesis rather than act as a primer. Jelinek (1980) proposed that the Alu (and Alu-like) middle repetitive dispersed DNA sequences

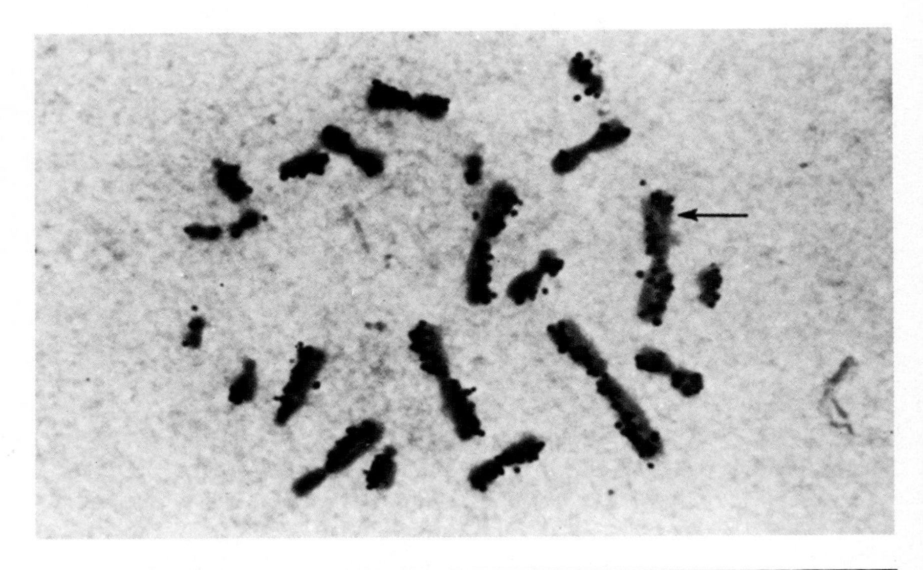

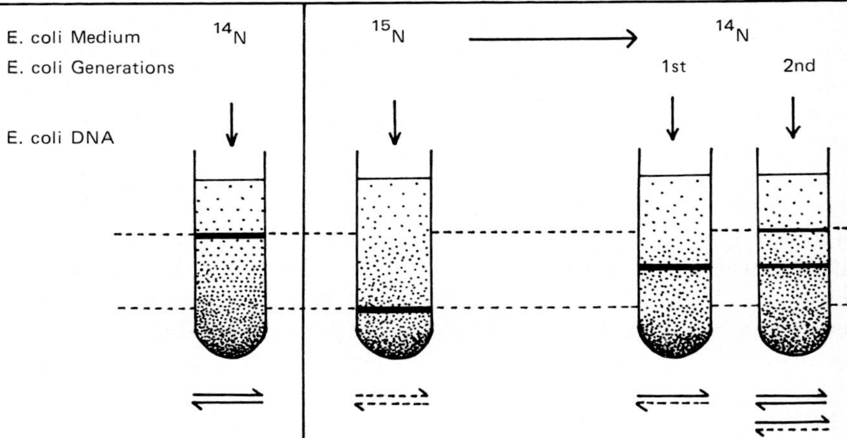


FIG. 3. (a) Autoradiograph of mitotic chromosomes of the Chinese hamster at the second division after labeling of the DNA with [$^{3}$H]thymidine. Only single chromatids are labeled. The arrow points to a sister chromatid exchange. [Preparation by D. M. Prescott, in Callan, 1972.] (b): The Meselson–Stahl experiment demonstrating semiconservative replication of the DNA molecules in the bacterium *Escherichia coli*. Bacteria were cultured in a medium containing heavy nitrogen ($^{15}$N). The DNA isolated from these bacteria has a high density. After transfer into normal medium ($^{14}$N), DNA acquires an intermediate ($^{14}$N-$^{15}$N) density after the first generation. After the second generation in normal medium, two DNA bands can be separated by gradient centrifugation. They correspond respectively to normal $^{14}$N-$^{15}$N double helices and $^{15}$N-$^{14}$N heavier molecules ⇌, light ($^{14}$N) strands; ⇌, heavy ($^{15}$N) strands. [Original drawing by P. Van Gansen.]

might be a universal origin sequence for DNA replication. It must be admitted, however, that we do not yet know for certain the "start signal(s)" for DNA replication.

A considerable amount of work (which lies outside the scope of this book) has been done on the biochemical mechanisms of DNA replication. Unwinding of the DNA double helix is believed to be due to a "nicking-rejoining" enzyme (thus, a DNA topoisomerase) that is almost inactive in $G_0$–$G_1$ and becomes very active in the S phase (Rosenberg *et al.*, 1976). A major role in DNA replication should be ascribed to DNA polymerase-α, since DNA synthesis almost immediately ceases in sea urchin eggs (Ikegami *et al.*, 1978) and other cells (Pedrali-Noy and Spadari, 1979) when its specific inhibitor, aphidicolin, is added. More recently, it has been shown that a primase activity is associated with DNA polymerase-α. It synthesizes the oligoribonucleotides (9–32 nucleotides) that prime DNA synthesis. This enzyme is particularly active in *Drosophila* embryos (Shioda *et al.*, 1982; Conaway and Lehman, 1982a,b; Kaguni *et al.*, 1983a,b) and *Xenopus* eggs (Tseng and Ahlem, 1982; Méchali and Harland, 1982; Riedel *et al.*, 1982; König *et al.*, 1983), but it is also present in large amounts in lymphocytes (Tseng and Ahlem, 1982). Primase activity requires the presence of DNA polymerase-α and ribonucleoside triphosphates; it is not inhibited by α-amanitin (Yoshida *et al.*, 1983). Its function is to synthesize the 8–10 (or multiples thereof) ribonucleotide chains covalently linked to the nascent DNA chains at the 5′ end of the replicating DNA molecule; this primer or initiator RNA is required for the initiation of Okazaki fragment synthesis in replicating chromosomal DNA. Primase and DNA polymerase-α are so tightly associated that it has been suggested that primase is one of the subunits of DNA polymerase-α (Yagura *et al.*, 1982, 1983; Kaguni *et al.*, 1983a,b; König *et al.*, 1983a,b); however, Tseng and Ahlem (1983) purified a DNA primase that was devoid of DNA polymerase activity, and Wang *et al.* (1984) concluded that the catalytic sites of the two enzymes are different. In addition to DNA polymerase-α and primase, a DNA ligase is involved in linking together the Okazaki fragments in order to form larger and larger DNA molecules [in cleaving sea urchin eggs, newly synthesized DNA is macromolecular within 15 min, according to Shimada and Terayama (1976)].

As already mentioned, Reddy and Pardee's (1980) replitase is a multi-enzymatic complex comprising at least six enzymes involved in DNA replication (DNA polymerase, DNA topoisomerase, DNA methylase, dihydrofolate reductase, thymidylate kinase, and ribonucleotide reductase). According to Noguchi *et al.* (1983), this arrangement allows rapid incorporation of deoxyribonucleoside diphosphates into DNA during the S phase, because of the presence of ribonucleotide reductase in the cytoplasm (Engström *et al.*, 1984). It also allows the selective methylation of cytidine into 5-methylcytidine residues during DNA

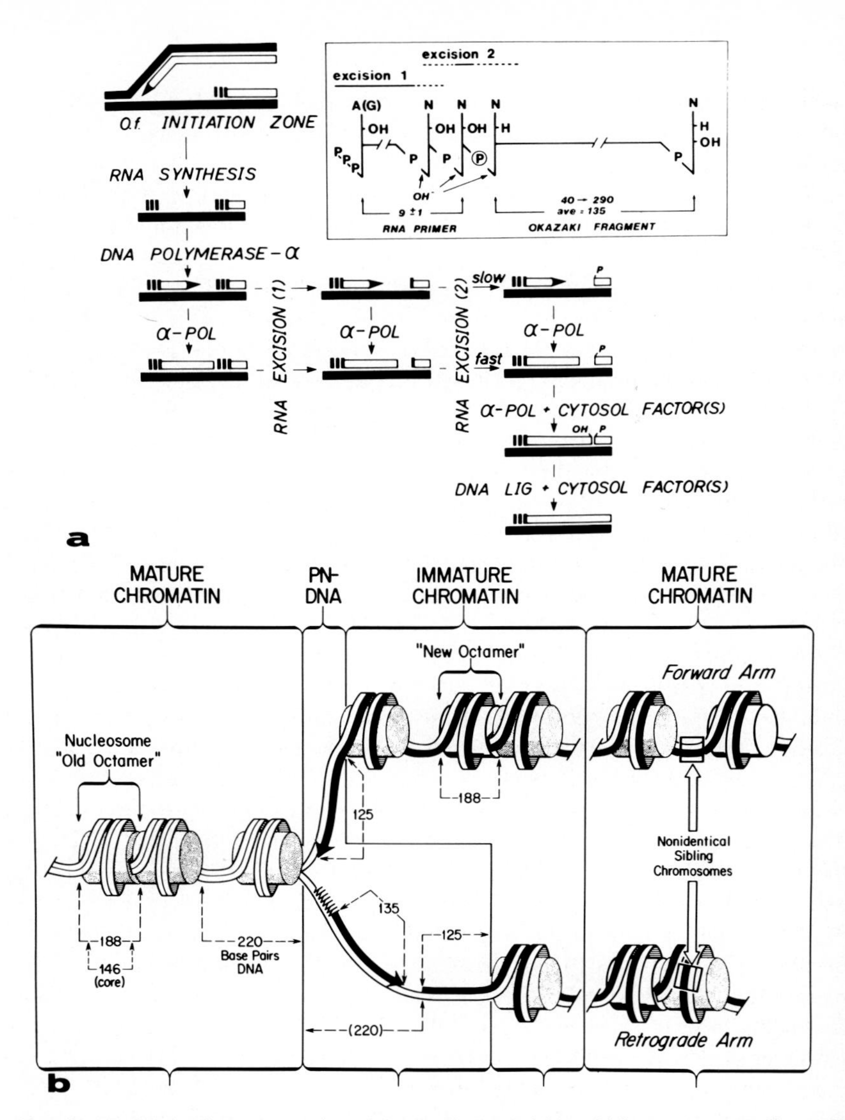


FIG. 4. (a) Schematic representation of the metabolism of Okazaki fragments, according to De Pamphilis and Wasserman (1980). Initiation of Okazaki fragments (O.f.) occurs stochastically within an ''initiation zone'' via *de novo* synthesis of RNA primers (III) complementary in the retrograde DNA template ( ▬ ). Within this zone, initiation events are promoted at preferred DNA sequences. The maturation of Okazaki fragments ( ▭▸ ) is represented on the vertical coordinate in four distinct steps: synthesis of RNA primers, elongation of DNA by DNA polymerase-α, incorporation of the final deoxyribonucleotides (gap filling), and joining of Okazaki fragments to growing DNA chains (ligation). Excision of RNA primers is represented on the horizontal coordinate in two distinct steps; excision of the bulk of the RNA, which is independent of Okazaki fragment synthesis, and removal of the p-rN-p-dN-(pdN) junction, which is facilitated by concomitant DNA synthesis. Excision of all RNA primers is presumed to take place concurrently. A single example is illustrated for simplicity.

replication by DNA methyltransferases. Reddy and Pardee (1983) have shown that one of the replitase (a spherical particle of 25 nm) enzymes, thymidylate kinase, is functionally associated with other enzymatic activities. It is inhibited by inhibitors of other enzymes present in the replitase complex: HU (an inhibitor of ribonucleotide reductase), novobiocin (an inhibitor of topoisomerase), and the classic inhibitor of DNA polymerase-α, aphidicolin. All of these inhibitors suppress thymidylate kinase activity in living S-phase cells, but not in soluble extracts of these cells. Another (or possibly the same) multienzyme complex of DNA replication has been isolated by Wickremasinghe *et al.* (1983); it consists of DNA polymerase, thymidine kinase, deoxythymidine monophosphate kinase, nucleoside diphosphate phosphatase, and thymidylate synthetase. There is no doubt that the numerous enzymes involved in the synthesis of the precursors required for DNA synthesis are closely associated *in vivo* with DNA polymerase and primase. Jazwinski and Edelman (1982, 1984) have given the name "replisome" to the replitase particle of yeast cells; the activity of this multiprotein complex, which participates in DNA replication in these cells, is not inhibited by α-amanitin or aphidicolin. It comprises DNA polymerase-α, DNA ligase, DNA primase, DNA topoisomerase II, and probably other unknown components.

Important progress in our understanding of DNA replication at the cellular level resulted when Huberman and Riggs introduced their fiber autoradiography technique in 1968. If, after [$^3$H]thymidine pulses of varying lengths of time, the isolated nuclei are spread and resolved in individual chromatin fibers and then subjected to autoradiography, individual units of replicating DNA segments can be visualized (Fig. 5). These small DNA-replicating units along chromatin fibers are called "replicons" (reviewed by Kapp and Painter, 1982). DNA synthesis starts at apparently fixed origins in each replicon and progresses by two divergent paths; DNA replication thus proceeds bidirectionally. The existence of replicons proves that DNA replication is a discontinuous process at both the molecular and cellular levels. Replicating DNA units can be seen under the electron microscope (Fig. 6) in the form of single-stranded DNA eyes and clusters of microbubbles; the latter correspond to multiple initiation sites within a single replicon and have 100–300 bp, like the Okazaki fragments (Micheli *et al.*, 1982).

The size and spacing of the replicons on the chromatin fibers may vary consid-

---

The inset shows the structure of RNA-primed Okazaki fragments, with sizes given in nucleotides. N represents A, G, C, or T. The cleavage sites during alkaline hydrolysis are indicated by arrows. Excisions 1 and 2 refer to regions excised (solid bars) during replication; possible extensions of those regions are indicated by broken bars. (b) Model of replication forks in SV40 chromosomes. Nascent DNA is represented by a black ribbon, with arrowheads indicating the growing 3′-ends and small bars indicating an RNA primer. An average of one Okazaki fragment per fork is represented on the retrograde arm. Numbers give the average distance in nucleotides. A nucleosome is a 110 Å × 22 Å cylinder consisting of 1.75 turns of duplex DNA (20 Å in diameter) coiled around a histone octamer. [Part (b) courtesy of Dr. De Pamphilis. Reproduced, with permission, from the Annual Review of Biochemistry, Volume **49.** © 1980 by Annual Reviews Inc.]

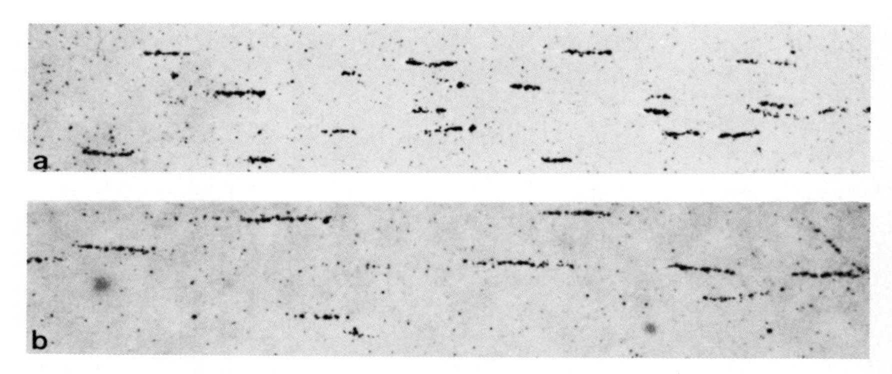


FIG. 5. Photographs of DNA fiber autoradiographs from [$^3$H]thymidine-labeled *Xenopus laevis* cells in culture. (a) Labeling for 2 hr. (b) Labeling for 2 hr and chase with nonradioactive medium for another 2-hr period. Heavily labeled segments are flanked on both sides by diminishing grain-density gradients, demonstrating that replication proceeds in opposite directions from each origin. [Callan (1972).]

erably. In general, their size is 15–25 μm, but replicons as small as 3–4 μm are found in cleaving eggs in which DNA replication is exceedingly rapid. For instance, during the first cleavages in *Drosophila,* interphase, which is limited to the S phase, lasts for only 3–4 min and mitosis for 6–9 min. At these early stages of development, the replicons are separated by only 200 nucleotides in 45% of the cells studied by Wolstenholme (1973), Kriegstein and Hogness (1974), and Micheli *et al.* (1982). Instead of clusters of small eyes, electron microscopy

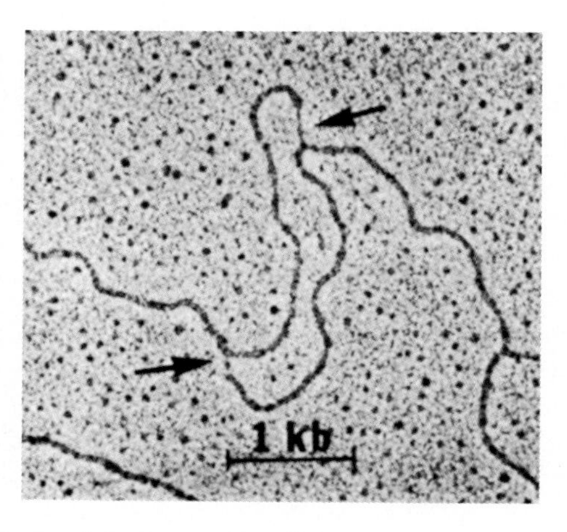


FIG. 6. Two replication forks form an "eye." Arrows indicate the single-stranded regions in each fork. [Kreigstein and Hogness (1974).]

shows the presence of macrobubbles (long replication eyes of about 10,000 DNA bases). Bidirectional replication takes place at a speed of 3000 bases per minute. For readers who like figures, we might add that the DNA of a completely uncoiled *Drosophila* chromosome measures 2.3 cm and is composed of $62 \times 10^6$ bases. There are no nucleosomes in *Drosophila* egg replicons, but they re-form as soon as DNA replication stops (McKnight and Miller, 1977).

In eggs, generally the rate of DNA replication slows down markedly after cleavage. As shown by Callan (1972) for amphibians and by Kriegstein and Hogness (1974) for *Drosophila,* this slowdown is due to a decrease in the number of replicons. The distance between adjacent replicons increases when the speed of DNA synthesis levels off; however, the speed of the progression of the replication forks is not reduced. The values given for the distance separating two replicons in mammals may vary from 4 μm (Taylor and Hozier, 1976) to 40–200 μm (Yurov and Liapunova, 1977); such differences are almost certainly correlated with the rate of DNA replication in the various cells under study.

The speed of fork progression in replicons varies from one cell type to another. According to the literature, it generally varies from 0.2 to 1.5 μm/min in mammalian cells; it may be as low as 0.02 μm/min in cultured amphibian cells. The speed of DNA fork movement may vary by a factor of 10 in a single cell type if individual replicons are compared; it may either increase or remain constant at different times during the S phase. This speed does not change greatly when rapidly and slowly dividing cells are compared (the S phase lasts for several hours in cultured *Drosophila* cells, as compared to 4 min in cleaving *Drosophila* eggs). Thus, the method used by cells for regulating the rate of DNA synthesis is to modify the spacing between replicons, and not to alter the speed of replication fork progression. How they succeed in doing that, however, remains a mystery.

Buongiorno-Nardelli *et al.* (1982) have pointed out that chromosomal DNA is arranged in loops of several tens of kilobases on the nuclear matrix, and that there is a correlation between loop size and replicon size in many animal and plant species. DNA synthesis apparently takes place at fixed sites on the nuclear matrix. In *Xenopus* kidney cells in culture the fork movement is 0.43 μm/min, and a loop is replicated in 30 min by bidimensional DNA replication. The size of a replicon is about twice that of a DNA loop; thus, a replicon might correspond to two adjacent DNA loops.

A few words should be said about unscheduled (repair) DNA synthesis (reviewed by Lehmann and Karran, 1981) and mitochondrial DNA synthesis. If DNA is injured by radiation or chemicals, the damaged segments must be repaired. If uv-irradiated cells are treated with [$^3$H]thymidine, autoradiography shows that nuclear DNA synthesis completely unrelated to the S phase of the normal cell cycle (unscheduled DNA synthesis) is taking place. The repair biochemical mechanisms are different from those at work during DNA semiconservative replication. For one thing, while the error rate of DNA polymerase when it is acting *in vivo* is very low [$10^{-8}$–$10^{-12}$ misincorporarion per base pair replicated, according to Kunkel and Loeb (1981)], the repair systems are more error prone. This may be partially due to the fact that the error rate of the replicating enzyme DNA polymerase-α is about 10 times less than that of the β-enzyme, which is a key enzyme in DNA repair, but the error rate of DNA polymerase-α *in vitro* (1/30,000) is much greater than the $10^{-8}$–$10^{-11}$ misincorporations by base pairs replicated *in vivo*. It follows that there are mechanisms for the correction of mistakes and the repair of spontaneous damage to the DNA during its replication. Possibly these correcting (proofreading) mechanisms do not work efficiently when the problem is the repair of damage (in general, single-stranded breaks in DNA molecules resulting from exposure to radiation or chemical mutagens). As pointed out by Kunkel and Loeb (1981), concerted action of several proteins, in addition to that of the DNA polymerases, is required for faithful DNA replication. It seems that, at least in bacteria, the repair systems have to work so fast that they are unable to do a perfect job [SOS repair of Witkin (1976) and Radman (1974)].

Many enzymes are involved in DNA repair (reviewed by Grossman, 1981; Friedberg *et al.*, 1981; Caradonna and Cheng, 1982), which occurs in several steps. First, DNA glycosylases recognize alkylated DNA and remove the damaged bases; second, in base excision repair, specific endonucleases excise the apurinic and apyrimidic sites; an exonuclease then produces a gap in the DNA molecule that is repaired by a DNA polymerase; finally, a DNA ligase seals the break by joining the 3′-OH and 5′-phosphate termini (Fig. 7).

There is no doubt that DNA polymerase-β is involved in DNA repair, but there has been a good deal of discussion about the possible involvement of DNA polymerase-α as well. An EM autoradiography study of uv-irradiated and aphidicolin-treated cells concluded that DNA polymerase-α does not play a significant part in DNA repair (Hardt *et al.*, 1981). But similar experiments using biochemical methods have led to results that vary according to the cells and the DNA-damaging agents involved. Miller and Chinault (1982) and Miller and Lui (1982) concluded that DNA polymerase-β plays a major role in the repair of damage induced by bleomycin and neocarzinostatin, but DNA polymerase-α is also involved in repairing the damage due to uv-radiation and alkylating agents. Dresler and Lieberman (1983) also concluded that, in fibroblasts, DNA excision repair is achieved by both the α- and β-enzymes but that their relative importance depends on both the DNA-damaging agent and the dose.

Another factor involved in DNA repair is the synthesis of poly(ADP) ribose (reviewed by Mandel *et al.*, 1982; Berger, 1985) at the expense of the coenzyme $NAD^+$. After DNA damage, the cellular $NAD^+$ content apparently decreases, and there is a concomitant increase in poly(ADP) ribose polymerase activity. A few papers indicate that ADP ribosylation of nuclear proteins stimulates DNA ligase activity (Creisson and Shall, 1982; James and Lehmann, 1982); according

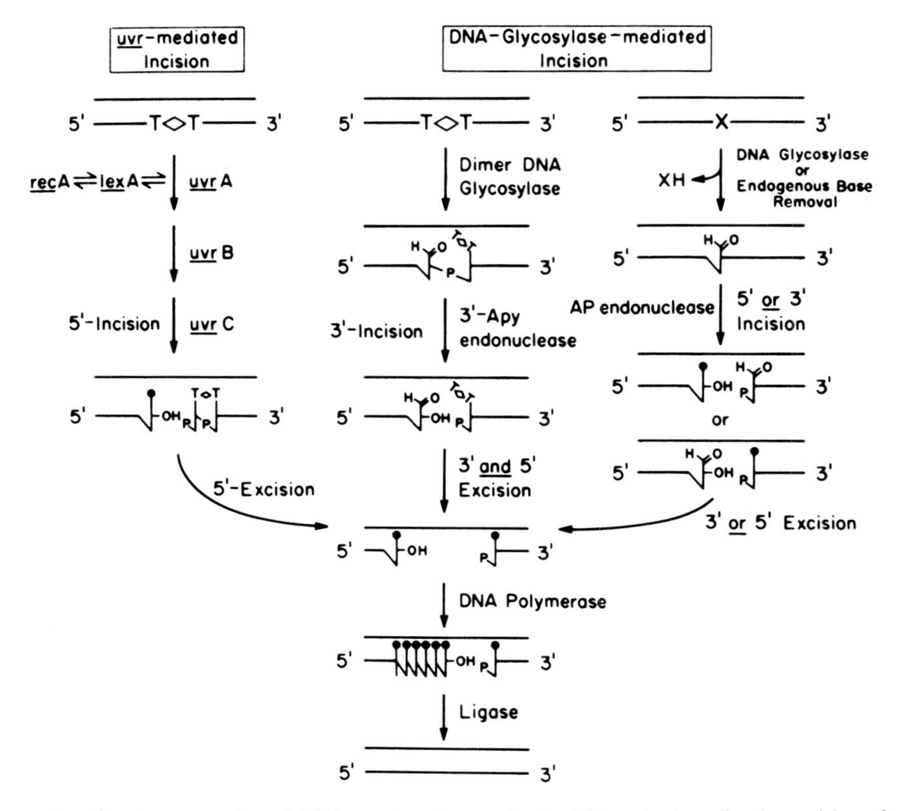


FIG. 7. A current view of DNA repair pathways. At the left mechanisms for the excision of thymidine dimers formed in uv-irradiated *E. coli;* successive repair stages require the products of several genes. At the right, removal of damaged bases (thymidine dimers T ◇ T, or others X) by DNA glycosylases and apurinic (AP) and a pyrimidinic (Apy) endonucleases; repair of the DNA molecule results from the successive actions of DNA polymerase-β and a DNA ligase. [Grossman (1981).]

to Ohashi *et al.* (1983), DNA ligase activity is inhibited by histones, and this inhibition is reversed by poly(ADP) ribose. However, if poly(ADP) ribosylation is necessary for DNA repair after alkylation, it does not appear to be required after uv-radiation (Cleaver *et al.*, 1983). Malik *et al.* (1983) found that only 10% of the nucleosomes are poly(ADP) ribosylated and associated with poly(ADP) ribose polymerase. These nucleosomes are the ones that have the highest number of single-stranded breaks in their DNA. This finding agrees with the concept that poly(ADP) ribosylation plays a role in DNA repair.

Except for trypanosomes (Van Assel and Steinert, 1971), mitochondrial DNA synthesis is independent of the cell cycle. The signals that control chromosomal

DNA synthesis are apparently not understood by mitochondrial DNA that is replicated continuously throughout the cell cycle by mitochondrial DNA polymerase-γ. As we have seen in Chapter 3, replicating mitochondrial DNA differs from nuclear DNA by the presence of displacement loops (D-loops). These D-loops are due to asynchronous DNA replication in which the two strands have a different origin (in contrast to nuclear replicons). Replication in mitochondria begins with the initiation of the heavy-strand (H-strand) DNA at its origin. Synthesis of the light strand (L-strand) does not begin until the nascent H-strand has been elongated to the origin of the L-strand, two-thirds of the way around the circular mitochondrial genome (Fig. 50, Chapter 3).

### 3. *Histone Synthesis and Nucleosome Assembly*

It was believed a few years ago that DNA synthesis and histone synthesis were always closely coordinated and took place side by side during the S phase. Stein *et al.* (1975) noted that histone mRNA was present in the cells only during the S phase. In chromatin isolated during this period of the cell cycle, but not during the others, histone genes could be transcribed *in vitro* by adding RNA polymerase. The control of transcription during the various phases of the cell cycle would be due to non-histone chromosomal proteins. Unfortunately, it is now known that the results of the experiments of Stein *et al.* could have been due to artifacts that were unknown at that time.

Our present views are somewhat different. According to Melli *et al.* (1977), histone mRNA is synthesized during the entire cell cycle in HeLa cells, but its concentration in the cytoplasm may vary due to posttranscriptional mechanisms. Groppi and Coffino (1980) concluded that the rate of histone synthesis was the same in G and S phases, but that the newly synthesized histones would not become associated with DNA until its replication had begun. In baby hamster kidney (BHK) cells, even the synthesis of the individual histone species is not synchronous. Histone H1 is synthesized in $G_1$ and the "core" histones (H2A, H2B, H3, H4) during the S phase (Tarnowka *et al.*, 1978). However, in other cells, synthesis of core histones and histone H1 undergoes parallel fluctuations during the cell cycle, with a peak correlated with DNA synthesis (Plumb *et al.*, 1984). Finally, Russev and Hancock (1981) observed that histone synthesis continues when DNA synthesis has been inhibited by HU. All of these experiments suggest that there is no strict correlation between histone synthesis and DNA replication. However, this view is not universally shared. According to a report by Wu and Bonner, basal histone synthesis (i.e., continuous synthesis at a constant rate throughout the cell cycle) amounts to only 10% of total histone synthesis; 90% takes place during the S phase and is much more HU sensitive than "basal" histone synthesis. The latter represents only 2–3% of histone production during the S phase (Waithe *et al.*, 1983). Close correlation between histone synthesis and DNA replication has also been reported in hamster fibroblasts and HeLa cells by Delegeane and Lee (1981), Marashi *et al.* (1982), and Heintz *et al.* (1983). In HeLa cells, the synthesis of nucleosome core histones increases 15 times during the S phase; this increase is due to both enhanced histone mRNA production and greater stability of the histone messages (Heintz *et al.*, 1983). On the other hand, Alterman *et al.* (1984) recently concluded that the regulation of mouse histone $H_3$ mRNA during the cell cycle largely depends on posttranscriptional processes that operate in the nucleus; they found no evidence for changes in the stability of the mRNA molecules during their decay after the S phase of the cell cycle. At the other extreme, there is the case of the sea urchin eggs, in which, as we shall see in Volume 2, Chapter 2, histone synthesis can take place in the complete absence of the nucleus. These apparent discrepancies show once more that biologists should never be too dogmatic and that the striking diversity of the living world also exists at the molecular level.

There is currently a good deal of discussion about the mechanisms of nucleosome assembly. Most investigators agree that assembly takes place as soon as DNA has replicated and has thus regained a double-stranded structure. However, the exact mechanisms of DNA–histone interaction after DNA replication are not yet clear. Different models (Prior *et al.*, 1980; Laskey and Earnshaw, 1980) have been proposed, but it seems too early to discuss them here.

For Russev and Hancock (1981), newly synthesized histones would be incorporated into "hybrid nucleosomes" containing both new and old histones. According to Jackson *et al.* (1981), the nucleosomes that contain newly synthesized DNA also possess 50% of the newly synthesized H2A and H2B histones; the other 50% of these histones, as well as 100% of H1, would be deposited elsewhere. In more recent papers, Jackson and Chalkley (1981a,b) concluded that only the newly synthesized H3 and H4 histones are specifically deposited on newly replicated DNA; histones H2A and H2B have only a partial preference for this DNA, while H1 is deposited on old DNA. During the cell cycle, all non-H1 histones remain attached to the same DNA; during the following cell cycle, the histones are equally distributed on both daughter DNA strands. For Annunziato *et al.* (1982), the newly synthesized histones are deposited on both replicating and nonreplicating regions of chromatin. Interestingly, these histone–DNA interactions are not affected by butyrate-induced histone acetylation. DNase sensitivity appears 3 min after replication of the DNA regions that are active in transcription. As pointed out by Laskey and Earnshaw (1980), the decision to assemble potentially active or inactive chromatin occurs at the time of DNA replication. One may guess that the level of DNA methylation and the addition of non-histone HMG-14 and -17 proteins (which increase during the S phase, according to Bhorjee, 1981) are important factors in reaching that decision. Another important factor is protein synthesis, since chromatin synthesized in the presence of cycloheximide is devoid of nucleosomes; this could be due to the small size of the histone pool (Riley and Weintraub, 1979). Although the problem of nucleosome assembly during the S phase of the cell cycle has not been completely solved, one may conclude, along with Russev and Hancock (1982), that the majority of the newly synthesized histones, together with newly replicated DNA, form new nucleosomes.

### D. The $G_2 \rightarrow M$ Transition[2]

During the $G_2$ phase, the cell is the site of important protein synthesis; at least 35 new proteins appear during this period (Al-Bader *et al.*, 1978). The following entry into prophase is characterized by two major morphological events: breakdown of the nuclear membrane and chromosome condensation.

There is ample evidence that phosphorylation of histone H1 plays an important role in chromosome condensation (reviewed by Laskey, 1983). This was first shown by the elegant experiments of Bradbury *et al.* (1974), who injected a purified preparation of H1 histone kinase into the myxomycete *Physarum polycephalum*. This interesting organism possesses thousands of nuclei in a common cytoplasm; all of these nuclei divide simultaneously, demonstrating that mitosis is induced by signals received from the cytoplasm. Bradbury's experiments, which were confirmed by Inglis *et al.* (1976), showed that injection of H1 histone kinase (isolated from Ehrlich ascites tumor cells) accelerates the entry into prophase of the *Physarum* interphase nuclei. That histone H1 phosphorylation is one of the factors that control chromosome condensation and mitosis in many, and possibly all, cells has been further shown by studies on temperature-sensitive (*ts*) cell mutants that are incapable of phosphorylating H1 at a nonpermissive (high) temperature. Only at a permissive (low) temperature can the cells condense their chromosomes and enter into mitosis (Matsumoto *et al.*, 1980). Furthermore, it has been found that, in synchronized hepatoma cells, the activity of H1 histone kinase increases four to six times between $G_2$ and metaphase, where chromosome condensation has reached its maximum (Zeilig and Langan, 1980).

For Gurley *et al.* (1974a,b, 1975), phosphorylation of histones might even play a broader role in the cell cycle. They have shown that phosphorylation of histone H3 might also be involved in chromosome condensation (Gurley *et al.*, 1974a). They found (Gurley *et al.*, 1974b) that histone H1 is superphosphorylated in prophase and concluded (Gurley *et al.*, 1975) that phosphorylation of H1 histone is linked to changes in the structure of chromatin in $G_1$, to DNA replication in S, and to chromosome condensation during mitosis. More recently, Gurley *et al.* (1978) also showed that both H1 and H3 histones are dephosphorylated after anaphase. HMG-14 also displays a phosphorytlation-dephosphorylation cycle during mitosis.

It is clear that chromosome condensation largely depends on conformational

[2]Reviewed by M. V. N. Rao (1980).

changes in chromatin organization. In addition to histone phosphorylation, histone acetylation might also play a role. Although nothing is known about such changes during $G_2$, it has been reported that histone acetylation strongly decreases at metaphase (Gómez-Lira and Bode, 1981). Changes in the conformation of DNA itself, catalyzed by the DNA topoisomerases, are also likely to play an important role in chromosome condensation. However, according to Paulson and Langmore (1983), nucleosomes have the same structure in condensed metaphase chromosomes and in interphase nuclei; the difference between the two must reside in higher orders of folding of the chromatin fibers.

Cell fusion experiments (reviewed by P. N. Rao, 1980), which will be discussed again later, have already provided us with evidence that chromosome condensation can be induced by factors present in the cytoplasm of cells that have condensed chromosomes. This was first shown by A. Brachet in 1921 in polyspermic sea urchin eggs. The results of similar experiments on eggs from many species will be discussed in Volume 2, Chapter 2. For example, the fusion of a mouse oocyte in meiotic metaphase II with a mouse blastomere induces the premature condensation of the chromosomes (PCC) (Fig. 8) in the interphase nucleus of the blastomere; there is no species specificity in this reaction of the interphase nuclei (Bałakier, 1979). In mammalian somatic cells, the fusion of cytoplasts separated from cells undergoing mitosis (the so-called mitoplasts) with cells in $G_1$ induces PCC in the latter (Sunkara *et al.*, 1980). These experiments clearly demonstrate that the cytoplasm of mitotic cells contains factors that can induce chromosome condensation in interphase nuclei. Whether these factors are enzymes that modify the conformation of histones of other proteins, or of DNA, or of something else is not yet known.

The other important event that takes place when a cell enters prophase is the breakdown of the nuclear membrane. The first observations of this event stem from work done on embryological material that will be discussed in more detail in Volume 2, Chapter 2. When amphibian oocytes are treated with progesterone, they undergo maturation after a few hours. Breakdown of the nuclear envelope and chromosome condensation take place. This is followed by the buildup of the meiotic spindle carrying the highly condensed chromosomes. If cytoplasm from such a progesterone-treated oocyte is removed and injected into an untreated oocyte, the same sequence of events characteristic of maturation takes place. The cytoplasm of oocytes that have undergone maturation thus contains a maturation-promoting factor (MPF) capable of inducing nuclear membrane breakdown and chromosome condensation in a normal oocyte (Smith and Ecker, 1971; Masui and Markert, 1971). More recently, Wasserman and Smith (1978) found that MPF, which is absent or inactive in unfertilized eggs, reappears during cleavage and shows a marked peak of activity during the $G_2$–M period; this cyclic MPF activity can be detected during several cleavage cycles. Interestingly, the same cyclic behavior of MPF formation and disappearance is found in both nucleate

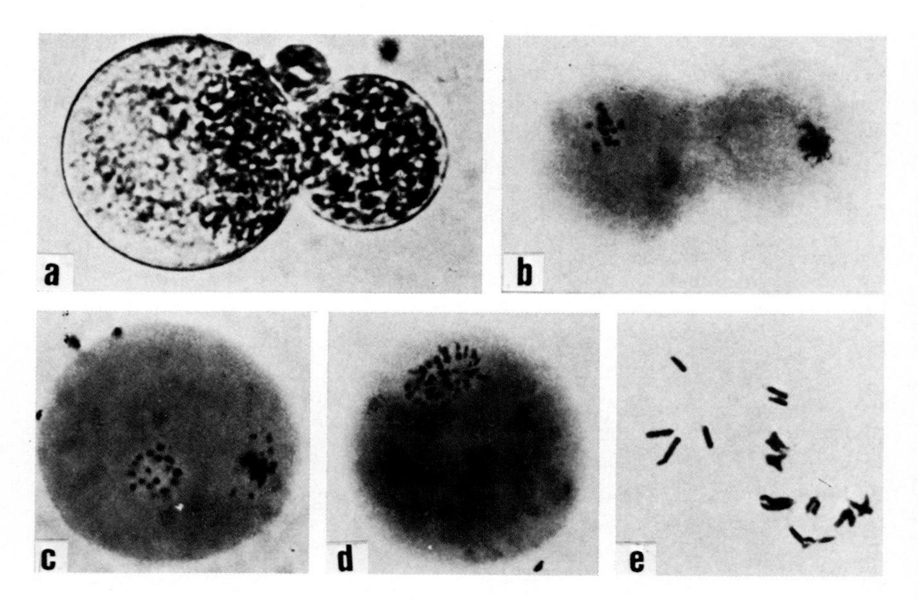


FIG. 8. Premature chromosome condensation (PCC) induced by fusing a maturing mouse oocyte with a mouse blastomere in interphase. (a) Shortly after fusion, the dark cytoplasm originating from the blastomere invades the oocyte cytoplasm; adhering to this large cell are the sister unfused blastomere and the two polar bodies. (b) The two contributing cells are still connected by a cytoplasmic bridge. The blastomere chromosomes (right) have begun to condense; on the left are the oocyte chromosomes. (c) One hour after fusion, chromosome condensation occurs; the chromosomes of the oocytes are in the center, and those of the fused blastomere are near the cell surface. (d) After 8 hr, chromosomes from the oocyte and the blastomere form a common group. (e) Bivalents of metaphase I and blastomere chromosomes from a hybrid cell fixed 2 hr postfusion in an air-dried preparation. [Tarkowski and Bałakier (1980).]

and anucleate eggs. These observations of Wasserman and Smith (1978) led Sunkara *et al.* (1979b) to a search for MPF activity in synchronized HeLa cells (human cancer cells). Extracts prepared at various stages of the cell cycle were injected into *Xenopus* oocytes. It was found that only the extracts prepared during the $G_2$–M period could induce the breakdown of the oocyte's nuclear membrane and the condensation of its chromosomes (Fig. 9).

It has been shown (Adlakha *et al.*, 1982a,b) that in mammalian cells, the MPF-like factors that induce the breakdown of the nuclear membrane after injection into the cytoplasm of *Xenopus* oocytes are concentrated in the chromosomes. The specific activity of the MPF-like factors in these cells is threefold higher in the chromosomes than in the cytoplasm. During the $G_2$ phase, MPF activity is found only in the nuclei. Later in this phase, increasing amounts of the factor are found in the cytoplasm. Mild digestion with DNase liberates MPF activity from isolated metaphase chromosomes or $G_2$ nuclei. The MPF-like factor of mammalian somatic cells, as in *Xenopus* oocytes, is a nondialyzable

$Ca^{2+}$-sensitive protein. In both cases, this protein is stabilized by the addition of protease and phosphatase inhibitors. However, there is a marked difference between *Xenopus* oocytes and mammalian somatic cells. In the former, MPF is localized in the cytoplasm and is produced by progesterone-treated, enucleated oocytes (see Volume 2, Chapter 2).

Interestingly, Adlakha *et al.* (1983, 1984) found that extracts from $G_1$ (but not $G_2$ or S) HeLa cells neutralize the activity of MPF-like factors. After mixing with $G_1$ cell extracts, the MPF-like factors no longer induce nuclear membrane break-

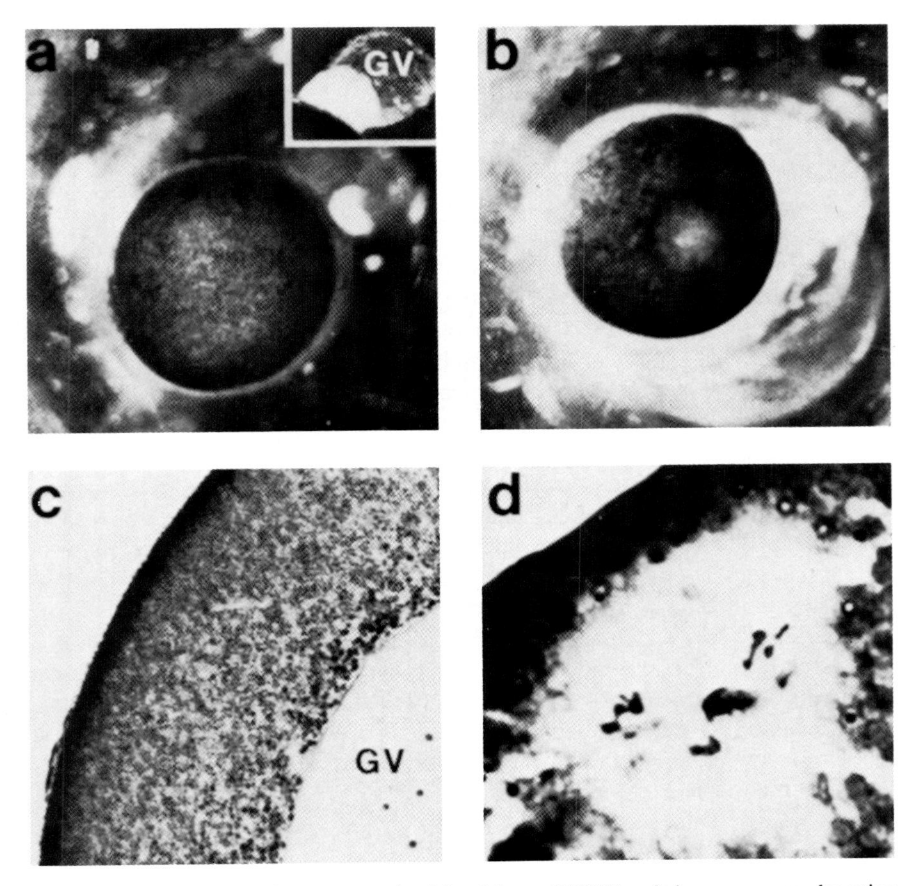


FIG. 9. Germinal vesicle (oocyte nucleus) breakdown (GVBD) and chromosome condensation after injection of *Xenopus* oocytes with mititic HeLa cell extracts. (a) Injection of extraction medium (control); the insert shows the GV dissected out of the oocyte. (b) Three hours after injection with mitotic HeLa cell extract, a white spot appears at the animal pole; it shows that GVBD took place. (c) After injection with an extract of S-phase HeLa cells, the GV remains intact. (d) One and one-half hours after injection with mitotic HeLa extract, the chromosomes are condensed, and a meiotic spindle is assembling [Sunkara *et al.* (1979b).]

down after injection into *Xenopus* oocytes. These inhibitors of MPF-like mitotic factors become active when cells enter into telophase; they might control the decondensation of the chromosomes at this late stage of mitosis. These non-histone chromosomal proteins are inactivated by radiation and have an $M_r$ higher than 12,000; they probably form an inactive complex with the MPF-like factors. However, pulses with protein synthesis inhibitors show that the MPF-like factors (mitotic proteins) disappear at anaphase-telophase in sea urchin eggs and must be resynthesized before prophase of the next cell division (Wagenaar, 1983).

Thus, it is clear that during the $G_2$ phase, somatic mammalian cells accumulate an MPF-like factor that is involved in the breakdown of the nuclear membrane. How universal the presence of an MPF-like activity in $G_2$ cells is remains to be seen. All that can be said is that we have been able to detect some MPF activity in cleaving sea urchin eggs (unpublished observations of A. Pays), and it is therefore likely that we are dealing with a very widespread phenomenon. It would be of particular interest to know whether an MPF is produced in protozoa and lower algae when the mitotic spindle is built up in a nucleus that still possesses an intact nuclear membrane.

Unfortunately, there has been only partial purification and characterization of MPF isolated from either *Xenopus* oocytes or HeLa cells. Present indications suggest that MPF is either directly or indirectly involved in protein phosphorylation. Interestingly, it has been proposed that the disassembly of the nuclear pore–lamina complex at prophase results from the phosphorylation of the lamins (laminoproteins). Kletzien (1981) has shown that protein kinase activity is present in isolated nuclear membrane preparations; it phosphorylates some of the nuclear membrane proteins, and its activity increases three- to five-fold when the cells are induced to proliferate. It is tempting to suggest that this increase is correlated with nuclear membrane breakdown; this hypothesis is reinforced by the recent finding that phosphorylation of the lamins increases 4 to 7 times at mitosis (Ottaviani and Gerace, 1985).

A paper by Jost and Johnson (1981) has shed new light on the behavior of lamin B during the cell cycle. This protein is synthesized during the S phase and accumulates at the nuclear surface. It is released into the cytoplasm during the $G_2$–prophase transition. At anaphase it associates with the chromosomes; in $G_1$, it associates with the nuclear membrane. In a more recent paper, Jost and Johnson (1983) described the results of experiments in which cells at different stages of the mitotic cycle were fused together. They concluded that lamina depolymerization is controlled by diffusible factors present in the cytoplasm of metaphase partners; repolymerization would result from the inactivation of these factors. We presume that they are closely related to the MPF of maturing *Xenopus* oocytes. The major glycoprotein constituent of the nuclear pore–lamina complex moves into the cytoplasm at prophase and returns to the nuclear surface at telophase (Gerace *et al.*, 1982). Its behavior during mitosis is similar to that of lamin B. Thus, it seems that the proteins that constitute the nuclear membrane are not broken down and resynthesized during mitosis; it is only their intracellular localization that changes. During the $G_1 \rightarrow M$ transition, modifications of the nuclear matrix should be expected to occur; a study by Henry and Hodge (1983) shows that nuclear matrix proteins undergo cyclic changes in phosphorylation. They are low in $G_1$, increase in $G_2$, and reach a maximum just before prophase; their phosphorylation is carried on by endogenous protein kinases. All of these results reinforce the idea that MPF-like factors are involved in protein phosphorylation, but direct evidence will be lacking until these factors have been purified and characterized. Further circumstantial evidence for the view that there is a link between protein phosphorylation and MPF activity comes from the recent work of Saharasbuddhe *et al.* (1984). They found that phosphorylation of eight nonhistone nuclear proteins increases 8–10 times between mid-$G_2$ and mitosis. After a peak during mitosis, phosphorylation drops. The authors point out that there is a close parallelism between this protein phosphorylation cycle and MPF activity.

One nonhistone nuclear protein, cyclin ($M_r$ = 36,000), should be mentioned here. Its synthesis is correlated with DNA replication when growth of fibroblasts is stimulated by addition of serum (Bravo, 1984a). Malignant cells, which divide continuously, synthesize much larger amounts of cyclin than normal cells (Celis *et al.*, 1984). In a human carcinoma cell line where the growth factor EGF paradoxically decreases proliferation, EGF inhibits cyclin synthesis (Bravo, 1984b). The conclusion that cyclin is somehow involved in cell division seems inescapable. Bravo and Macdonald-Bravo (1985) have shown that the synthesis of cyclin precedes that of DNA, that it is not inhibited by the HCl, and that its nuclear distribution changed during the cell cycle.

The intensive protein synthesis that characterizes $G_2$ is essential for all of the events we have just discussed. Cells previously synchronized in $G_1$ can be synchronized (and thus arrested) in $G_2$ if they are treated with fluorophenylalanine (an analog of phenylalanine). Such $G_2$-arrested cells inhibit chromosome condensation after cell fusion; extracts of these cells have no MPF activity and do not induce premature chromosome condensation. One can conclude that synthesis of phenylalanine-containing polypeptides is a prerequisite for the production, during $G_2$, of the factors that control nuclear membrane breakdown and chromosome condensation. The identification of these factors and the elucidation of their mode of action are awaited with particular interest.

## III. MITOSIS

### A. Metaphase Chromosomes[3]

It is possible to isolate and separate chromosomes of various sizes and morphologies from dividing cells arrested in metaphase. This is done, in general,

[3]See Lima-de-Faria (1983).

after treatment of the cells with colchicine or colcemid, which has the advantage that the metaphase chromosomes, not being held together on the mitotic spindle, are dispersed in the cytoplasm. Partial large-scale separation of large from small chromosomes is often possible by centrifugation or cell-sorting methods.

It is common knowledge that the number of chromosomes ($2n$ in diploid cells, $n$ in haploid gametes) is constant in a given species. However, it should be mentioned that there exists, in many plant and animal species, so-called B-chromosome systems (reviewed by Jones, 1975). These additional (supernumerary) chromosomes, (whose number varies both among individuals of the same species and among the various tissues) increase at meiosis. The significance of these additional chromosomes remains unknown; it would be worthwhile to analyze them using the cytological methods now available. B-chromosomes might be analogous to the already mentioned "double-minute" chromosomes. These extrachromosomal gene copies appear in unstable cell mutants that are resistant to various chemicals, such as colchicine or methotrexate (Baskin *et al.*, 1981). B-chromosomes might represent amplified gene copies that have left their chromosomal loci, or they might have something in common with mobile genetic elements. The existence of B-chromosomes is often overlooked.

The rule of chromosome number constancy in a given species is, of course, valid only for the autosomes. The heterochromosomes (sex chromosomes) are not homologous, and in many species the genetically inactive Y chromosome is greatly reduced in size or has completely disappeared.

Figure 10 is a schematic representation of the structure of metaphase chromosomes as accepted by most cytologists; Fig. 11 shows how they actually appear in squashed preparations under the light microcope. They are composed of two chromatids (the future daughter chromosomes at anaphase), often linked together at metaphase by the kinetochore (the spindle attachment region, which is also called the "centromere"). The position of the kinetochore on the chromosome may vary from one chromosome to another. One can distinguish (Fig. 10) metacentric, acrocentric, and telocentric chromosomes according to the kinetochore's position. The kinetochore is usually recognizable as a clear, constricted zone called the "primary constriction." In certain species, one pair of

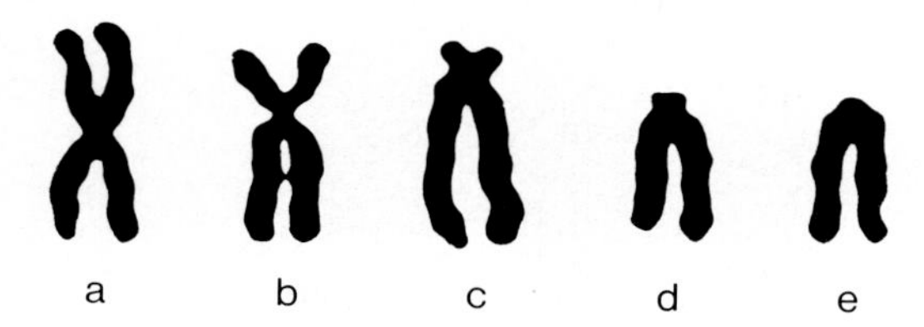


FIG. 10. Types of metaphasic chromosomes, depending on the position of the centromere. (a) Metacentric, (b) submetacentric, (c) subtelocentric, (d) acrocentric, (e) telocentric.

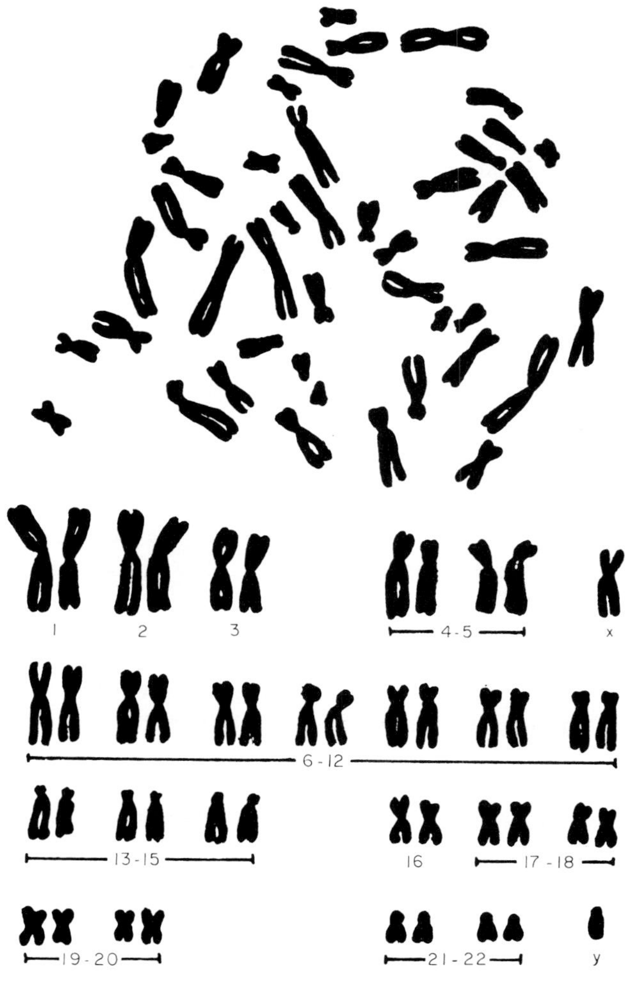


FIG. 11. The normal human karyotype. Above, karyotype obtained by squashing a metaphase cell. Below, identification of the various chromosomes. The autosomes are numbered from 1 to 22 according to their decreasing size. The heterochromosomes, X and Y, show that the squashed cell came from a man. [Redrawn from a photograph published in Srb *et al.* (1965).]

chromosomes displays a "secondary constriction" separating the nucleolar organizers from the bulk of these two chromosomes. They give rise to the nucleoli at the end of telephase.

As mentioned in Chapter 2, cytogeneticists now have at their disposal an array of "banding" techniques. After appropriate treatments involving, in general, protein and DNA denaturation or selective extraction, a banded structure (Fig. 3, Chapter 1) can be seen after staining with Giemsa or a fluorescent dye such as quinacrine. These reproducible banding patterns help to identify all of the chromosomes present in a metaphase plate. The chromosomal map established in this way is called the "karyotype" of the species (Fig. 11). Analysis of the karyotype is very important for many studies. Precise identification of the chromosome that bears a given gene helps to establish a map of the genome. In somatic intraspecies hybrids, some of the chromosomes are lost when the hybrid cells undergo repeated divisions; they can be identified with banding techniques. Several hereditary diseases can be detected in humans by careful examination of the fetal karyotype during pregnancy.

As we have seen in Chapter 2, differential staining can be due to non-histone proteins or to differences in base composition along the chromosome. Improvements in this technology have shown that individual bands replicate separately. It is, therefore, possible to distinguish "subphases" in the S phase in a more precise way than by what had been previously called "early"- and "late"-replicating DNAs (Cawood, 1981). It has also been proposed that, in both euchromatin and heterochromatin, "chromosomal replicons" allow the ordered sequence of chromosome replication. Analogous to the DNA replicons at the molecular level, they would be the organizational units of chromosomal replication (Lau and Arrighi, 1981). Finally, banding techniques are very helpful in attempts to detect by *in situ* hybridization $^{3}$H-labeled, single-copy DNA sequences (Harper and Saunders, 1981). Using such a method, Harper *et al.* (1981) have localized the insulin gene on the distal end of the short arm of human chromosome 11. With Harper's technique, great progress in the localization of individual genes on chromosome segments has already been made.

There is now universal agreement that metaphase chromatids are unineme, i.e., are formed from a single DNA fiber, as was first proposed by Wolff and Perry (1975) and by Schwartz (1975). If so, the $M_r$ of the DNA present in the largest *Drosophila* chromosome would be as high as $41 \times 10^9$; this implies that a metaphase chromosome is "contracted" 400 times (Bak and Zeuthen, 1978). There is also general agreement that metaphase chromosomes, like chromatin in the interphase nucleus, are organized in nucleosomes (Wigler and Axel, 1976; Paulson and Langmore, 1983). More open to discussion, because the facts and ideas are more recent, is the existence of a chromosome or chromatid scaffold analogous to the nuclear matrix of interphase nuclei (Fig. 12). According to Satya-Prakash *et al.* (1980), metaphase chromosomes, after digestion with nucleases and extraction of the histones, retain a central core that can be stained with silver. This scaffold would be some kind of proteinaceous skeleton for the chromosome. However, for Hadlaczky *et al.* (1981), Nasedkina and Slesinger (1982), and Burkholder (1983), chromosomal scaffolds after histone extraction would be an artifact and DNA would be necessary to maintain the shape of a

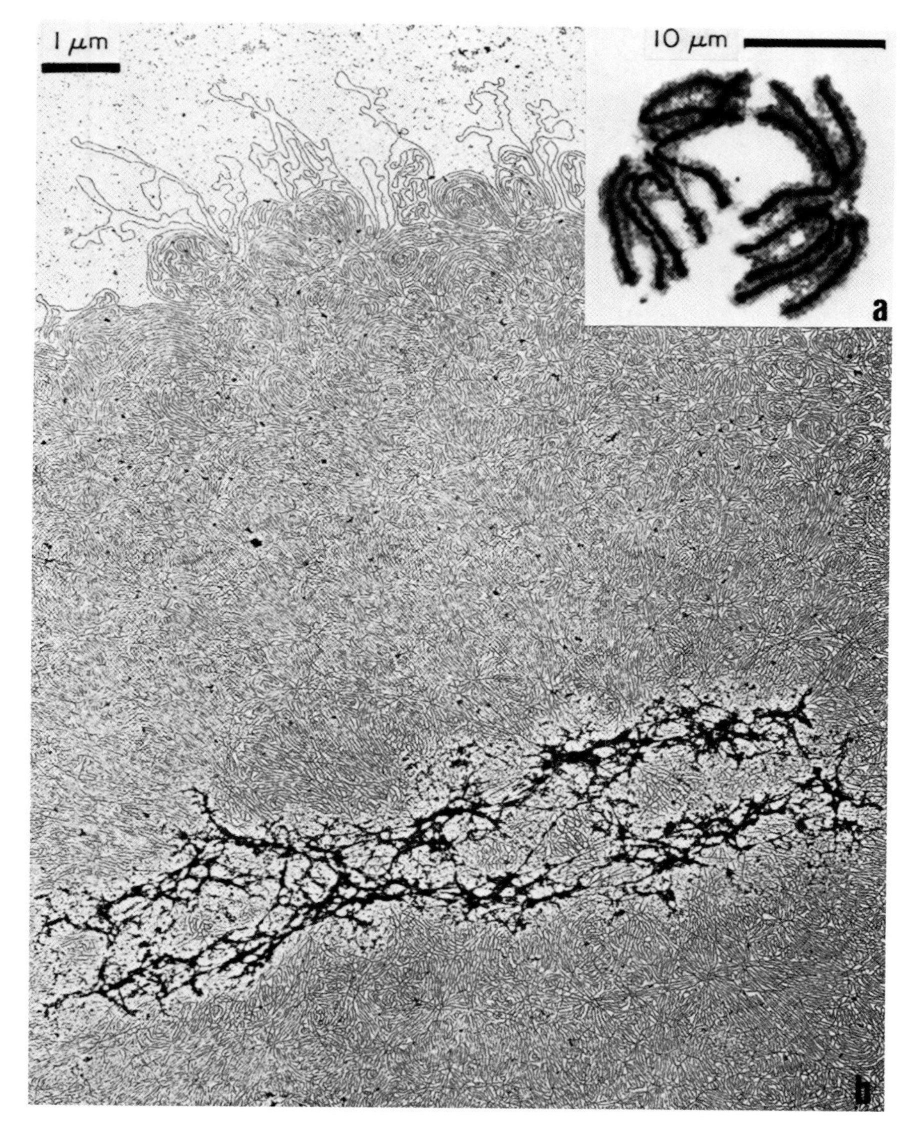


FIG. 12. The chromosome scaffold (core structure). (a) Anaphase chromosomes of the Indian muntjac; silver staining after a 50-min hypotonic treatment. (b) Electron micrograph of a histone-depleted metaphase chromosome from an HeLa cell showing the central protein scaffold and part of the surrounding DNA fibers. [(a) Howell and Hsu (1979). (b) Laemmli *et al.* (1978).]

chromosome. Finally, on the molecular level, the non-histone scaffold remaining after digestion of metaphase chromosomes with various DNases retains 0.1–0.5% of the initial DNA content. These DNA segments do not seem to have any specificity and have the same base composition as total DNA (Bowen, 1981).

Recently, Kerem *et al.* (1984) have studied the effects of DNase I digestion on fixed metaphase chromosomes and have looked for possible correlations with the classical Giemsa banding. As expected, they found that the inactive, strongly heterochromatic X chromosome of mammalian females displays only a weak sensibility to DNase. Constitutive heterochromatin is insensitive to DNase digestion. In the other regions of the chromosomes, a banding pattern, corresponding in general to the light bands after Giemsa staining, was obtained.

Attempts to elucidate the chemical composition of the residual scaffold remaining after nucleohistone removal have been made. According to Earnshaw and Laemmli (1983, 1984), the scaffold of metaphase chromosomes, which can be stained with silver, is composed of non-histone, fibrous proteins; the number of these proteins is small, and one of them is vimentin (Gooderham and Jeppesen, 1983). Earnshaw and Laemmli (1983) think that the residual scaffold of mitotic chromosomes plays a dynamic role in chromosome condensation; however, Labhart *et al.* (1982) found that removal of H1 histone by extraction with 0.5 *M* NaCl is sufficient to destroy the higher-order structure of chromatin and the structure of mitotic chromosomes. Since no chromosomal scaffold was left after this treatment, the authors think that the chromosome structure is maintained by a tight assembly of the 23- to 30-nm-thick chromatin fibers. It is obviously too early to draw strong conclusions; all we can do is to wait and see. However, it is hard to visualize DNA condensation in metaphase chromosome without wrapping of the DNA fibers around a proteinaceous scaffold; the latter, according to Detke and Killer (1982), is biochemically very similar to the nuclear matrix. However, two proteins of the nucleoskeleton are modified when the cells enter mitosis. On the biochemical level, it is known that histone deacetylase activity is highest in metaphase and probably plays a role in chromosome condensation (Wateborg and Matthews, 1982).

When metaphase chromosomes are spread on a grid and examined under the electron microscope, they break down into innumerable fibrils; their diameter is about 25 nm, and they form loops that seemingly arise from a pin cushion. A comparison of Figs. 8 and 9 in Chapter 4 provides an interesting demonstration of the progress made in EM techniques during the last 25 years. However, the problem of DNA packing in a metaphase chromosome remains unsolved, despite the discovery of nucleosomes. There would be little point in rediscussing what has already been said in Chapter 4 about the possible organization of supranucleosomal units, the probable role of nucleoplasmin and DNA topoisomerase I (nicking-closing enzymes that produce single-stranded breaks in DNA), and the importance of histone phosphorylation, etc., in the packing of chromatin into highly condensed metaphase chromosomes. If the chromosome scaffold is real, one can imagine that the nucleohistone loops wind around fibrous proteins. At this point it might be best—and certainly safer—to admit that we do not really know.

Some mention must be made of kinetochores. In 1956, Lima-de-Faria found that the kinetochore was composed of three different zones, and this has been confirmed by more recent EM studies. For Ris and Witt (1981), the kinetochore is, in general, a trilaminar disk (Fig. 13). Bundles of microtubules (MTs) are attached to the electron-dense outer layer. This outer plate is formed from hairpin loops of chromatin stacked together to form a solid layer. On its external surface, one can distinguish a "fibrous corona" where the polymerization of the MTs would be initiated. Both the inner and outer layers (or plates) are electron dense, while the middle layer is less opaque. The darker inner layer is immediately adjacent to centromeric heterochromatin. In the rat, this heterochromatin (which is also present in the telomeres, i.e., at the distal end of the chromosomes) is made up of a highly reiterated DNA sequence of 93 bp; this sequence is transcribed in the nucleus, but the transcripts do not cross the nuclear membrane and thus remain in the nucleus (Sealy *et al.*, 1981).

The presence of highly reiterated sequences (satellite DNAs) in kinetochores seems to be the rule. Brenner *et al.* (1981) discovered that serum of patients suffering from scleroderma, (an autoimmune disease) contains antibodies that, if made fluorescent or if bound to peroxidase, stain the kinetochores (reviewed by

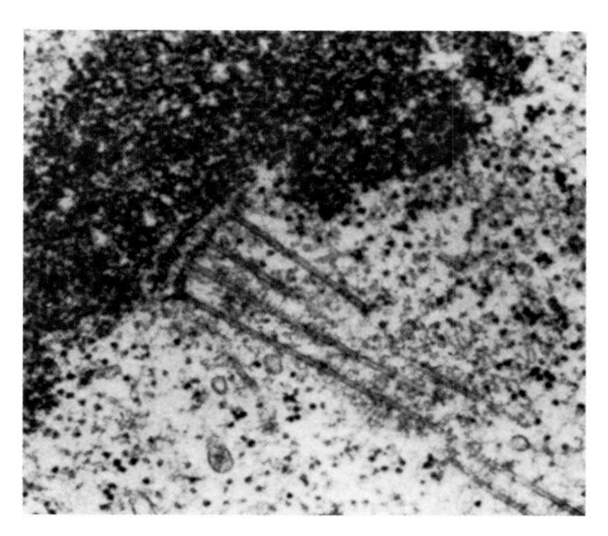

FIG. 13. Trilaminar structure of the kinetochore in a metaphase chromosome. The inner layer is in close contact with condensed centromeric heterochromatin; MTs are attached to the outer layer. The middle layer is much less electron dense. [De Brabander *et al.* (1981b).]

Hyams, 1984). By using immunoperoxidase staining and EM observation, they confirmed the trilaminar structure of the kinetochores. More surprising is the fact that according to the same authors, kinetochores replicate in $G_2$, later than the late-replicating chromatin. Immunofluorescence shows a number of fluorescent dots in interphase nuclei; these are presumed to be kinetochores. In general, these fluorescent dots are attached to the nuclear membrane or the nucleoli in interphase nuclei (Moroi *et al.*, 1981). These conclusions have been confirmed by Cox *et al.* (1983), who also used human anticentromere antibodies to stain the kinetochores of metaphase chromosomes and the discrete "prekinetochores" of interphase nuclei. Staining with the fluorescent antibodies is retained by the proteinaceous matrix in both nuclei and chromosomes; it is due to four distinct antigens. According to Earnshaw *et al.* (1984), at least one of them is an integral protein of the mitotic chromosomal scaffold. Finally, Rieder (1979a,b) found that the inner layer of the kinetochores contains ribonucleoprotein (RNP) particles; after colchicine treatment, the trilaminar structure of the kinetochore disappears, and all that remains is a large band of RNP particles that (hypothetically) might be the MT-organizing centers responsible for the formation of spindle fibers in contact with the kinetochores.

So far, no unexpected findings have emerged from the chemical analysis of isolated metaphase chromosomes. Their main components are, of course, DNA and histones. They contain fewer non-histone chromosomal proteins than interphase chromatin and, as one would expect from their highly condensed structure, are inactive in DNA or RNA synthesis. This could be due to the fact that metaphase chromosomes can phosphorylate their own H1 and H2 histones, as well as HMG-14 (Paulson and Taylor, 1982). Interestingly, genes that were active in interphase retain their DNAse sensitivity in the condensed metaphase chromosomes. The chromatin conformation responsible for nuclease sensitivity is thus retained even when the gene is silent at metaphase (Gazit *et al.*, 1982). DNAse sensitivity also allows one to distinguish between active and inactive X chromosomes (Kerem *et al.*, 1983; Riley *et al.*, 1984), at least at the level of individual genes carried by the active X chromosome.

We will conclude this section by mentioning the provocative and exciting experiments reported by Murray and Szostak (1983). They constructed, in yeast, artificial chromosomes containing cloned individual genes (as markers), replicators, centromeres (kinetochores), and telomeres from *Tetrahymena* (in which the telomeres are tandem arrays of single repetitive sequences). If the artificial chromosomes have fewer than 20 kb, the kinetochores do not function properly and the chromosomes are quickly lost; however, artificial chromosomes of 55 kb have normal behavior at both mitosis and meiosis, although they tend to be lost during successive cell divisions. The future will tell whether artificial chromosomes will provide a breakthrough in our understanding of mitosis or remain an interesting curiosity.

## B. The Lampbrush Chromosomes of Amphibian Oocytes[4]

Lampbrush chromosomes will be discussed in more detail in Volume 2, Chapter 2, but a brief description of their morphology and structure will be given here because it is now generally agreed that they can be used as a model for all chromosomes. They are, however, meiotic, not mitotic, chromosomes. When oocytes begin to grow, their chromosomes are at the diplotene stage of meiosis and are linked by chiasmata (synaptinemal complexes), which will be discussed briefly at the end of this chapter. The largest lampbrush chromosomes are found in the oocytes of urodeles (newts, salamanders); they differ from metaphase chromosomes by their extended configuration and by the fact that they are present in an intact nucleus.

Figure 14 is a phase-contrast photograph of two homologous lampbrush chromosomes linked by a chiasma. Loops extending from a continuous axis that originate from DNA-containing granules called ''chromomeres'' can be seen. Callan and Macgregor (1958) have shown that a continuous DNA molecule forms the axis of both chromomeres and loops, proving that chromatids are unineme. If lampbrush chromosomes isolated from the oocyte's nucleus (*germinal vesicle*) are treated with DNase, their entire structure (chromomeres and loops) collapses. The total length of one of the *Triturus* lampbrush chromosomes, if completely stretched, would be about 50 cm. The genetically active parts of the lampbrush chromosomes are the loops, where the DNA axis is very actively transcribed; the chromomeres, which have a more condensed structure, are not transcribed. Figure 15 is a diagram of the structure of lampbrush chromosomes as deduced by Gall and Callan (1962) from autoradiography studies on RNA synthesis. Electron microscopy has not greatly modified the Callan–Gall model, as we shall see in Volume 2, Chapter 2, but has merely added more detail. According to Mott and Callan (1975), chromomeres are composed of DNA–protein particles of 30 nm. The DNA-containing fibrils that link adjacent chromomeres measure 10 nm and show no signs of transcription. The transcribed matrix of the loops is composed of 30-nm particles, which associate in aggregates of RNP particles measuring 200–300 nm.

The DNA content of the lampbrush chromosomes varies greatly from one amphibian species to another; there is almost 10 times more DNA in the nuclei of the urodele *Triturus* than in those of the anuran *Xenopus*. This is the so-called C-paradox (Callan, 1972). Thus, the number of chromomeres in the lampbrush chromosomes of the two species differs widely. Since these species have approximately the same number of active genes, it was a mistake to believe, as was done a few years ago, that one gene corresponds to one chromomere. Although the length of the loops is correlated with the c-value, there is no correlation between this value and the hnRNA content (Scheer and Sommerville, 1982).

[4]Reviewed by Callan (1982).

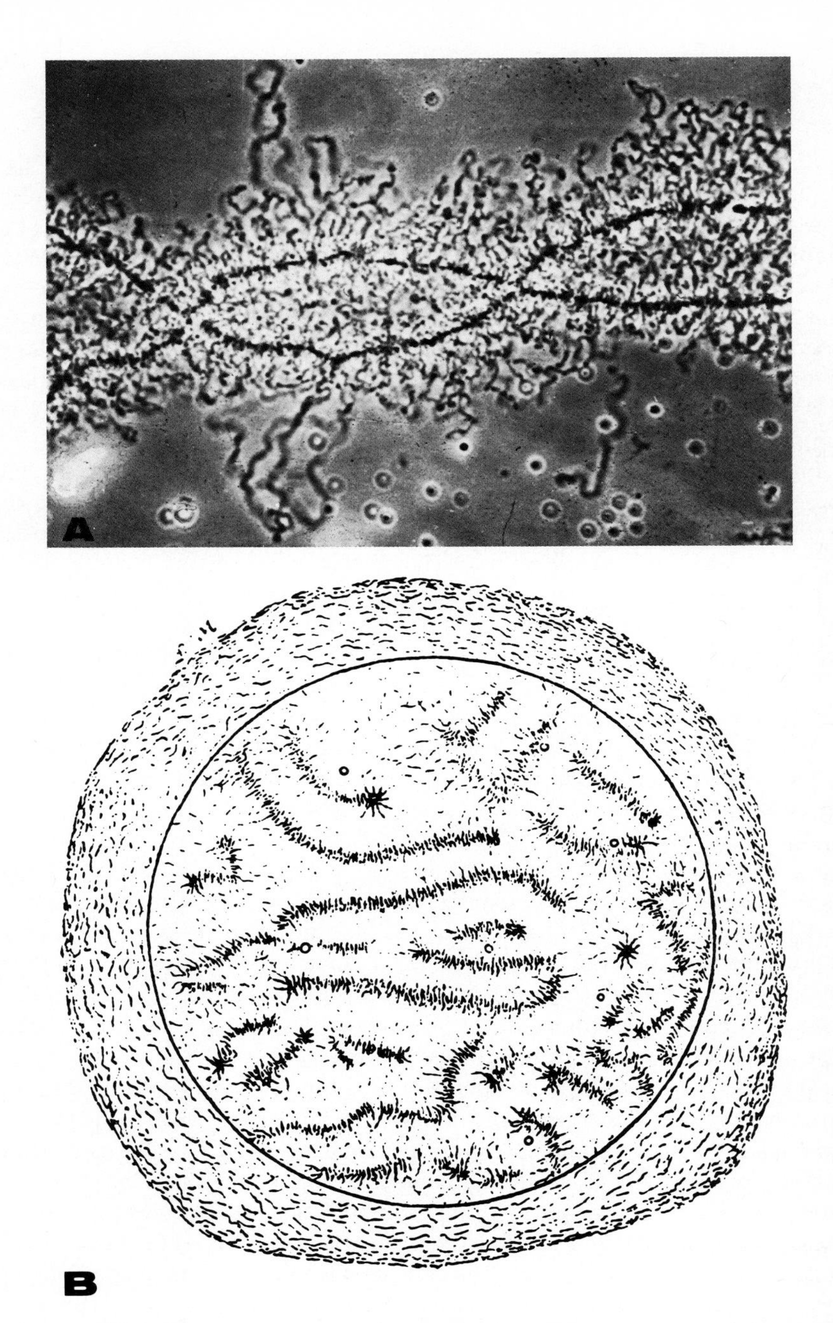


FIG. 14. (*Top*) Phase-contrast photograph of paired lampbrush chromosomes. [Courtesy of Prof. H. G. Callan.] (*Bottom*) Lampbrush chromosomes observed in stained sections of urodele ovary by Flemming (1982).

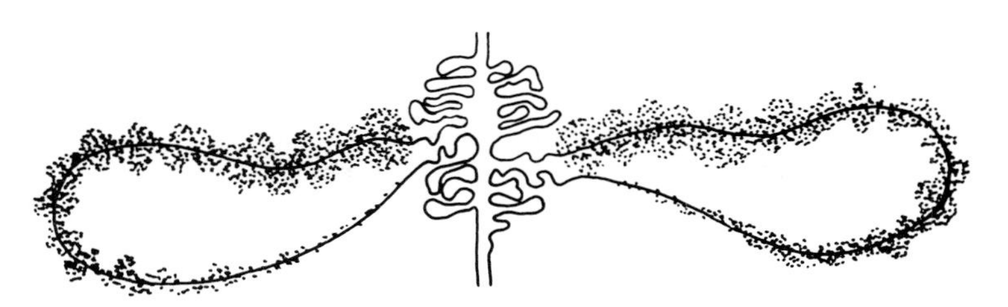

FIG. 15. Structure of chromomeres and loops according to J. G. Gall (1956). Two DNA threads undergo extensive coiling and form a chromomere; loops result from localized uncoiling of a DNA thread in the chromomere. Uncoiling allows transcription to take place. In this model, transcription is believed to proceed unidirectionally, leading to a progressive accumulation of ribonucleoprotein particles along the loop.

The main reason for using the lampbrush chromosomes as a model was to show that the bulk of DNA, which is present in the condensed chromomeres, is genetically inactive; in order to be active, a gene should be in an extended configuration, as are the loops. We shall see that this conclusion is reinforced by studies done on other giant chromosomes: the "polytene" chromosomes present in the salivary glands of *Drosophila* and *Chironomus* larvae.

### C. POLYTENE CHROMOSOMES OF *DROSOPHILA* AND *CHIRONOMUS* LARVAL SALIVARY GLANDS[5]

Balbiani (1881) found long ago that chromosomes in the salivary glands of *Chironomus* larvae are very large and have a characterized banded structure. The two very conspicuous rings surrounding them have been called the "Balbiani rings" (BR1 and BR2). Figure 16 compares Balbiani's original drawing and a micrograph of spread *Chironomus* chromosomes. The dark bands, which are visible in the intact cell by phase microscopy, are very rich in DNA; between these bands are interbands that, as was shown by Caspersson and Schultz (1938) in *Drosophila,* have proteins as their main constituent. There are at least 3200 bands in the giant chromosomes of *Drosophila* (Grond *et al.,* 1982).

The morphology of the giant chromosomes in *Chironomus* and *Drosophila* is essentially the same (Fig. 17); there is the same alternation of bands and interbands. However, in *Drosophila,* the four chromosomes are held together by a deeply staining, strongly heterochromatic centromere region (chromocenter, Fig. 17). Painter (1934) showed that the banded structure in the giant chromosomes corresponds to specific genes. The linear organization of the genome in the chromosomes has been known since the work of Morgan and his school, but one had not made a correlation between the bands present in the giant chromosomes

[5]Reviewed by Zegarelli-Schmidt and Goodman (1981) and Zhimulev *et al.* (1981).

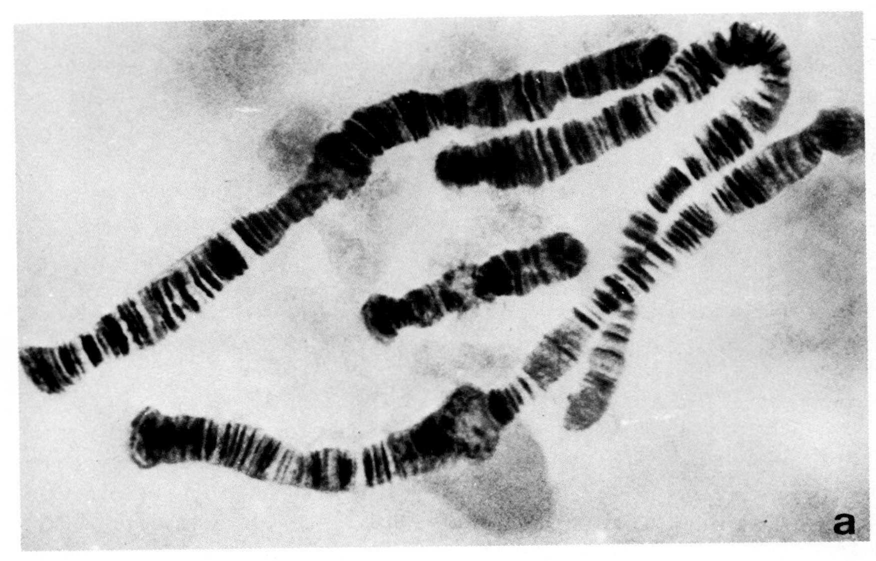


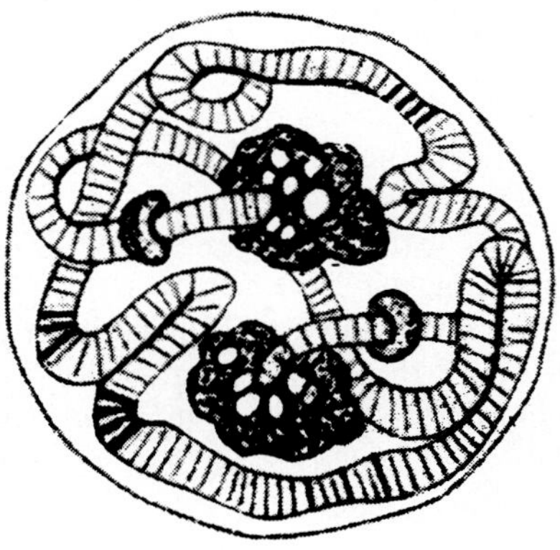


FIG. 16. (a) The four giant chromosomes of *C. tentans* after spreading and staining. Two of the long chromosomes bear nucleoli. Chromosome IV is the shortest and has a Balbiani ring (giant puff). The banding pattern is the same in all tissues, but is particularly clear in larval salivary glands. (b) Salivary gland chromosomes of *Chironomus,* as seen by Balbiani in 1881. [(a) Beermann and Clever (1964).]

and the genes that had been located in the chromosomes by crossing-over analysis. Painter was certainly a pioneer in the now broad field of cytogenetics.

In both *Drosophila* and *Chironomus* salivary glands, the four giant chromosomes undergo somatic pairing. Thus, the diploid number of chromosomes in these species is eight, but in salivary glands the homologous chromosomes are paired band by band. This is clearly shown in hybrids between two different

*Drosophila* species. In nonhomologous regions, the chromosomes are separated from each other (Fig. 18). The total length of the giant *Drosophila* chromosomes is about 2 mm compared to about 7.5 μm for the chromosomes of *Drosophila* cells in culture. The length of the giant chromosomes can be increased 176 times by stretching (Gruzder and Reznik, 1981). The giant chromosomes are also called "polytene" chromosomes because they are composed of many parallel fibrils that result from the repeated replication of the four chromatids initially present in the two paired chromosomes. This process is called "endoreduplication," "endomitosis," or "polytenization" because DNA replication takes place in an intact nucleus and is not followed by mitotic division. Insect salivary glands, after secreting larval saliva, degenerate during metamorphosis. Large chromosomes, which have a banding pattern similar to that of salivary gland chromosomes, are found in the intestine and Malpighian tubes of the larvae.

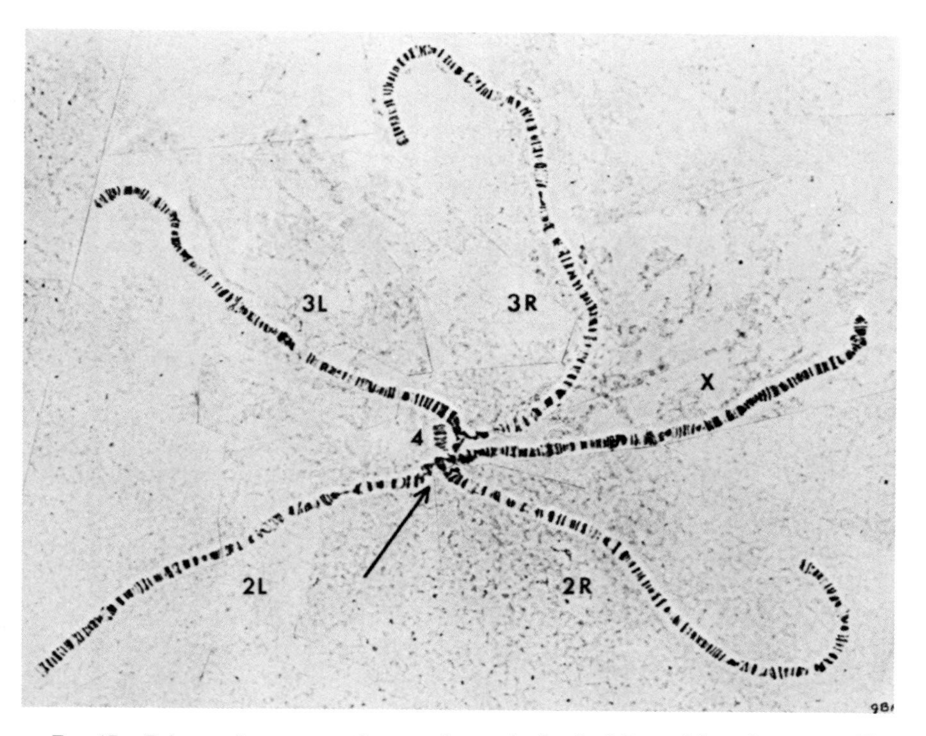


FIG. 17. Polytene chromosomes from a salivary gland cell of *Drosophila melanogaster*. The banded giant chromosomes are held together at the chromocenter (arrow). Homologous chromosomes are initimately paired, so that only four chromosomes are seen. The X chromosome and the small chromosome 4 have single arms; chromosomes 2 and 3 have left (L) and right (R) arms. [Lefevre (1976). With permission from "The Genetics and Biology of *Drosophila*," (M. Ashburner and E. Novitski, eds.). Copyright: Academic Press Inc. (London) Ltd.]

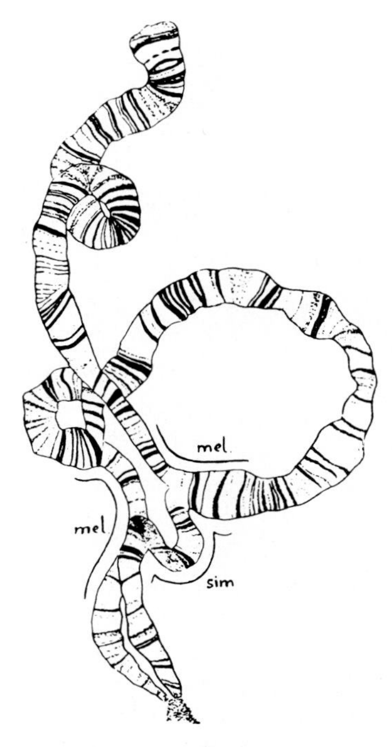


FIG. 18. Chromosome 3 of a hybrid between *D. melanogaster* ♀ × *D. simulans* ♂. Due to an inversion in the region between the genes *scarlet* and *peach,* pairing of the two polytene chromosomes 3 is incomplete and a large loop is seen between the two nonhomologous regions. [Pätau (1935).]

However, their banding pattern is not easily analyzed. In salivary glands of *Drosophila* and *Chironomus,* the degree of polyteny can reach 1.024 C in the former and 2.048 in the second (with 256 C and 512 C as intermediaries during the polytenization process, where C = 2 chromatids). According to Beermann and Bohr (1954), about 10 cycles of endoreduplication of the initial chromatids occur in *Chironomus*. A more recent report by Laird (1980) indicates that there are local variations in polyteny among interbands in *Drosophila* giant chromosomes, but Spierer and Spierer (1984) found no differences in polyteny between bands and interbands.

It is generally believed that repetitive sequences are underreplicated during polytenization. Thus, differential underreplication (reviewed by Brutlog, 1980) has been observed for satellite DNA and for ribosomal DNA (Spear and Gall, 1973; Spear, 1974). Figure 19, taken from Gall, summarizes the results obtained on the localization (by *in situ* hybridization, according to Gall and Pardue, 1969), replication (by thymidine incorporation), and autoradiography of highly repetitive DNA in giant and metaphase chromosomes of *Drosophila*. As proposed by Heitz (1934) this scheme distinguishes an α- and a β-heterochromatin. The former is more condensed, is formed from satellite DNA sequences, and constitutes the compact center of the chromocenter; it is not replicated during polytenization (Lakhotia, 1984). In contrast, β-heterochromatin is less compact, forms the major part of the chromocenter, and extends into euchromatic chromosome areas; it is replicated like the adjacent chromatin and is transcribed. There is no underreplication of unique sequences, whatever their size (Lifschytz, 1983).

Although differential underreplication during polytenization is generally considered an established fact, it is only fair to say that it has been questioned by Dennhöfer (1981). On the basis of cytophotometric measurements, he has concluded that the entire DNA is completely replicated and that there is no heterochromatic underreplication.

The chromatids of the polytene chromosomes are unineme and are, therefore, composed of a single DNA double helix (Laird, 1971; Gruzdev and Resnick, 1981). Thus, as in lampbrush chromosomes, the DNA molecules are continuous in polytene chromosomes. The bands, like the chromomeres of the lampbrush chromosomes, are regions where the chromatin fibrils have undergone extensive coiling; in interbands, where non-histone proteins are much more abundant than in bands (as was first shown by Caspersson and Schultz, 1939), the DNA-histone threads would be much less coiled (Crick, 1977; Fig. 20). A more elaborate model has been worked out by Sass (1980) on the basis of EM studies. The basic constituent of bands, interbands, and heterochromatin is the chromatid, a fiber of 11–25 nm; its coiling varies in interbands and bands, where it forms superbeads of 0.1 μm. The chromatid subunits are the classic 11-nm nucleosomes (the DNA molecule is only 2 nm wide). Unfolding of the chromosome fibers would produce loops, as in the lampbrush chromosomes (Fig. 21). More original is the model proposed by Mortin and Sedat (1982) for *Drosophila* polytene chromo-

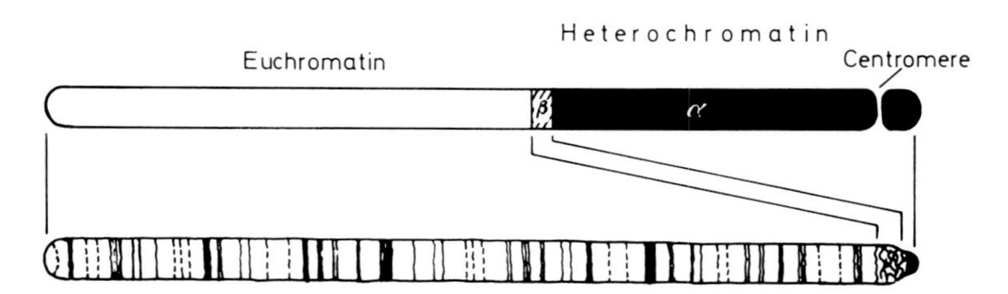


FIG. 19. Comparison of mitotic (above) and polytene (below) chromosomes of *Drosophila*. α-Heterochromatin (α) is composed of highly repetitive DNA sequences (satellites) and is probably not replicated during polytenization. β-Heterochromatin (B) contains repetitive sequences, at least some of which are replicated during polytenization. Euchromatin is composed of unique sequences and interspersed repetitive sequences; both are replicated during polytenization. [Gall *et al.* (1971).]

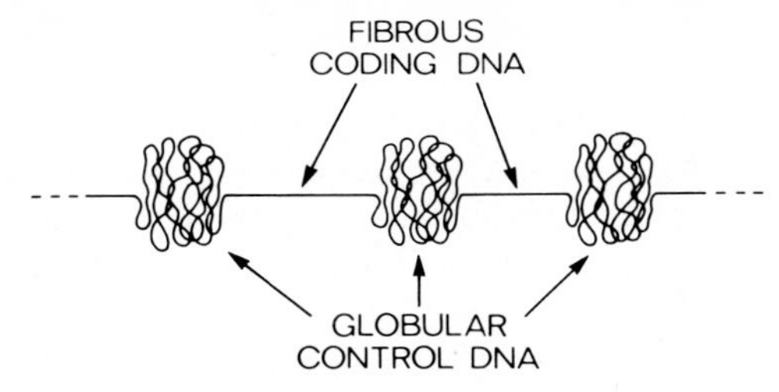


FIG. 20. General structure of the DNA of a chromatid, according to F. H. C. Crick (1971). The line represents part of the continuous DNA in the uninemic chromatid. The straight portions correspond to interband regions and the intricated folded regions to the bands. Both globular and fibrous DNA are complexed with proteins. [Reprinted by permission from *Nature*, **24,** 26. Copyright © 1971 Macmillan Journals Limited.]

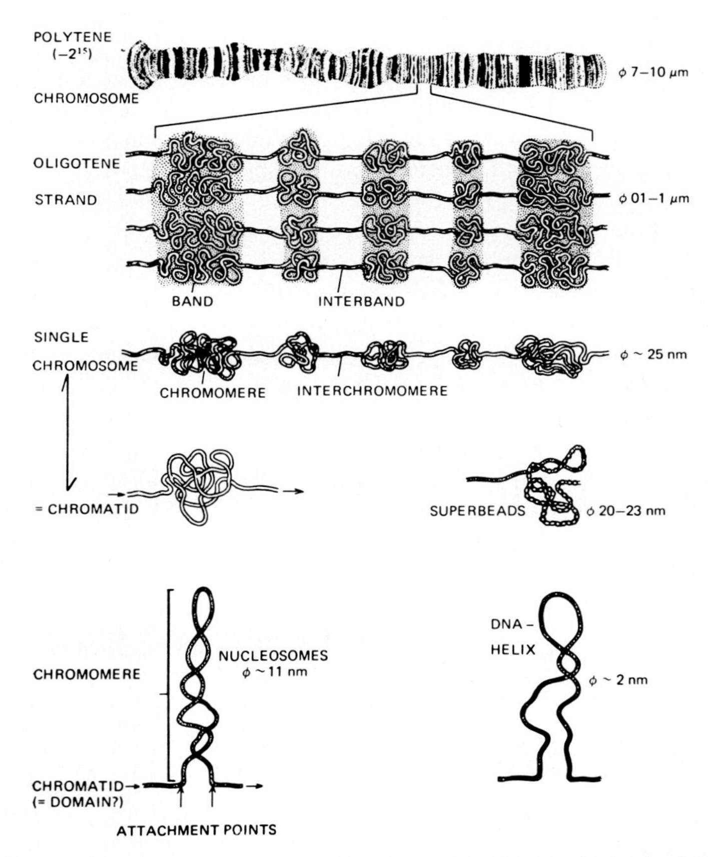


FIG. 21. A schematic representation of the hierarchy of fibrillar organization levels in polytene chromosomes of *Chironomus* larval salivary glands as seen in thin sections and spread preparations. [Sass (1980).]

somes. Torus-shaped DNA molecules would be confined to the rim of the bands, whose center would be hollow; the bands would thus have a radially symmetric structure. For Sorsa (1983), these chromosomes have a cable-like organization with disk-like bands containing toroidal subunits. The fact that there there are no differences in polyteny between bands and interbands is taken by Spierer and Spierer (1984) as evidence for the classic folded fiber model for the organization of *Drosophila* polytene chromosomes. Finally, we should mention the recent observations of Mathog *et al.* (1984) on these chromosomes. They found that each arm of the polytene chromosomes folds up in a characteristic way, contacts the nuclear surface, and is topologically isolated from the other arms. There is thus a nonrandom organization of the chromosomes in the nucleus; the chromocenter sticks to the nuclear membrane, while the telomeres congregate around the opposite pole.

When Crick originally proposed his new model of polytene chromosome organization in 1971, he boldly pointed out that the genes might be located in the interbands rather than in the bands, where extensive coiling of the chromatin fibers might prevent their transcription. This question of band versus interband transcription remains controversial. What seems certain, however, is that the average DNA content of a band (around 30 kb pairs) is about 20 times higher than the amount of DNA coding for a protein of average size; it is, therefore, no longer possible to accept Painter's classic idea that one band corresponds to one gene. Indeed, it has been shown that four genes, having similar functions (the production of heat shock proteins, which will be discussed later), are clustered at the same locus in an 11-kb DNA segment (the band corresponding to this locus has 28 kb of DNA) (Corces *et al.*, 1980b). This suggests that the bands might correspond not to single genes but to genetic functional units; thus, genes having the same function would be located in the same band (Craig and McCarthy, 1980). Another possibility that has often been suggested is that a structural gene occupies both an interband and the adjacent region of a band. Our difficulties in deciding whether the genes are localized in bands, interbands, or both are very apparent when one reads the critical review published in 1981 by Zhimulev *et al.* They think that polytene chromosomes have a dynamic, not a static, organization; interbands would be active in transcription, and when their activity ends, they would condense into the neighboring band, whose size would increase. The bands are supposedly fragments of temporarily or constantly inactive DNA, but there is still no proof that the genes are localized in the interbands. Hall *et al.* (1983) analyzed the correlation in *Drosophila* between transcription units, chromomeric units (bands), and complementation groups (genes). They found a good correlation, but with exceptions. Two large bands have several transcription units, but the major part of a large band has none. They concluded that some genes are in the bands and others in the interbands. In a recent study, Bossy *et al.* (1984), using a similar methodology (''chromosome walking'') analyzed a 315-

kb segment of *Drosophila* chromosomal DNA and found that the transcripts were associated with the chromomeres. If one recalls that many genes have been assigned to definite chromosome bands in *Drosophila* polytene chromosomes by *in situ* hybridization, it seems difficult to avoid the conclusion that the bulk of the *Drosophila* genes are located in the bands. However, one cannot exclude the fact that part of the genetic material extends into the interbands.

Another possible way to correlate active genes with bands or interbands might be the immunofluorescent study of the localization of RNA polymerase II, the enzyme required for gene transcription. Early work by Jamrich *et al.* (1977) led to the conclusion that this enzyme can be detected in both bands and interbands. But according to Krämer *et al.* (1980), Christensen *et al.* (1981), Sass (1982), and Sass and Bautz (1982), the main localization of RNA polymerase II is in the interbands, not in the condensed loci; this distribution fits well with Crick's (1971) model shown in Fig. 21. RNA polymerase II is always abundant in the Balbiani rings and in similar structures called "puffs," to which we shall soon return.

Several investigators have been interested in the localization of left-handed Z-DNA, which can be detected by immunofluorescence and is believed to play a role in the control of gene activity. Unfortunately, the few data at our disposal are confusing. Lemeunier *et al.* (1982) found that, in *Chironomus,* Z-DNA is restricted to two heterochromatic bands and one interband, all located in the area of the nuclear organizers and the centromere. For Arndt-Jovin *et al.* (1983), Z-DNA is in the bands, where it could play a role in the higher-order organization of the chromosome and modulate the functional status of the genes. But a paper by Hill and Stollar (1983) led to unexpected conclusions: fluorescent antibodies against Z-DNA do not stain unfixed *Drosophila* polytene chromosomes; after classic fixation with acetic acid, strong staining of either the bands or the interbands can be obtained, depending upon the length of fixation.

Beermann (1952) found that certain bands in *Chironomus* polytene chromosomes undergo marked swelling at certain stages of larval development; this swelling takes place in a programmed sequence during development. This local increase in the volume of individual bands is called "puffing" (Fig. 22), and can be induced in isolated salivary glands by the addition of the molting hormone ecdysone. Puffs can also be induced by a number of nonspecific agents, such as, sugars (Beermann, 1973), inhibitors of respiration (Leenders and Berendes, 1972), changes in the $Na^+/K^+$ ratio (Lezzi, 1966), cycloheximide, vitamin $B_6$, and heat. The addition of actinomycin D, cordycepin [which inhibits poly(A) synthesis], HU, or cortisone causes their regression. That they correspond to localized gene activation (increased transcription) is shown by autoradiography studies using [$^3$H]uridine in which only the puffs and nucleoli are labeled. Beermann (1971) showed that α-amanitin (the specific inhibitor of RNA polymerase II) inhibits RNA synthesis in the puffs, but not in the nucleoli; it also

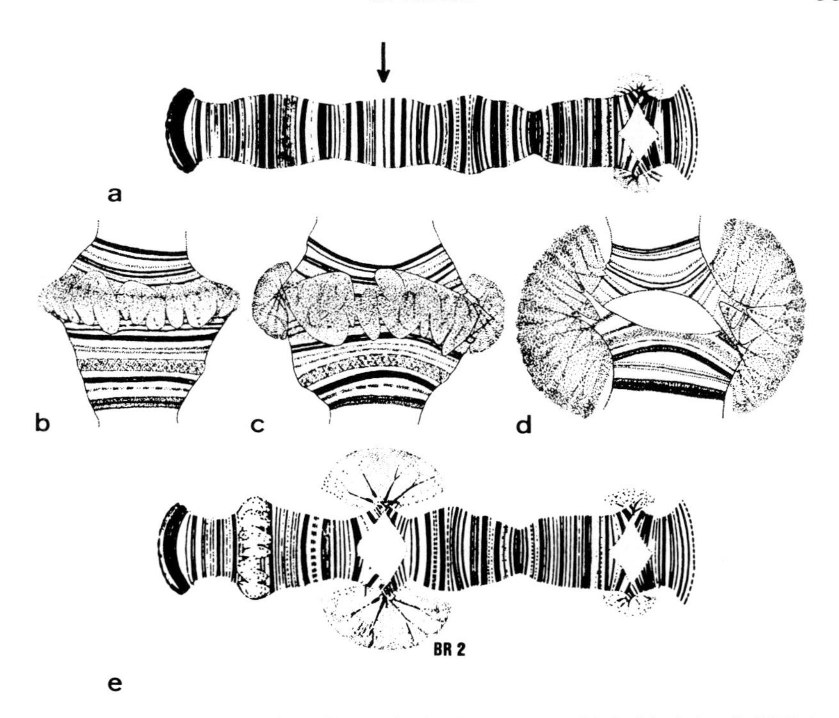


FIG. 22. Chromosome 4 in the salivary glands of *C. tentans* with Balbiani ring 2 (BR2) in an unpuffed (a) and a puffed (e) state. The position of the BR2 band is indicated by an arrow. In (a) BR2 is the largest puff in *Chironomus*. (b,c,d) These represent schematically intermediate stages in the puffing process. [Modified from Beermann (1952, 1973).]

inhibits puffing itself. The Balbiani ring, therefore, is nothing but a giant puff sitting on *Chironomus* chromosome II.

Puffs are chromosomal bands in which chromatin has undergone local decondensation, as shown schematically in Fig. 23. They are, therefore, equivalent to the loops in lampbrush chromosomes. Uncoiling of the chromatin fibers in a puff loosens the band structure, facilitates access to RNA polymerase II molecules, and allows mRNA synthesis.

It should be recalled that, in addition to these more common RNA puffs, DNA puffs can be found in some insect species (*Rhynchosciara, Sciara*) at given stages of larval development. Certain bands selectively replicate their DNA, as demonstrated by autoradiography after labeling with [$^3$H]thymidine (Ficq and Pavan, 1957) (Fig. 24). In such cases, we are probably witnessing selective gene amplification, as was shown by Glover *et al.* in 1982.

It appears that the first stage of puff induction is the accumulation, in a given band, of non-histone proteins (Berendes, 1968). In *Drosophila,* ecdysone treat-

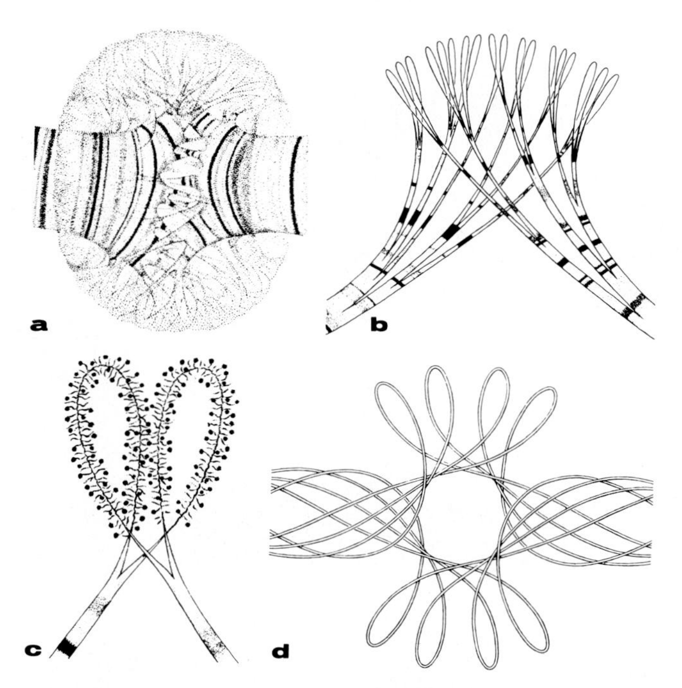


FIG. 23. Diagram of the structure of a large puff (BR2). (a) Balbiani ring 2 as seen under the light microscope. (b) At very high magnification under the light microscope, a few fibrils can be seen. (c) The electron microscope shows two puff fibrils with granules composed of mRNPs. (d) This schematic representation of the formation of a large puff shows fibrils untwisted and ''popped out'' of the cable-like structure. A giant chromosome contains thousands of fibrils, but they are tightly coiled when they are in the form of bands. [Beermann and Clever (1964).]

ment or an increase in temperature induces the penetration of non-histone proteins into the salivary gland nuclei; these proteins are concentrated in specific bands when puffing begins, and it is probable that they play an important role in loosening the compact structure of the band.

Our understanding of RNA puffs has been greatly increased by the work done in J. E. Edström's laboratory. Using delicate microdissection methods, Edström dissected the *Chironomus* chromosome III from nuclei isolated from larval salivary glands. Further micromanipulation allowed the isolation of one of the Balbiani rings, the giant puff BR2 (Edström and Daneholt, 1967; Fig. 25). This allowed Daneholt (1972) to show that the BR2 gene is transcribed into a giant RNA; although its sedimentation constant is as high as 75 S, it is transferred as such from the nucleus to the salivary gland cell cytoplasm. Case and Daneholt (1978) have pointed out that this 75 S RNA corresponds to a gene of 37 kb. Since the band adjacent to BR2 has only 5 kb, it follows that the major part of the BR2 gene is located in the 100-kb chromomere. A more recent estimate by Derksen *et al.* (1980) gives still higher values for the BR2 chromomere: it contains $5 \times 10^4$ pg of DNA, corresponding to 470 kb, and its compaction is 380 times. These authors think that, in the case of the giant BR2 puff, transcription might occur without unfolding of the chromomere into loops. Finally, the 75 S RNA synthesized by the BR2 locus can be translated *in vitro*. Its primary product is a very large protein ($M_r$ 850,000) related to larval saliva proteins (Rydlander *et al.*, 1980). Similar results have been obtained with *Drosophila hydei,* except that the mRNA produced by the gene coding for salivary proteins is smaller (45 S) than that found in *Chironomus* (75 S). However, a study by Hertner *et al.* (1983) reduces the giant size of BR2 to more modest dimensions. They found that the BR DNA genes occur mainly in the form of tandem repeats of 200–240 bp; there is also an 18,000-dalton repeat unit in the giant secretory protein.

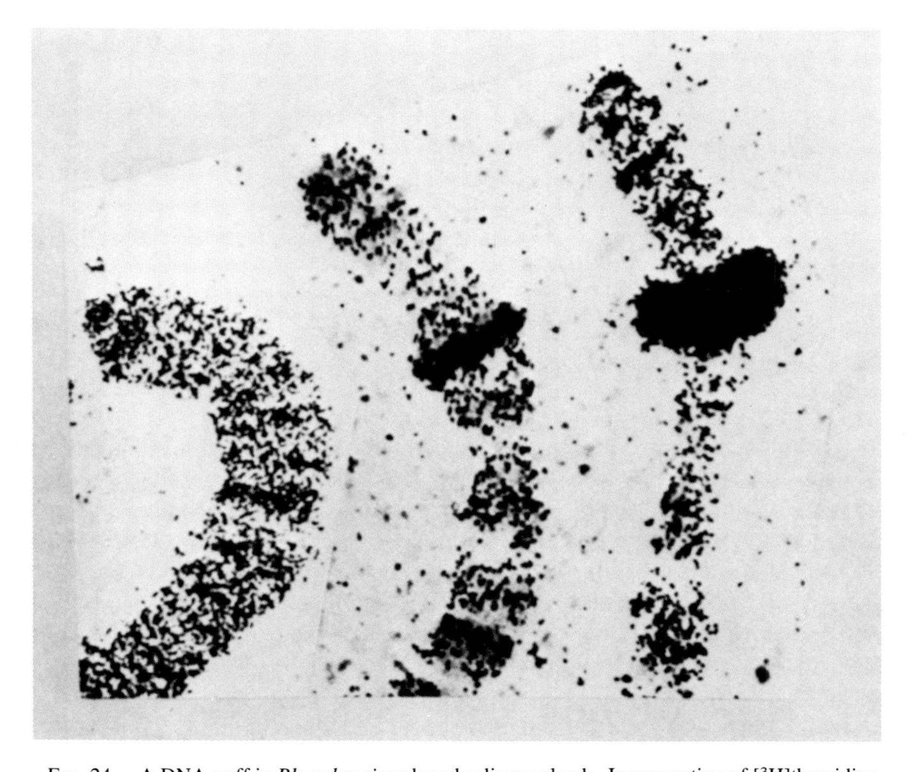

FIG. 24. A DNA-puff in *Rhynchosciara* larval salivary glands. Incorporation of [$^3$H]thymidine, followed by autoradiography, shows selective DNA amplification of a single band during larval development. [Ficq and Pavan (1957). Reprinted with permission from *Nature* **180,** 983. Copyright © 1957 Macmillan Journals Limited.]

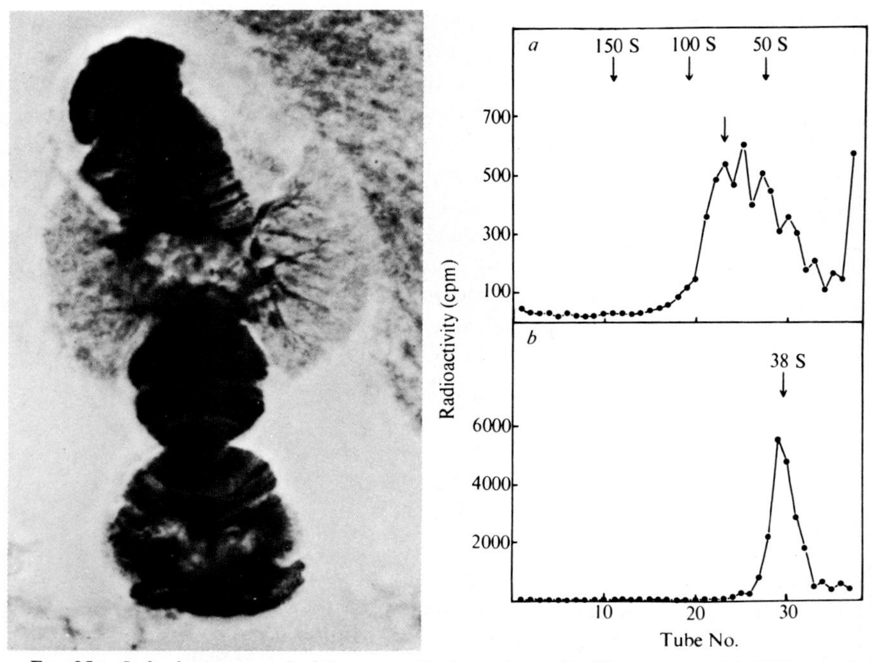


FIG. 25. *Left:* chromosome 4 of *C. tentans* displays a large, fan-like structure; it is BR2, seen in phase contrast after staining. *Right:* Balbiani rings and nucleoli were collected by micromanipulation from 30 [$^3$H]uridine-treated *Chironomus* larval salivary gland cells. The labeled RNAs have sedimented in sucrose gradients. (a) BR2 RNA has a sedimentation constant as high as 75 S. (b) The sedimentation constant of nucleolar RNA is only 38 S. [(a) Pelling (1978). (b) Daneholt (1972). Reprinted by permission from *Nature New Biology,* **240,** 231. Copyright © 1972 Macmillan Journals Limited.]

Electron microscopy has given some insight into the primary transcripts of the BR locus. According to Skoglund *et al.* (1983), the transcripts are packed in a series of structures of growing complexity: 10-nm fibers, 19-nm fibers, 26-nm fibers, and finally, 50-nm granules, which become rod-shaped when they move into the cytoplasm through the nuclear pores. An independent study by Olins *et al.* (1982), who used a new technique (EM tomography), also concludes that the BR transcription units are 45- to 50-nm granules; they are connected by short stems to the chromatin axes. The compaction factor, in the 2-μm-long transcription unit is about 8.

Finally, a few words are warranted about the heat shock proteins (reviewed by Ashburner and Bonner, 1979), discovered by Ritossa in 1964. When *Drosophila* larvae are heated at 37°C, nine new puffs appear in the salivary gland giant chromosomes. Simultaneously, the already existing puffs regress. Induction of heat shock puffs in *Drosophila* occurs rapidly. The new puffs are detectable after a 1-min heat treatment and increase in size after 30–40 min at 37°C. Appearance of these new puffs is also observed when stress is produced by agents other than excessive temperature, namely, by anaerobiosis and by inhibitors of energy production in the mitochondria (dinitrophenol, rotenone). In *Chironomus,* a heat shock at 33–40°C leads to the appearance of four to five new puffs in 5 min, which reach their maximum size in 30–60 min; the Balbiani rings are greatly reduced in size simultaneously (Lezzi *et al.*, 1981). Concomitant with the morphological appearance of the new puffs, heat treatment induces the synthesis of several RNAs and a limited number of heat shock proteins (six major bands, according to Tissières *et al.*, 1976). We have already mentioned that four heat shock genes are clustered in an 11- to 12-kb DNA segment located in "band 67B" (Corces *et al.*, 1980b; Craig and McCarthy, 1980); the possible significance of this finding has already been discussed. It is likely that non-histone chromosomal proteins play an important role in the control of heat shock protein synthesis (reviewed by Bienz, 1985). Dangli and Bautz (1983) have reported that heat shocks remove non-histone proteins from the inactivated loci and cause them to accumulate in the activated loci; the various heat shock puffs differ in protein composition, at least in *Drosophila*.

While heat shocks induce the synthesis of a small set of proteins, they repress the preexisting overall protein synthesis. Possibly correlated with protein synthesis inhibition is the dephosphorylation of one of the ribosomal proteins (called S6) (Glover, 1982).

The main *Drosophila* heat shock protein, hsp 70 ($M_r$ = 70,000), accumulates in the nuclei during a heat shock (or during anoxia); it moves into the cytoplasm after recovery from heating and migrates back into the nuclei during a second heat shock (Velazquez and Lindquist, 1984).

One should add that the heat shock response is a very widespread phenomenon. In *Tetrahymena,* heat shock, deciliation, and release from anoxia all induce the synthesis of the same set of 16 "stress proteins," one of which is associated with the scaffolding proteins of the nuclear matrix (Guttman *et al.*, 1980). In chicken and mammalian (human and hamster) cells, heat shock leads to the appearance of three new proteins. One of them (called "thermin") is associated with MTs and MFs. Although these proteins exist before heat shock, they are present in only very small amounts. Heat shocks have not activated inactive genes, but they have stimulated the activity of genes working at a low rate (C. Wang *et al.*, 1981). Particularly interesting are two systems in which the production of heat shock proteins seems to be somehow correlated with cell differentiation. Heat shock induces the synthesis of four major proteins in undifferentiated myoblasts but fails to do so in differentiated myotubes (Atkinson, 1981). In sea urchin embryos, heat shock induces the synthesis of several proteins and decreases overall protein synthesis, but this does not occur at stages earlier than the blastula (Roccheri *et al.*, 1981). It is too early to draw conclusions from these studies, but heat shock protein synthesis seems to be a promising field for the analysis of gene expression.

*Drosophila* is an exceptional biological material. It is the only organism for which we have a very extensive knowledge of genome organization, due to the combined approaches of formal genetics and cytological technology. Because of the large size of the bands and their identification as specific genetic loci, salivary gland chromosomes are an ideal material for the detection, by *in situ* molecular hybridization, of individual genes. One example is the detection by Pardue *et al.* (1977) of 100 copies of histone genes extending on 12 bands of chromosome 2L.

## D. The Mitotic Apparatus: Spindle and Asters (Achromatic Apparatus)

It is something of a paradox that in many cells the so-called achromatic apparatus (spindle and asters) is more basophilic than the rest of the cell. As shown by the ribonuclease test, this is due entirely to RNA (Brachet, 1942). Electron microscopy shows that the spindle and asters are essentially composed of MTs and that the basophilic material detected in the spindles of certain cells corresponds to ribosomes trapped between arrays of MTs (Fig. 21, Chapter 3). One should distinguish between achromatic apparatus and mitotic apparatus (Mazia and Dan, 1952). Chromosomes are a constituent of the mitotic apparatus. Since, in a dividing cell, the chromosomes are tightly bound to spindle fibers and since, as we shall see, they play an important role in spindle MT assembly, they are essential constituents of the whole system. The term "mitotic apparatus" is, therefore, preferred to the old-fashioned achromatic apparatus.

The mitotic apparatus and the mechanisms of its assembly have been the subject of several reviews since Mazia's (1961) now classic treatment of the subject. Among the most complete and recent reviews on the subject, are those of Sakai (1978), Raff (1979), Rebhun (1977), Peterson and Berns (1980), Petzelt (1979), and Berns *et al.* (1981).

As was previously mentioned, centrioles (see Chapter 3) are not always present at the opposite poles of the mitotic apparatus. They are absent in plant cells and in the meiotic spindles of eggs. Debec *et al.* (1982) have reported that a *Drosophila* cell line lacks centrioles; nevertheless, the cells divide in a normal way. Conversely, there are numerous centrioles in a mouse neuroblastoma cell line; they continue to organize MTs during mitosis. Nevertheless, cell division is normal and bipolar (Ring *et al.*, 1982). All of this shows that two centrioles are not necessarily required for the establishment of bipolar cell division. However, a "cloud" of osmiophilic granules and fibrous material is always present at the poles. Despite the fact that this cloud may be devoid of centrioles or diplosomes, we shall call it the "pericentriolar cloud," since this is the accepted name; mysterious virus-like particles have been observed in this pericentriolar cloud by Gould and Borisy (1977). Pericentriolar structures of fibrillar nature (called "satellites" or "centriolar appendages") are found only around the parent centriole; this suggests that the daughter centriole does not function as an MT-organizing center before the following centriolar cycle. These so-called satellites are thus simply condensed forms of the pericentriolar cloud (Rieder and Borisy, 1982). The latter can exist, disperse, and reorganize independently of the centriole during the first cleavages of mouse eggs (Calarco-Gillam *et al.*, 1983).

The centriole and the pericentriolar cloud surrounding it form a centrosome. As pointed out by Mazia (1984), centrioles have the same structure in all cells. In contrast, centrosomes have a flexible structure. They are not corpuscles like the centrioles; instead they can change their shape during cell division. It is likely that the centrioles help in the organization and distribution of the pericentriolar clouds that are necessary for the formation of the spindle poles (Keryer *et al.*, 1984).

We know little about the chemical composition of the centrioles and the pericentriolar cloud. Kuriyama (1984) recently succeeded in isolating centrosomes from chinese hamster ovary cells. If tubulin is added to isolated centrosomes, from 200 to 250 microtubules polymerize around each centriole. This polymerization is suppressed by proteolytic digestion, while DNase and RNase have no inactivating effects. Vandre *et al.* (1984) found that all MTOCs (including the centrosomes) contain phosphoproteins that are also present in the cytoplasm of mitotic cells but not in that of interphase cells.

Elegant experiments by Karsenti *et al.* (1984) have shed a new light on the role of the centrosomes in microtubule assembly. They worked on nucleate and anucleate fragments of cells (karyoplasts and cytoplasts, respectively) (see Chapter 2, Volume 2) treated with nocodazole, a powerful microtubule inhibitor. Microtubular arrays are reformed only in the small, centrosome-containing karyoplasts. Their number increases with the density of the culture. In cytoplasts devoid of centrosomes, microtubules are formed at confluence, but their distribution is abnormal: almost all of them are located at the cell periphery. The conclusion of these experiments was that centrosomes stabilize the microtubules.

Mitchison and Kirschner (1984a) recently analyzed the nucleation of microtubule assembly by isolated centrosomes and found unusual dynamic properties: the number of microtubules assembled by a centrosome increases with the tubulin concentration until it reaches a plateau value. At certain tubulin concentrations some of the microtubules detach from the centrosome, but the others continue to grow; microtubule loss results from depolymerization at the ends. Shrinking and growing microtubules are not directly interconvertible, and it is likely that free microtubules are unstable in the cell. The pericentriolar material has a structural autonomy independent of the centrioles, which represent only about 5% of the centrosomes.

In a second paper, Mitchison and Kirschner (1984b) emphasize the importance of this dynamic instability of the microtubules in both interphase and dividing cells. Growth of certain microtubules, while others are lost by depolymerization, can be demonstrated, even in the absence of nucleating centrosomes. It is suggested that microtubules lacking a GTP cap and having instead a GDP end are unstable. Only the GTP-capped lattice would be stable, and GTP hydrolysis would lead to instability. This model, which is in opposition to the treadmilling model discussed in Chapter 3, has the advantage of explaining how microtubules can grow or shrink quickly during cell locomotion or mitosis. The mitotic spindle is a much more dynamic structure than was previously believed: exchanges between microtubules and soluble subunits may take place with a few seconds as shown by experiments where fluorescent tubulin was injected into sea urchin eggs (Salmon *et al.*, 1985) or somatic cells (Saxton *et al.*, 1985): incorporation was much faster in the mitotic apparatus than in cytoskeletal MTs. Whether incorporation of tubulin takes place all along the MT tubules or at their distal end remains controversial (Soltys and Borisy, 1985).

The aster and spindle MTs, as shown in Fig. 27, have the same ultrastructure as the cytoskeletal MTs described in Chapter 3. However, in the spindle they are organized in almost parallel bundles; some MTs extend through the spindle from one pole to the other, while others link the kinetochores of the chromosomes to the polar regions. According to Borisy (1978), the growth of the MTs is unidirectional. Microtubules originating from the centriolar region and the kinetochores are disposed in an antiparallel direction. The almost parallel disposition of the MTs results in strong spindle birefringence (Inoué, 1953).

Surrounding the mitotic spindle in both starfish eggs (Harris, 1975) and HeLa cells (Paweletz and Finze, 1981) are numerous vesicles. They are related to the endoplasmic reticulum cisternae and are believed to sequester calcium ions. Indeed, many fluorescent dots are seen around the poles of the mitotic apparatus in HeLa cells and sea urchin eggs when free $Ca^{2+}$ is detected with chlorotetracycline (Schatten and Schatten, 1980). Aubin *et al.* (1980) have shown that the mitotic apparatus in somatic cells is surrounded by a cage of intermediate filaments composed of vimentin. This cage, which forms when the cells enter mitosis and presumably determines the orientation of the mitotic apparatus, has been isolated by Zieve *et al.* (1980). These alterations in the organization of the intermediate filaments seem to be linked to changes in the phosphorylation of vimentin (Evans and Fink, 1982) and three other "cytoarchitectural" proteins (Bravo *et al.*, 1982).

As one can see, the entire cytoskeleton undergoes extensive change during mitosis; this change is particularly striking in regard to the cytoplasmic MTs, which disappear during mitosis (Brinkley *et al.*, 1975), and the intermediate filaments (Franke *et al.*, 1982). As we shall see, thick bundles of cortical MFs accumulate in the cortex, where they play a role in cytokinesis (furrow formation). At metaphase the endoplasmic reticulum (Porter, 1954) and the acid phosphatase–containing (GERL) network (see Chapter 3) break down (Garvin *et al.*, 1981), and the usually active displacements of the mitochondria are brought to a standstill; some of the mitochondria may even break down into a string of heads

(Chèvremont and Frédéric, 1952; Frédéric, 1954). When mitosis is induced in hepatocytes by partial hepatectomy, the lysosomes migrate toward the nucleus at prophase; at metaphase, they concentrate in the polar regions; at telophase, they form perinuclear clusters. These changes in the localization of the lysosomes do not occur if the cells have been treated with colchicine, suggesting that their movement is a consequence of rearrangements in the MT network (Mori *et al.*, 1982). It has also been shown that within a few minutes, the Golgi complex is dispersed in small vacuoles in dividing cells; these vacuoles reassociate and reform dictyosomes at telophase (Burke *et al.*, 1982). This brief description shows that the organization of the entire cytoplasm changes dramatically in dividing cells.

There is continuing interest in the behavior of the centrioles during the cell cycle. Kuriyama and Borisy (1981a,b) and Kuriyama (1982) have described a centriole cycle. In the $G_1$ phase, the mitotic centrioles undergo disorientation and nucleation of a daughter centriole; in $G_2$, elongation of the daughter centriole takes place. During mitosis, diplosomes are positioned at right angles. The centriole cycle is independent of DNA synthesis. In cytoplasts, elongation and disorientation occur normally, but there is no nucleation; a procentriole does not form in the absence of the nucleus. Cytoplasts may contain free centrioles, which can nucleate MTs under *in vitro* conditions (in a tubulin-polymerizing medium). The *in vitro* nucleating activity of centrioles increases five times during the $G_2$–M transition (Kuriyama and Borisy, 1981a,b). The behavior of the centrioles during the cell cycle has also been described by Vorobjev and Chentsov (1982), who pointed out that, in $G_0$ cells, mother centrioles often form a cilium.

The relationship between the centriole cycle and nuclear DNA synthesis is not yet clear. We have just seen that Kuriyama and Borisy found that the centriole cycle is independent from DNA synthesis, but experiments where cells were treated with either mitogens or inhibitors of mitotic activity have led Sherline and Mascardo (1984) to the conclusion that there is a close correlation between centrosomal separation and DNA synthesis.

A major breakthrough occurred when, in 1952, Mazia and Dan isolated the entire mitotic apparatus of cleaving sea urchin eggs. These preparations are presented in Fig. 26. Mazia (1954a,b, 1955) found that the mitotic apparatus represents as much as 12% of the total proteins of the sea urchin egg and contains a major protein. The astral rays and spindle fibers (which retain their birefringence after isolation of the mitotic apparatus) result from the linear polymerization of this protein, which was later identified as tubulin. Treatment of the isolated mitotic apparatus with reducing agents, for example, mercaptoethanol, leads to its breakdown, which shows that —SS— bonds are essential in retaining the integrity of the structure. Interestingly, Mazia (1958) found that the treatment of dividing sea urchin eggs with mercaptoethanol abolishes mitotic apparatus birefringence; however, mercaptoethanol does not inhibit centriole division. Therefore, if fertilized eggs are treated with mercaptoethanol and then released

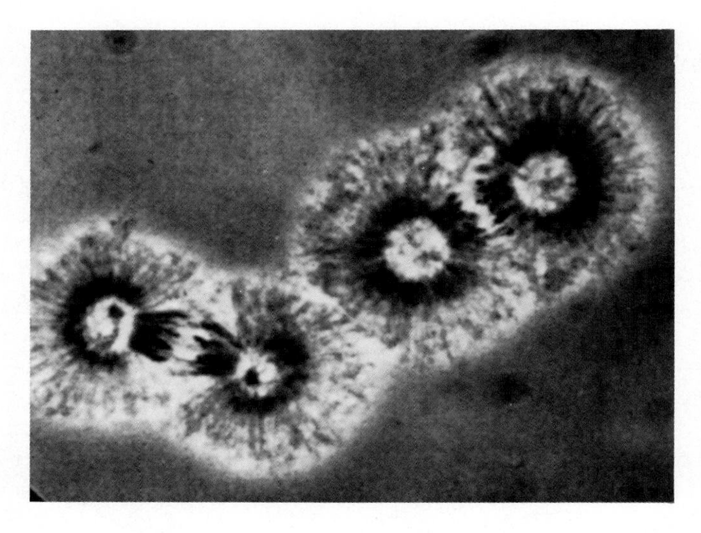

FIG. 26. Two mitotic apparatuses isolated from sea urchin eggs. Asters and anaphase chromosomes are very conspicuous; they are surrounded by a thin layer of cytoplasmic amorphous material. [Courtesy of Prof. D. Mazia.]

from mitotic arrest by simple washing, they cleave directly into four cells instead of two (Fig. 27). These experiments clearly show that mitotic apparatus assembly and centriole duplication are independent events. Much more recently, Mazia *et al.* (1981), using *in vivo* mercaptoethanol treatment at the correct stage, obtained unipolar mitotic apparatuses. One pole produced a half-spindle where one kinetochore faced the pole, while its sister kinetochore faced away from the pole.

The effects of other agents known to influence tubulin polymerization *in vivo* and *in vitro* have been extensively studied by Mazia and his group [see Mazia (1961) for details]. For instance, urea or formamide, which break down hydrogen bonds, separate the spindle fibers from each other; this indicates that these fibers were apparently held together by hydrogen bonds both *in vivo* and *in vitro*. However, cross-linking via a MAP now seems a more likely possibility. Mazia further showed that the mitotic apparatus of a colchicine-treated sea urchin egg may be isolated as an amorphous gel (Fig. 28) and concluded (with great insight)

FIG. 27. Mitotic figures isolated at successive times after blocking sea urchin eggs at metaphase by means of mercaptoethanol. The sequence of stages in the generation of a four-polar figure from a two-polar figure is shown in (a) through (d). (e) The first mitosis following four-way division after blockage with mercaptoethanol. In this egg, one of the furrows failed, so that the upper cell received two centers, while the lower two received one each. The latter form mitotic figures with only one pole. These cells cannot divide but will return to interphase, undergo another cycle of replication of centriolar units, and reenter mitosis with two poles. [Mazia (1961).]

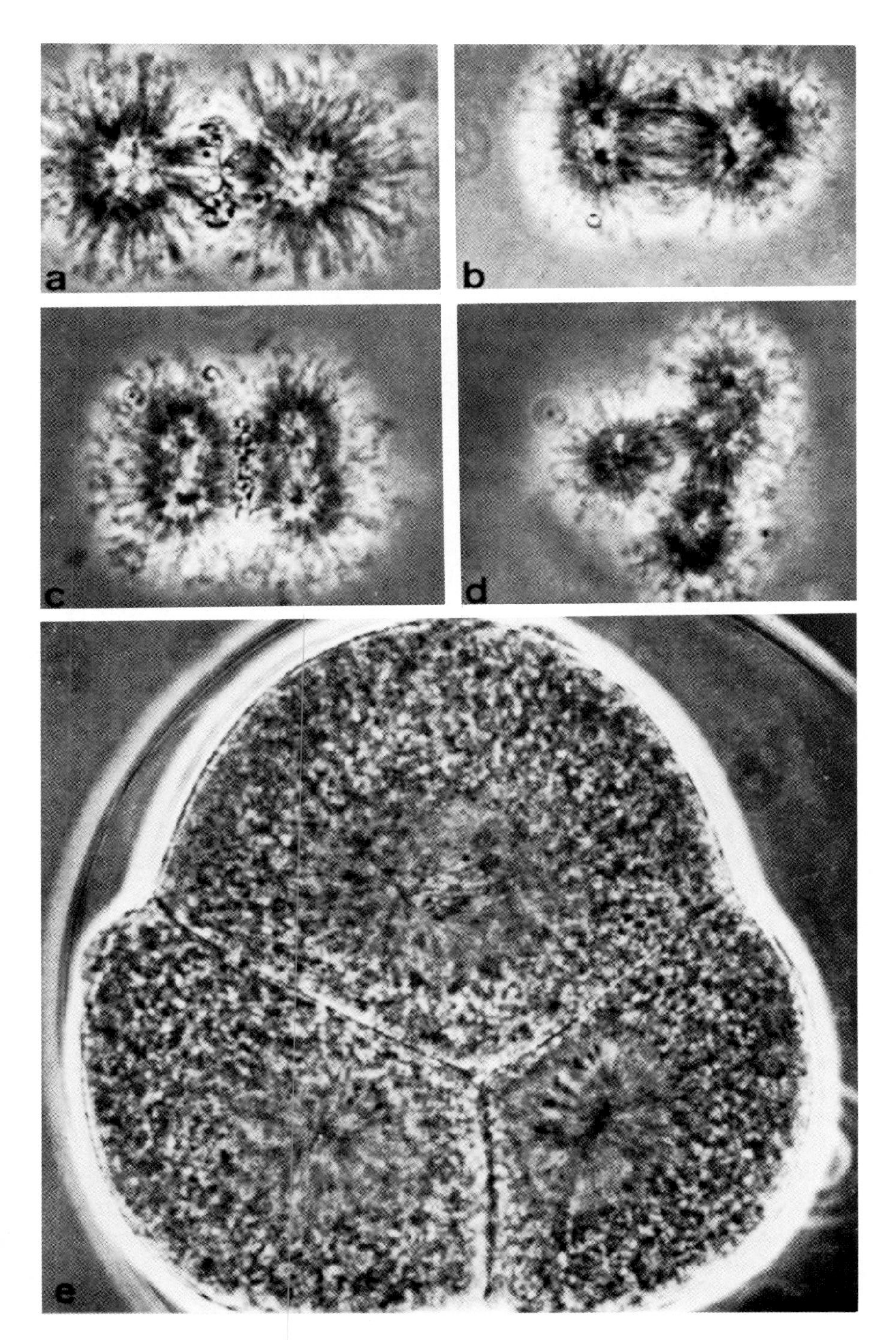
a
b
c
d
e

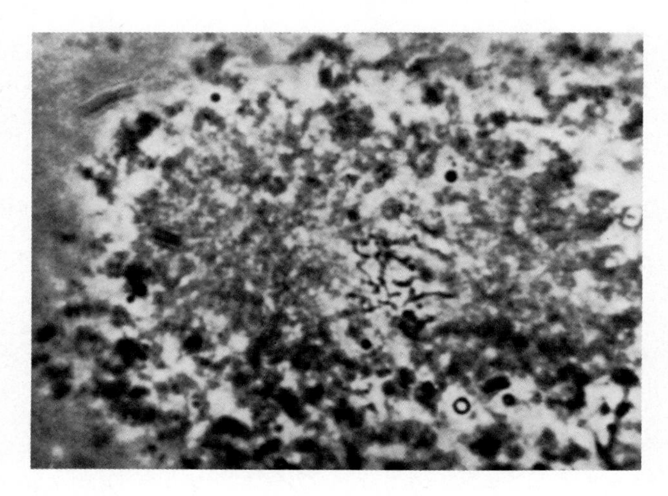

FIG. 28. Mitotic apparatus isolated from a colchicine-arrested sea urchin egg. An amorphous gel surrounds the chromosomes. [Courtesy of Prof. D. Mazia.]

that the chromosomes (probably the kinetochores) as well as the centrioles contributed to the organization of the microscopically visible spindle (Mazia, 1955, 1956a). Since the time of this pioneer work, a number of improved methods for isolating the mitotic apparatus from sea urchin eggs as well as other cells have been devised. These methods rely on the use of agents that, like heavy water ($D_2O$) or glycerol, increase MT stability.

So many papers have been devoted to the chemical composition of the mitotic apparatus, using either the isolated organelle or immunofluorescence methods, that only a brief account will be given here; details will be found in the reviews by Sakai (1978), Raff (1979), and Rebhun (1977).

It has been demonstrated that tubulin is always a major constituent of asters and spindle (Fuller *et al.*, 1975); in sea urchin eggs, tubulin present in the mitotic apparatus (which is quite different from bovine brain tubulin, according to Keller *et al.*, 1982) is not neosynthesized, but is assembled from a preexisting cytoplasmic pool (Bibring and Baxandall, 1977). In other cells, it is likely that a similar situation prevails since, as we have seen, the cytoplasmic MTs disappear at mitosis (Brinkley *et al.*, 1975). However, in tissue culture cells, tubulin increases in S and reaches a maximum in $G_2$, just prior to mitotic apparatus assembly (Forrest and Klevecz, 1978). If brain tubulin is added to an isolated mitotic apparatus or a meiotic spindle, it is incorporated into the spindle fibers (Rebhun, 1977; Sakai *et al.*, 1975). As a result, both the size and birefringence of the mitotic apparatus increase. This shows that the major event at mitosis, in eggs at any rate, is spindle MT assembly, and not tubulin synthesis.

Although tubulin is a constant and necessary constituent of all mitotic apparatuses, it is by no means the only one. In mitotic apparatuses isolated from sea urchin eggs, Sakai *et al.* (1977) found that more than 16 proteins were associated with tubulin; in those isolated from HeLa cells, tubulin represented only 25% of the total proteins (Chu and Sisken, 1977). Among the proteins present in the spindle and asters are, as one might expect, the MAPs, which were discussed in Chapter 3. Connolly *et al.* (1977) showed that the mitotic spindle is stained by a fluorescent serum directed against the tau proteins; Sherline and Schiavone (1978) showed that the MAPs have the same distribution as tubulin in the mitotic apparatus. However, MAP-2 is not detectable in the spindle and seems to be specific for brain MTs (Izant and McIntosh, 1980). In HeLa cells, where the MAPs differ in $M_r$ from those of the brain, two MAPs ($M_r$ 210,000 and 125,000) are associated with the MTs during both interphase and mitosis (Bulinski and Borisy, 1980). A 200,000-dalton MAP from HeLa cells is associated with the spindle during mitosis and with the nucleus during interphase (Izant *et al.*, 1982). The mitotic spindle of mammalian cells has been reported to possess, in addition to MAPs, a 150,000-dalton mitosis-specific protein of still unknown function (Zieve and Solomon, 1982). A spindle-specific antigen has been detected in HeLa cells by Izant *et al.* (1983); its chemical identity is not known, but it seems to play an important role, since microinjection into a mitotic cell of antibodies against this antigen disorganizes the spindle.

We have seen in Chapter 3 that tubulin polymerization is inhibited by low concentrations of $Ca^{2+}$. It is likely that the free $Ca^{2+}$ concentration plays a very important role in MT assembly *in vivo*. Kiehart (1981) made the interesting observation that microinjection of $CaCl_2$ into cleaving sea urchin eggs locally abolishes spindle birefringence; the free $Ca^{2+}$ concentration would "orchestrate" the assembly and disassembly of the spindle. It has already been mentioned that the vesicles that surround the mitotic apparatus (Harris, 1975; Paweletz and Finze, 1981) in sea urchin eggs and in HeLa cells are believed to sequester free calcium ions; these vesicles might thus play an important role in the control, via free $Ca^{2+}$, of mitotic apparatus assembly and disassembly. Injection of $CaCl_2$ into somatic cells decreases spindle birefringence, as it does in sea urchin eggs, but this does not affect anaphase and cytokinesis. Therefore, an increase in free $Ca^{2+}$ does not directly induce the separation of the kinetochores, but a spontaneous increase in free $Ca^{2+}$ at metaphase might stimulate the onset of anaphase (Izant, 1983).

One would also expect calmodulin to be one of the proteins present in the spindle and asters. Welsh *et al.* (1978), using immunofluorescence, were the first to show the presence of calmodulin in mitotic spindles. Its distribution is not the same as that of tubulin, which has led to the conclusion that, in the presence of $Ca^{2+}$, calmodulin might play a role in the anaphase movement of the chromosomes. Marcum *et al.* (1978) noted that calmodulin is mainly localized in the chromosomes and the centrioles. It inhibits *in vitro* MT assembly in the presence of 10 $\mu M$, but not of 1 $\mu M$, $Ca^{2+}$, which suggests that it might be the endogenous regulator of spindle disassembly. In a more recent paper, Welsh *et al.* (1979) reported that calmodulin is mainly associated with the kinetochore-pole fibers. At the end of metaphase, it moves toward the poles, in contrast to tubulin. At telophase, it accumulates in the midbody that links the two daughter cells. Finally, according to De Mey *et al.* (1980), antibodies against calmodulin stain the mitotic, but not the interphase, MTs. The detectable amount of calmodulin quickly increases at prometaphase at the time when spindle fibers first appear. At this stage, calmodulin displays a gradient distribution in the direction of the poles (Fig. 29). Electron microscopy shows that, at anaphase, it is present selectively in the kinetochore-pole fibers and is absent from the pole-to-pole fibers. It is abundant in the pericentriolar cloud but seems to be absent in the centriole itself. Later it is accumulated in the midbody, as previously shown by Welsh *et al.* (1979). De Mey *et al.* (1980) conclude, from their observations, that calmodulin is somehow involved in the migration of the chromosomes toward the poles and in cytokinesis. Similar observations on the distribution of calmodulin in dividing cells have been made by Willingham *et al.* (1983). Zavortink *et al.* (1983) injected a fluorescent derivative of calmodulin into the cytoplasm of interphase and dividing cells. It is incorporated into the mitotic spindles and remains diffuse in interphase cells. Rather unexpectedly, injection of calmodulin had no effect on the progression of mitosis.

If the presence of tubulin, MAPs, and calmodulin in mitotic apparatus fibers seems to be established beyond any doubt, the situation regarding contractile proteins (which might theoretically pull the chromosomes toward the poles at anaphase) is less clear. According to Sanger (1975), actin is present in the kinetochores, centrioles, and spindle, but not in the asters. The presence of actin in the mitotic spindle has been confirmed by Cande *et al.* (1977) and by Schloss *et al.* (1977). It is said to be localized in the poles, the kinetochores, and the fibers that link them together. But a paper by Barak *et al.* (1981) throws considerable doubt on these earlier results. According to these authors, there are only very small amounts of F-actin (the polymerized form of actin) in the pole-kinetochore region, but F-actin is present around the spindle, where it might play a role in chromosome movement. Finally, it has been reported that, in sea urchin eggs, tropomyosin, like actin, is associated with the mitotic apparatus and with the equatorial cortex region where the furrow will form (Ishimoda-Takagi, 1979).

In recent years, there has been great interest in the mechanisms of *in vivo* MT assembly in dividing cells (reviewed by Raff, 1979; Berns *et al.*, 1981). Mazia's idea that both centrioles and kinetochores are centers of mitotic MT assembly has received considerable experimental support. For instance, McGill and Brinkley (1975) showed that both centromeres and centrioles act as nucleation centers for

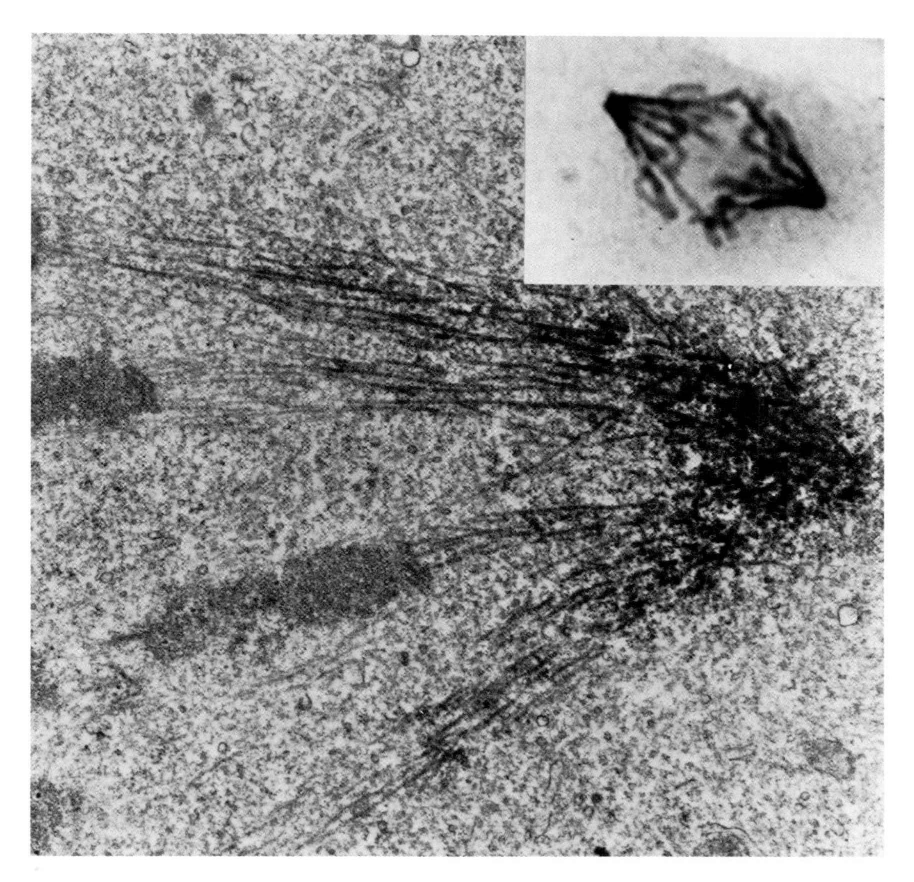

FIG. 29. Distribution of calmodulin in the mitotic apparatus of a dividing somatic cell; immunocytochemical detection. [Courtesy of Dr. J. De Mey.]

brain tubulin; at the same time, Telzer *et al.* (1975), using electron microscopy, elegantly demonstrated that isolated chromosomes polymerize brain tubulin at the level of their kinetochores (Fig. 30). More recently, Gould and Borisy (1978) showed that 98% of isolated chromosomes form about eight MT, starting from their kinetochores. There is convincing evidence (Wiche *et al.*, 1978; Corces *et al.*, 1978, 1980a; Villasante *et al.*, 1981; Avila *et al.*, 1983) that MAPs, but not tubulin itself, bind to the highly repeated DNA sequences (satellite DNA) present in the kinetochores. But it is likely that the proteins associated with kinetochore DNA also play a role in the organization of the mitotic spindle. We have seen that the kinetochores contain at least four different antigens; interestingly, the serum raised against these antigens inhibits specifically the

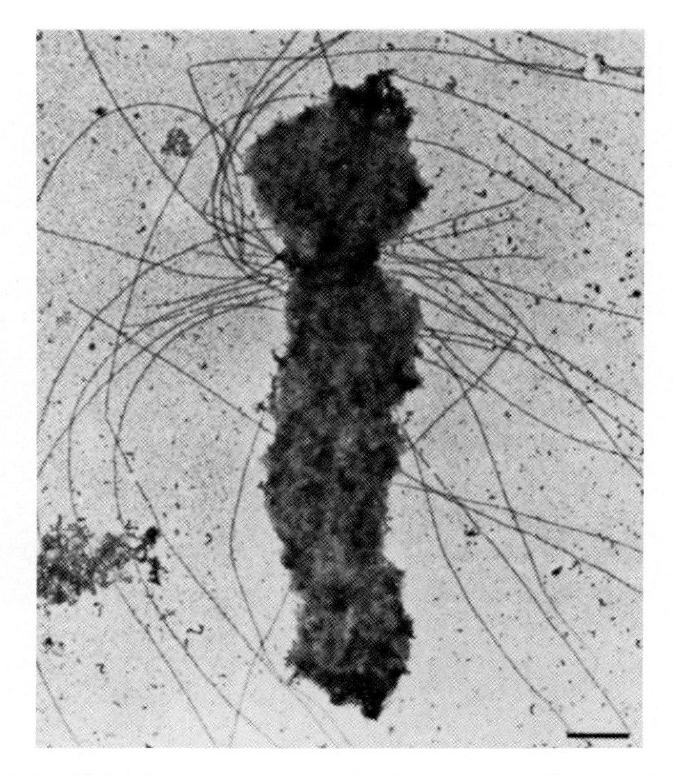

FIG. 30. Isolated HeLa chromosomes have been incubated in brain tubulin; electron microscopy shows that polymerization of the tubulin dimers and formation of MTs has taken place at the kinetochore. [Telzer *et al.* (1975).]

organization of MTs at the kinetochores in lysed cells (Cox *et al.*, 1983; Hyams, 1984).

It can be concluded from all of these experiments that high molecular weight MAPs bind to repetitive sequences present in the pericentromeric zone of metaphase chromosomes, which allows binding of the MTs to the kinetochores. On the other hand, there is some disagreement about the direction in which the kinetochore-originated MTs grow, since MTs are clearly polarized and might grow from either their basal (i.e., from the kinetochore) or their distal end. Summers and Kirschner (1979) believe that growth of the MTs would be three times faster at their distal ends (thus, distal to the kinetochore). However, Euteneuer and McIntosh (1981), using a new and reliable method for detecting the direction of MT growth, have come to the opposite conclusion, namely, that growth of kinetochore MTs is initiated in such a way that the fast-growing end would be proximal to the kinetochore.

According to classic cytology textbooks, MTs arise around the centrioles (although this is not necessarily the case in eggs, as we shall see later). However, the work of Berns and his Colleagues casts strong doubts on the role played by the centriole as an MT-organizing center (MTOC). This skepticism is based on elegant experiments in which a centriole was selectively destroyed by laser microbeam irradiation (Berns *et al.*, 1977, 1981; Brenner *et al.*, 1977). These experiments have led to the conclusion that it is the pericentriolar cloud, and not the centriole itself, that is important in MT assembly. In fact, according to Berns *et al.* (1977), irradiation of the centrioles in prophase does not prevent mitosis. Telzer and Rosenbaum (1979) isolated centrioles and the surrounding clouds from lysed, metaphase-arrested HeLa cells and found that, in the presence of tubulin, MT assembly occurs not only on the pericentriolar cloud but also (although to a lesser extent) on the centriole itself. More recently, Kuriyama (1984) added brain tubulin to isolated centrioles and observed the polymerization of MTs around them; isolated centrioles surrounded by MTs strikingly resemble asters. Treatment of the isolated centrioles by DNase or RNase had no effect on MT formation; however, proteases inactivated them, suggesting that some centriolar proteins play an essential role in the induction of aster fibers.

We know little about the chemical composition of the MTOCs. They contain no tubulin, according to De Brabander *et al.* (1979), who were also unable to detect this protein in interphase nuclei, chromosomes, kinetochores, and pericentriolar material. There is some evidence that RNA-containing structures might be involved in MTOC activity (although RNA itself, in *in vitro* experiments, inhibits brain tubulin polymerization). According to Rieder (1979a,b), both centrioles and kinetochores give positive cytochemical reactions for RNA at the EM level. More suggestive, perhaps, are the experiments by Peterson and Berns (1978). They treated cells with psoralens, which are substances that bind to nucleic acids, making them sensitive to light, and then laser-irradiated these cells. Sensitivity to light occurred in cells treated with an RNA-specific psoralen; no effect was seen after treatment with a DNA-specific psoralen. Destruction of the RNA-containing structure present in the centriolar cloud could account for these results, which agree with the results of older experiments by Brachet and Ledoux (1955). They treated frog eggs with RNase and observed definite abnormalities of the mitotic apparatus (in particular, fading of the spindle and asters) in the cells that were dividing when the enzyme penetrated the eggs. However, the great majority of the cells were blocked in interphase and had enormously swollen nuclei (Fig. 31), suggesting that the primary action of RNase might be on the formation or reduplication of the centrosomes. The effects of RNase on fibroblasts were also studied at about the same time, by Chèvremont and Chèvremont-Comhaire (1955) and by Chèvremont *et al.* (1956). They found that RNase does not inhibit DNA replication, but arrests the cells in $G_2$. Again, an effect of the enzyme on the centrioles or the pericentriolar cloud is a good possibility. It

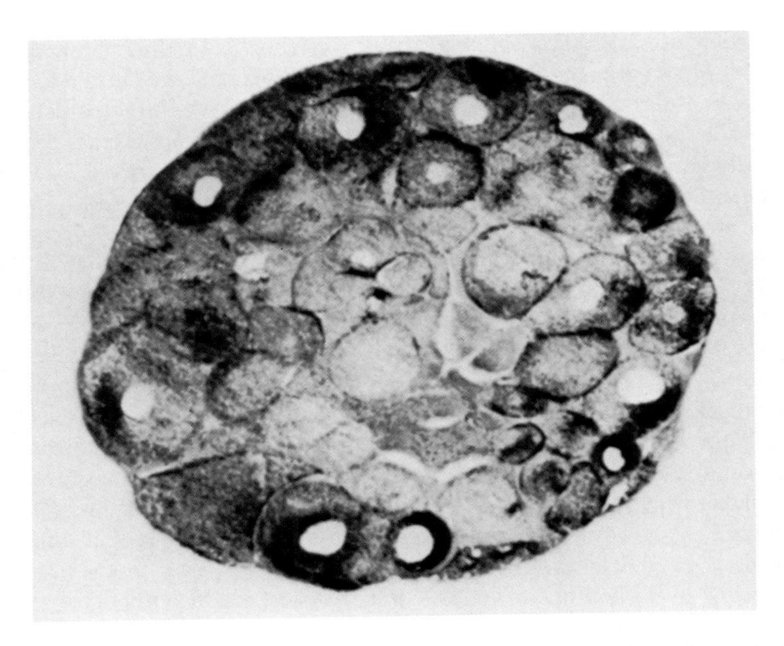

FIG. 31. Swollen interphase nuclei in a frog morula treated *in vivo* with RNase. Arrest of mitosis might be due to inhibition of the centriole-organizing activity. [Brachet and Ledoux (1955).]

might be worthwhile to repeat these older experiments, but at the ultrastructural level. We have already seen that the pericentriolar cloud contains high concentrations of calmodulin (De Mey *et al.*, 1980), and Petzelt (1980) has shown that a $Ca^{2+}$-activated ATPase is localized in the centriolar region and around the mitotic apparatus of both sea urchin eggs and tissue cultured cells. There is thus circumstantial evidence for the view that RNA and calcium ion fluxes are involved in MTOC activity.

How MTs originating from the centriolar region and from the kinetochores come together in the spindle is an interesting question. It has been the subject of a paper by Rieder and Borisy (1981). As is already known, MTs undergo disassembly at low temperatures (ca.5°C); this is true of both mitotic and cytoplasmic MTs. Rieder and Borisy (1981) have studied the re-formation, at 18°C, of MTs disassembled in intact cells at 5°C. They found that the centrosome-nucleated MTs interact either directly with the kinetochores or with kinetochore-nucleated MTs.

A very important question for our understanding of the dynamics of mitosis is the molecular mechanism of chromosome movement at anaphase. Sakai *et al.* (1976) found that the migration of chromosomes toward the poles can be induced in isolated mitotic apparatuses by the addition of ATP. This process is inhibited by colchicine and by an antiserum against dynein, the $Mg^{2+}$-ATPase. On the other hand, chromosome migration is not inhibited by an antimyosin serum (Kiehart *et al.*, 1982). These experiments suggest that chromosome migration is not due to the contraction of actomyosin fibers linking poles to kinetochores, but rather (like ciliary bending and flagellar movement) to a dynein-induced sliding of adjacent MTs or to $Ca^{2+}$-dependent depolymerization of the spindle MTs. That dynein is indeed involved in chromosome movements at anaphase is confirmed by the fact that vanadate inhibits both chromosome migration and dynein activity (Cande and Wolniak, 1978). According to Pratt *et al.* (1980), isolated mitotic apparatuses of sea urchin eggs possess both a dynein-like $Mg^{2+}$-ATPase and a $Ca^{2+}$-ATPase [found earlier in sea urchin eggs by Petzelt and von Ledebur-Villiger (1973), by Petzelt and Auel (1977), and reviewed by Pratt (1984)]. The dynein-like enzyme is sensitive to vanadate, but not to ouabain. It has also been shown that there is no species specificity in the interactions between dynein and the mitotic apparatus. Isolated meiotic spindles from eggs of the mollusk *Spisula* bind dynein isolated from the axonemes of the ciliate *Tetrahymena* flagella (Telzer and Haimo, 1981). Simultaneously, the ATPase activity of the isolated spindles increases seven times. However, it has not been possible to detect dynein from mammalian spermatozoa flagella in the mitotic apparatus of fibroblasts (Zieve and McIntosh, 1981) using immunofluorescence. It thus seems that dynein and the mitotic apparatus can bind together without species specificity in *in vitro* reconstitution experiments, but that there is a tissue specificity *in vivo*. Finally, in experiments by Cande *et al.* (1981), dividing cells made permeable to proteins were treated with a number of substances. It was found that chromosome movement is inhibited by tubulin, MAPs, and taxol (a chemical that, as we have seen, prevents MT depolymerization). None of these substances had any effect on cytokinesis, which is inhibited by cytochalasin B and phalloidin; these two poisons of the MFs do not affect chromosome migration (Cande *et al.*, 1981; Aubin *et al.*, 1981). The obvious conclusion is that chromosome movement is not due to actomyosin contraction, but requires MT depolymerization at the equator of the spindle. It is very likely that Petzelt's $Ca^{2+}$-ATPase (reviewed by Petzelt, 1979) plays a role, possibly in conjunction with calmodulin and the $Ca^{2+}$-sequestering vesicles that surround the spindle, in the localized and progressive polewise disassembly of the mitotic apparatus, which seems to be necessary for chromosome migration at anaphase. Finally, the interested reader is referred to a new model of chromosome translocation at anaphase proposed by Margolis and Wilson (1981). This model is based on the treadmilling of MTs, i.e., the unindirectional flow of tubulin molecules from one end of the MTs to the other. The efficiency of this GTP-dependent process is increased by ATP, which would increase the phosphorylation of MAP-2 by a cAMP-dependent protein kinase.

Cande (1982a,b) has distinguished between two processes that take place at anaphase: movement of the chromosomes (anaphase A) and elongation of the spindle (anaphase B). The use of a new inhibitor of dynein ATPase (an adenine derivative) has shown that this enzyme is involved in spindle elongation, but not in chromosome movement. Experiments on permeabilized cells have shown (Cande, 1982a) that only anaphase B requires ATP and is inhibited by vanadate and -SH reagents. The conclusion is that a dynein-like ATPase plays a role in spindle elongation, while dynein, tubulin, and actomyosin are not involved in chromosome movements; actomyosin would have an effect on the collapse of an elastic component of the spindle. The view that chromosome movement at anaphase results from a localized depolymerization of the kinetochore MTs is shared by Jarosch and Foissner (1982).

In a more recent paper, Cande (1983) has emphasized the role that creatine kinase might play in anaphase. This enzyme catalyzes the following reaction: ADP + phosphocreatine → ATP + creatine; it thus generates ATP, and its function in muscle biochemistry is well known. Cande (1983) found that creatine kinase is associated with the intermediate filaments and stress fibers in interphase, and with the spindle during mitosis. He observed that, in permeabilized cells, the addition of creatine kinase + ADP induces continuous elongation of the spindle. He concluded that anaphase should be divided into two steps (A and B). The first (A) is chromosome migration toward the poles; the second (B), the ATP-requiring step, is spindle elongation. The function of the spindle-associated creatine kinase would be to generate ATP to be used by a dynein-like ATPase. Silver *et al.* (1983) have demonstrated the presence of creatine kinase in mitotic apparatuses isolated from cleaving sea urchin eggs.

This discussion on the mechanisms of spindle assembly, spindle elongation, chromosome movement, etc. perhaps lacks clarity. The author is not the only one to be baffled by the still unsolved problems raised by the dynamics of mitosis. In an authoritative review of our present knowledge of mitosis, Pickett-Heaps *et al.* (1982) adopted a highly critical attitude. They pointed out that there is no evidence for sliding between kinetochore and polar MTs and for the treadmilling theory of Margolis and Wilson (1981); The presence of actin in the spindle remains doubtful, and there is no proof that this protein is involved in chromosome migration. A dynein-like ATPase might play a role in establishing the chromosomes at metaphase, but not their migration. More recently, Paweletz (1983) summarized the findings of a workshop on the dynamics of mitosis. The tentative conclusion was that chromosome movements might be due to sliding assembly and disassembly and lateral interactions (Zipping) between spindle MTs. The same mechanism would cooperate in the separation and migration of the daughter chromosomes at mitosis. This phenomenon is of fundamental importance for all eukaryotes; we do not yet understand how it takes place at the molecular level.

Before concluding this very long section, a brief mention should be made of the many agents that allow us to modify the integrity of the mitotic apparatus. As we have seen in Chapter 3, many drugs other than colchicine and colcemid (vinblastine, vincristine, maytansin, podophyllotoxin, nocodazole, etc.) inhibit tubulin polymerization; all prevent the building up of a mitotic apparatus and arrest the cell in metaphase. Mercaptoethanol dissolves the mitotic apparatus in living sea urchin eggs; in contrast, dithiodiglycol, glycerol, and heavy water ($D_2O$) have a stabilizing effect such that cell division becomes impossible. More interesting are the results obtained with taxol. Schiff *et al.* (1979) have found that taxol inhibits mitosis by favoring the assembly of MTs, arresting the cell cycle in either $G_2$ or M. Microtubules assembled in the presence of taxol are resistant to cold or to high $Ca^{2+}$ treatments. More recently, De Brabander *et al.* (1981a) observed that taxol induces the assembly of MTs in interphase cells and inhibits the organizing capacity of the centrioles and kinetochores during mitosis.

### E. Cytokinesis (Furrow Formation)

Among the various theories proposed to explain furrow formation leading to cell division, the contractile ring theory was once (30 years ago) the most popular (Mitchison, 1952). It assumed that cleavage was caused by the contraction of an equatorial ring to the furrow (Fig. 32). Currently, this theory is still accepted, although alternative models are still occasionally proposed.

The cortex of eggs and cells contains proteins similar to the contractile proteins of muscles, thus requiring ATP and $Ca^{2+}$ for contraction. This has been known since the classic experiments of Hoffmann-Berling (1954a,b, 1956) on cells (fibroblasts or spermatocytes) that had been treated with glycerol in order to remove most of the soluble substances present in the cells without extracting the (then unknown) fibrous cytoskeletal proteins. The addition of ATP in the presence of $Ca^{2+}$ produced marked contraction of the cells. In particular, glycerol-extracted cells in telophase divided into two daughter cells as a result of ATP-induced contraction. Even in intact cells (amebas, myxomycetes, tissue culture cells), the addition of ATP produced strong cytoplasmic contraction (for a review of the older literature, see Brachet, 1957). The existence of contractile proteins, similar to actomyosin, in the cell cortex was thus widely accepted 30 years ago.

Definite progress was made when Schroeder, (1972) demonstrated, by ultrastructural studies on cleaving sea urchin eggs, the existence of a ring of filaments (3.5–6.0 nm wide) at the equator of the egg, at the exact place where the furrow would appear. This ring can be observed for 6–7 min at telophase; it is 8 μm wide and about 0.2 μm thick. Since formation of the ring was suppressed when eggs were treated with cytochalasin B (which also suppresses cytokinesis), it was obvious that actin was a major constituent of the contractile ring; this was soon established by ultrastructural cytochemistry. Fujiwara and Pollard (1976, 1978) later showed that, in HeLa cells, myosin too is strongly accumulated in the contractile ring region. The same holds true for some, if not all, of the actin-

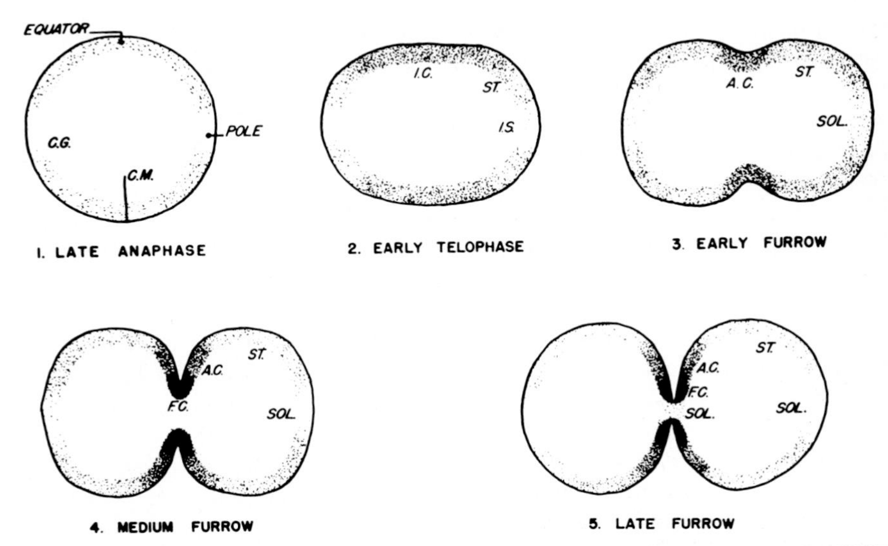


FIG. 32. A schematic representation of sea urchin egg cleavage based on Mitchison's (1952) contractile ring theory (cortical gel contraction theory) of cytokinesis. It shows progressive changes in the gelational state of cell cortex and how these changes may be related to the furrowing process. C. G., cortical gel; C. M., cell membrane; I.C., region of incipient contraction; ST, region subjected to stretching; I.S., region of incipient solution; A.C., actively contracting region; SOL, region of solation; F.C., fully contracted region. [Marsland and Landau (1954).]

associated proteins. According to Fujiwara *et al.* (1978), both α-actinin and myosin accumulate in the furrow at cytokinesis. Similar observations have been reported for tropomyosin (Ishimoda-Takagi, 1979), filamin, α-actinin, and myosin (Nunnally *et al.,* 1980), whose presence in the cell cortex seems to be necessary for cytokinesis. Previous observations of the presence of actin, myosin, and tropomyosin in the mitotic spindle have been negated by Aubin *et al.* (1979), who concluded that these proteins are primarily located in the contractile ring.

Innumerable experiments have shown that cytochalasin B inhibits furrowing in all types of cells. In fact, cytochalasins are now classic tools for obtaining binucleated or multinucleated cells. Suppression of cytokinesis is selective, and the entire karyokinetic process proceeds normally, resulting in two daughter nuclei at each cell cycle. Work by Aubin *et al.* (1981) with cytochalasin B led to the conclusion that MF activity is not required for the S → M progression in the cell cycle, and that it is not involved in spindle formation and chromosome movement. Thus, the drug produces binucleate cells. Experiments with cytochalasin thus provide a major argument against a significant intervention of actin in chromosome movement at anaphase. Ingenious experiments by Meeusen *et al.* (1980) have provided additional evidence for a major role of actomyosin in furrowing. Meromyosin treated with the −SH reagent *N*-ethylmaleimide (NEM) binds to actin. Such a modified actin molecule can no longer bind to myosin and is, therefore, unable to contract. Injection of NEM-treated meromyosin into fertilized frog eggs suppresses cleavage and wound closure, which require egg cortex contractility involving an actin–myosin interaction. The same conclusion can be drawn from the experiments of Kiehart *et al.* (1982), who found that injection of actomyosin antibodies into sea urchin eggs blocks their cleavage.

The mechanisms of furrow formation differ in small cells, such as fibroblasts, and very large ones, such as amphibian eggs. In the former, the two daughter cells are held together at the conclusion of cytokinesis by a midbody, and a 49,000-dalton protein, which is soluble at metaphase and particulate at the end of telophase when the midbody becomes microscopically visible. This protein, which has not been characterized further, disappears during the $G_1$ phase of the cell cycle (Krystal *et al.*, 1978). More recently, Mullins and McIntosh (1982) found that isolated midbodies have a chemical composition very similar to that of mitotic spindles: α- and β-tubulins represent 30% of the total proteins, and about 35 other polypeptides are present in isolated midbodies. Using their method for establishing the polarity of MTs (based on the ''decoration'' of MTs with curved ribbons of partially polymerized tubulin), Euteneuer and McIntosh (1980) showed that the midbody is formed by interdigitation of the aster MTs. As already mentioned, midbodies also contain large quantities of calmodulin (Welsh *et al.*, 1979; De Mey *et al.*, 1980), suggesting that calcium ions are involved in the ultimate stages of cytokinesis; a plausible role for this calmodulin–$Ca^{2+}$ system would be the depolymerization of the midbody tubulin that holds together the two daughter cells.

In the case of large cells, such as amphibian eggs, the furrow begins from the animal pole, proceeding to the vegetal pole, which is 1–2 mm away from its origin. This unusual situation leads to an asynchrony between karyokinesis and cytokinesis. When the furrow reaches the vegetal pole, the nuclei of the two blastomeres are not in telophase, but are already in prophase. The yolk mass should be a very serious obstacle to furrow progression, but this obstacle is greatly reduced by a gelification of the cytoplasm surrounding the asters, leaving a clear, almost yolk-free ''diastema'' in the middle (Fig. 33). If periastral gelification is suppressed by phenylurethane, there is no diastema formation and the furrow is unable to progress. Thus, treatment with phenylurethane results in the formation of undivided but multinucleated eggs, since nuclear division proceeds normally for several cycles despite the continuous presence of the drug (Brachet, 1934). In addition to the presence of the diastema (or perhaps correlated with its formation) is a cleavage-inducing factor discovered by Sawai (1972). Using very delicate microsurgery, he showed that the material lying at the tip of the progressing furrow can induce furrowing if it is injected elsewhere in the egg cortex.

Progression of the furrow through these large amphibian eggs requires an

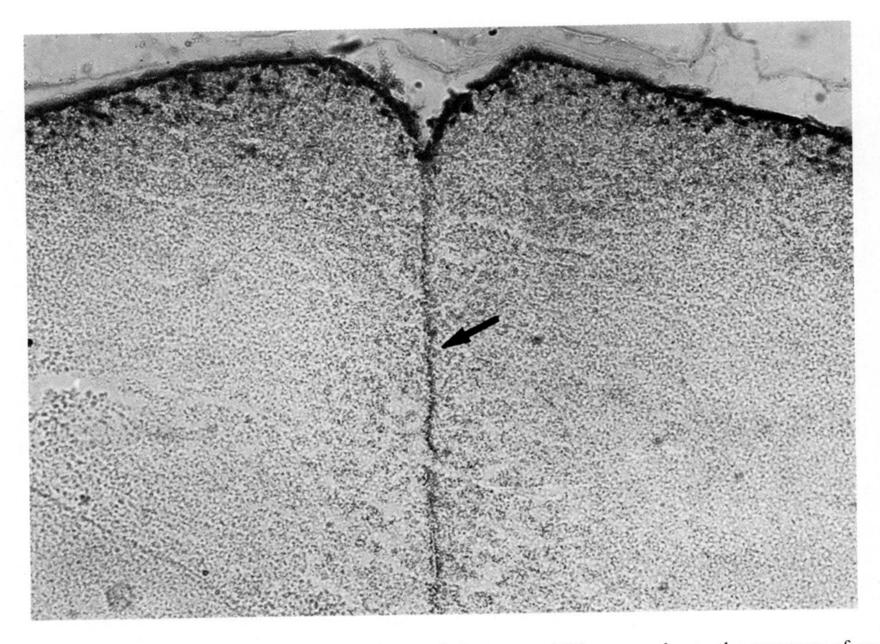

FIG. 33. Cytokinesis is possible in large, yolk-laden amphibian eggs due to the presence of an almost yolk-free diastema (arrow) between the animal and vegetal poles.

interaction between two factors: contraction of the egg cortex and apposition of new membranes to the progressing furrow (Sawai, 1976). The importance of continuous cortical contraction is shown by the fact that an injection of cytochalasin B at the tip of the growing furrow inhibits cytokinesis in *Xenopus* eggs. Electron microscopy shows that cytochalasin B causes a disorganization in the MF bundles without causing complete destruction of the MFs (Luchtel *et al.*, 1976). During cytokinesis, there is, of necessity, an increase in cell surface. In amphibian eggs, at least, this increase is due mainly to the apposition of membranes already present in the cytoplasm to the growing furrow. This is shown by the fact that the furrow's membrane contains very few intramembrane particles compared to the plasma membrane surrounding the rest of the egg (Bluemink *et al.*, 1976). However, the plasma membrane affects the formation and progression of the furrow, since addition of some lectins suppresses cleavage in *Xenopus* eggs (Tencer, 1978). The old and new membranes obviously form a continuum, which is in a highly dynamic state.

One problem that is often controversial is the role of the mitotic apparatus, particularly the asters, in cytokinesis. It has been suggested that the asters release factors that induce furrowing. This question was reviewed in 1971 by Rappaport. The main objection to this hypothesis is that, in sea urchin eggs, removal by micromanipulation of the entire mitotic apparatus at anaphase or telophase does *not* inhibit cell cleavage; furrowing apparently depends on autonomous contractility of the egg cortex (Hiramoto, 1956). A reinvestigation of the question by Rappaport (1981) has shown that sand dollar eggs cleave normally if one removes the mitotic apparatus 4 min before furrowing takes place, but not if removal takes place 5 min before the furrow forms. It is, therefore, argued by Rappaport (1981) that an interaction between asters and the cell surface is required for furrowing. This interpretation is in agreement with a hypothesis presented by Schroeder (1981). According to his ''global contraction-polar relaxation'' hypothesis, the entire egg would first undergo general contraction, followed by localized relaxation of the cortex at the egg equator prior to furrowing (Fig. 34). This would lead to localized contraction between the two relaxed hemispheres. This hypothesis is more elaborate than the ititial contractile ring theory, for which Schroeder himself had provided experimental evidence in 1972. It is, indeed, easily conceivable that something originating from the asters at anaphase is responsible for polar relaxation, since so many factors affect actin polymerization and depolymerization.

In another model, proposed by White and Borisy (1983), cytokinesis would occur in two steps. In the first step, contractile cortical MFs would be activated, inducing the development of surface tension. In the second step, the asters (via their MTs) would modulate the tension activity and produce gradients of tension

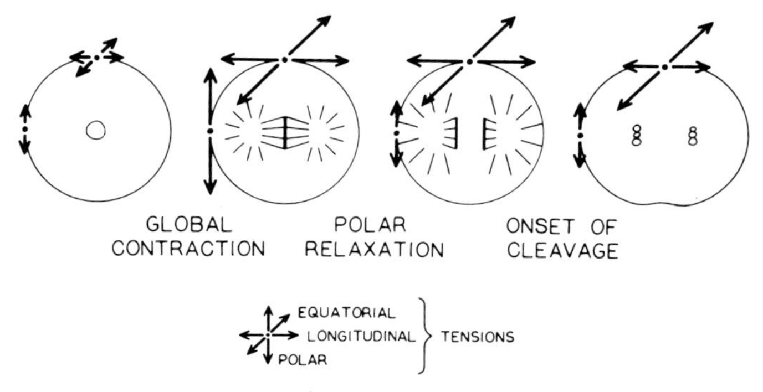


FIG. 34. Schematic representation of the global contraction–polar relaxation hypothesis. A sea urchin egg with a still intact nucleus (*left*) exhibits low tangential tension (arrows). By midmitosis, tangential tension has increased as the result of global contraction, but remains equal in all directions so that the egg remains spherical. This change is independent of mitosis and is intrinsic to the cortex. At anaphase, enlarging mitotic asters interact with the cortex at the poles and convey the cleavage stimulus; this causes relaxation of the polar cortex without affecting the equatorial cortex. Once the poles relax, there is an imbalance in tensile forces over the spherical egg leading to elongation of the egg. Tension in the equatorial direction is greatest. Thus, a cleavage furrow forms at the equator, while the poles distend and stretch to accommodate the change in shape while preserving the volume of the egg. This constitutes the onset of cleavage (right). [Schroeder (1981).]

with a maximum at the equator. The tension-generating elements would finally accumulate at the equator, forming a contractile ring.

This wealth of theories, hypotheses, and models shows that we do not yet understand completely the mechanisms that lead to cytokinesis. More work on the interactions that may take place between the full-grown asters and the cell cortex is obviously needed before a general theory of cytokinesis can be proposed.

We have not yet discussed the strange differentiation, without cleavage, in *Chaetopterus* eggs or the curious axolotl mutant (nc), which is unable to assemble a mitotic apparatus and to undergo cleavage. These topics will be discussed in Volume 2, Chapter 2.

Telophase is not only the time when a furrow divides the cytoplasm of the dividing cell into two daughter cells. It is also the time when the swollen anaphase chromosomes fuse together and reconstitute a nuclear envelope, and when the reactivated nucleolar organizer genes build up new nucleoli. The first attempt to analyze nuclear reassembly at the conclusion of mitosis was made by Lydersen and Pettijohn (1980). They found that one of the protein constituents of the nuclear matrix is associated, during mitosis, with the poles of the mitotic apparatus. This protein is not found in the bridge between the two daughter cells (midbody) but rather in their nuclei, where it presumably participates in nuclear matrix reorganization. More recently, Van Ness and Pettijohn (1983) described a protein (called NUMA, for nuclear mitotic apparatus) located in the interphase cell nuclei and metaphase chromosomes of primates; its role is to bind the mitotic chromosomes to the nuclear membrane when nuclei re-form at telophase. In regard to nuclear membrane formation at this stage of mitosis, Fuchs *et al.* (1983) found that, in *Drosophila* embryos, a nuclear membrane antigen can be detected by immunocytochemistry as dots in the cytoplasm at metaphase-anaphase. This antigen then associates with the inner nuclear membrane, forming the outer nuclear membrane.

## F. Biochemical Changes during Mitosis

In addition to the dramatic changes described in the two preceding sections, a number of more discrete changes occur during mitosis. For instance, endocytosis is greatly reduced. Thus, cells grown in culture are less sensitive to changes in the protein composition of the surrounding medium during mitosis than during interphase (Berlin *et al.*, 1978).

Since it is impossible to give even a very brief account of all of the papers in which various biochemical parameters have been followed during mitosis, we shall limit ourselves to a summary of the results obtained in three fields: energy production, macromolecular synthesis, and cyclic nucleotide metabolism.

In "Biochemical Cytology," several pages were dedicated to the energy requirements during mitosis, a subject in which there is little current interest. We will only summarize here what was already known 25 years ago. Generally, the presence of oxygen and of an oxidizable substrate is required for mitosis. Inhibitors of cytochrome oxidase (KCN, CO, $N_3Na$) and uncouplers (dinitrophenol) quickly arrest cell division in both cultured cells and sea urchin eggs. A striking exception is the frog egg, which can reach the early blastula stage in complete anaerobiosis and in the presence of KCN at concentrations that inhibit oxygen consumption by 90% (Brachet, 1934). Since dinitrophenol arrests frog egg development at a much earlier stage (morulae with large interphase nuclei), it was concluded that ATP formation is more important than respiration for division of the nucleus (Brachet, 1954) and for RNA synthesis (Steinert, 1953). If one compares the ATP content of sea urchin and frog eggs under anaerobic conditions, one finds that mitotic activity ceases in both cases when the ATP content has decreased to about 40% of its initial value. This occurs rapidly in sea urchin eggs and more slowly in frog eggs, explaining the different results obtained with the two species.

In the past, many attempts were made to find out whether there are cyclic changes in oxygen consumption during cell division. By far the best work in this field was that of E. Zeuthen, who used a very delicate method (the cartesian diver technique) to measure the respiration of single sea urchin or frog eggs during cleavage. In 1957, it was concluded that large changes in oxygen consumption during cell division did not exist and that the variations observed by Zeuthen were always small. The most recent paper on the subject by Zeuthen and Hamburger (1972) shows that oxygen consumption is at a minimum and that the respiratory quotient ($CO_2/O_2$) is at a maximum shortly before the appearance of the first cleavage furrow (Fig. 35). The fact that, in sea urchin eggs, respiration is somewhat higher during interphase than during mitosis has been taken by Zeuthen (1953) as indicating that a correlation exists between oxygen consumption and DNA synthesis, an idea proposed independently by Comita and Whiteley (1953). It should be possible to test this suggestion by observing cyclic changes in respiration in sea urchin eggs treated with aphidicolin (an inhibitor of DNA polymerase), which almost completely inhibits DNA synthesis and cleavage in sea urchin eggs (Ikegami *et al.*, 1978). Unfortunately, with the recent death of Zeuthen, the experiments on aphidicolin-treated eggs are not likely to be repeated.

Louis Rapkine (1931) emphasized the importance of thiols in cell division. He showed, for instance, that if sea urchin eggs are treated with small amounts of $HgCl_2$, cleavage is irreversibly arrested unless an $-SH$-containing compound is added to the eggs. He believed that, during sea urchin egg cleavage, there are cyclic changes in the content of glutathione and protein-SH groups, leading to the hypothesis that a reversible denaturation of proteins takes place during mitosis, which results in the formation of the spindle and asters. Mazia's (1955, 1956a) experiments on the effects of mercaptoethanol and dithiodiglycol on the

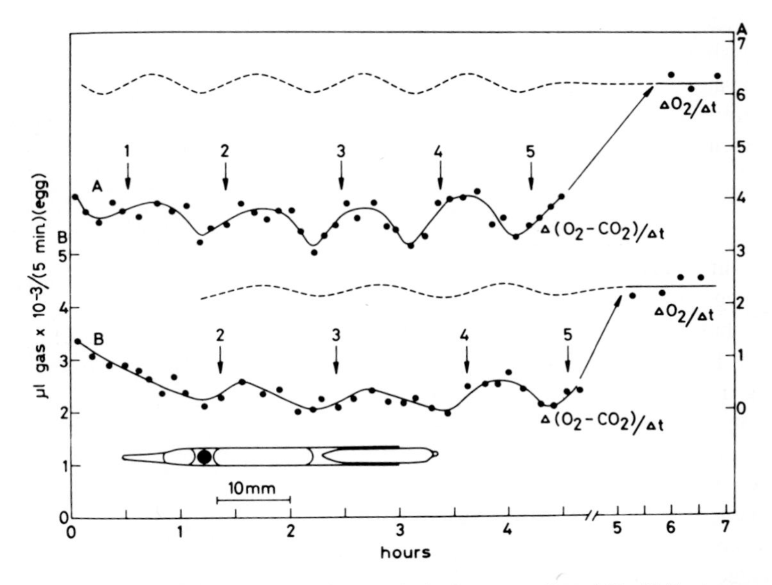


FIG. 35. Rates of gaseous exchanges in two single frog eggs (A and B). Fully drawn curves show $\Delta(O_2 - CO_2)/\Delta t$ and $\Delta O_2/\Delta t$ measured at 0–5 and 5–7 hr, respectively. The dashed lines show extrapolated values for $\Delta O_2/\Delta t$. Arrows indicate when cleavages 1–5 begin. Below is the diver used; it floats "egg up." [Zeuthen and Hamberger (1972).]

isolated mitotic apparatus have confirmed that the $-SH \rightleftharpoons SS$- equilibrium plays an essential role in the assembly and disassembly of the mitotic apparatus. The question of the "Rapkine cycle" has been reviewed by Mazia (1961) and, more recently, by Rebhun (1977). Undoubtedly, Rapkine's bold and original views on mitosis have been the basis of important research. However, more recent work does not support the idea that cyclic changes in the glutathione content occur during sea urchin egg cleavage. The most recent paper on the subject (Fahey *et al.*, 1976) concludes that there are no changes in reduced and oxidized glutathione, as well as in protein -SS- groups, at fertilization and during first cleavage in sea urchin eggs. The existence of reversible protein denaturation is a concept that was suggested at a time when macromolecules were unknown.

One enzyme associated with energy utilization clearly displays cyclic activity during sea urchin egg cleavage, namely, ATPase; it shows activity peaks during interphase and metaphase (Petzelt, 1972). According to Petzelt and Avel (1977), a $Ca^{2+}$-activated ATPase is synthesized during interphase and activated during mitosis. Its role would be to control the free calcium ion concentration at the site of spindle formation; local changes in $Ca^{2+}$ concentration would, in turn, regulate mitotic MT assembly or disassembly.

Most of the work summarized in the last pages deals with eggs, and it would be hazardous to extrapolate the results obtained with them to all cells. Skog *et al.* (1982) showed that this warning should be taken seriously; they studied the content and production of ATP in Ehrlich ascites tumor cells and came to the following conclusions: these cells contain $19 \times 10^{12}$ molecules of ATP; 60% of them result from aerobic metabolism; ATP production is increased in the S and G + M phases of the cell cycle. The energy-requiring processes are at a maximum at the beginning of the $G_1$ phase, and there is no correlation between these processes and DNA synthesis. Of course, the situation may be different in cancer cells (which have high aerobic glycolysis) and in eggs in which aerobic glycolysis is negligible and DNA replication proceeds at an exceptionally fast rate.

We shall now briefly discuss macromolecular synthesis during mitosis. As we already know, nuclear DNA synthesis takes place during interphase, but not during mitosis. The only DNA synthesis that can be detected during actual cell division is of mitochondrial origin (as well as of chloroplastic origin in green plant cells).

RNA synthesis (reviewed by M. V. N. Rao, 1980) greatly decreases during prophase, falling to 10–15% of the interphase value during metaphase-anaphase (Stein and Baserga, 1970). However, synthesis of the small 4 and 5 S RNAs, catalyzed by RNA polymerase III, continues during metaphase, according to Zylber and Penman (1971). It has also been reported that the addition of terminal-CCA to the tRNAs continues in mitotic cells. Synthesis of mitochondrial (and chloroplastic) RNAs continues during mitosis. One must therefore conclude, with M. V. N. Rao (1980), that the highly condensed metaphase chromosomes are not the site of measurable RNA synthesis. They are comparable, in this respect, to the strongly condensed nuclei of the spermatozoa or bird erythrocytes that do not synthesize RNA. One could even say that, from the viewpoint of RNA transcription, metaphase cells are equivalent to anucleate cytoplasts and that their genome is temporarily completely repressed. Only when the chromosomes have undergone sufficient decondensation will the proteins present in the surrounding cytoplasm penetrate the daughter nuclei. This protein intake is accompanied by the renewal of RNA synthesis and, concomitantly, by gene derepression. Among the proteins that move into the daughter nuclei are those that were already present in the initial interphase nucleus and had been dispersed in the cytoplasm during mitosis. It is not known for certain whether these "shuttle" proteins play a more important role than newly synthesized proteins in the resumption of transcription in the daughter nuclei.

Overall protein synthesis also decreases during mitosis, with a minimum at metaphase-anaphase. This decrease is usually not as great as it is for RNA, but Tarnowska and Baglioni (1979) found a 75% decrease during mitosis in HeLa cells. A disaggregation of the polyribosomes, due to the arrest of mRNA synthesis, is responsible for the decrease in protein synthesis that occurs during metaphase and anaphase. During mitosis, protein synthesis is directed by preexisting, stable mRNAs. Consequently, protein synthesis at metaphase is insensitive to actinomycin D, the classic inhibitor of RNA synthesis. It seems that in certain cells the ribosomes undergo changes during mitosis. Their capacity for artificial mRNA [poly(U), for instance] *in vitro* translation is decreased, but it can be restored by mild digestion with trypsin. Tarnowska and Baglioni (1979) found that the marked inhibition of protein synthesis would be due to the presence in mitotic cells of small inhibitory RNAs. The general conclusion based on this work is that mitotic cells are much less active in the biosynthesis of macromolecules than interphase cells due to the suppression of DNA transcription in the condensed metaphase chromosomes. A cell in metaphase very closely approximates an enucleated cell in which protein synthesis continues due to the presence of stable mRNA molecules in the cytoplasm. To view an interphase nucleus as a "resting" nucleus, as was done for many years, is obviously erroneous.

As we shall see later in this chapter, it is widely accepted that cyclic nucleotides (cAMP and cGMP) play an important role in the control of cell division. Here we shall limit ourselves to a few papers dealing with cyclic changes during mitosis in the cyclic nucleotide content and with enzymes involved in their metabolism. Burger *et al.* (1972) reported that the cAMP level is minimal during mitosis. This was confirmed by Zeilig *et al.* (1976), who found that, in HeLa cells, cAMP is at a maximum during the $G_1 \rightarrow S$ transition and at a minimum at metaphase. Gray et al. (1980) found that cAMP increases in $G_1$ and decreases during mitosis; cAMP-dependent protein kinase activity follows a parallel course, and it was concluded that cAMP controls mitosis via this protein kinase activity. In agreement with the view that the cAMP level is low during mitosis, Howard and Sheppard (1981) found that the activity of adenylate cyclase, the enzyme that synthesizes cAMP, decreases during mitosis. While there is substantial agreement that cAMP decreases during mitosis, it should be mentioned that Oleinick *et al.* (1981) failed to detect any changes in cAMP and cGMP content during the cell cycle of the naturally synchronized myxomycete *Physarum*.

Changes in protein kinase activity during the cell division cycle imply changes in cellular phosphoproteins. Indeed, Davis *et al.* (1983) detected three major and several minor mitosis-specific phosphoproteins in synchronized HeLa cells using monoclonal antibodies. They concluded that condensation and decondensation of chromatin do not depend on the phosphorylation and dephosphorylation of H1 histone alone; phosphorylation and dephosphorylation of non-histone proteins are presumably also involved. Laskey (1983) mentioned that, in addition to the strong phosphorylation of H1 histone, H3 histone, HMG-14, and the A, B and C lamins also undergo reversible changes in phosphorylation during mitosis. Several protein kinases are located in the nucleus: in particular, H1 histone kinase increases in activity during mitosis. Finally, Celis *et al.* (1983) have reported that cytokeratins and vimentin are phosphorylated during the mitotic cycle; this is probably correlated with the changes in the intermediate filament network that occur in dividing cells. It is likely that research on protein phosphorylation during the cell cycle will lead to important conclusions in the years to come.

### G. Description of a Few Mitotic Abnormalities

Physical agents (temperature, radiation, etc.), as well as innumerable chemical substances (narcotics, enzyme inhibitors, etc.) easily produce a number of mitotic abnormalities; these will be briefly described below.

#### *1. Formation of Cytasters*

The formation of numerous asters in the cytoplasm of unfertilized eggs is a frequent phenomenon. They were first obtained by E. B. Wilson (1901), who treated unfertilized eggs with hypertonic sea water. The formation of many small asters in the cytoplasm is a frequent occurrence at the time of maturation in the eggs of marine invertebrates. The mixing of nuclear sap and cytoplasm seems to be necessary for the formation of asters (this is discussed in greater detail in Volume 2, Chapter 2). This abnormality is particularly interesting because the formation of an aster in eggs does not necessarily require the presence of a centriole. Remarkable experiments by Lorch (1952) have shown that, if an aster is removed with a micromanipulator from a sea urchin egg, it re-forms, after a long delay, even if the nucleus has also been removed; thus, the aster is entirely of cytoplasmic origin. An easy way to induce the formation of hundreds of cytasters in amphibian eggs is to treat them with $D_2O$. Cytaster formation does not require DNA, RNA, or protein synthesis, since once formed, they never divide (Van Assel and Brachet, 1968) (Fig. 36).

#### *2. Multipolar and Catenar Mitoses*

The production of multipolar mitoses (Fig. 37) is also a common phenomenon. Mitoses involving several spindles and centrioles are found in many protozoa and in the vitelline syncytium of fish eggs. They are also common in cancer cells and in eggs after treatment with physical or chemical agents. Multipolarity is apparently caused by an abnormal division of the centrioles. The distribution of the chromosomes on the different spindles is highly irregular, and their number in the daughter cells, if the cell succeeds in dividing, is often uneven. The outcome is the formation of cells with a strongly unbalanced chromosome complement. Such aneuploidy will, sooner or later, prove to be lethal for the cell.

Catenar mitoses have been described by Dalcq and Simon (1932b) in irradiated amphibian eggs. The dividing cell forms a large number of asters without spindles or chromosomes to link them together; these asters line up in a row.

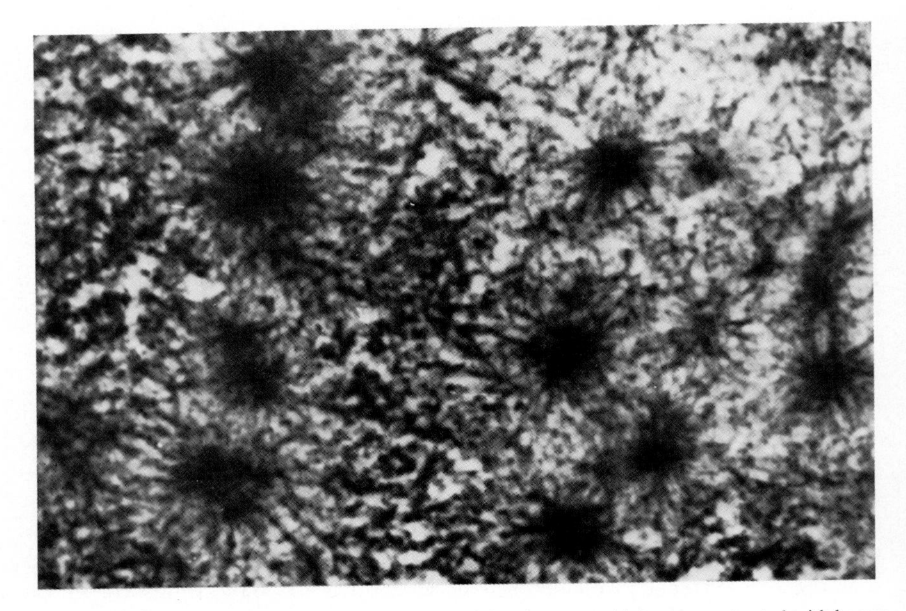

FIG. 36. Cytasters form when unfertilized amphibian (or sea urchin) eggs are treated with heavy water ($D_2O$). The condensed material that can be seen in their center probably corresponds to a "procentriole." [Van Assel and Brachet (1966).]

### 3. *Achromosomal Mitoses*

That large asters surrounding a centriole in eggs are capable of dividing independently has been proved by the fact that repeated cleavage can occur in the total absence of the nucleus. Cleavage of nonnucleated amphibian eggs was first obtained by Fankhauser (1934) and by Dalcq and Simon (1932a,b). In Dalcq and Simon's (1932b) experiments, frog eggs were fertilized with trypaflavine-treated or irradiated sperm (the sperm nucleus soon degenerates completely), and the maturation spindle was removed by pricking. A similar technique was used by Briggs *et al.* (1951), who fertilized frog eggs with heavily irradiated sperm and removed the egg maturation spindle by pricking and sucking; partial nonnucleate blastulae (Fig. 38) were obtained in this manner. An almost normal nonnucleate axolotl blastula was obtained by Stauffer (1945), who destroyed the maturation spindle and the sperm nuclei by pricking in recently fertilized axolotl eggs.

Similar results were described by E. B. Harvey (1933, 1936) with sea urchin eggs. She separated the unfertilized eggs, by centrifugation, into nucleate and nonnucleate halves; when the latter were treated with parthenogenetic agents, asters were formed and repeated cleavage of the cytoplasm took place. More will be said about these parthenogenic merogones in Volume 2, Chapter 2.

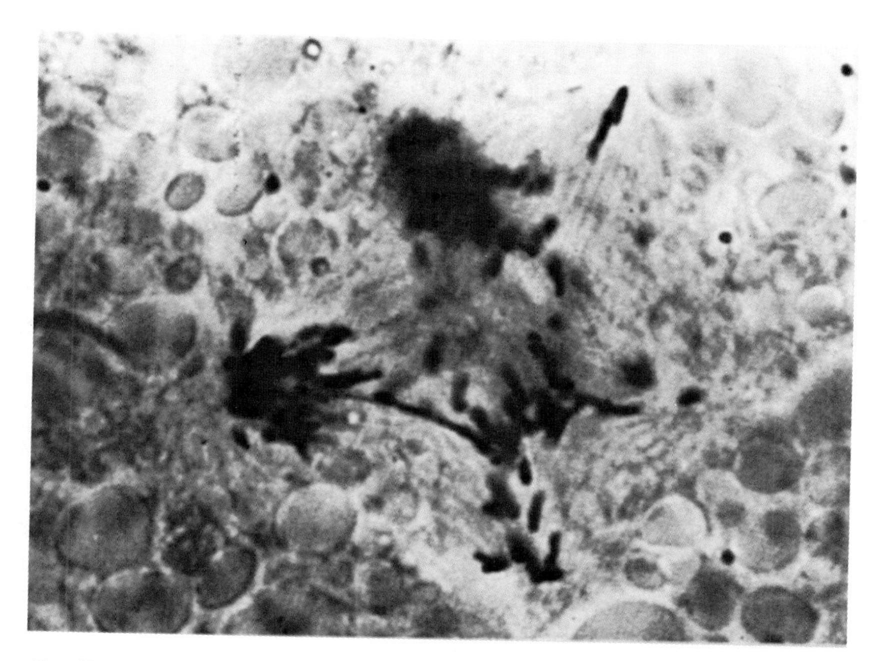

Fig. 37. Pluricentric mitosis in a cell from an 8-hr *Xenopus* blastula. Soon after fertilization the egg cortex had been slightly damaged by pricking, but development until the blastula stage had been normal. [Brachet and Hubert (1973).]

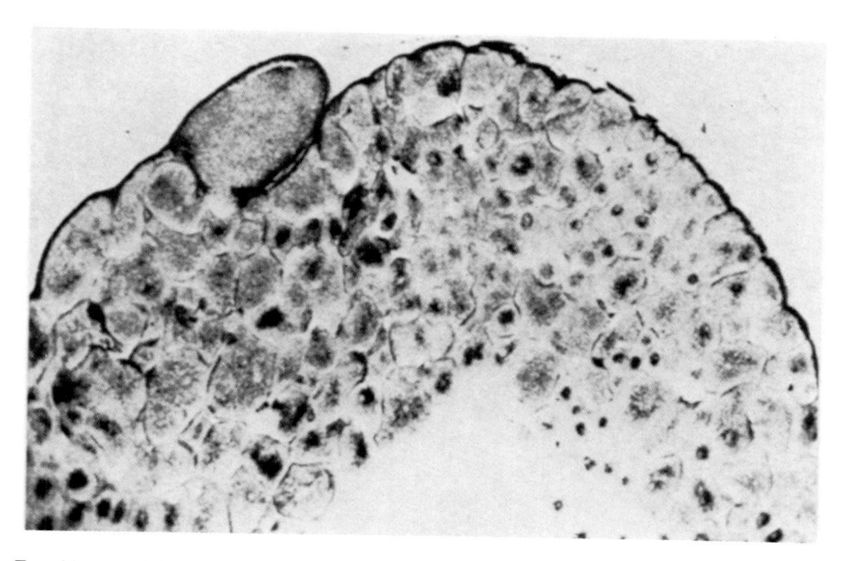

Fig. 38. Partial nonnucleate blastula obtained by fertilizing a frog egg with x-irradiated sperm and removing manually the maturation spindle. The dark spots are not nuclei, but local accumulations of pigment (melanin) granules. [Courtesy of the late Prof. R. Briggs.]

Thus, in sea urchins as well as in frogs, cleavage of nonnucleated eggs is the consequence of repeated aster division. Needless to say, the so-called purpose of mitosis, i.e., the equal distribution of the chromosomes between the two daughter cells, is completely defeated in such cases.

### 4. *Suppression of Cytokinesis*

Suppression of cleavage furrow formation, leading to the formation of binucleated cells, is easy to obtain in many cells. Phenylurethane-type drugs are very effective in inhibiting cytoplasmic division (cytokinesis) in both amphibian eggs (Brachet, 1934) and tissue culture cells (Frédéric, 1954). However, as already mentioned, the classic agent used for the suppression of cytokinesis is now cytochalasin B, which disrupts the MF cytoskeleton and prevents the contraction of the contracting ring.

### 5. *Anastral Mitosis*

Asters and centrioles (anastral mitosis) are absent in plant cells and oocytes during the maturation divisions. A similar condition can often be found, as an abnormality, in amphibian eggs during cleavage (Fig. 39). Anastral mitoses have been described by Bataillon and Tchou Su (1930, 1933) in amphibian in-

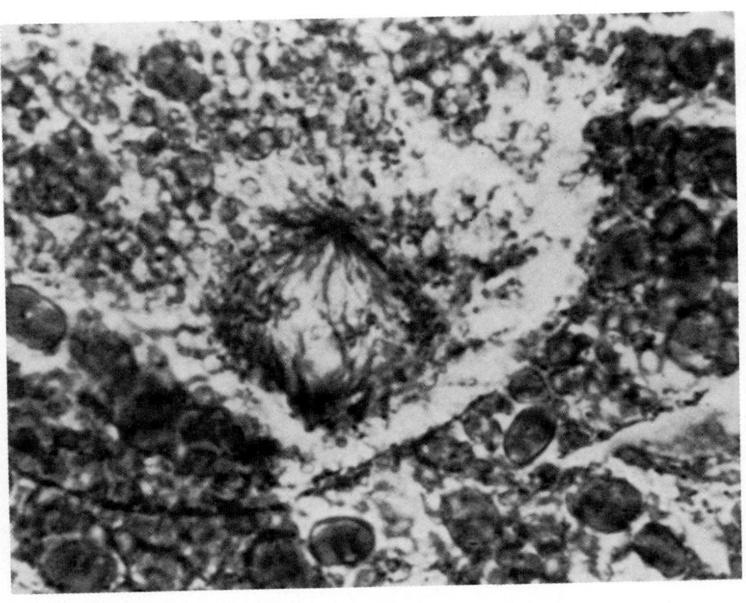

FIG. 39. An anastral mitosis in an RNase-treated morula. The spindle is broad, and there are no asters; this metaphase spindle resembles the metaphase I meiotic spindle during amphibian oocyte maturation.

terspecific hybrids. They can also be observed occasionally in amphibian eggs treated with a variety of inhibitors. In such experimentally obtained anastral mitoses, the spindle is often barrel-shaped; the missing ends of the spindle, as well as the asters, might require the presence of a normal centriole for their formation. Such abnormal anastral mitoses never lead to cleavage, and their chromosomes remain blocked in the equatorial plate stage. Chromosome migration toward the ends of the barrel-shaped spindle is never observed, suggesting that the centrioles (or the pericentriolar clouds) are necessary for chromosome migration at anaphase and telophase, at least in experimentally induced anastral mitoses.

#### 6. *Amitosis*

Amitosis is the fragmentation of the nucleus into two pieces. The phenomenon is very frequent in the endosperm of seeds (Darlington, 1955). Amitosis, if followed by cytokinesis, inevitably leads to the formation of aneuploid daughter cells, since no mechanism is provided for equal distribution of the chromosomes. For this reason, it is usually believed that amitosis is an agonic phenomenon and that it can no longer be followed by normal mitosis.

#### 7. *Abnormal Spindle Formation: C-Mitosis*

It has been known since the time of A. Dustin's (1934) pioneer work that colchicine inhibits mitosis by causing disorganization of spindle formation (Fig. 40). The result, especially in plant cells, is the formation of polyploid cells (Blakeslee, 1939). Thus, replication of the chromosomes occurs normally, but the latter are unable to migrate toward the poles, since the spindle is absent. If the cell survives and a nuclear membrane is formed, a polyploid cell results. Many substances, in addition to colchicine, can induce similar mitotic abnormalities (colchicine mitosis or c-mitosis). An extensive study of c-mitosis and of the agents that produce it can be found in a book by P. Dustin (1978).

In eggs, where the spindle is very sensitive to temperature changes, cooling or warming may easily induce poly- or aneuploidy (Fankhauser and Godwin, 1948; Costello and Henley, 1950). Heat shock immediately following fertilization inhibits polar body emission and thus provides a useful means for the production of triploids (Fankhauser and Godwin, 1948). The disruption of the spindle by heating to about 37°C for 1 hr at later stages (morulas, blastulas, or gastrulas) commonly causes polyploidy and aneuploidy in frog embryos (Brachet, 1949a,b).

As Mazia's (1955) work has already shown, c-mitosis is due to disorganization of the spindle rather than to the complete prevention of its formation.

#### 8. *Chromosomal Abnormalities*

The cytogenetic importance of chromosomal aberrations is too well known to require a detailed review here; precise information about chromosomal linkage and its genetic effects may be found in all textbooks dealing with genetics.

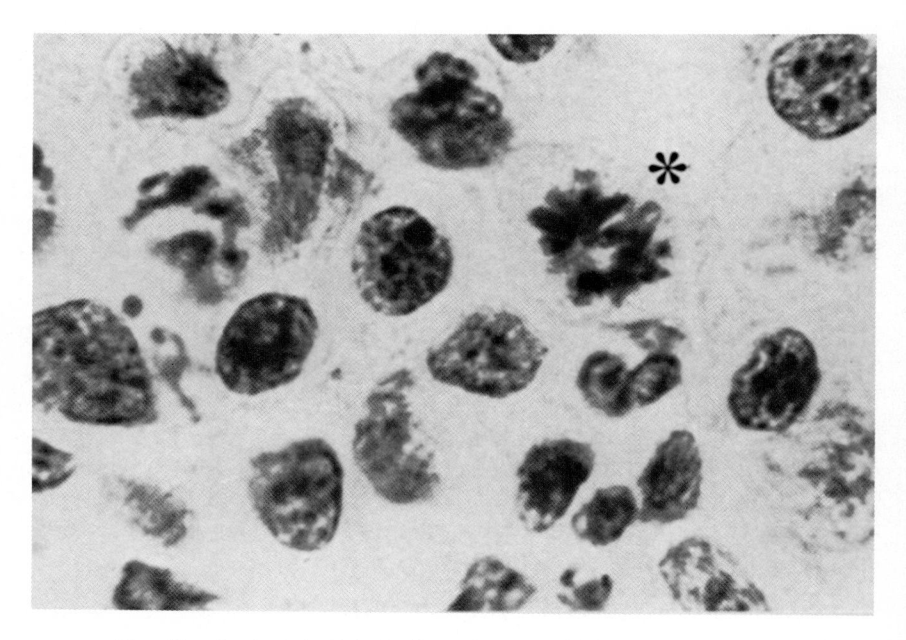


FIG. 40. *C-mitosis* (star) in the liver of an axolotl injected with colchicine.

Chromosomal aberrations always result from damage to DNA that can be induced by a host of physical (ionizing radiations, uv light, and even light from fluorescent lamps) and chemical (mutagens, carcinogens, chemicals used in cancer chemotherapy, etc.) agents. As pointed out by Woodcock *et al.* (1982), prolonged inhibition of DNA synthesis almost inevitably leads to chromosomal aberrations, changes in gene control, DNA hypermethylation, and, finally, cell death.

The most frequent chromosomal abnormalities are chromosome breakages, in which chromosome fragments devoid of a kinetochore degenerate in the cytoplasm (Fig. 41) and the lagging behind of anaphase chromosomes (Fig. 42). One of the chromosomes is stretched like a rubber band between the two poles of the spindle and finally breaks into pieces. These chromosomal aberrations lead to aneuploidy. Rearrangements may follow, providing these aberrations are limited.

In recent years, a new type of chromosomal abnormality has been discovered due to the development of sophisticated cytological techniques: sister chromatid exchanges (SCE). Interest in SCE stems from the fact that they are the earliest cytological sign of DNA damage known so far. SCEs (Fig. 4, Chapter 2) are usually detected by allowing the cells to replicate from one to three times in the presence of BrdUrd, staining with the fluorescent dye Hoechst 33858 and uv irradiating or by analyzing [$^3$H]thymidine incorporation autoradiographically. According to Gutierrez and Calvo (1981) SCEs form near the replication fork following DNA synthesis and result from postreplicative exchange of DNA fragments and errors in unraveling the daughter double helices by topoisomerases (Cleaver, 1981). There are about 0.065 SCE/pg DNA/cell cycle in normal cells. They occur, by polymerization or recombination, when error-prone systems are at work. Curiously, the inhibitors of (ADP)ribose polymerase markedly increase the frequency of both SCEs and nonsister chromatid exchanges (Hori, 1981; Althaus *et al.*, 1982). Although the precise role of this enzyme in chromatid exchanges is not yet fully understood, it certainly plays a role in excision-repair processes. It has been found that all of the agents that damage DNA increase the conversion of $NAD^+$ to poly(ADP)ribose.

### 9. *Elimination of Chromosomes*

Elimination of complete chromosomes may occur in normal insect development (Metz, 1938) and is a fairly common phenomenon in intergeneric or in-

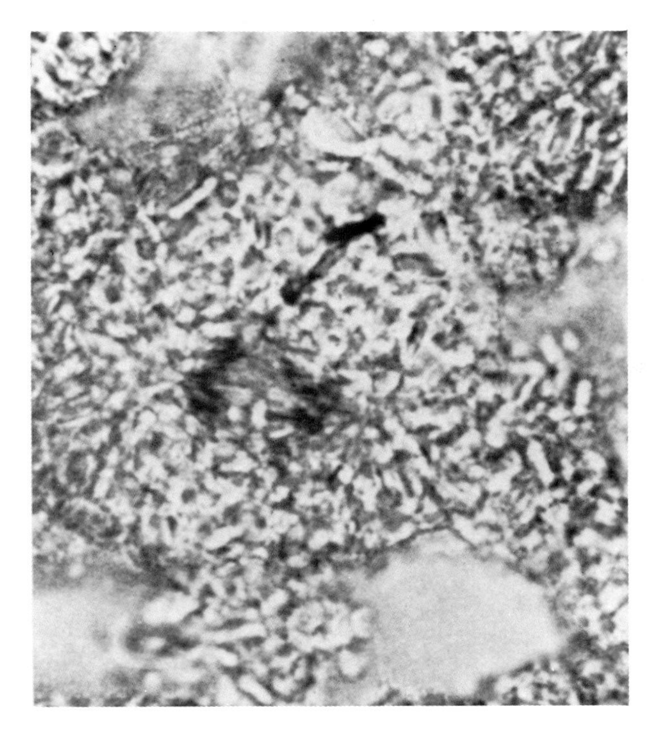

FIG. 41. Elimination of the male chromatin (right to the anaphase mitotic figure) bearing the egg chromosomes) in a *Rana esculenta* ♀ × *R. temporaria* ♂ hybrid.

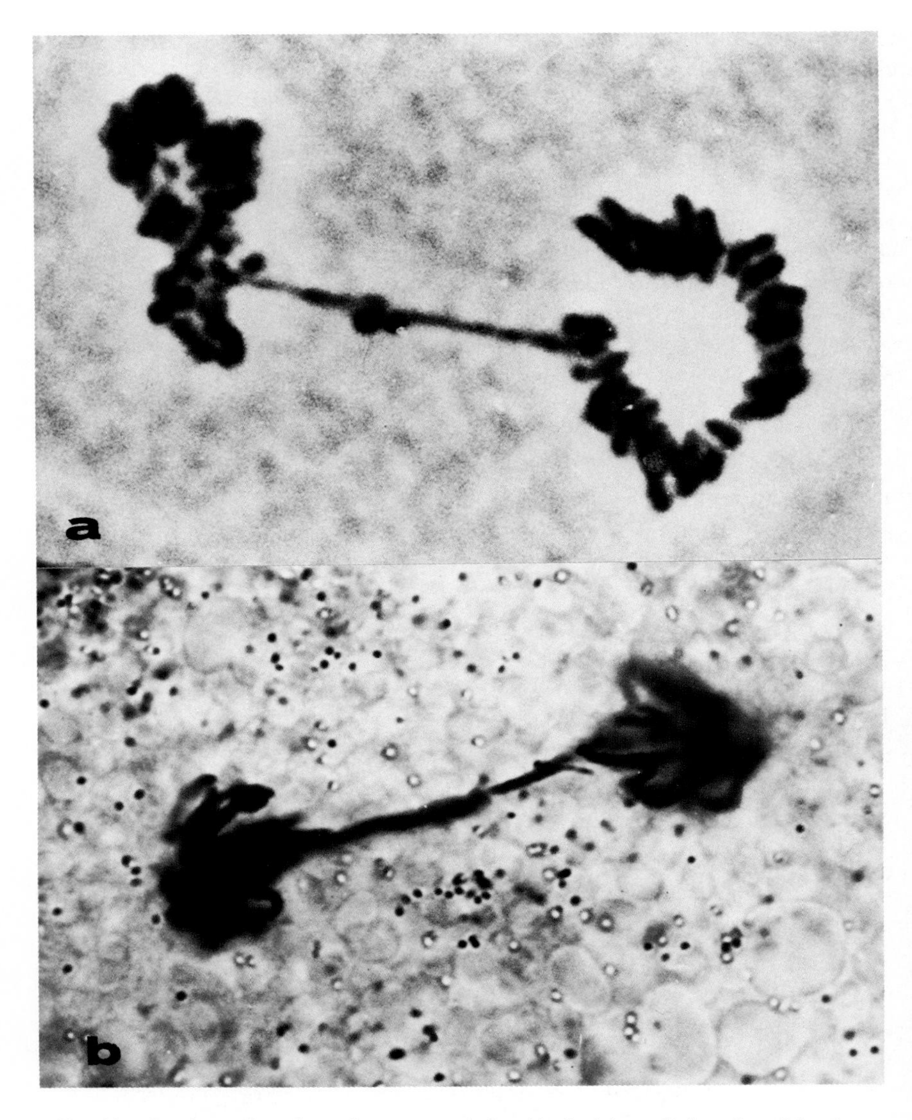


FIG. 42. Breakage of anaphase chromosomes induced by ionizing radiation. One of the chromosomes is stretched between the two poles of the spindle and finally will break into pieces. (a) Cultivated human fibroblast. (b) Blastomere of *Pleurodeles*. [(a) Courtesy of Dr. F. Zampetti; (b) courtesy of Dr. H. Alexandre.]

terspecific hybrids; the male chromatin is eliminated during one of the first cleavages (Baltzer, 1910; Bataillon and Tchou Su, 1930, 1933) (Fig. 41). Similar observations have been made on frog eggs that were fertilized with uv-irradiated sperm (Dalcq and Simon, 1932a,b; Brachet, 1954). This treatment provides a very convenient and reliable method for obtaining haploid (parthenogenic) frog embryos. The eliminated chromosomes sooner or later degenerate in the cytoplasm. Chromosome elimination is also very widespread in somatic interspecific hybrids. For instance, in most human–mouse somatic hybrid cell lines, human chromosomes are preferentially lost at each mitotic cycle.

The finer cytological details of chromosome elimination have not yet been investigated, although it is probable that the abnormality is a result of some problem in the kinetochore region, which is no longer capable of attaching itself to the spindle fibers.

#### *10. Pycnosis and Karyorrhexis*

Death of the cell nucleus is a frequent event. Usually the nucleus rounds up in a single spherical compact mass (pycnosis), which progressively disappears. The pycnotic chromatin, once it lies free in the cytoplasm, is usually rapidly digested by cytoplasmic enzymes. Sometimes the chromatin of the dying nucleus breaks down into a large number of spherical bodies (karyorrhexis), which soon degenerate (Fig. 43). Although pycnotic nuclei stain very strongly with Feulgen and basic dyes, their actual DNA content steadily decreases (Leuchtenberger, 1950). Thus, the intense staining of the pycnotic nuclei is simply a result of their marked condensation.

### H. Mitotic Poisons

The list of agents, physical or chemical, that interfere with mitosis is almost endless (even changes in the ionic balance of the medium can arrest mitosis). Only a brief discussion of the subject will be presented here because the question is obviously an important one in the field of cancer therapy.

There are numerous spindle poisons that adversely affect MT assembly. The classic spindle poisons—colchicine, colcemid, vinblastin, and vincristine—are alkaloids of plant origin; the *Vinca* alkaloids, which form crystals in the presence of tubulin, are less toxic than colchicine and are frequently used in cancer chemotherapy. As we have seen, spindle formation is affected by –SH reagents and temperature changes. Taxol inhibits mitosis through its effects on MT assembly. An herbicide, chloro-isopropyl-*N*-phenyl carbamate (CIPC), produces multiple mitoses leading to multinucleate cells, acting on the MTOC, but not on MTs, even in plant cells devoid of centrioles (Lloyd, 1979). While colchicine stops mitosis in metaphase, the tranquilizer diazepam arrests it in prometaphase (Andersson *et al.*, 1981). Its target, at the cellular and molecular levels, is not yet known.

Any agent that inhibits DNA replication is a potential chromosome poison

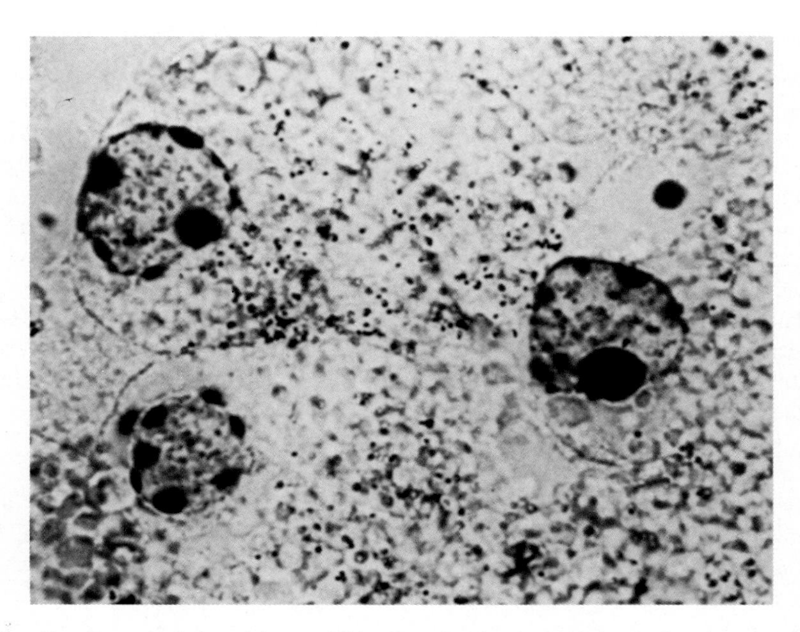

FIG. 43. Karyorhexis in a dying amphibian blastula. The chromatin has broken down into large DNA-containing spherules.

and, thus, a potential mutagen and carcinogen. Nevertheless, such agents must be used for cancer therapy to stop malignant growth. Although prolonged radiotherapy or chemotherapy may increase the risk of developing cancer a few years later, it may also be the only way to save the patient's life today. In addition to x-rays and radiomimetic agents, there are hundreds of substances that damage DNA molecules, such as the nitrogen mustards. Inhibitors of DNA synthesis such as HU and thymidine analogs (FrdUrd, for instance) are used widely for cell synchronization. Particularly interesting for the cell biologist are aphidicolin, a specific inhibitor of DNA polymerase-α, and methotrexate, which, as we have seen, induces the amplification of dihydrofolate reductase genes in resistant cells.

Both spindle and chromosome poisons arrest mitosis at metaphase, while a few other inhibitors block mitosis at other stages. For instance, RNase and fluorophenylalanine, as we have seen, arrest cells at the $G_2 \rightarrow M$ transition. Finally, cytokinesis can be inhibited by cytochalasin B treatment. According to Sunkara *et al.* (1979a), the same effect is obtained by treating cells with inhibitors of polyamine synthesis.

Clearly, physicians and cell biologists are well provided with dangerous, often lethal, drugs. Every step of cell division can be arrested at will. But we now have to face a more important problem: how is cell division controlled in normal, untreated cells? A great deal of investigation awaits us.

## I. The Regulation of Cell Division[6]

The cell cycle, which culminates in cell division, is so complex that its accurate regulation is a matter of life or death for the cell. It is therefore not surprising that cell division is not controlled by a single, simple regulatory factor. Instead, many different control factors (positive or negative) work together to regulate the rate of cell proliferation.

This section begins with an analysis of experiments in which cells at various steps of cell cycle progression have been fused together (this type of experiment has already been briefly mentioned in this chapter). It then examines the role played by stimuli (signals) received at the cell membrane in inducing mitogenesis. The role of endogenous, ubiquitous molecules that are believed to be important in growth regulation (in particular c-nucleotides and polyamines) will be studied next. Finally, the effects of a number of factors of cellular origin that stimulate or inhibit cell proliferation will be considered.

### *1. Cell Fusion Experiments (Hybrids and Cybrids)*

The important fact that the cytoplasm contains factors (unfortunately, of unknown nature) that control the cell cycle stems from the fundamental work of Henry Harris (summarized in book form in 1974). He fused bird erythrocytes, whose nuclei are completely inert in DNA transcription and replication, with different types of cells. The result of these experiments was that the bird nucleus was reactivated by fusion with the other cells. The erythrocyte will begin to synthesize RNA (this aspect of nuclear reactivation will be examined in more detail in the next chapter) if it has been fused with an RNA-synthesizing cell; if it has been fused with a cell that is synthesizing DNA—and only in that case—its nucleus will be induced to synthesize DNA (Fig. 44). Therefore, cells that synthesize DNA contain, in their cytoplasm, diffusible factors that can induce DNA synthesis in an inactive nucleus; these factors are not present in the cytoplasm of cells that are unable to synthesize DNA.

With the discovery of the cell cycle and the possibility of working with synchronized cell cultures, many cell fusion experiments were performed to analyze the cellular mechanisms that control DNA replication and cell division. We shall now examine the results obtained from these experiments. They have provided us with more information about the control of DNA replication during the S phase of the cell cycle than about the regulation of mitotic division.

In 1970, Johnson *et al.* discovered premature chromosomal condensation (PCC), which takes place in interphase nuclei after the fusion of a variety of cells with HeLa cells that have been arrested in metaphase by colcemid treatment. Later, Matsui *et al.* (1972) found that fusion of cytoplasm from cells in metaphase induces prophase in interphase cells. These experiments suggest that the

[6]Reviewed by Holley (1980).

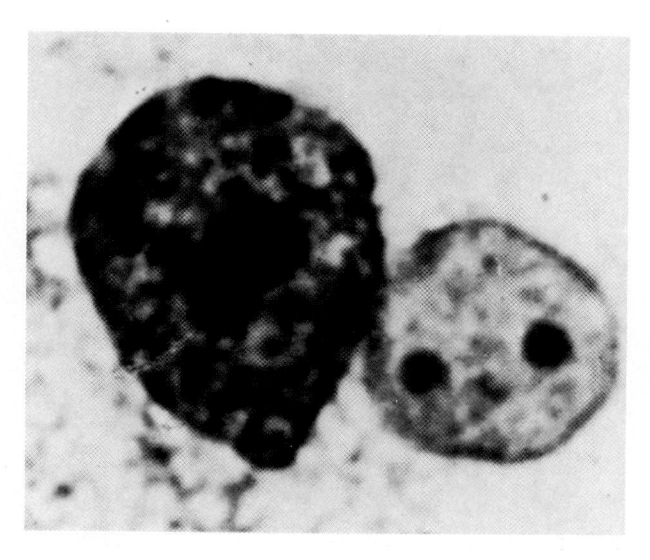

FIG. 44. Reactivation of a chick erythrocyte nucleus by fusion between chick erythrocytes and active cells (rat hepatoma cells in the present case). The figure shows a heterokaryon containing one hepatoma nucleus (left) and one reactivated chick erythrocyte nucleus (right). This nucleus has swollen, its chromatin has undergone decondensation, and two nucleoli have appeared. [Szpirer (1974).]

cytoplasm of cells in metaphase possesses the factors needed for the $G_2 \rightarrow M$ transition (histone H1 protein kinase, MPF, etc.).

A discussion of the control of DNA synthesis initiation during interphase (a process that is necessary but not sufficient for mitosis, as shown by polytenization and gene amplification) is appropriate. Marshall Graves (1972) showed that, in cell hybrids, the cytoplasm controls the initiation of DNA synthesis, but not the length of the S period. That the control factors are actually located in the cytoplasm is shown by the fact that in cybrids (resulting from the fusion of a cell with an anucleate cytoplast), reactivation of a chicken erythrocyte nucleus occurs, provided that the cytoplast has been separated from a cell that was actually replicating its DNA; otherwise there is no reactivation of DNA replication in the erythrocyte nucleus (Lipsich *et al.*, 1978). Cell fusion experiments by Rao *et al.* (1977, 1978) have shown that the agents that induce DNA synthesis are accumulated in $G_1$. They reach a peak during the S phase and then decrease. The accumulation of these inducers of DNA synthesis would therefore occur in the $G_1$ phase. Tsutsui *et al.* (1978), however, did not obtain the reactivation of chicken erythrocyte nuclei after fusion with cells in $G_1$. According to these authors, the factors that reactivated DNA synthesis did not appear before the S phase. That the $G_1$ phase may be reduced to a minimum is, as we have seen, also shown by the cell fusion experiments of Yanishevsky and Prescott (1978). If a cell in $G_1$ is fused with a cell in S, DNA synthesis is quickly induced in the $G_1$ nucleus. In a more detailed analysis of the cell cycle by cell fusion, Rao and Smith (1981) reached the following conclusions: induction of DNA synthesis by fusion with a cell in S is slower for a cell in $G_0$ than for a cell in $G_1$; this might be due to the absence, in $G_0$ cells, of some non-histone proteins required for the initiation of DNA synthesis that would be present in $G_1$ cells.

In fusion experiments, a cell does not always exert a positive effect on its partner, since cases of negative control have also been recorded. For instance, Rao *et al.* (1975) observed that, after fusion of cells in S and in $G_2$, the $G_2$ chromosomes failed to undergo condensation and the nucleus did not enter mitosis. The presence of the S cell nucleus thus inhibited the entry into mitosis of the $G_2$ cell nucleus. Rao and Smith (1981) have shown that cells in $G_0$, in contrast to cells in $G_1$, are incapable of inhibiting the progression of a $G_2$ cell toward mitosis. Thus, cells in $G_1$, but not those in $G_0$ or in S, possess cytoplasmic factors that delay mitosis in cells that are ready for prophase.

A factor that certainly plays a role in the outcome of cell fusion experiments is the age of the cell. Cell aging will be discussed in Volume 2, Chapter 3, but a paper by Yanishevsky and Prescott (1978) and a review by Yanishevsky and Stein (1981) should be mentioned here. According to the hybrid combination between young and aged cells, the senescent phenotype (i.e., the suppression of DNA synthesis) can be dominant or recessive. If a quiescent or senescent cell is fused with a cell in the S phase, entry into mitosis of the hybrid is inhibited. However, DNA synthesis in the S phase cell does not cease, indicating that mitosis could not occur because some event linked to the $G_2 \rightarrow M$ progression did not take place after fusion with the quiescent or senescent cell.

Finally, a few words should be said about the interactions between the two nuclei present in a binucleate cell obtained after cytochalasin B treatment. Fournier and Pardee (1975) found that there is cooperation between the two nuclei. The $G_1$ phase is accelerated and DNA synthesis is synchronous in the two nuclei. However, according to a report by M. V. N. Rao (1980), there is no such synchronism in 7% of the population. In these cases, one of the two nuclei seems to have "retired" from the cell cycle and to have become refractory to the cytoplasmic factors that induce DNA synthesis. Celis and Celis (1985) also found that in fused amniotic cells, the individual nuclei control DNA synthesis and the intranuclear distribution of cyclin.

There is no doubt that all of the work done on cell fusion has clearly demonstrated the presence in the cytoplasm of positive and negative factors controlling DNA synthesis and mitosis. However, after reading the papers devoted to this very interesting subject, the reader remains somewhat confused, since the chemical nature and mode of action of these still mysterious factors are unidentified. It will not be easy to identify them, because the yield of successful cell hybridization remains low. Thus, no meaningful biochemical work can be done. In fact, practically all of the work just summarized was done by autoradiography after

[$^{3}$H]thymidine incorporation. For this reason, we have learned far more about the control of DNA replication than about the other phases of the cell cycle, including mitosis, from cell fusion experiments. However, as mentioned before, we know that the breakdown of the nuclear membrane (lamina depolymerization) is controlled by diffusible factors present in the cytoplasm of the metaphase partners in hybrids between cells taken at different stages of the mitotic cycle (Jost and Johnson, 1983). We shall certainly learn much more from such experiments when we succeed in markedly increasing the percentage of successful cell fusions, or when suitable ultramicromethods are applied to the biochemical analysis of cell hybrids and cybrids.

### 2. *Cell Surface and Mitogenesis*

When still undifferentiated embryonic cells (fibroblasts, for instance) are seeded in a serum-containing medium, they first attach to the substratum and then begin to proliferate. After a period of logarithmic growth, the culture becomes confluent and the cells stop dividing. This arrest of cell proliferation is called "contact inhibition of growth"; a better term for this phenomenon is "density-dependent inhibition of growth." The fact that normal cells require a solid substratum for growth, and that they no longer synthesize DNA and divide when they come in contact in confluent cultures, shows that signals received at the cell membrane level must play an important role in the control of cell division. In cancer cells, as we shall see in more detail in Volume 2, Chapter 2, normal growth controls are lost. Malignant cells do not require a solid substratum for multiplication and do not behave like "well-educated" normal cells when they bump against each other. Thus, there are multiple cell divisions and the malignant cells pile up in multilayers. The loss of density-dependent inhibition of growth in malignant cells is probably due to deletion or alteration of plasma membrane components, since the addition of membranes from normal fibroblasts arrests DNA synthesis in malignant fibroblasts (Peterson and Lerch, 1983). Density-dependent inhibition of growth results from cell-to-cell contact and is certainly due to events that take place at the level of the cell membrane. It has been shown that the addition of plasma membrane preparations to sparse cultures of cells can mimic contact inhibition of growth (reviewed by Lieberman and Glaser, 1981).

In the remainder of this section, we shall deal essentially with tissue culture cells, particularly fibroblasts. The cell lines used for studies of the regulation of cell proliferation are diploid, and are thus genetically normal, at least at the chromosomal level. We have developed many mutant or variant cell lines that have been selected for resistance to some chemical. These cells are not malignant—but are they really normal? This point has already been raised in Chapter 3, and it is necessary to recall here that culture conditions, particularly the presence of serum, can profoundly modify cell physiology. In fact, it has been reported that hepatocytes, when cultured in a chemically defined (serum-free) medium, secrete a number of proteins in the medium (fibrinogen, albumin, and one of the globulins). The addition of small amounts of serum is enough to induce an increased secretion of these proteins within 1 hr. The authors (Plant *et al.*, 1981) correctly conclude that the effects of growth-stimulating factors should be studied in serum-free medium. This warning was needed because some contradictions in the literature were probably due to the use of different cell lines and different culture conditions. However, *in vitro* cultured cells divide, have a normal chromosomal complement, and can thus be taken as a model for normal cells.

One of the first strong indications of the importance of the cell surface in the control of cell division was the discovery that lectins (phytohemagglutinins, concanavalin A, etc.), which bind to the sugar residues of the cell surface glycoproteins, are mitogenic for lymphocytes (no objection to the use of serum in the culture medium can be raised in this case!). Binding of lectins to their membrane receptors induces RNA synthesis, followed by DNA synthesis and, finally, cell division. Lymphocytes respond in the same way to the addition of antigens, although in this case the mitogenic reaction is limited to immunocompetent cells. In contrast, lectins (ConA, in particular) induce DNA synthesis in many cell types—for instance, in chick embryo retina cells (Kaplowitz and Moscona, 1973, 1976). They may also have the opposite effects. As we have seen, certain lectins inhibit cleavage in *Xenopus* eggs (Tencer, 1978), in which DNA replication is so rapid that lectin binding to the egg surface cannot increase its rate. It is highly probable that, in cleaving eggs, lectins interfere mainly with cytokinesis. Another mitogenic stimulus, particularly in lymphocytes, is oxidation of the cell surface sugar residues by treatment with periodate (reviewed by O'Brien and Parker, 1976). In the same cells, a mitogenic response can also be induced by the calcium ion ionophore A23187 (Toyoshima *et al.*, 1976).

There is some evidence that the lectin-binding sites undergo modification during mitosis. Their lateral mobility changes during the mitotic cycle (Shoham and Sachs, 1974), and ConA binding increases three times during mitosis (Noonan *et al.*, 1973).

Finally, work done mainly in M. Burger's laboratory has shown that, for many cells, treatment with proteolytic enzymes is another mitogenic stimulus. Limited proteolytic digestion of cell membrane proteins induces mitosis and modifies the cell surface in such a way that cell agglutinability by ConA is greatly increased. Trypsin treatment increases RNA and DNA synthesis in the treated cells (Burger *et al.*, 1972; Noonan and Burger, 1973). Inhibitors of protease activity have no effect on the growth of normal cells but slow the multiplication of malignant transformed cells, which secrete a protease, plasminogen activator (Schnebli and Burger, 1979). According to Chen and Buchanan (1975), thrombin is also a potent mitogenic agent. Zetter *et al.* (1976) confirmed that trypsin and thrombin are mitogenic for fibroblasts, but found pepsin to be inactive.

We have seen that $Ca^{2+}$ is probably involved in the response of the cells to mitogens; we shall return to this point later. Are other ions also involved in the stimulation of DNA synthesis and cell division? According to Rubin and Koide (1976), DNA synthesis in tissue cultures is stimulated by both calcium and magnesium ions. However, they conclude from their analysis that the free $Mg^{2+}$ concentration is the main factor stimulating DNA synthesis. More recently, Rubin *et al.* (1981) have repeated their previous conclusion that $Mg^{2+}$ plays an important role in the regulation of normal cell growth, but they found that this role is lost in malignant transformed cells. Finally, Sanui and Rubin (1982) conclude that the main factor in stimulating cell growth by the addition of serum is the intracellular ionic concentration and the $pH_i$.

We have already mentioned that there is strong evidence for an important role of monovalent ions in the control of cell proliferation; no detailed discussion is needed here. We have seen that $Na^+$ entry into the cells is a mitogenic signal; when it results, as was first found in sea urchin eggs, from an $Na^+/H^+$ exchange, the $pH_i$ is increased in cells treated by mitogenic stimuli. It is believed that this increase leads to the initiation of DNA synthesis, but we do not know how this is achieved. In addition, Mummery *et al.* (1981) and Lopez-Rivas *et al.* (1982) have reported that a transitory influx of potassium ions is required for the $G_1 \rightarrow S$ transition in neuroblastoma cells. Thus, a high $K^+$ intracellular concentration is needed for the initiation of DNA synthesis. Maintenance of a sufficient $K^+$ influx is also required for the next phases of the cell cycle.

It is obvious from this summary that a great variety of stimuli received at the cell surface act as signals for the induction of DNA replication and cell division. How these signals are transduced from the cell membrane to the cell nucleus remains unknown.

### 3. *Transduction of Signals from the Cell Surface to the Nucleus*

Edelman (1976) suggested that the microtubular cytoskeletal system is the link between the cell membrane and the nucleus. This conclusion was based on experiments showing that the mitogenic effect of ConA on lymphocytes is inhibited by colchicine (Edelman and Yahara, 1976). More recent experiments by Friedkin *et al.* (1979) have shown that colchicine, on the contrary, enhances the response of tissue culture cells to the various growth factors (insulin, for instance); this will be discussed in the next section. The MTs would thus slow down the stimuli received at the cell membrane instead of facilitating the transfer of positive signals from the membrane to the nucleus. A further analysis of the question by Friedkin *et al.* (1980), besides confirming their earlier results, has shown that the positive effect of colchicine on DNA replication occurs during the $G_1$ phase of the cell cycle. A reinvestigation of the problem by McClain and Edelman (1980) led to the conclusion that colchicine may either inhibit or stimulate cell growth, depending on the cell type and the density of the cell population.

Finally, Otto *et al.* (1981) have found that the MTs must be disrupted for a long time (i.e., colchicine should be added in early $G_1$) in order to increase the rate of DNA synthesis initiation. They also observed additive effects between colchicine and one of the growth factors (the so-called EGF). According to Wang and Rozengurt (1983), colchicine and various agents that increase the cAMP content of the cells act synergistically to induce DNA synthesis in fibroblasts. Colchicine also accelerates cytokinesis in cultured cells, suggesting that MTs restrict the rate of furrowing (Hamilton and Snyder, 1983). According to Thyberg (1984), partial disassembly of the MT network might be an inherent step in the reactions that precede DNA synthesis and mitosis. However, the possibility that MT inhibitors have side effects should not be overlooked. Chou *et al.* (1984) found that colchicine increases the UTP pool and suggested that this might be the reason for the stimulation of RNA and DNA synthesis in colchicine-treated cells. Taxol blocks these increases in UTP pool size and nucleic acid synthesis.

In summary, it seems unlikely that the transduction of signals received by the cell surface from the nucleus occurs through the MTs. Although MTs probably play a role in this transfer, they seem to exert a negative rather than a positive control (reviewed by Otto, 1982). We know very little about the possible role of MFs and IFs in the initiation of DNA synthesis; according to a report by Maness and Walsh (1982), dihydrocytochalasin B, which disrupts the actin cytoskeleton, inhibits the initiation of DNA synthesis in fibroblasts; this suggests that, at least in these cells, the integrity of the actin cytoskeleton is required for entry into the S phase. If the transfer of signals from the membrane to the nucleus has a morphologically visible substratum, MFs and IF-sized filaments are as good candidates as MTs for such a role.

The necessity of a morphological substratum for signal transfer is not certain, since soluble "second messengers" such as calcium ions and cyclic nucleotides are probably involved in the control of cell proliferation (reviewed by Whitfield *et al.*, 1979).

According to Parker (1974), all of the mitogenic agents (lectins, proteases, periodate) increase the intake of $Ca^{2+}$ into lymphocytes. As we have seen, Toyoshima *et al.* (1976) induced mitotic activity in lymphocytes by adding the $Ca^{2+}$ ionophore A23187; this leads to an increase in free $Ca^{2+}$ in the cell, which might, in turn, increase the cell's cGMP content (Otani *et al.*, 1982). Evidence for a possible role of calmodulin was presented by Whitfield *et al.* (1979) in the aforementioned review article. It should be mentioned, however, that Rink *et al.* (1980) were unable to detect any fluctuation in the free $Ca^{2+}$ content during the cleavage of *Xenopus* egg (in which free $Ca^{2+}$ amounts to only 0.1 $\mu M$ compared to a total calcium store of 7 $\mu M$). It is possible that eggs may not behave like ordinary cells.

Evidence for the existence of cyclic changes in the cAMP content was presented in the discussion of biochemical changes during mitosis. In 1970–1975, strong evidence was obtained to indicate that a high cAMP content slowed down or prevented cell multiplication and that a decrease in cAMP was correlated with DNA replication and cell division. The reverse situation would hold for cGMP, although the experimental evidence is not as convincing as one might wish. Nevertheless, it is now customary to consider cAMP as a "stop" signal and cGMP as a "go" signal. This conclusion was based on a number of factors. When cultured cells reach confluence and stop dividing, their cAMP content increases. When DNA replication is induced by diluting the cells in fresh medium, for instance, their cAMP content decreases. Trypsin treatment activates DNA synthesis and decreases cAMP, whereas prostaglandins increase cAMP and decrease the rate of cell growth (Otten *et al.*, 1972). The conclusion that contact inhibition of growth is due to an increase in the cAMP content was confirmed by D'Armiento *et al.* (1973), who found an increase in cAMP when cells reached confluence. Addition of dibutyryl cAMP (which penetrates cells better than cAMP) slows down proliferation in many cells; the same effect is obtained in hamster cells with butyrate alone (Wright, 1973).

That treatment with trypsin, which stimulates cell proliferation, decreases the cAMP content has been observed by Burger *et al.* (1972) and by Noonan and Burger (1973). The decrease in cAMP, according to these authors, precedes the stimulation of RNA synthesis and the initiation of DNA synthesis. This has led Bombik and Burger (1973) to conclude that cell proliferation is regulated by fluctuations in the cAMP content. These fluctuations do not necessarily cause a decrease in cAMP, as shown by the work of Rozengurt and his colleagues on fibroblasts. As mentioned in Section II,B, they have repeatedly presented experimental evidence showing that an increase in the cAMP content of confluent, density-inhibited fibroblasts is a mitogenic signal (Rozengurt *et al.*, 1981c, 1983a,b).

Addition of cAMP to malignant transformed cells slows down their growth and modifies their phenotypic appearance. They appear more normal than they were before cAMP treatment (reverse transformation), although they have not lost their malignancy.

In a very general way, all of the effects of cAMP are mediated by cAMP-dependent protein kinases. This is probably also true for the control of mitotic activity since, as was already mentioned, both cAMP content and cAMP-dependent kinase activity decrease during mitosis and increase in the $G_1$ phase (Gray *et al.*, 1980). Protein phosphorylation may profoundly modify a cell; it is probably directly involved in the $G_2 \rightarrow M$ transition. As previously discussed, it seems well established that histone H1 phosphorylation plays a role in chromosome condensation, and there is also a good possibility that protein phosphorylation is involved in nuclear membrane breakdown at prophase. Several proteins, including the ribosomal S6 protein, are phosphorylated when DNA synthesis is reinitiated by the addition of serum (Pouysségur *et al.*, 1982; Martin-Pérez *et al.*, 1984).

We should mention that diadenosine tetraphosphate (A $P_4A$) is present in relatively large amounts in proliferating cells. If this adenine derivative is added to $G_1$-arrested cells permeabilized by an osmotic shock, DNA synthesis is induced (Grummt, 1978). Whether A $P_4A$ is an important factor in the regulation of cell growth or a mere biochemical curiosity remains to be seen.

If the case for intervention of cAMP in a positive or negative control of cell proliferation is very convincing and rests on strong evidence, the situation regarding cGMP is far from clear, since there are contradictory reports in the literature. While increases in cGMP have been observed in lymphocytes treated with mitogens (reviewed by Abell and Monahan, 1973), addition of cGMP failed to stimulate the growth of fibroblasts in the experiments of Nesbitt *et al.* (1976).

Besides cAMP, another group of substances is believed to play a key role in the control of cell growth. They are the polyamines, in particular, putrescine and spermidine [the role of polyamines in cell multiplication was reviewed by Tabor and Tabor (1976), Jänne *et al.* (1978), and Heby (1981)]. The key enzyme for polyamine biosynthesis (Fig. 45) is ornithine decarboxylase (ODC), which converts *l*-ornithine into putrescine. The next step is catalyzed by *S*-adenosylmethionine decarboxylase (SAM-DC); the reaction product is used for the synthesis of spermidine and spermine by spermidine and spermine synthases, respectively. In many instances, stimulation of cell growth by a great variety of processes is followed by a rapid and marked increase in ODC activity; its peak in enzymatic activity precedes RNA and DNA synthesis. Generally, ODC activity decreases after a few hours, because the enzyme is under very strict control (a very short half-life of only about 10 min, binding by an "antizyme," inhibition by polyamines, which might be due to enzyme modification by transglutaminase) (Russell, 1981). Some examples of ODC activity stimulation in connection with induction of growth are liver regeneration (partial hepatectomy induces a mitotic wave), malignant transformation of the liver into hepatoma, dilution of confluent cells, addition of serum to serum-deprived cells, treatment with phorbol esters endowed with cocarcinogen activity, and embryonic development [see Heby (1981) for other examples and more details]. Inhibitors of ODC (analogs of *l*-ornithine, such as α-methylornithine or difluoromethylornithine) arrest mitotic activity in cell cultures after one round of DNA replication (Mamont *et al.*, 1976). Analysis of control and treated cells showed that prior to treatment with the inhibitors, the cells possessed a polyamine store sufficient to allow one mitotic cycle to take place; the polyamine content then dropped markedly, and mitotic actitity stopped. In all of the cases studied so far, polyamine depletion by treatment with ODC inhibitors leads to the arrest of mitotic activity; biochemical studies have shown that, in general, DNA replication is more af-

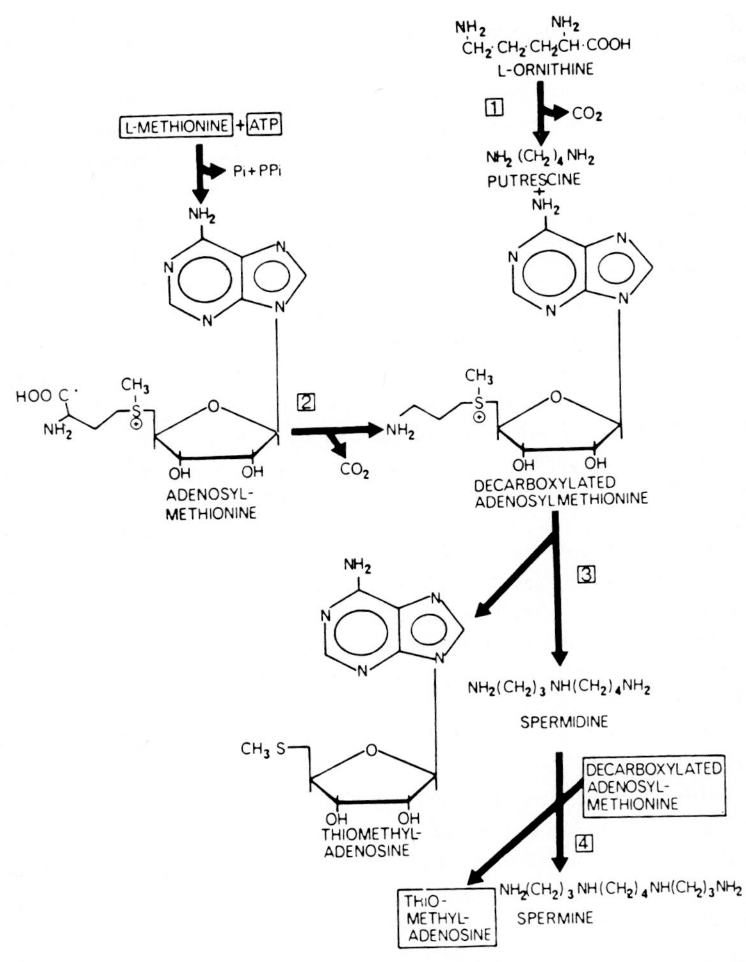


FIG. 45. Pathway of polyamine biosynthesis. (1) *L*-ornithine is decarboxylated into putrescine by the enzyme ornithine decarboxylase (ODC); ODC activity is the limiting factor in polyamine biosynthesis. (2) *S*-adenosylmethionine (SAM, a general donor of methyl groups) is decarboxylated by *S*-adenosylmethionine decarboxylase (SAM-DC). (3) Putrescine reacts with decarboxylated SAM to form spermidine under the action of spermidine synthase. (4) Spermidine, in the presence of spermine synthase, reacts with another molecule of decarboxylated SAM to form spermine. [Jänne *et al.* (1978).]

fected by polyamine depletion than RNA and protein synthesis. Finally, DNA replication and mitotic activity are resumed when putrescine or spermidine is added to polyamine-depleted cells. Careful analysis of the effects of the ODC inhibitors and of polyamine addition to polyamine-depleted cells shows that a sufficient polyamine content—spermidine in particular—is required for the $G_1 \rightarrow S$ transition (Pegg *et al.*, 1981). It has been shown that cells depleted of polyamines by treatment with difluoromethylornithine (DFMO) have a decreased ability to induce premature chromosome condensation in cell fusion experiments; addition of polyamines restores this ability, suggesting that putrescine and spermidine somehow control chromosome condensation (Sunkara *et al.*, 1983). Unfortunately, we do not know precisely why polyamine synthesis is required for DNA replication and mitosis, although it is well known that polyamines (like magnesium ions) bind to DNA, tRNAs, and ribosomes (see Tabor and Tabor, 1976, for details). However, *in vitro,* spermine (at relatively high concentrations) is more effective than spermidine and, especially, putrescine; the reverse is generally true *in vivo.* A report by Tomita *et al.* (1981) indicates that the increase in ODC activity induced by the growth factor EGF is not necessary for the stimulation of DNA synthesis in hepatocytes. A report by Koenig *et al.* (1983) furnishes a possible explanation of the favorable effects of polyamines on DNA synthesis and mitosis. Experiments with ODC inhibitors have shown that synthesis of ODC and polyamines is indispensable for the $Ca^{2+}$ fluxes that occur when quiescent cells are induced to proliferate.

ODC, unlike adenylcyclase, is a soluble enzyme. Since an important part of adenylcyclase is bound to the plasma membrane, one could imagine a direct response of the enzyme to signals received at the cell surface. This can hardly be the case for ODC. There is at least one report (O'Brien *et al.*, 1976) showing that the very large increase in ODC activity induced by phorbol esters is suppressed in the presence of colchicine. It would be interesting to know whether this is a general situation and whether MTs are really involved in ODC activation and polyamine synthesis.

### 4. *Cell Growth-Stimulating Factors*[7]

A great variety of substances present in the cells or excreted from them stimulate DNA synthesis and cell division. In general, the stimulating factors that have been isolated from actively dividing cells are glycoproteins with a high $M_r$; in contrast, the growth factors excreted by the cells into the medium or those present in serum have a much lower $M_r$ (15,000 or less). These factors are similar to the pancreatic hormone insulin. From many, if not all, cells that rapidly proliferate, it is possible to isolate factors that induce or stimulate DNA synthesis in nuclei isolated from adult livers (which hardly incorporate labeled

[7]Reviewed by Pardee *et al.* (1978), Holley (1980), and Rozengurt (1983b).

thymidine). Such a DNA synthesis "initiating factor" has been isolated from fertilized *Xenopus* eggs by Benbow and Ford (1975). They suggested that it might be an endonuclease that recognizes palindromic sequences. A similar factor has been isolated from the nuclei of sea urchin embryos (Murakami-Murofushi and Mano, 1977). In this case, the initiating factor stimulates DNA synthesis in isolated nuclei from sea urchin embryos and has an $M_r$ of 220,000. Such DNA synthesis-stimulating factors are not limited to eggs and embryos and display no tissue specificity. Thompson and McCarthy (1973) and Jazwinski *et al.* (1976) found that extracts from all proliferating cells increase DNA synthesis in isolated liver nuclei; the $M_r$ of these factors is higher than 50,000. Similar large, nondialyzable factors have been found in HeLa cells (Fraser and Huberman, 1978), Chinese hamster ovary cells (Reinhard *et al.*, 1979), and *Physarum* (Brewer, 1979), in which the DNA synthesis-stimulating factor is somewhat smaller; it is a glycoprotein of $M_r$ 30,000.

A large number of growth factors of small $M_r$ are present in serum and in "conditioned media" (i.e., media in which cells have been cultured for some time). The main natural growth-promoting factors are, in addition to the already mentioned EGF and PDGF (Section II,B, fibroblast-derived growth factor (FDGF), fibroblast growth factor (FGF), and the two insulin-like growth factors (IGF-1 and IGF-2) (somatomedins). But many other substances also stimulate cell proliferation and may be considered as artificial growth factors. Among them are the pituitary hormones vasopressin and oxytocin; the phorbol esters and teleocidin, which promote tumor growth (see Volume 2, Chapter 3); and retinoic acid, melittin, cholera toxin, dibutyryl-cAMP, and colchicine, according to Rozengurt's (1983b) review article.

Curiously, a rich source of EGF (reviewed by Carpenter and Cohen, 1979) is the submaxillary salivary gland of the male rat. It is curious because this gland secretes two other low molecular weight substances endowed with biological activity. One is the nerve growth factor (NGF), which stimulates the differentiation of neurons and will be discussed in Volume 2, Chapter 2; the other is renin, a hypertensive substance that is also produced by the kidney.

EGF induces mitosis in fibroblasts (Rose *et al.*, 1975) after binding to species-specific receptors; the EGF–receptor complex is then endocytosed and degraded (Carpenter and Cohen, 1976). The EGF receptor [reviewed by Hunter (1984)] possesses a signal peptide of 24 amino acids. The mature protein is made of 1186 amino acids and possesses three distinct domains: the N-terminal region is external to the cell and contains the EGF-binding site; the central domain spans the cell membrane; and the C-terminal domain has probably a tyrosine kinase activity. The gene coding for the EGF receptor is amplified 3–100 times in a malignant cell line (A 431) which possesses an unusually large number of receptors on the cell surface. It encodes two RNAs of 10 and 6 kb and a small RNA of 3 kb. Internalization and degradation of the growth hormone receptor complexes are required for mitogenesis and are inhibited by lysosomotropic amines (King *et al.*, 1981). EGF, which is composed of 53 amino acids and has a $M_r$ of 6000, forms a complex with an arginine esteropeptidase (Lembach, 1976). It is derived from a much larger precursor by limited degradation through this arginine esterase (Server *et al.*, 1976). NGF is also linked to an arginine esterase; the two enzymes are similar, but not identical. Interaction of EGF with its receptor activates a cAMP-independent phosphorylation system. The substrates for phosphorylation are several membrane proteins, including the EGF receptor itself (Cohen, 1981; Ehrhart *et al.*, 1981). Interestingly, EGF stimulates the phosphorylation of a tyrosine residue in the receptor; as a rule, protein kinases phosphorylate serine and threonine residues, but not tyrosine residues. Tyrosine phosphorylation was believed to be specific for cells infected with cancer viruses, as we shall see in Volume 2, Chapter 3. Activation of tyrosine-specific protein kinases closely associated with specific plasma membrane receptors is observed not only for EGF [see Cohen *et al.* (1982) for details about the EGF receptor protein kinase] but also for PDGF (Nishimura *et al.*, 1982; Ek and Heldin, 1982), IGF-1 (Rubin *et al.*, 1983; Jacobs *et al.*, 1983), and insulin (Kasuga *et al.*, 1982a,b; Petruzzelli *et al.*, 1982; Van Obberghen *et al.*, 1983). It seems to be an absolute prerequisite for stimulation of cell proliferation.

Another protein that is phosphorylated in cells treated with EGF (or serum) is the S6 ribosomal protein (Nilsen-Hamilton *et al.*, 1982). It is believed that S6 protein phosphorylation is required for entry of the cell into $G_1$; it might modify the affinity of the ribosomes for some, but not all, mRNAs (Decker, 1981). If so, phosphorylation of the S6 ribosomal protein might play a role in the control of translation specificity. As already mentioned, EGF also stimulates ODC activity leading to increased polyamine synthesis; this increase is not required for DNA replication, at least in hepatocytes (Tomita *et al.*, 1981).

We have already mentioned that EGF is not a fast-acting mitogen and that it exerts pleiotropic effects (activation of the plasma membrane [$Na^+$,$K^+$]–ATPase, increased transport of amino acids and glucose, influx of $Na^+$ ions, etc.) before it induces DNA replication; this multiplicity of signals makes it impossible to pinpoint the biochemical changes that are directly responsible for the initiation of DNA synthesis. Interesting observations have been reported by Sherline and Mascardo (1982a,b). They found that EGF speeds up the separation of the centromeres; this effect of EGF is inhibited by trifluoperazine, cytochalasin, and taxol, and is enhanced by the MT poisons colchicine and nocodazole. Sherline and Mascardo (1982a,b) suggest that EGF first increases the free $Ca^{2+}$ content of the cells. This would activate calmodulin, which would stimulate the contraction of MFs attached to the centrosomes. This contraction would finally result in the separation of the centrosomes.

PDGF (reviewed by Antoniades and Williams, 1983) is a heat-stable growth factor that is released from blood platelets during blood coagulation. It exists in two different forms (PDGF-1 and PDGF-2, with weights of 32,000 and 34,000 daltons, respectively; like insulin, they are composed of two polypeptide chains linked by —SS— bridges. Like EGF, PDGF exerts many effects. It stimulates growth and migration in many cell types, and increases amino acid transport, protein synthesis, phospholipid turnover, etc. Today there is a growing interest in PDGF because, as we shall see in Volume 2, Chapter 3, it is very similar to the products of genes involved in malignant transformation (oncogenes). Interestingly, cytoplasts isolated from PDGF-treated cells transfer the growth response to untreated cells after fusion (Smith and Stiles, 1981). A cytoplasmic step is therefore involved in the response of the nucleus (DNA replication) to the growth factor.

Warden and Friedkin (1984) have shown that many, and perhaps all, growth factors increase the synthesis of phosphatidyl choline in fibroblasts, suggesting an effect on the lipids of the cell membrane. An article by Marx (1984), dealing with still unpublished work, reported that when the receptors of the growth factors are activated by tyrosine phosphorylation, phosphatidyl inositol-4 phosphate is hydrolyzed with the production of diacylglycerol and inositol triphosphate. This may be important, since diacyl glycerol activates the phospholipid, $Ca^{2+}$-dependent protein kinase C, while inositol triphosphate is involved in the release of calcium ions from internal stores (in particular, the endoplasmic reticulum). There is no doubt that our understanding of the changes induced by the growth factors in the plasma membrane and the cytoskeleton will increase rapidly in the near future.

In short, there are many early cell responses to EGF and PDGF: phosphorylation of membrane and S6 proteins, stimulation of $NA^{+},K^{+}$–ATPase and ODC activities, and changes in the organization of the MFs leading to changes in morphology. It seems, however, that these early changes are not sufficient to induce DNA synthesis and that later changes, including an increase in cellular pH (Burns and Rozengurt, 1983), are required (Yarden *et al.*, 1982).

The somatomedins (reviewed by Rothstein, 1982) are insulin-like peptides produced by the liver in response to pituitary growth hormone; this family of growth factors is composed of somatomedins A and C, the insulin-like factors IGF-1 and -2, and multiplication-stimulating activity (MSA). The last was first detected in liver cell–conditioned medium by Dulak and Temin in 1973; an analysis by Marquardt *et al.* (1981) shows that MSA has an $M_r$ of 7484 and that its amino acid sequence shows 93% homology with IGF-2. MSA stimulates the proliferation of hepatocytes, the cells that release it in the conditioned media (autocrine control of cell growth). According to a report by Fryklund *et al.* (1974), blood plasma contains a somatomedin B, which is formed from four polypeptides; it has an $M_r$ of 5000, and it stimulates DNA synthesis and cell division. Like FDF and FDGF, the somatomedins are internalized by endocytosis after binding to specific receptors.

There are also other growth-stimulating factors. For instance, Gospodarowicz (1975) discovered an FDGF ($M_r$ 14,300) that strongly stimulates DNA synthesis; it has similarities to PDGF (Dicker *et al.*, 1981). Rozengurt *et al.* (1981b) found that vasopressin, a nonapeptide pituitary hormone, increases ODC activity and is mitogenic for fibroblasts.

Immunologists are very interested in growth factors called "lymphokines," because cell division of lymphocytes requires two factors called "interleukin-1" and "interleukin-2"; interleukin-1 is produced by macrophages and induces the production of interleukin-2 by the lymphocytes themselves. Proliferation of T cells requires transferrin (the protein that allows the uptake of plasma iron in the cells) in addition to interleukin-2 (called the "T-cell growth factor" by Gallo *et al.*, 1983), according to Neckers and Cossman (1983).

Other growth factors, which also act selectively on certain cell types, have been discovered. The serum of partially hepatectomized animals contains a hepatopoietin that induces DNA synthesis in adult hepatocytes (Michalopoulos *et al.*, 1982). A glial growth factor present in the brain and pituitary stimulates the division of Schwann cells (Lemke and Brockes, 1983). Gilchrest *et al.* (1984) discovered a keratinocyte growth factor of 1700 daltons in the hypothalamus; Gajdusek (1984) found that endothelial cells produce endothelial cell–derived growth factors (ECDGF) which support DNA synthesis in smooth muscle cells and fibroblasts. Some ECDGFs, but not all, display similarities with PDGF. A PDGF-like factor is secreted by aortic smooth muscle cells in young rats, but not in the adults (Seifert *et al.*, 1984).

Also deserving of mention is the colony-stimulating factor (CSF) discovered by Sachs (1980). It is released in the medium and specifically stimulates the proliferation of the granulocyte and macrophage progenitor cells. This factor, which is as important for cell differentiation and cancer studies as for cell proliferation, has been shown to be a sialoglycoprotein (Tsuneoka *et al.*, 1981).

Other powerful mitogenic agents are the phorbol esters present in croton oil. Since they are mainly used as cocarcinogens, they will be discussed in Volume 2, Chapter 3, as will those factors that induce a wave of cell divisions prior to differentiation (hematopoietin, for instance).

There are some contradictory reports about the effects of histones on cell division. In 1964, the author found that injection of a poorly defined mixture of histones into amphibian eggs arrested cleavage. According to Pehrson and Cole (1980), histone $H_1^\circ$ inhibits replication in cell cultures. However, in a more recent paper, Kundahl *et al.* (1981) reported that the addition of histone H1 or poly-lysine increases RNA synthesis and cell division in lymphocytes and mouse erythroleukemic cells; fiber autoradiography shows a decrease in replicon size in the chromatin of the treated cells. More work is obviously needed before these discrepancies can be explained.

Finally, plant cells have their own growth factors called "cytokinins." They are purine derivatives, the main representative of which is kinetin (dimethylallyl aminopurine), which is believed to act in the $G_2$ phase and to accelerate the entry into prophase.

Cells contain many agents that stimulate DNA synthesis and mitotic activity; many of them are present in the serum that is added to cultured cells. Thus, serum is not only a nutrient for the cells but also a rich source of growth factors. Since normal cells do not grow forever, negative controls must also exist; the substances that specifically slow the mitotic rate are called ''chalones.''

### 5. *Cell Growth-Inhibitory Factors: Chalones*[8]

It is, of course, much easier to inhibit cell division than to promote it. Any toxic agent will arrest mitosis before it kills the cell. Therefore, in order to demonstrate that cells possess their own inhibitory factors, more stringent criteria than the mere arrest of cell division are needed. These criteria were established by Bullough (1962) when he found that skin epidermis cells contained a factor that inhibited mitotic activity in these cells, but not in others; this factor displayed no species specificity. Bullough (1962) called these tissue-specific, but non-species-specific, antimitotic agents ''chalones''. Since that time, chalones have been found in many tissues and organs, and some progress has been made in their isolation and characterization. Like the growth factors, they seem to belong to two different chemical categories: small glycopeptides or large proteins with an $M_r$ between 30,000 and 50,000.

The epidermis has two different chalones, according to Thornley and Laurence (1976). One of these is called the ''$G_2$-epidermal chalone'' because it arrests cell division in $G_2$ and not at the $G_1 \rightarrow S$ transition, like the other epidermal chalones. It has been partially characterized by Isaksson-Forsén *et al.* (1981), who showed that its weight is lower than 10,000 daltons.

In 1970, Bullough and Laurence discovered a lymphocyte chalone and described it as thermostable, tissue specific, and devoid of species specificity. According to Houck *et al.* (1971) and Attallah and Houck (1977), it has an $M_r$ of about 40,000, and it inhibits the stimulation of DNA synthesis induced in erythrocytes by photohemagglutinins; it is not species specific. In 1977, Houck *et al.* found, in the conditioned medium of fibroblasts, a chalone ($M_r$ 40,000) that selectively inhibits DNA synthesis in fibroblasts. More recently, Steck *et al.* (1982) confirmed that the conditioned medium of density-inhibited fibroblasts inhibits the growth of fibroblasts in sparse cultures; this fibroblast growth regulator is a protein with an $M_r$ of 10,000–30,000.

But Houck *et al.* (1977) later noted that the lymphocyte and fibroblast chalones have an $M_r$ of only 5000, are basic, and probably contain mannose. In their last report, however, Patt and Houck (1980) concluded that the $M_r$ of lymphocyte and granulocyte chalones was only 600–700, while that of the fibroblast chal-

[8]Reviewed by Houck and Hennings (1973).

ones was below 10,000. If one can draw any conclusion from these rather confusing results, it is that there are probably two kinds of chalones: small peptides and average-sized proteins.

Chalones have been found in cells and tissues other than epidermis, fibroblasts, and lymphocytes. They are also present in liver, kidney, mammary glands, heart, and, presumably, in other organs as well. For instance, Chopra and Simnett (1971) demonstrated the existence of kidney and liver chalones by transplantation experiments in *Xenopus* embryos. Adult kidney contains two growth-inhibitory factors that arrest the cell cycle in $G_1$; their effect can be reversed by the addition of EGF. These factors act on epithelial cells, but not on fibroblasts (Holley *et al.*, 1980); cultured epithelial cells from Simian kidneys also produce a chalone (Holley *et al.*, 1983). The liver chalone is a polypeptide of low $M_r$ that inhibits DNA synthesis in regenerating liver and has no species specificity (Verly *et al.*, 1971; Sekas *et al.*, 1979). Another chalone has been isolated from mammary glands by Gonzalez and Verly (1976). More recently, Kriek *et al.* (1981) isolated a chalone from cardiac muscle, which was a large glycoprotein ($M_r$ 715,000) and which displayed tissue, but not species, specificity.

Part of the current interest in growth-promoting factors, such as EGF or the somatomedins, and in the growth-inhibitory chalones stems from their potential applications in medicine. Growth factors might help to heal wounds, especially in the aged. However, excess growth factor might prove carcinogenic in clinical trials, and this possibility should not be overlooked, particularly for EGF and PDGF; their ability to induce phosphorylation of tyrosine residues, such as the protein kinases produced by tumor viruses, make them somewhat suspect. Chalones might slow down malignant growth. Thus, if one or the other of the factors that stimulate or inhibit cell division can be produced by genetic engineering, it would provide new avenues of treatment and unlimited potential applications.

Another substance is now being manufactured by bacteria that also inhibits cell proliferation (and virus multiplication): interferon. Although it slows the growth of malignant cells *in vitro,* it does not seem very likely to become a general and complete cure for cancer. Interferon synthesis is induced in fibroblasts and leukocytes by treatment using double-stranded RNA molecules. We know little about its mode of action on cells, although a paper by C. Wang *et al.* (1981) reports that interferon inhibits the multiplication of malignant carcinoma cells and increases both membrane rigidity and the number of polymerized actin filaments in the cell cortex. Virologists interested in the antiviral properties of interferon have found that one of its early biochemical effects is the synthesis of an enzyme (2′–5′ oligoadenylate synthetase) and the accumulation of its product [an oligoadenylate called, in abbreviated form, (2′–5′)A]. According to Smekens-Etienne *et al.* (1983), (2′–5′)A synthetase is located in the nuclei of liver cells and decreases in activity after partial hepatectomy. The oligonucleotide (2′–5′)A is present in liver cells of animals that have neither been treated with interferon nor infected with a virus. Since there is a decrease of (2′–5′)A in the liver after partial hepatectomy, Smekens-Etienne *et al.* have concluded that this oligoadenylate represses DNA synthesis in normal cells. Similar results have been obtained by Creasey *et al.* (1983), who found an inverse relationship between the (2′–5′) oligoadenylate synthetase level and the percentage of S-phase cells. Since they also found that sera raised against interferon stimulate cell proliferation, they concluded that interferon is a negative growth factor. It should be added that it is unlikely that (2′–5′)A is a universal regulator of DNA synthesis. Its role seems to be limited to mammalian cells capable of responding to interferon, since it is not detectable in amphibian cells (personal communication by Dr. J. Content).

## J. A Brief Comparison between Mitosis and Meiosis

Figure 46 shows schematically the well-known events that lead from a spermatogonium to four haploid spermatozoa and from an oogonium to a single

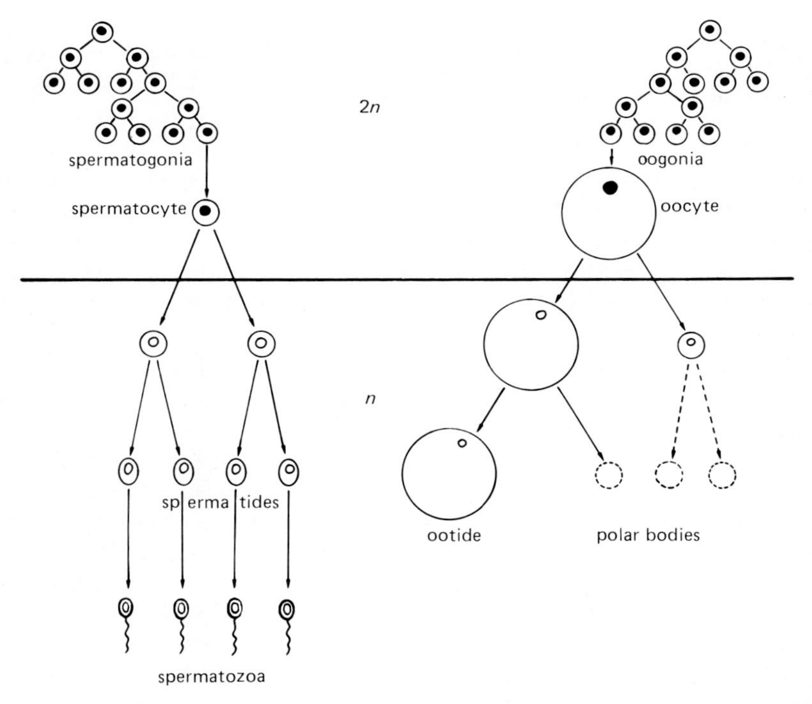


Fig. 46. A classic diagram of spermatogenesis (left) and oogenesis (right). [Original drawing by P. Van Gansen.]

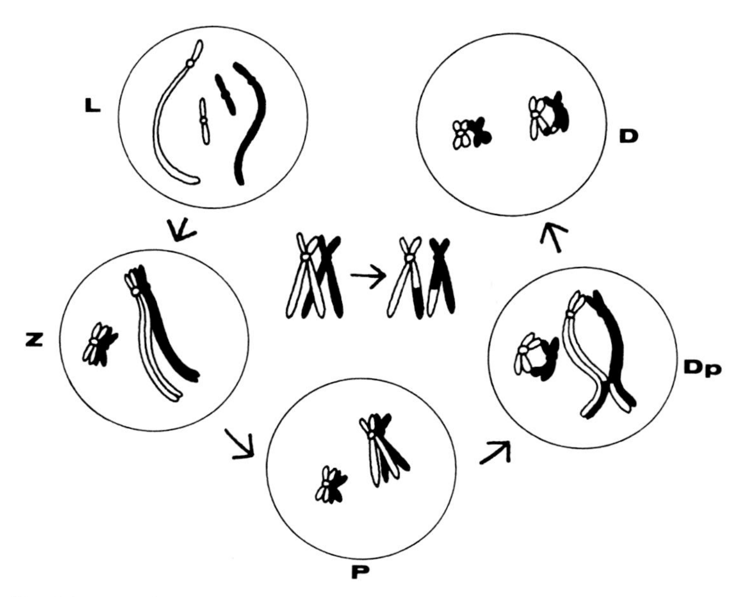


FIG. 47. Diagram of meiotic prophase I. L, leptotene stage; Z, zygotene stage; P, pachytene stage; Dp, diplotene stage; D, diakinesis. The crossing-overs (shown in the middle) take place in the pachytene stage. [Original drawing by P. Van Gansen.]

haploid ootid and two polar bodies. In comparing this schematic representation of meiosis with that of mitosis, one is struck by the fact that meiotic division is reductional (i.e., it leads to haploidy), while mitotic division is equational. Reductional division is due to the fact that, during meiosis, the kinetochores do not divide at metaphase-anaphase. The molecular reasons for this remain unknown.

There is another major difference between meiosis and mitosis. Genetic recombination, through crossing-over, occurs between homologous chromosomes at the pachytene stage (Fig. 47). An equivalent to crossing-over, in somatic cells, is chromatid (sister or nonsister) exchange (SCE). As we have seen, SCE is an extremely rare event during normal cell division. It is a sensitive symptom of DNA damage, and we have therefore described it as one of the chromosomal abnormalities that can occur during mitosis in injured cells. Even in somatic hybrids, in which recombination events might have been expected to take place at a higher rate than in normal cells, mitotic recombination remains a very rare event (Tarrant and Holliday, 1977). Since DNA damage is followed by DNA repair, one would expect DNA repair mechanisms to be at work during the pachytene stage of meiosis. This has been conclusively shown by the work of Stern and Hotta (reviewed in 1980) on DNA metabolism during meiosis.

Finally, another characteristic of meiosis is the fact that the homologous chromosomes are held together by the so-called synaptinemal complexes (Fig. 48) that have been recently isolated, at the pachytene stage of meiosis, from hamster testis by Walmsley and Moses (1981). Their structure can be studied under the light microscope after silver staining (Dresser and Moses, 1980; Fig. 49) or under the electron microscope. Their main components are two lateral elements (40–60 nm thick) and one central element (20–40 nm thick). The central element is a DNA fiber, while the lateral elements are composed of proteins and RNA (Solari, 1972; Esponda and Giménez-Martín, 1973). Transverse filaments link the lateral elements, which end in ''attachment plates'' and fibrous material associated with the annuli of the nuclear membrane pore–lamina complexes (Walmsley and Moses, 1981). The synaptinemal complexes thus constitute a device for holding the paired chromosomes firmly together. This should facilitate the exchange of DNA segments during genetic recombination. However, the synaptinemal complexes are so long that precise association between paired chromosomes required for crossing-over is difficult to visualize. Carpenter (1975, 1981) has shown that genetic recombination may take place in *Drosophila* mutants that lack synaptinemal complexes; the chromosomes are held together by electron-dense structures called ''recombination nodules'' (Carpenter, 1975). These nodules become associated with synaptinemal complexes at pachytene. It has been suggested by Holliday (1977) that, at the molecular level, a fibrous protein of the synaptinemal complex might bind to specific DNA sites during meiosis. Synaptinemal complexes have been studied in rat spermatocytes at the pachytene stage of meiosis. It was shown that the synaptinemal complexes are composed of nonhistone proteins and DNA (Li *et al.*, 1983) and that they are integral components of the nuclear matrix (Ierardi *et al.*, 1983). In the ovary of the insect *Ephestia,* the synaptinemal complexes are also associated with the nuclear matrix (Raveh and Ben-Zéev, 1984). This seems to be a general feature of both spermatogenesis and oogenesis.

A general model for genetic recombination at the molecular level has been proposed by Meselson and Radding (1975). This model (Fig. 50) is accepted by a majority of those who work in the field. It proposes that genetic recombination begins with an asymmetrical transport of single-stranded DNA that, after isomerization, would become a symmetrical, double-stranded exchange. The model implies that DNA recombination involves DNA synthesis; this corollary of the model fits perfectly with the experimental data published by Hotta and Stern.

Hotta and Stern showed the existence of DNA repair replication at pachytene in *Lilium;* it is linked to the synthesis of a repair enzyme (1971a) and of a DNA-binding protein that catalyzes the renaturation of DNA (thus, the formation of double-stranded DNA) and that would facilitate genetic recombination. Hotta and Stern (1971b) have suggested that this DNA-binding protein allows precise pairing of chromosomal genetic loci at pachytene. By the zygotene stage, the

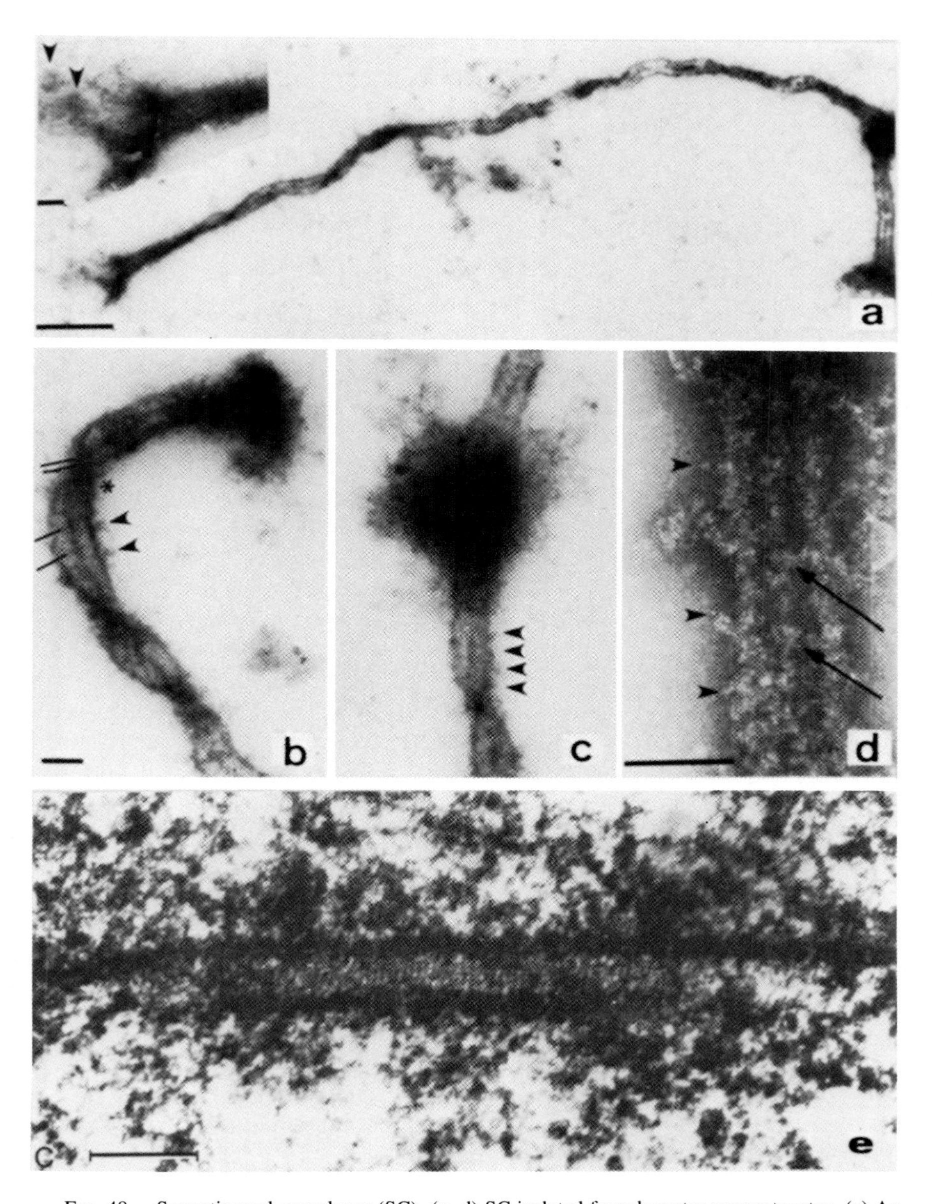


FIG. 48. Synaptinemal complexes (SC). (a–d) SC isolated from hamster spermatocytes. (a) An acrocentric SC; note the tuft extending from the long arm attachment plaque containing annuli (inset: arrowheads). (b) A portion of SC showing stubs (arrowheads), fine transverse filaments (lines), and a twist (asterisk). (c) Kinetochore region showing a large, dense mass of associated fibrillar material. Stubs (arrowheads) are again visible. (d) A portion of SC from a similar preparation, but negatively stained with ammonium molybdate; stubs (arrowheads) and transverse filaments (arrows) are visible. (e) SC observed *in situ* in mid-pachytene *Bombyx* spermatocytes. [(a–d) Walmsley and Moses (1981); (e) Holm and Rasmussen (1980).]

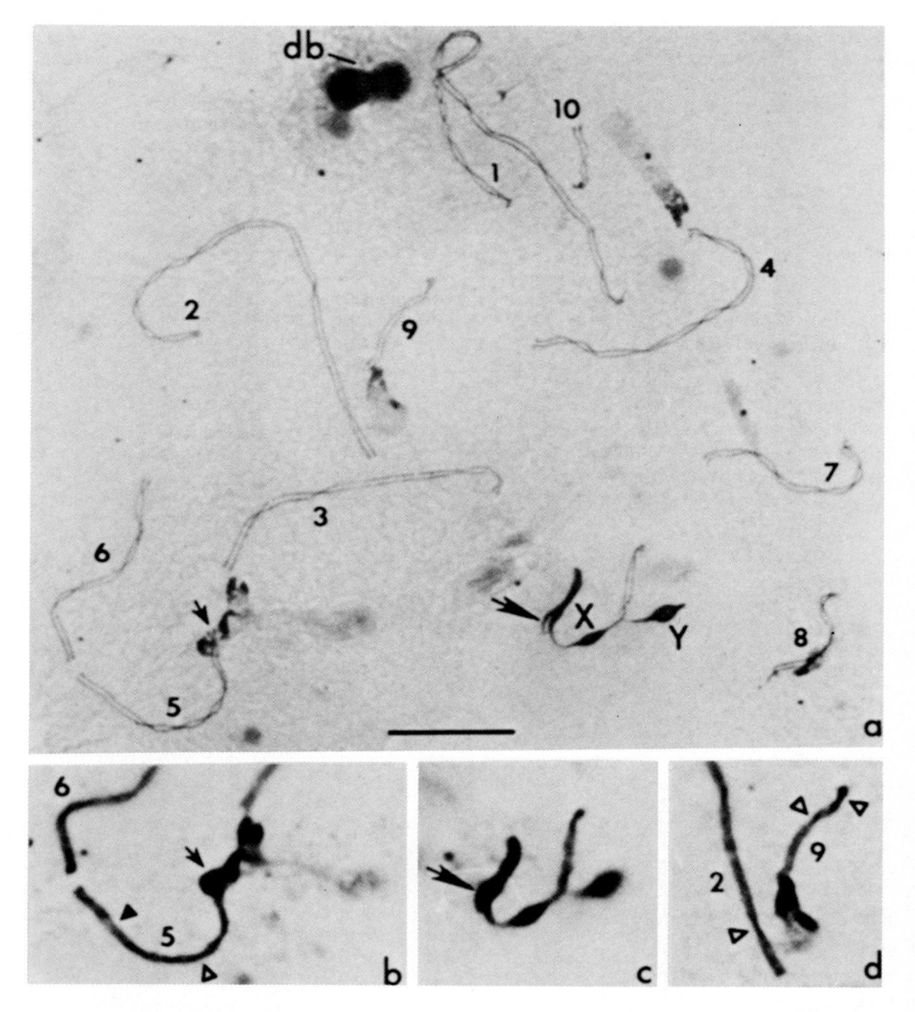


FIG. 49. Synaptinemal karyotyping in spermatocytes of the Chinese hamster. Light (b–d) and electron microscopy (a) of synapsis and nuclear development shown by a silver method. (a) Comet-shaped nucleoli are associated with the ends of synaptinemal complexes 3–5, 8, and 9. (b–d) Light micrographs of synaptinemal complex 5, the XY pair, and synaptinemal complex 9, respectively. [Dresser and Moses (1980).]

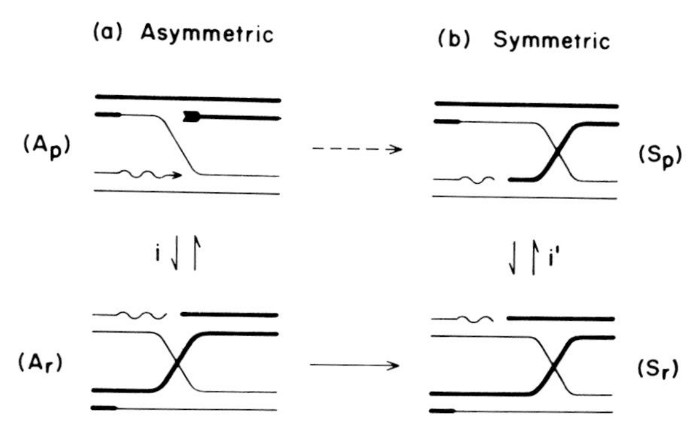


FIG. 50. The Meselson–Radding (1975) general model for genetic recombination. In the asymmetric phase (a), heteroduplex DNA is formed on only one chromatid. In the first specified intermediate, $A_p$, the flanking arms are in a parental configuration. By isomerization (i), the structure Ar is produced, in which heteroduplex DNA is still restricted to one chromatid, but the flanking arms have acquired the recombinant configuration. Branch migration driven by rotary diffusion converts $A_r$ to $S_r$, and heteroduplex DNA is subsequently formed symmetrically on both chromatids. The configuration of the flanking arms in (b), the symmetric phase, can be rearranged by the isomerization i′. Structure $S_p$ might also be produced directly from Ap by rotary diffusion (dashed arrow). The interruption shown in one strand in Ar is closed by a polynucleotide ligase at some specified time.

synaptinemal complexes have already allowed a coarse pairing between homologous chromosomes. Stern and Hotta (1977) have suggested that DNA synthesis during meiosis is of the repair type. They found that an endonuclease removes 100–200 bp at precise sites of the DNA molecules at pachytene; this local nicking would be followed by DNA repair during meiosis. The universal mechanism for crossing-over proposed by Hotta *et al.* in 1977 is as follows: after the appearance of a protein (called "protein R") that favors the reannealing of DNA, single-stranded nicks would be formed and then repaired, particularly at the level of moderately repeated sequences (present in 1000–2000 copies and representing 0.1% of the total DNA). Further work showed the existence, in meiotic cells, of an ATP-dependent, DNA-unwinding enzyme that is also a DNA-dependent ATPase. The extent of unwinding is 50 bp at the end of the DNA molecule and 400–500 bp at each nick (Hotta and Stern, 1978). Thus, the function of protein R in *Lilium* anthers is to destabilize the DNA double helix; protein R is activated when it is phosphorylated by a protein kinase present in meiotic lily cells (Hotta and Stern, 1979). Finally, Hotta and Stern (1981) found that snRNAs were synthesized at the time of pairing, which would allow accessibility of specific DNA sequences to endonuclease nicking. More recently, Hotta *et al.* (1984) found that a 70-kilodalton protein (called "leptotene protein") is present in the nuclear membrane of preleptotene, leptotene, and zygotene cells. Its function is to stop the replication of zygotene DNA sequences at the leptotene stage. This protein binds specifically to 90 bp of zygotene DNA sequences and has an ATP-dependent endonucleolytic activity, which could play a role in the initiation of synapsis. This protein might be responsible for the irreversible commitment to meiosis at the end of the preleptotene S phase. Finally, Hotta *et al.* (1985) found that the DNA which replicates at the zygotene stage (0.2% of the genome) is transcribed in an RNA (*zyg* RNA) which is absent before leptotene and disappears at mid-pachytene.

There is good reason to believe that the enzyme involved in the repair of the nicks during meiosis is DNA polymerase-β. Grippo *et al.* (1978) found, in cells isolated from mouse testes, that both DNA polymerases-α and -β are present in spermatogonia and young spermatocytes; but at pachytene, only the β enzyme can be detected, suggesting that it is implied in genetic recombination. Orlando *et al.* (1984) have also found that meiotic male cells possess a high-uracil-DNA glycosylase activity. This enzyme catalyzes the excision of uracil from DNA in which it has been misincorporated; it is involved in postreplicative DNA repair processes (Lehmann and Karran, 1981).

As one can see, genetic recombination at meiosis is a highly complex biochemical process, and is very different from the semiconservative DNA replication (catalyzed by DNA polymerase-α) that takes place during the S phase of the somatic cell cycle. The main biochemical events at pachytene seem to be the appearance of DNA polymerase-β activity; an endonuclease that produces nicks in the DNA molecules; an ATP-dependent, unwinding protein; a protein R that catalyzes the reassociation of single-stranded DNA and stimulates DNA polymerase-β activity; snRNAs that recognize "meiotically active" DNA sequences; and *zyg* RNA specific for the zygotene stage.

How do the cell nucleus and the cytoplasm interact? How do somatic cells originate from gametes? How do cells differentiate, become malignant, grow old, and die? These topics will be discussed in Volume 2.

## REFERENCES

Abell, C. W., and Monahan, T. W. (1973). *J. Cell Biol.* **59,** 549.

Adlakha, R. C., Sahasrabuddhe, C. G., Wright, D. A., Lindsey, W. F., and Rao, P. N. (1982a). *J. Cell Sci.* **54,** 193.

Adlakha, R. C., Sahasrabuddhe, C. G., Wright, D. A., Lindsey, W. F., Smith, M. L., and Rao, P. N. (1982b). *Nucleic Acids Res.* **10,** 4107.

Adlakha, R. C., Sahasrabuddhe, C. G., Wright, D. A., and Rao, P. N. (1983). *J. Cell Biol.* **97,** 1707.

Adlakha, R. C., Wang, Y. C., Wright, D. A., Sahasrabuddhe, C. G., Bigo, H., and Rao, P. N. (1984). *J. Cell Sci.* **65,** 279.

Al-Bader, A. A., Orengo, A., and Rao, P. N. (1978). *Proc. Natl. Acad.Sci. U.S.A.* **75,** 6064.

Alterman, R. B. M., Ganguly, S., Schulze, D. H., Marzluff, W. F., Schidkraut, C. L., and Skoultchi, A. I. (1984). *Mol. Cell. Biol.* **4,** 123.

Althaus, F. R., Lawrence, S. D., Sattler, G. L., and Pitot, H. C. (1982). *J. Biol. Chem.* **257,** 5528.

Andersson, L. C., Lehto, Y. P., Stenman, S., Badley, R. A., and Virtanen, I. (1981). *Nature (London)* **291,** 247.

Annunziato, A. T., Schindler, R. K., Riggs, M. G., and Seale, R. L. (1982). *J. Biol. Chem.* **257,** 8507.
Antoniades, H. N., and Williams, L. T. (1983). *Fed. Proc., Fed. Am. Soc. Exp. Biol.* **42,** 2630.
Arndt-Jovin, D. J., Robert-Nicoud, M., Zarling, D. A., Greider, C., Weimer, E., and Jovin, T. M. (1983). *Proc. Natl. Acad. Sci. U.S.A.* **80,** 4344.
Ashburner, M., and Bonner, J. J. (1979). *Cell (Cambridge, Mass.)* **17,** 241.
Atkinson, B. G. (1981). *J. Cell Biol.* **89,** 666.
Attallah, A. M., and Houck, J. C. (1977). *Exp. Cell Res.* **105,** 137.
Aubin, J. E., Weber, K., and Osborn, M. (1979). *Exp. Cell Res.* **124,** 93.
Aubin, J. E., Osborn, M., Franke, W. W., and Weber, K. (1980). *Exp. Cell Res.* **129,** 149.
Aubin, J. E., Osborn, M., and Weber, K. (1981). *Exp. Cell Res.* **136,** 63.
Avila, J., Montejo de Garcini, E., Wandosell, F., Villasante, A., Sogo, J. M., and Villanueva, N. (1983). *EMBO J.* **2,** 1229.
Bak, A. L., and Zeuthen, J. (1978). *Philos. Trans. R. Soc. London, B Ser.* **283,** 415.
Bałakier, H. (1979). *J. Exp. Zool.* **209,** 323.
Baldari, C. T., Amaldi, F., and Buongiorno-Nardelli, M. (1978). *Cell (Cambridge, Mass.)* **15,** 1095.
Baltzer, F. (1910). *Arch. Zellforsch.* **5,** 498.
Barak, L. S., Nothnagel, E. A., De Marco, E. F., and Webb, W. W. (1981). *Proc. Natl. Acad. Sci. U.S.A.* **78,** 3034.
Baskin, F., Rosenberg, R. N., and Dev, V. (1981). *Proc. Natl. Acad. Sci. U.S.A.* **78,** 3654.
Bataillon, E., and Tchou Su (1930). *Arch. Biol.* **40,** 439.
Bataillon, E., and Tchou Su (1933). *Arch. Anat. Microsc.* **29,** 285.
Beermann, W. (1952). *Chromosoma* **5,** 139.
Beermann, W. (1971). *Chromosoma* **34,** 152.
Beermann, W. (1973). *Chromosoma* **41,** 297.
Beermann, W., and Bahr, G. F. (1954). *Exp. Cell Res.* **6,** 195.
Beermann, W., and Clever, U. (1964). "The Living Cell," p. 234. Freeman, San Francisco, California.
Beermann, W., Daneholt, B., and Hosick, H. (1973). *Cold Spring Harbor Symp. Quant. Biol.* **38,** 629.
Benbow, R. M., and Ford, C. C. (1975). *Proc. Natl. Acad. Sci. U.S.A.* **72,** 2437.
Berendes, H. D. (1968). *Chromosoma* **24,** 418.
Berezney, R., and Buchholtz, L. A. (1981). *Exp. Cell Res.* **132,** 1.
Berger, N. A. (1985). *Rad. Res.* **101,** 4.
Berlin, R. D., Oliver, J. M., and Walter, R. J. (1978). *Cell (Cambridge, Mass.)* **15,** 327.
Berns, M. W., and Richardson, S. M. (1977). *J. Cell Biol.* **75,** 977.
Berns, M. W., Rattner, J. B., Brenner, S., and Meredith, S. (1977). *J. Cell Biol.* **72,** 351.
Berns, M. W., Aist, J., Edwards, J., Strahs, K., Girton, J., McNeill, P., Rattner, J. B., Kitzes, M., Hammer-Wilson, M., Liaw, L. H., Siemens, A., Koonue, M., Peterson, S., Brenner, S., Burt, J., Walter, R., Bryant, P. J., van Dijk, D. Coulomb, J., Cahill, T., and Berns, G. S. (1981). *Science* **213,** 505.
Berridge, M. J. (1984). *Biochem. J.* **220,** 345.
Berridge, M. J., Heslop, J. P., Irvine, R. F., and Brown, K. D. (1984). *Biochem. J.* **222,** 195.
Bhorjee, J. S. (1981). *Proc. Natl. Acad. Sci. U.S.A.* **78,** 6944.
Bibring, T., and Baxandall, J. (1977). *Dev. Biol.* **55,** 191.
Bienz, M. (1985). *Trends Biochem. Sci.* **10,** 157.
Blakeslee, A. F. (1939). *Ann. J. Bot.* **26,** 163.
Bluemink, J. G., Tertoolen, L. G. J., Ververgaert, P. H., and Verkleij, A. J. (1976). *Biochim. Biophys. Acta* **443,** 143.
Bombik, B. M., and Burger, M. M. (1973). *Exp. Cell Res.* **80,** 88.

Borisy, G. G. (1978). *J. Mol. Biol.* **124,** 565.
Bossy, B., Hall, L. M. C., and Spierer, P. (1984). *EMBO J.* **3,** 2537.
Bostock, C. J., and Prescott, D. M. (1971). *Exp. Cell Res.* **64,** 267.
Bowen, B. C. (1981). *Nucleic Acids Res.* **9,** 5093.
Boynton, A. M., Whilfield, J. F., and Kleine, L. P. (1983). *Biochem. Biophys. Res Commun.* **115,** 383.
Brachet, A. (1910). *Arch. Entwicklungsmech. Org.* **30,** 261.
Brachet, A. (1922). *Arch. Biol.* (Liège) **32,** 505.
Brachet, J. (1934). *Arch. Biol.* **45,** 611.
Brachet, J. (1942). *Arch. Biol.* **53,** 207.
Brachet, J. (1949a). *Experientia* **4,** 353.
Brachet, J. (1949b). *Pubbl. Stn. Zool. Napoli* **21,** 71.
Brachet, J. (1954). *Arch. Biol.* **65,** 1.
Brachet, J. (1957). ''Biochemical Cytology.'' Academic Press, New York.
Brachet, J. (1964). *Nature (London)* **204,** 1218.
Brachet, J., and Donini-Denis, S. (1978). *Differentiation* **11,** 19.
Brachet, J., and Hubert, E. (1973). *J. Embryol. Exp. Morphol.* **27,** 121.
Brachet, J., and Ledoux, L. (1955). *Exp. Cell Res., Suppl.* **3,** 27.
Bradbury, E. M., Inglis, R. J., and Matthews, H. R. (1974). *Nature (London)* **247,** 257.
Bravo, R. (1984a). *FEBS Lett.* **169,** 185.
Bravo, R. (1984b). *Proc. Natl. Acad. Sci. USA* **81,** 4848.
Bravo, R., and Macdonald-Bravo, H. (1985). *EMBO J.* **4,** 655.
Bravo, R., Small, J. V., Fey, S. J., Larsen, P. M., and Celis, J. E. (1982). *J. Mol. Biol.* **154,** 121.
Brenner, S., Branch, A., Meredith, S., and Berns, M. W. (1977). *J. Cell Biol.* **72,** 368.
Brenner, S., Pepper, D., Berns, M. W., Tan, E., and Brinkley, B. R. (1981). *J. Cell Biol.* **91,** 95.
Brewer, E. N. (1979). *Biochim. Biophys. Acta* **564,** 154.
Briggs, R., Green, E. H., and King, T. J. (1951). *J. Exp. Zool.* **116,** 455.
Brinkley, B. R., Fuller, G. M., and Highfield, D. P. (1975). *Proc. Natl. Acad. Sci. U.S.A.* **72,** 4981.
Brutlag, D. L. (1980). *Annu. Rev. Genet.* **14,** 121.
Bulinski, J. C., and Borisy, B. G. (1980). *J. Cell Biol.* **87,** 792.
Bullough, W. S. (1962). *Biol. Rev.* **37,** 307.
Bullough, W. S., and Laurence, E. B. (1970). *Eur. J. Cancer* **6,** 525.
Buongiorno-Nardelli, M., Michéli, G., Carri, M. T., and Marilley, M. (1982). *Nature (London)* **298,** 100.
Burger, M. M., Bombik, B. M., Breckenridge, B., and Seppard, J. R. (1972). *Nature (London) New Biol.* **239,** 161.
Burke, B., Griffiths, G., Reggio, H., Louvard, D., and Warren, G. (1982). *EMBO J.* **1,** 1621.
Burkholder, G. D. (1983). *Exp. Cell Res.* **147,** 287.
Burns, C. P., and Rozengurt, E. (1983). *Biochem. Biophys. Res. Commun.* **116,** 931.
Calarco-Gillam, P. D., Siebert, M. C., Hubble, R., Mitchison, T., and Kirschner, M. (1983). *Cell (Cambridge, Mass.)* **35,** 621.
Callan, H. G. (1963). *Int. Rev. Cytol.* **15,** 1.
Callan, H. G. (1972). *Proc. R. Soc. London, Ser. B* **181,** 19.
Callan, H. G. (1982). *Proc. R. Soc. London, Ser. B* **214,** 417.
Callan, H. G., and Macgregor, H. C. (1958). *Nature (London)* **181,** 1479.
Campisi, J., Medrano, E. E., Morreo, G., and Pardee, A. B. (1982). *Proc. Natl. Acad. Sci. U.S.A.* **79,** 436.
Cande, W. Z. (1982a). *Cell (Cambridge, Mass.)* **28,** 15.
Cande, W. Z. (1982b). *Nature (London)* **295,** 700.
Cande, W. Z. (1983). *Nature (London)* **304,** 557.

Cande, W. Z., and Wolniak, S. M. (1978). *J. Cell Biol.* **79,** 573.
Cande, W. Z., Lazarides, E., and McIntosh, J. R. (1977). *J. Cell Biol.* **72,** 552.
Cande, W. Z., McDonald, K., and Meeusen, R. L. (1981). *J. Cell Biol.* **88,** 618.
Caradonna, S. J., and Cheng, Y. C. (1982). *Mol. Cell. Biochem.* **46,** 49.
Carpenter, A. T. C. (1975). *Proc. Natl. Acad. Sci. U.S.A.* **72,** 3186.
Carpenter, A. T. C. (1981). *Chromosoma* **83,** 59.
Carpenter, G., and Cohen, S. (1976). *J. Cell Biol.* **71,** 159.
Carpenter, G., and Cohen, S. (1979). *Annu. Rev. Biochem.* **48,** 193.
Case, S. T., and Daneholt, B. (1978). *J. Mol. Biol.* **124,** 223.
Caspersson, T., and Schultz, J. (1938). *Nature (London)* **142,** 294.
Caspersson, T., and Schultz, J. (1939). *Nature (London)* **143,** 602.
Cassel, D., Rothenberg, P., Zhuang, Y. X., Deuel, T. F., and Glaser, L. (1983). *Proc. Natl. Acad. Sci. U.S.A.* **80,** 6224.
Cawood, A. H. (1981). *Chromosoma* **83,** 711.
Celis, J. E., and Celis, A (1985). *EMBO J.* **4,** 1187.
Celis, J. E., Larsen, P. M., Fey, S. J., and Celis, A. (1983). *J. Cell Biol.* **97,** 1429.
Celis, J. E., Fey, S. J., Mose Larsen, P., and Celis, A. (1984). *Proc. Natl. Acad. Sci. U.S.A.* **81,** 3128.
Chabanas, A., Lawrence, J. J., Humbert, J., and Eisen, H. (1983). *EMBO J.* **2,** 833.
Chafouleas, J. G., Lagace, L., Bolton, W. E., Boyd, A. E., III, and Means, A. R. (1984). *Cell (Cambridge, Mass.)* **36,** 73.
Chen. L. B., and Buchanan, J. M. (1975). *Proc. Natl. Acad. Sci. U.S.A.* **72,** 131.
Chèvremont, M., and Chèvremont-Comhaire, S. (1955). *Nature (London)* **176,** 1075.
Chèvremont, M., and Frédéric, J. (1952). *Arch. Biol.* **63,** 259.
Chèvremont, M., Chèvremont-Comhaire, S., and Firket, H. (1956). *Arch. Biol.* **67,** 635.
Chopra, D. P., and Simnett, J. D. (1971). *J. Embryol. Exp. Morphol.* **25,** 321.
Chou, I. N., Zeiger, J., and Rapaport, E. (1984). *Proc. Natl. Acad. Sci. USA* **81,** 2401.
Christensen, M. E., LeStourgeon, W. M., Jamrich, M., Howard, G. C.,Serunian, L. A., Silver, L. M., and Elgin, S. C. (1981). *J. Cell Biol.* **90,** 18.
Chu, L. K., and Sisken, J. E. (1977). *Exp. Cell Res.* **107,** 71.
Cleaver, J. E. (1981). *Exp. Cell Res.* **136,** 27.
Cleaver, J. E., Bodell, W. J., Morgan, W. F., and Zelle, B. (1983). *J. Biol. Chem.* **258,** 9059.
Cohen, S. (1981). *Fed. Proc., Fed. Am. Soc. Exp. Biol.* **40,** 1654.
Cohen, S., Fava, R. A., and Sawyer, S. T. (1982). *Proc. Natl. Acad. Sci. U.S.A.* **79,** 6237.
Collins, A. R. S., Squires, S., and Johnson, R. T. (1982). *Nucleic Acids Res.* **10,** 1203.
Comita, J. J., and Whiteley, A. H. (1953). *Biol. Bull. (Woods Hole, Mass.)* **105,** 412.
Conaway, R. C., and Lehman, I. R. (1982a). *Proc. Natl. Acad. Sci. U.S.A.* **79,** 2523.
Conaway, R. C., and Lehman, I. R. (1982b). *Proc. Natl. Acad. Sci. U.S.A.* **79,** 4585.
Connolly, J. A., Kalnins, V. I., Cleveland, D. W., and Kirschner, M. W. (1977). *Proc. Natl. Acad. Sci. U.S.A.* **74,** 2437.
Cook, P. R., and Lang, J. (1984). *Nucleic Acids Res.* **12,** 1069.
Cooper, J. A., Sefton, B. M., and Hunter, T. (1984). *Mol. Cell. Biol.* **4,** 30.
Corces, V. G., Salas, J., Salas, M. L., and Avila, J. (1978). *Eur. J. Biochem.* **86,** 473.
Corces, V. G., Manso-Martínez, R., Torre, J., Avila, J., Nasr, A., and Wiche, G. (1980a). *Eur. J. Biochem.* **105,** 7.
Corces, V., Holmgren, R., Freund, R., Morimoto, R., and Meselson, M. (1980b). *Proc. Natl. Acad. Sci. U.S.A.* **77,** 5390.
Costello, D. P., and Henley, C. (1950). *Biol. Bull. (Woods Hole, Mass.)* **99,** 386.
Cottrell, S. F., and Lee, L. H. (1981). *J. Cell Biol.* **91,** 277a.
Cox, J. V., Schenk, E. A., and Olmsted, J. B. (1983). *Cell (Cambridge, Mass.)* **35,** 331.
Craig, E. A., and McCarthy, B. J. (1980). *Nucleic Acids Res.* **8,** 4441.

Creasey, A. A., Eppstein, D. A., Marsh, V., Khan, Z., and Merigan, T. C. (1983). *Mol. Cell. Biol.* **3,** 780.
Creissen, D., and Shall, S. (1982). *Nature (London)* **296,** 271.
Crick, F. H. C. (1971). *Nature (London)* **234,** 25.
Crossin, K. L., and Carney, D. H. (1981). *Cell (Cambridge, Mass.)* **27,** 341.
Dalcq, A., and Simon, S. (1932a). *Protoplasma* **14,** 497.
Dalcq, A., and Simon, S. (1932b). *Arch. Biol.* **43,** 343.
Daneholt, B. (1972). *Nature (London) New Biol.* **240,** 229.
Dangli, A., and Bautz, E. K. F. (1983). *Chromosoma* **88,** 201.
D'Anna, J. A., Gurley, L. R., and Tobey, R. A. (1982). *Biochemistry* **21,** 3991.
Daoust, R., Leblond, C. P., Nadler, N. J., and Enesco, M. (1956). *J. Biol. Chem.* **221,** 727.
Darlington, C. D. (1955). *Nature (London)* **176,** 1139.
D'Armiento, M., Johnson, G. S., and Pastan, I. (1973). *Nature (London), New Biol.* **242,** 78.
Darzynkiewicz, Z., Sharpless, T., Staiano-Coico, L., and Melamed, M. R. (1980). *Proc. Natl. Acad. Sci. U.S.A.* **77,** 6696.
Das, M. (1982). *Int. Rev. Cytol.* **78,** 233.
Davis, F. M., and Rao, P. N. (1982). *Exp. Cell Res.* **137,** 381.
Davis, F. M., Tsao, T. Y., Fowler, S. K., and Rao, P. N. (1983). *Proc. Natl. Acad. Sci. U.S.A.* **80,** 2926.
Debec, A. Szöllösi, A., and Szöllösi, D. (1982). *Biol. Cell.* **44,** 133.
De Brabander, M., Geuens, G., De Mey, J., and Joniau, M. (1979). *Biol. Cell.* **34,** 213.
De Brabander, M., Geuens, G., Nuydens, R., Willebrords, R., and De Mey, J. (1981a). *Proc. Natl. Acad. Sci. U.S.A.* **78,** 5608.
De Brabander, M., Geuens, G., Nuydens, R., Willebrords, R., and De Mey, J. (1981b). *Cold Spring Harbor Symp. Quant. Biol.* **46,** 238.
Decker, S. (1981). *Proc. Natl. Acad. Sci. U.S.A.* **78,** 4112.
Delegeane, A. M., and Lee, A. S. (1981). *Science* **215,** 79.
De Mey, J., Moeremans, M., Geuens, G., Nuydens, R., Van Belle, H., and De Brabander, M. (1980). *Eur. J. Cell Biol.* **22,** 297a.
Dennhöfer, L. (1981). *Wilhem Roux's Arch. Dev. Biol.* **190,** 237.
De Pamphilis, M. L., and Wasserman, P. H. (1980). *Annu. Rev. Biochem.* **49,** 629.
Derksen, J., Wieslander, L., van der Ploeg, M., and Daneholt, B. (1980). *Chromosoma* **81,** 65.
Detke, S., and Keller, J. M. (1982). *J. Biol. Chem.* **257,** 3905.
Dicker, P., and Rozengurt, E. (1981). *Biochem. Biophys. Res. Commun.* **100,** 433.
Dicker, P., Pohjanpelto, P., Pettican, P., and Rozengurt, E. (1981). *Exp. Cell Res.* **135,** 221.
Dresler, S. L., and Lieberman, M. W. (1983). *J. Biol. Chem.* **258,** 9990.
Dresser, M. E., and Moses, M. J. (1980). *Chromosoma* **76,** 1.
Dulak, N. C., and Sing, Y. W. (1977). *J. Cell. Physiol.* **90,** 127.
Dulak, N. C., and Temin, H. M. (1973). *J. Cell. Physiol.* **81,** 153.
Dustin, A. P. (1934). *Bull. Cl. Sci., Acad. R. Belg.* [5] **14,** 487.
Dustin, P. (1984). "Microtubules," 2nd ed. Springer-Verlag, Berlin and New York.
Earnshaw, W. C., and Laemmli, U. K. (1983). *J. Cell Biol.* **96,** 84.
Earnshaw, W. C., and Laemmli, U. K. (1984). *Chromosoma* **89,** 186.
Earnshaw, W. C., Halligan, N., Cooke, C., and Rothfield, N. (1984). *J. Cell Biol.* **98,** 352.
Edelman, G. M. (1976). *Science* **192,** 218.
Edelman, G. M., and Yahara, I. (1976). *Proc. Natl. Acad. Sci. U.S.A.* **73,** 2047.
Edmunds, L. N., Jr., and Adams, K. J. (1981). *Science* **211,** 1002.
Edström, J. E., and Daneholt, B. (1967). *J. Mol. Biol.* **28,** 331.
Ehrhart, J. C., Creuzet, C., Rollet, E., and Loeb, J. (1981). *Biochem. Biophys. Res. Commun.* **102,** 602.
Ek, B., and Heldin, C. H. (1982). *J. Biol. Chem.* **257,** 10486.

Elsevier, S. M., and Ruddle, F. H. (1976). *Chromosoma* **56,** 227.
Engström, Y., Rozell, B., Hansson, H. A., Stemme, S., and Thelander, L. (1984). *EMBO J.* **3,** 863.
Esponda, P., and Giménez-Martín, G. (1973). *Chromosoma* **42,** 335.
Etienne-Smekens, M., Vandenbussche, P., Content, J., and Dumont, J. E. (1983). *Proc. Natl. Acad. Sci. U.S.A.* **80,** 4609.
Euteneuer, U., and McIntosh, J. R. (1980). *J. Cell Biol.* **87,** 509.
Euteneuer, V., and McIntosh, J. R. (1981). *J. Cell Biol.* **89,** 338.
Evans, R. M., and Fink, L. M. (1982). *Cell (Cambridge, Mass.)* **29,** 43.
Fahey, R. C., Mikolajczyk, S. D., Meier, G. P., Epel, D., and Carroll, E. J., Jr. (1976). *Biochim. Biophys. Acta* **437,** 445.
Fankhauser, G. (1934). *J. Exp. Zool.* **67,** 159.
Fankhauser, G., and Godwin, D. (1948). *Proc. Natl. Acad. Sci. U.S.A.* **34,** 544.
Fehlmann, M., Canivet, B., and Freychet, P. (1981). *Biochem. Biophys. Res. Commun.* **100,** 254.
Ficq, A., and Pavan, C. (1957). *Nature (London)* **180,** 983.
Flemming, W. (1882). "Zellsubstanz, Kern und Zellteilung." Leipzig.
Forrest, G. L., and Klevecz, R.R. (1978). *J. Cell Biol.* **78,** 441.
Fournier, R. E., and Pardee, A. B. (1975). *Proc. Natl. Acad. Sci. U.S.A.* **72,** 869.
Fox, R. M., Mendelsohn, J., Barbosa, E., and Goulian, M. (1973). *Nature New Biol.* **245,** 234.
Franke, W. W., Deumling, B., Zentgraf, H., Falk, H., and Rae, P. M. M. (1973). *Exp. Cell Res.* **81,** 365.
Franke, W. W., Schmid, E., Grund, C., and Geiger, B. (1982). *Cell (Cambridge, Mass.)* **30,** 103.
Fraser, J. M. K., and Huberman, J. A. (1978). *Biochim. Biophys. Acta* **520,** 271.
Frédéric, J. (1954). *Ann. N.Y. Acad. Sci.* **58,** 1246.
Friedberg, E. C., Bonura, T., Love, J. D., MacMillan, S., Radany, E. H., and Schultz, R. A. (1981). *J. Supramol. Struct. Cell. Biochem.* **16,** 91.
Friedkin, M., Legg, A., and Rozengurt, E. (1979). *Proc. Natl. Acad. Sci. U.S.A.* **76,** 3909.
Friedkin, M., Legg, A., and Rozengurt, E. (1980). *Exp. Cell Res.* **129,** 23.
Fryklund, L., Uthne, K., Sievertsson, H., and Westermark, B. (1974). *Biochem. Biophys. Res. Commun.* **61,** 950.
Fuchs, J. P., Giloh, H., Kuo, C. H., Saumweber, H., and Sedat, J. (1983). *J. Cell Biol.* **64,** 331.
Fujiwara, K., and Pollard, T. D. (1976). *J. Cell Biol.* **71,** 848.
Fujiwara, K., and Pollard, T. D. (1978). *J. Cell Biol.* **77,** 182.
Fujiwara, K., Porter, M. E., and Pollard, T. D. (1978). *J. Cell Biol.* **79,** 268.
Fuller, G. M., Brinkley, B. R., and Boughter, J. M. (1975). *Science* **187,** 948.
Fusijawa, Y., and Sibatani, A. (1954). *Experientia* **10,** 178.
Gajdusek, C. M. (1984). *J. Cell. Physiol.* **121,** 13.
Gall, J. G. (1956). *J. Biophys. Biochem. Cytol., Suppl.* **2,** 393.
Gall, J. G., and Callan, H. G. (1962). *Proc. Natl. Acad. Sci. U.S.A.* **48,** 562.
Gall, J. G.,and Pardue, M. L. (1969). *Proc. Natl. Acad. Sci. U.S.A.* **63,** 378.
Gall, J. G., Cohen, E. H., and Polan, M. L. (1971). *Chromosoma* **33,** 319.
Gallo, R. C., Sarngadharan, M. G., Popovic, M., and Sarin, P. (1983). *Cell Biol. Int. Rep.* **7,** 515.
Garvin, A. J., Lyubsky, S., and Poore, C. M. (1981). *Exp. Cell Res.* **133,** 297.
Gazit, B., Cedar, H., Lere, I., and Voss, R. (1982). *Science* **217,** 648.
Gelfant, S. (1981). *Int. Rev. Cytol.* **70,** 1.
Gerace, L., Ottaviano, Y., and Kondor-Koch, C. (1982). *J. Cell Biol.* **95,** 826.
Gerson, D. F., and Kiefer, H. (1982). *J. Cell Physiol.* **112,** 1.
Gilchrest, B. A., Marshall, W. L., Karassik, R. L., Weinstein, R., and Maciag, T. (1984). *J. Cell. Physiol.* **120,** 377.
Glover, C. V. C. (1982). *Proc. Natl. Acad. Sci. U.S.A.* **79,** 1781.
Glover, D. M., Zaha, A., Stocker, A. J., Santelli, R. V., Pueyo, M. T., De Toledo, S. M., and Lara, F. J. S. (1982). *Proc. Natl. Acad. Sci. U.S.A.* **79,** 2947.

Goldman, M. A., Holmquist, G. P., Gray, M. C., Caston, L. A., and Nag, A. (1984). *Science* **224,** 686.
Gómez-Lira, M. N., and Bode, J. (1981). *FEBS Lett.* **127,** 228.
Gonzalez, R., and Verly, W. G. (1976). *Proc. Natl. Acad. Sci. U.S.A.* **73,** 2196.
Gooderham, K., and Jeppesen, P. (1983). *Exp. Cell Res.* **144,** 1.
Gospodarowicz, D. (1975). *J. Biol. Chem.* **250,** 2515.
Gould, R. R., and Borisy, G. G. (1977). *J. Cell Biol.* **73,** 601.
Gould, R. R., and Borisy, G. G. (1978). *Exp. Cell Res.* **113,** 369.
Gray, A., Dull, T. J., and Ullrich, A. (1983). *Nature (London)* **303,** 722.
Gray, J. P., Johnson, R. A., and Friedman, D. L. (1980). *Arch. Biochem. Biophys.* **202,** 259.
Grippo, P., Geremia, R., Locorotondo, G., and Monesi, V. (1978). *Cell Differ.* **7,** 237.
Grond, C. J., Derksen, J., and Brakenhoff, G. J. (1982). *Exp. Cell Res.* **138,** 458.
Groppi, V., and Coffino, P. (1980). *Cell (Cambridge, Mass.)* **21,** 195.
Grossman, L. (1981). *Arch. Biochem. Biophys.* **211,** 511.
Grummt, F. (1978). *Proc. Natl. Acad. Sci. U.S.A.* **75,** 371.
Gruzdev, A. D., and Reznik, N. A. (1981). *Chromosoma* **82,** 1.
Guerriero, V., Jr., Rowley, D. R., and Means, A. R. (1981). *Cell (Cambridge, Mass.)* **27,** 449.
Gurley, L. R., Walters, R. A., and Tobey, R. A. (1974a). *J. Cell Biol.* **60,** 356.
Gurley, L. R., Walters, R. A., and Tobey, R. A. (1974b). *Arch. Biochem. Biophys.* **164,** 469.
Gurley, L. R., Walters, R. A., and Tobey, R. A. (1975). *J. Biol. Chem.* **250,** 3936.
Gurley, L. R., D'Anna, J. A., Barham, S. S., Deaven, L. L., and Tobey, R. A. (1978). *Eur. J. Biochem.* **84,** 1.
Gutiérrez, C., and Calvo, A. (1981). *Chromosoma* **83,** 685.
Guttman, S. D., Glover, C. V. C., Allis, C. D., and Gorovsky, D. (1980). *Cell (Cambridge, Mass.)* **22,** 299.
Hadlaczky, G., Sumner, A. T., and Ross, A. (1981). *Chromosoma* **81,** 557.
Hall, L. M. C., Mason, P. J., and Spierer, P. (1983). *J. Mol. Biol.* **169,** 83.
Hamilton, B. T., and Snyder, J. A. (1983). *Exp. Cell Res.* **144,** 345.
Hardt, N., Pedrali-Novy, G., Focher, F., and Spadari, S. (1981). *Biochem. J.* **199,** 453.
Harel, L., Blat, C., and Chatelain, G. (1983). *Biol. Cell.* **48,** 11.
Harper, M. E., and Saunders, G. F. (1981). *Chromosoma* **83,** 431.
Harper, M. E., Ulrich, A., and Saunders, G. F. (1981). *Proc. Natl. Acad. Sci. U.S.A.* **78,** 4458.
Harris, H. (1974). "Nucleus and Cytoplasm." Oxford Univ. Press (Clarendon), London and New York.
Harris, P. (1975). *Exp. Cell Res.* **94,** 409.
Harvey, E. B. (1933). *Biol. Bull. (Woods Hole, Mass.)* **64,** 125.
Harvey, E. B. (1936). *Biol. Bull. (Woods Hole, Mass.)* **71,** 101.
Heby, O. (1981). *Differentiation* **19,** 1.
Heintz, N., Sive, H. L., and Roeder, R. G. (1983). *Mol. Cell. Biol.* **3,** 539.
Heitz, E. (1934). *Biol. Zentralbl.* **54,** 588.
Helmsing, P. J., and Berendes, H. D. (1971). *J. Cell Biol.* **50,** 893.
Hendrickson, S. L., and Scher, C. D. (1983). *Mol. Cell Biol.* **3,** 1478.
Henry, S. M., and Hodge, L. D. (1983). *Eur. J. Biochem.* **133,** 23.
Hersketh, T. R., Moore, J. P., Morris, J. D. H., Taylor, M. V., Rogers, J., Smith, G. A., and Metcalfe, J. C. (1985). *Nature (London)* **313,** 481.
Hertner, T., Eppenberger, H. M., and Lezzi, M. (1983). *Chromosoma* 88, 194.
Hill, R. J., and Stollar, B. D. (1983). *Nature (London)* **305,** 338.
Hiramoto, Y. (1956). *Exp. Cell Res.* **11,** 630.
Hiramoto, Y. (1971). *Exp. Cell Res.* **68,** 291.
Hochhauser, S. J., Stein, J. L., and Stein, G. S. (1981). *Int. Rev. Cytol.* **71,** 95.

Hoffmann-Berling, H. (1954a). *Biochim. Biophys. Acta* **14,** 182.
Hoffmann-Berling, H. (1954b). *Biochim. Biophys. Acta* **15,** 226.
Hoffmann-Berling, H. (1956). *Biochim. Biophys. Acta* **19,** 453.
Holley, R. L. (1980). *J. Supramol. Struct.* **13,** 191.
Holley, R. W., Böhlen, P., Fava, R., Baldwin, J. H., Kleeman, G., and Armour, R. (1980). *Proc. Natl. Acad. Sci. U.S.A.* **77,** 5989.
Holley, R. W., Armour, R., Baldwin, J. H., and Greenfield, S. (1983). *Cell Biol. Int. Rep.* **7,** 525.
Holliday, R. (1977). *Philos. Trans. R. Soc. London, Ser. B.* **277,** 359.
Holm, P. B., and Rasmussen, S. W. (1980). *Carlsberg Res. Commun.* **45,** 483.
Hori, T.-a. (1981). *Biochem. Biophys. Res. Commun.* **102,** 38.
Hotta, Y., and Stern, H. (1971a). *Dev. Biol.* **26,** 87.
Hotta, Y., and Stern, H. (1971b). *J. Mol. Biol.* **55,** 337.
Hotta, Y., and Stern, H. (1978). *Biochemistry* **17,** 1872.
Hotta, Y., and Stern, H. (1979). *Eur. J. Biochem.* **95,** 31.
Hotta, Y., and Stern, H. (1981). *Cell (Cambridge, Mass.)* **27,** 309.
Hotta, Y., Chandley, A. C., and Stern, H. (1977). *Nature (London)* **269,** 240.
Hotta, Y., Tabata, S., and Stern, H. (1984). *Chromosoma* **90,** 243.
Hotta, Y., Tabata, S., Stubbs, L., and Stern, H. (1985). *Cell (Cambridge, Mass.)* **40,** 785.
Houck, J. C., and Hennings, H. (1973). *FEBS Lett.* **32,** 1.
Houck, J. C., Irausquin, H., and Leikin, S. (1971). *Science* **173,** 1139.
Houck, J. C., Sharma, V. K., and Cheng, R. F. (1973). *Nature (London) New Biol.* **246,** 111.
Houck, J. C., Kanagalingam, K., Hunt, C., Attallah, A., and Chung, A. (1977). *Science* **196,** 896.
Howard, A., and Pelc, S. R. (1953). *Heredity, Suppl.* **6,** 261.
Howard, R. F., and Sheppard, J. R. (1981). *J. Cell Biol.* **90,** 169.
Howell, W. M., and Hsu, T. C. (1979). *Chromosoma* **73,** 64.
Huberman, J. A., and Riggs, A. D. (1968). *J. Mol. Biol.* **32,** 327.
Hunter, T. (1984). *Nature* **311,** 414.
Hunter, T., and Francke, B. (1974). *J. Mol. Biol.* **83,** 123.
Hyams, J. S. (1984). *Nature (London)* **308,** 604.
Ierardi, L. A., Moss, S. B., and Bellvé, A. R. (1983). *J. Cell Biol.* **96,** 1717.
Iida, H., and Yahara, I. (1984). *J. Cell Biol.* **99,** 199.
Ikegami, S., Taguchi, T., Ohashi, M., Oguro, M., Nagano, H., and Mano, Y. (1978). *Nature (London)* **275,** 458.
Inglis, R. J., Langan, T. A., Matthews, H. R., Hardie, D. G., and Bradbury, E. N. (1976). *Exp. Cell Res.* **97,** 418.
Inoué, S. (1953). *Chromosoma,* **5,** 487.
Isaksson-Forsén, G., Elgjo, K., Burton, D., and Iversen, O. H. (1981). *Cell Biol. Int. Rep.* **5,** 195.
Ishimoda-Takagi, T. (1979). *Exp. Cell Res.* **119,** 423.
Izant, J. G. (1983). *Chromosoma* **88,** 1.
Izant, J. G., and McIntosh, J. R. (1980). *Proc. Natl. Acad. Sci. U.S.A.* **77,** 4741.
Izant, J. G., Wheatherbee, J. A., and McIntosh, J. R. (1982). *Nature (London)* **295,** 248.
Izant, J. G., Wheatherbee, J. A., and McIntosh, J. R. (1983). *J. Cell Biol.* **96,** 424.
Jackson, V., and Chalkley, R. (1981a). *Cell (Cambridge, Mass.)* **23,** 121.
Jackson, V., and Chalkley, R. (1981b). *J. Biol. Chem.* **256,** 5095.
Jackson, V., Marshall, S., and Chalkley, R. (1981). *Nucleic Acids Res.* **9,** 4563.
Jacobs, S., Kull, F. C., Jr., Earp, H. S., Svoboda, M. E., Van Wyk, J. J., and Cuatrecasas, P. (1983). *J. Biol. Chem.* **258,** 9581.
James, M. R., and Lehmann, A. R. (1982). *Biochemistry* **21,** 4007.
Jamrich, M., Greenleaf, A. L., and Bautz, E. K. F. (1977). *Proc. Natl. Acad. Sci. U.S.A.* **74,** 2079.
Jänne, J., Pösö, H., and Raina, A. (1978). *Biochim. Biophys. Acta* **473,** 241.

Jarosch, R., and Foissner, I. (1982). *Eur. J. Cell Biol.* **26,** 295.
Jazwinski, S. M., and Edelman, G. M. (1982). *Proc. Natl. Acad. Sci. U.S.A.* **79,** 3428.
Jazwinski, S. M., and Edelman, G. M. (1984). *J. Biol. Chem.* **259,** 6852.
Jazwinski, S. M., Wang, J. L., and Edelman, G. M. (1976). *Proc. Natl. Acad. Sci. U.S.A.* **73,** 2231.
Jelinek, W. R., Toomey, T. P., Leinwand, L., Duncan, C. H., Biro, P. A., Choudary, P. V., Weissman, S. N., Rubin, C. M., Houck, C. M., Deininger, P. L., and Schmid, C. W. (1980). *Proc. Natl. Acad. Sci. U.S.A.* **77,** 1398.
Jimenez de Asua, L., Foecking, K., and Otto, A. M. (1983). *Cell Biol. Int. Rep.* **7,** 499.
Johnson, L. F., Rao, L. G., and Muench, A. J. (1982). *Exp. Cell Res.* **138,** 79.
Johnson, R. T., Rao, P. N., and Hughes, H. D. (1970). *J. Cell Physiol.* **76,** 151.
Jones, R. N. (1975). *Int. Rev. Cytol.* **40,** 1.
Jost, E., and Johnson, R. T. (1981). *J. Cell Sci.* **47,** 25.
Jost, E., and Johnson, R. T. (1983). *Eur. J. Cell Biol.* **30,** 295.
Kaguni, L. S., Rossignol, J. M., Conaway, R. C., and Lehman, I. R. (1983a). *Proc. Natl. Acad. Sci. U.S.A.* **80,** 2221.
Kaguni, L. S., Rossignol, J. M., Conaway, R. C., Banks, G. R., and Lehman, I. R. (1983b). *J. Biol. Chem.* **258,** 9037.
Kaplowitz, P. B., and Moscona, A. A. (1973). *Biochem. Biophys. Res. Commun.* **55,** 1326.
Kaplowitz, P. B., and Moscona, A. A. (1976). *Exp. Cell Res.* **100,** 177.
Kapp, L. N., and Painter, R. B. (1982). *Int. Rev. Cytol.* **80,** 1.
Karsenti, E., Kobayashi, S., Mitchison, T., and Kirschner, M. (1984). *J. Cell Biol.* **98,** 1763.
Kasuga, M., Zick, Y., Blithe, D. L., Crettaz, M., and Kahn, C. R. (1982a). *Nature (London)* **298,** 667.
Kasuga, M., Zick, Y., Blith, D. L., Karlsson, F. A., Häring, H. U., and Kahn, C. R. (1982b). *J. Biol. Chem.* **257,** 9891.
Keller, T. C. S., III, Jemiolo, D. K., Burgess, W. H., and Rebhun, L. I. (1982). *J. Cell Biol.* **93,** 797.
Kelley, P. M., and Schlesinger, M. J. (1982). *Mol. Cell. Biol.* **2,** 267.
Kerem, B. S., Goitein, R., Richler, C., Marcus, M., and Cedar, H. (1983). *Nature (London)* **304,** 88.
Kerem, B. S., Goitein, R., Diamond, G., Cedar, H., and Marcus, M. (1984). *Cell (Cambridge, Mass.)* **38,** 493.
Keryer, G., Ris, H., and Borisy, G. G. (1984). *J. Cell Biol.* **98,** 2222.
Key, J. L., Lin, C. Y., and Chen, Y. M. (1981). *Proc. Natl. Acad. Sci. U.S.A.* **78,** 3526.
Kiehart, D. P. (1981). *J. Cell Biol.* **88,** 604.
Kiehart, D. P., Mabuchi, I., and Inoué, S. (1982). *J. Cell Biol.* **94,** 165.
Kihara, H. K., Amano, M., and Sibatani, A. (1956). *Biochim. Biophys. Acta* **21,** 489.
King, A. C., Hernaez-Davis, L., and Cuatrecasas, P. (1981). *Proc. Natl. Acad. Sci. U.S.A.* **78,** 717.
Kletzien, R. F. (1981). *Biochem. J.* **196,** 853.
Kobayashi, A., Kurokawa, M., Murata, M., Tsukada, K., and Sugano, N. (1981). *J. Biochem. (Tokyo)* **90,** 341.
Koch, K. S., and Leffert, H. L. (1979). *Nature (London)* **279,** 104.
Koenig, H., Goldstone, A., and Lu, C. Y. (1983). *Nature (London)* **305,** 530.
König, H., Riedel, H. D., and Knippers, R. (1983). *Eur. J. Biochem.* **135,** 435.
Krämer, A., Haars, R., Kabisch, R., Will, H., Bautz, F. A., and Bautz, E. K. F. (1980). *Mol. Gen. Genet.* **180,** 193.
Kriegstein, H. J., and Hogness, D. S. (1974). *Proc. Natl. Acad. Sci. U.S.A.* **71,** 135.
Kriek, J. A., van der Walt, B. J., and Bester, A. J. (1981). *Proc. Natl. Acad. Sci. U.S.A.* **78,** 4161.
Krystal, G., Rattner, J. R., and Hamkalo, B. A. (1978). *Proc. Natl. Acad. Sci. U.S.A.* **75,** 4977.

Kundahl, E., Richman, R., and Flickinger, R. A. (1981). *J. Cell Physiol.* **108,** 291.
Kunkel, T. A., and Loeb, L. A. (1981). *Science* **213,** 765.
Kuo, M. T. (1982). *J. Cell Biol.* **93,** 278.
Kuriyama, R. (1982). *J. Cell Sci.* **53,** 155.
Kuriyama, R. (1984). *J. Cell Sci.* **66,** 277.
Kuriyama, R., and Borisy, G. G. (1981a). *J. Cell Biol.* **91,** 814.
Kuriyama, R., and Borisy, G. G. (1981b). *J. Cell Biol.* **91,** 822.
Labhart, P., Koller, T., and Wunderli, H. (1982). *Cell (Cambridge, Mass.)* **30,** 115.
Laemmli, U. K., Cheng, S. M., Adolph, K. W., Paulson, J. R., Brown, J. A., and Baumbach, W. R. (1978). *Cold Spring Harbor Symp. Quant. Biol.* **42,** 355.
Laird, C. D. (1971). *Chromosoma* **32,** 378.
Laird, C. D. (1980). *Cell (Cambridge, Mass.)* **22,** 869.
Lakhotia, S. C. (1984). *Chromosoma* **89,** 63.
L'Allemain, G., Paris, S. and Pouysségur, J. (1984). *J. Biol. Chem.* **259,** 5809.
Laskey, R. A. (1983). *Philos. Trans. R. Soc. London, Ser. B* **302,** 143.
Laskey, R. A., and Earnshaw, W. C. (1980). *Nature (London)* **286,** 763.
Lau, Y. F., and Arrighi, F. E. (1981). *Chromosoma* **83,** 721.
Leader, D. P., Thoma, A., and Voorma, H. O. (1981). *Biochim. Biophys. Acta* **656,** 69.
Leenders, H. J., and Berendes, H. D. (1972). *Chromosoma* **37,** 433.
LeFevre, G., Jr. (1976). *In* "The Genetics and Biology of Drosophila" (M. Ashburner and E. Novitski, eds.), Vol. 1A, p. 36. Academic Press, New York.
Lehmann, A. R., and Karran, P. (1981). *Int. Rev. Cytol.* **72,** 101.
Lembach, K. J. (1976). *Proc. Natl. Acad. Sci. U.S.A.* **73,** 183.
Lemeunier, F., Derbin, C., Malfoy, B., Leng, M., and Taillandier, E. (1982). *Exp. Cell Res.* **141,** 508.
Lemke, G. E., and Brockes, J. P. (1983). *Fed. Proc., Fed. Am. Soc. Exp. Biol.* **42,** 2627.
Leuchtenberger, C. (1950). *Chromosoma* **3,** 449.
Lewis, C. D., and Laemmli, U. K. (1982). *Cell (Cambridge, Mass.)* **29,** 171.
Lezzi, M. (1966). *Exp. Cell Res.* **43,** 571.
Lezzi, M., Meyer, B., and Mähr, R. (1981). *Chromosoma* **83,** 327.
Li, S., Meistrich, M. L., Brock, W. A., Hsu, T. C., and Kuo, M. T. (1983). *Exp. Cell Res.* **144,** 63.
Lieberman, M. A., and Glaser, A. (1981). *J. Memb. Biol.* **63,** 1.
Lifschytz, E. (1983). *J. Mol. Biol.* **164,** 17.
Lima-de-Faria, A. (1956). *Hereditas* **42,** 85.
Lima-de-Faria, A. (1983). "Molecular Evolution and Organization of the Chromosomes." Elsevier, Amsterdam.
Lipsich, L. A., Lucas, J. J., and Kates, J. R. (1978). *J. Cell Physiol.* **97,** 199.
Littlefield, B. A., Cidlowski, N. B., and Cidlowski, J. A. (1982). *Exp. Cell Res.* **141,** 283.
Lloyd, C. (1979). *Nature (London)* **277,** 515.
Lopez-Rivas, A., Adelberg, E. A., and Rozengurt, E. (1982). *Proc. Natl. Acad. Sci. U.S.A.* **79,** 6275.
Lopez-Rivas, A., Stroobant, P., Waterfield, M. D., and Rozengurt, E. (1984). *EMBO J.* **3,** 939.
Lorch, I. J. (1952). *Q. J. Microsc. Sci.* **93,** 475.
Luchtel, D., Bluemink, J. G., and De Laat, S. W. (1976). *J. Ultrastruct. Res.* **54,** 406.
Lydersen, B. K., and Pettijohn, D. E. (1980). *Cell (Cambridge, Mass.)* **22,** 489.
McClain, D. A., and Edelman, G. M. (1980). *Proc. Natl. Acad. Sci. U.S.A.* **77,** 2748.
McCurry, L. S., and Jacobson, M. K. (1981a). *J. Biol. Chem.* **256,** 551.
McCurry, L. S., and Jacobson, M. K. (1981b). *J. Supramol. Struct. Cell Biochem.* **17,** 87.
McGill, M., and Brinkley, B. R. (1975). *J. Cell Biol.* **67,** 189.
McKnight, S. L., and Miller, O. L. (1977). *Cell (Cambridge, Mass.)* **12,** 795.

McMahon, J. B., Malan-Shibley, L., and Iype, P. T. (1984). *J. Biol. Chem.* **259,** 1803.
Malik, N., Miwa, M., Sugimura, T., Thraves, P., and Smulson, M. (1983). *Proc. Natl. Acad. Sci. U.S.A.* **80,** 2554.
Mamont, P. S., Böhlen, P., MacCann, P. P., Bey, P., Schuber, F., and Tardif, C. (1976). *Proc. Natl. Acad. Sci. U.S.A.* **73,** 1626.
Mandel, P., Okazaki, H., and Niedergang, C. (1982). *Prog. Nucleic Acid Res. Mol. Biol.* **27,** 1.
Maness, P. F., and Walsh, R. C., Jr. (1982). *Cell (Cambridge, Mass.)* **30,** 253.
Marashi, F., Baumbach, L., Rickles, R., Sierra, F., Stein, J. L., and Stein, G. S. (1982). *Science* **215,** 683.
Marcum, J. M., Dedman, J. R., Brinkley, B. R., and Means, A. R. (1978). *Proc. Natl. Acad. Sci. U.S.A.* **75,** 3771.
Margolis, R. L., and Wilson, L. (1981). *Nature (London)* **293,** 705.
Mariani, B. D., Slate, D. L., and Schimke, R. T. (1981). *Proc. Natl. Acad. Sci. U.S.A.* **78,** 4985.
Marquardt, H., Todaro, G. J., Henderson, L. E., and Oroszlan, S. (1981). *J. Biol. Chem.* **256,** 6859.
Marshall Graves, J. A. (1972). *Exp. Cell Res.* **72,** 393.
Marsland, D., and Landau, J. V. (1954). *J. Exp. Zool.* **125,** 529.
Martin-Pèrez, J., Siegmann, M., and Thomas, G. (1984). *Cell (Cambridge, Mass.)* **36,** 287.
Marx, J. L. (1984). *Science* **224,** 271.
Mastropaolo, W., and Henshaw, E. C. (1981). *Biochim. Biophys. Acta* **656,** 246.
Masui, Y., and Markert, C. L. (1971). *J. Exp. Zool.* **177,** 129.
Mathog, D., Hochstrasser, M., Gruenbaum, Y., Saumweber, H., and Sedat, J. (1984). *Nature (London)* **308,** 414.
Matsui, S. I., Yoshida, H., Weinfeld, H., and Sandberg, A. A. (1972). *J. Cell Biol.* **54,** 120.
Matsui, S. I., Sandberg, A. A., Negoro, S., Seon, B. K., and Goldstein, G. (1982). *Proc. Natl. Acad. Sci. U.S.A.* **79,** 1535.
Matsukage, A., Yamamoto, S., Yamaguchi, M., Kusakabe, M., and Takahashi, T. (1983). *J. Cell Physiol.* **117,** 266.
Matsumoto, Y.-i., Yasuda, H., Mita, S., Marunouchi, T., and Yamada, M.-A. (1980). *Nature (London)* **284,** 181.
Mazia, D. (1954a). *Proc. Natl. Acad. Sci. U.S.A.* **40,** 521.
Mazia, D. (1954b). *In* "Glutathione" (S. Colowick, A. Lazarow, E. Racker, D. R. Schwarz, E. Stadtman, and H. Waelsch, eds.), p. 209. Academic Press, New York.
Mazia, D. (1955). *Symp. Soc. Exp. Biol.* **9,** 335.
Mazia, D. (1956). *Adv. Biol. Med. Phys.* **4,** 70.
Mazia, D. (1958). *Exp. Cell Res.* **14,** 486.
Mazia, D. (1961). *In* "The Cell" (J. Brachet and A. E. Mirsky, eds.), Vol. 3, p. 77. Academic Press, New York.
Mazia, D. (1977). "Mitosis. Facts and Questions," pp. 196–213. Springer-Verlag, Berlin and New York.
Mazia, D. (1978). *In* "Cell Reproduction: Essays in Honor of Daniel Mazia" (E. R. Dirksen, D. M. Prescott, and C. F. Fox, eds.), pp. 1–14. Academic Press, New York.
Mazia, D. (1984). *Exp. Cell Res.* **153,** 1.
Mazia, D., and Dan, K. (1952). *Proc. Natl. Acad. Sci. U.S.A.* **38,** 826.
Mazia, D., Paweletz, N., Sluder, G., and Finzi, E. M. (1981). *Proc. Natl. Acad. Sci. U.S.A.* **78,** 377.
Méchali, M., and Harland, R. M. (1982). *Cell (Cambridge, Mass.)* **30,** 93.
Meeusen, R. L., Bennett, J., and Cande, W. Z. (1980). *J. Cell Biol.* **86,** 858.
Melli, M., Spinelli, G., and Arnold, E. (1977). *Cell* **12,** 167.
Mercer, W. E., and Schlegel, R. A. (1982). *J. Cell Physiol.* **110,** 311.

Mercer, W. E., Nelson, D., DeLeo, A. B., Old, L. J., and Baserga, R. (1982). *Proc. Natl. Acad. Sci. U.S.A.* **79,** 6309.
Meselson, M. S., and Radding, C. M. (1975). *Proc. Natl. Acad. Sci. U.S.A.* **72,** 358.
Meselson, M. S., and Stahl, F. W. (1958). *Proc. Natl. Acad. Sci. U.S.A.* **44,** 671.
Metz, C. W. (1938). *Am. Nat.* **72,** 485.
Michalopoulos, G., Cianciulli, H. D., Novotny, A. R., Kligerman, A. D., Strom, S. C., and Jirtle, R. L. (1982). *Cancer Res.* **42,** 4673.
Micheli, G., Baldari, C. T., Carri, M. T., Di Cello, G., and Buongiorno-Nardelli, M. (1982). *Exp. Cell Res.* **137,** 127.
Miller, M. R., and Chinault, D. N. (1982). *J. Biol. Chem.* **257,** 10204.
Miller, M. R., and Lui, L. H. (1982). *Biochem. Biophys. Res. Commun.* **108,** 1676.
Mitchison, J. M. (1952). *Symp. Soc. Exp. Zool.* **6,** 105.
Mitchison, T., and Kirschner, M. (1984a). *Nature* 312, 232.
Mitchison, T., and Kirschner, M. (1984b). *Nature* **312,** 237.
Moolenaar, W. H., Mummery, C. L., van Der Saag, P. T., and de Laat, S. W. (1981). *Cell (Cambridge, Mass.)* **23,** 789.
Moolenaar, W. H., Yarden, Y., de Laat, S. W., and Schlessinger, J. (1982). *J. Biol. Chem.* **257,** 8502.
Moolenaar, W. H., Tsien, R. Y., van Der Saag, P. T., and de Laat, S. W. (1983). *Nature (London)* **304,** 645.
Moolenaar, W. H., Tertoolen, L. G. J., and de Laat, S. W. (1984a). *J. Biol. Chem.* **259,** 8066.
Moolenaar, W. H., Tertoolen, L. G. J., and de Laat, S. W. (1984b). *Nature* **312,** 371.
Morgan, J. F., Morton, H. J., Campbell, M. E., and Guérin, L. E. (1956). *J. Natl. Cancer Inst. (U.S.)* **16,** 1405.
Mori, M., Satoh, M., and Novikoff, A. B. (1982). *Exp. Cell Res.* **141,** 277.
Moroi, Y., Hartman, A. L., Nakane, P. K., and Tan, E. M. (1981). *J. Cell Biol.* **90,** 254.
Mortin, L. I., and Sedat, J. W. (1982). *J. Cell Sci.* **57,** 73.
Moser, G. C., Fallon, R. J., and Meiss, H. K. (1981). *J. Cell. Physiol.* **106,** 293.
Mott, M. R., and Callan, H. G. (1975). *J. Cell Sci.* **17,** 241.
Mullins, J. M., and McIntosh, J. R. (1982). *J. Cell Biol.* **94,** 654.
Mummery, C. L., Boonstra, J., van Der Saag, P. T., and de Laat, S. W. (1981). *J. Cell. Physiol.* **107,** 1.
Mummery, C. L., Boonstra, J., van Der Saag, P. T., and de Laat, S. W. (1982). *J. Cell. Physiol.* **112,** 27.
Murakami-Murofushi, K., and Mano, Y. (1977). *Biochim. Biophys. Acta* **475,** 254.
Murray, A. W., and Szostak, J. W. (1983). *Nature (London)* **305,** 189.
Nasedkina, T. V., and Slesinger, S. I. (1982). *Chromosoma* **86,** 239.
Neckers, L. M., and Cossman, J. (1983). *Proc. Natl. Acad. Sci. U.S.A.* **80,** 3494.
Nesbitt, J. A., Anderson, W. B., Miller, Z., Pastan, I., Russell, T. R., and Gospodarowicz, D. (1976). *J. Biol. Chem.* **251,** 2344.
Nilsen-Hamilton, M., Hamilton, R. T., Allen, W. R., and Potter-Perigo, S. (1982). *Cell (Cambridge, Mass.)* **31,** 237.
Nishimura, J., Huang, J. S., and Deuel, T. F. (1982). *Proc. Natl. Acad. Sci. U.S.A.* **79,** 4303.
Noguchi, H., Reddy, G. P., and Pardee, A. B. (1983). *Cell (Cambridge, Mass.)* **32,** 443.
Noonan, K. D., and Burger, M. M. (1973). *Exp. Cell Res.* **80,** 405.
Noonan, K. D., Levine, A. J., and Burger, M. M. (1973). *J. Cell Biol.* **58,** 491.
Nunnally, M. H., D'Angelo, J. M., and Craig, S. W. (1980). *J. Cell Biol.* **87,** 219.
O'Brien, R. L., and Parker, J. W. (1976). *Cell (Cambridge, Mass.)* **7,** 13.
O'Brien, T. G., Simsiman, R. C., and Boutwell, R. C. (1976). *Cancer Res.* **36,** 3766.
O'Farrell, M. K. (1983). *Cell Biol. Int. Rep.* **7,** 549.

Ohashi, Y., Ueda, K., Kawaichi, M., and Hayaishi, O. (1983). *Proc. Natl. Acad. Sci. U.S.A.* **80,** 3604.
Olashaw, N. E., and Pledger, W. J. (1983). *Nature (London)* **306,** 272.
Oleinick, N. L., Daniel, J. W., and Brewer, E. N. (1981). *Exp. Cell Res.* **131,** 373.
Olins, D. E., Olins, A. L., Levy, H. A., Durfee, R. C., Margle, S. M., Tinnel, E. P., and Dover, S. D. (1982). *Science* **220,** 498.
Oren, M., and Levine, A. J. (1983). *Proc. Natl. Acad. Sci. U.S.A.* **80,** 56.
Orlando, P., Grippo, P., and Geremia, R. (1984). *Exp. Cell Res.* **153,** 499.
Otani, S., Kuramoto, A., Matsui, I., and Morisawa, S. (1982). *Eur. J. Biochem.* **125,** 35.
Ottaviane, Y., and Gerace, L. (1985). *J. Biol. Chem.* **260,** 624.
Otten, J., Johnson, G. S., and Pastan, I. (1972). *J. Biol. Chem.* **247,** 7082.
Otto, A. M. (1982). *Cell Biol. Int. Rep.* **6,** 1.
Otto, A. M., Ulrich, M. O., Zumbé, A., and Jimenez De Asua, L. (1981). *Proc. Natl. Acad. Sci. U.S.A.* **78,** 3063.
Owen, N. E., and Villereal, M. L. (1983a). *Cell (Cambridge, Mass.)* **32,** 979.
Owen, N. E., and Villereal, M. L. (1983b). *J. Cell Physiol.* **117,** 23.
Painter, T. S. (1934). *J. Hered.* **25,** 1165.
Pardee, A. B., Dubrow, R., Hamlin, J. L., and Kletzien, R. F. (1978). *Annu. Rev. Biochem.* **47,** 715.
Pardoll, D. M., Vogelstein, B., and Coffey, D. S. (1980). *Cell (Cambridge, Mass.)* **19,** 527.
Pardue, M. L., and Dawid, I. B. (1981). *Chromosoma* **83,** 29.
Pardue, M. L., Kedes, L. H., Weinberg, E. S., and Birnstiel, M. L. (1977). *Chromosoma* **63,** 135.
Parker, C. W. (1974). *Biochem. Biophys. Res. Commun.* **61,** 1180.
Pätau, K. (1935). *Naturwissenschaften* **23,** 537.
Patt, L. M., and Houck, J. C. (1980). *FEBS Lett.* **120,** 163.
Paulson, J. R., and Langmore, J. P. (1983). *J. Cell Biol.* **96,** 1132.
Paulson, J. R., and Taylor, S. S. (1982). *J. Biol. Chem.* **257,** 6064.
Paweletz, N. (1983). *Nature (London)* **305,** 389.
Paweletz, N., and Finze, E. M. (1981). *J. Ultrastruct. Res.* **76,** 127.
Pedrali-Noy, G., and Spadari, S. (1979). *Biochem. Biophys. Res. Commun.* **88,** 1194.
Pedrali-Noy, G., Spadari, S., Miller-Faurès, A., Miller, A. O. A., Kruppa, J., and Koch, G. (1980). *Nucleic Acids Res.* **8,** 377.
Pegg, A. E., Borchardt, R. T., and Coward, J. K. (1981). *Biochem. J.* **194,** 79.
Pehrson, J., and Cole, D. (1980). *Nature (London)* **285,** 43.
Pelling, C. (1978). *Cold Spring Harbor Symp. Quant. Biol.* **35,** 522.
Peterson, S. P., and Berns, M. W. (1978). *J. Cell Sci.* **34,** 289.
Peterson, S. P., and Berns, M. W. (1980). *Int. Rev. Cytol.* **64,** 81.
Peterson, S. W., and Lerch, V. (1983). *J. Cell Biol.* **97,** 276.
Petruzzelli, L. M., Ganguly, S., Smith, S. J., Cobb, M. H., Rubin, C. S., and Rosen, O. M. (1982). *Proc. Natl. Acad. Sci. U.S.A.* **79,** 6792.
Petzelt, C. (1972). *Exp. Cell Res.* **70,** 333.
Petzelt, C. (1979). *Int. Rev. Cytol.* **60,** 53.
Petzelt, C. (1980). *Eur. J. Cell Biol.* **22,** 314a.
Petzelt, C., and Auel, D. (1977). *Proc. Natl. Acad. Sci. U.S.A.* **74,** 1610.
Petzelt, C., and von Ledebur-Villiger, M. (1973). *Exp. Cell Res.* **81,** 87.
Pickett-Heaps, J. D., Tippit, D. H., and Porter, K. R. (1982). *Cell (Cambridge, Mass.)* **29,** 729.
Plant, P. W., Liang, T. J., Pindyck, J., and Grieninger, G. (1981). *Biochim. Biophys. Acta* **655,** 407.
Plumb, M., Marashi, F., Green, L., Zimmerman, A., Zimmerman, S., Stein, J., and Stein, G. (1984). *Proc. Natl. Acad. Sci. U.S.A.* **81,** 434.
Polunovsky, V. A., Setkov, N. A., and Epifanova, O. I. (1983). *Exp. Cell Res.* **146,** 377.

Porter, K. R. (1954). *Union Int. Sci. Biol., Ser. B.* **21,** 236.

Pouysségur, J., Chambard, J. C., Paris, S., and Van Obberghen-Schilling, E. (1982). *Proc. Natl. Acad. Sci. U.S.A.* **79,** 3935.

Pouysségur, J., Sardet, C., Franchi, A., L'Allemain, G., and Paris, S. (1984). *Proc. Natl. Acad. Sci. U.S.A* **81,** 4833.

Pratt, M. M. (1984). *Int. Rev. Cytol.* **87,** 83.

Pratt, M. M., Otter, T., and Salmon, E. D. (1980). *J. Cell Biol.* **86,** 738.

Prescott, D. M. (1976). "Reproduction of Eukaryotic Cells." Academic Press, New York.

Prior, C. P., Cantor, C. R., Johnson, E. M., and Allfrey, V. G. (1980). *Cell (Cambridge, Mass.)* **20,** 597.

Purrello, F., Burnham, D. B., and Goldfine, I. D. (1983). *Science* **221,** 462.

Radman, M. (1974). "Molecular and Environmental Aspects of Mutagenesis," p. 128. Thomas, Springfield, Illinois.

Raff, E. C. (1979). *Int. Rev. Cytol.* **59,** 1.

Raff, E. C., Brothers, A. J., and Raff, R. A. (1976). *Nature (London)* **260,** 615.

Rao, M. V. N. (1980). *Int. Rev. Cytol.* **67,** 291.

Rao, P. N. (1980). *Mol. Cell. Biochem.* **29,** 47.

Rao, P. N., and Smith, M. L. (1981). *J. Cell Biol.* **88,** 649.

Rao, P. N., Hittleman, W. N., and Wilson, B. A. (1975). *Exp. Cell Res.* **90,** 40.

Rao, P. N., Sunkara, P. S., and Wilson, B. A. (1977). *Proc. Natl. Acad. Sci. U.S.A.* **74,** 2869.

Rao, P. N., Wilson, B. A., and Sunkara, P. S. (1978). *Proc. Natl. Acad. Sci. U.S.A.* **75,** 5043.

Rapkine, L. (1931). *Ann. Physiol. Physicochim. Biol.* **7,** 382.

Rappaport, R. (1971). *Int. Rev. Cytol.* **31,** 169.

Rappaport, R. (1981). *J. Exp. Zool.* **217,** 365.

Raveh, D., and Ben-Ze'ev, A. (1984). *Exp. Cell Res.* **153,** 99.

Rebhun, L. I. (1977). *Int. Rev. Cytol.* **49,** 1.

Rechler, M. M., Nissley, S. P., King, G. L., Moses, A. C., Van Obberghen-Schilling, E. E., Romanus, J. A., Knight, A. B., Short, P. A., and White, R. M. (1981). *J. Supramol. Struct. Cell. Biochem.* **15,** 253.

Reddy, G. P. V., and Pardee, A. B. (1980). *Proc. Natl. Acad. Sci. U.S.A.* **77,** 3312.

Reddy, G. P. V., and Pardee, A. B. (1983). *Nature (London)* **304,** 86.

Reinhard, P., Maillart, P., Schluchter, M., Gautschi, J. R., and Schindler, R. (1979). *Biochim. Biophys. Acta* **564,** 141.

Riedel, H. D., König, H., Stahl, H., and Knippers, R. (1982). *Nucleic Acids Res.* **10,** 5621.

Rieder, C. L. (1979a). *J. Cell Biol.* **80,** 1.

Rieder, C. L. (1979b). *J. Ultrastruct. Res.* **66,** 109.

Rieder, C. L., and Borisy, G. G. (1981). *Chromosoma* **82,** 693.

Rieder, C. L., and Borisy, G. G. (1982). *Biol. Cell.* **44,** 117.

Riley, C., and Weintraub, H. (1979). *Proc. Natl. Acad. Sci. U.S.A.* **76,** 328.

Riley, D. E., Canfield, T. K., and Gartler, S. M. (1984). *Nucleic Acids Res.* **12,** 1829.

Ring, D., Hubble, R., and Kirschner, M. (1982). *J. Cell Biol.* **94,** 549.

Rink, T. J., Tsien, R. Y., and Warner, A. E. (1980). *Nature (London)* **283,** 658.

Ris, H., and Witt, P. L. (1981). *Chromosoma* **82,** 153.

Ritossa, F. M. (1964). *Exp. Cell Res.* **35,** 601.

Roccheri, M. C., Di Bernardo, M. G., and Giudice, G. (1981). *Dev. Biol.* **83,** 173.

Rønning, Ø. W., and Lindmo, T. (1981). *Exp. Cell Res.* **134,** 113.

Rønning, Ø. W., and Lindmo, T. (1983). *Exp. Cell Res.* **144,** 171.

Rose, S. P., Pruss, R. M., and Herschman, H. R. (1975). *J. Cell. Physiol.* **86,** 593.

Rosenberg, B. H., Ungers, G., and Deutsch, J. F. (1976). *Nucleic Acids Res.* **3,** 3305.

Rothenberg, P., Reuss, L., and Glaser, L. (1982). *Proc. Natl. Acad. Sci. U.S.A.* **79,** 7783.

Rothenberg, P., Glaser, L., Schlesinger, P., and Cassel, D. (1983). *J. Biol. Chem.* **258,** 12644.

Rothstein, H. (1982). *Int. Rev. Cytol.* **78,** 127.
Rozengurt, E. (1982a). *Exp. Cell Res.* **139,** 71.
Rozengurt, E. (1982b). *J. Cell. Physiol.* **112,** 243.
Rozengurt, E. (1983a). *Cell Biol. Int. Rep.* **7,** 495.
Rozengurt, E. (1983b). *Mol. Biol. Med.* **1,** 169.
Rozengurt, E., Gelehrter, T. D., Legg, A., and Pettican, P. (1981a). *Cell (Cambridge, Mass.)* **23,** 781.
Rozengurt, E., Brown, K. D., and Pettican, P. (1981b). *J. Biol. Chem.* **256,** 716.
Rozengurt, E., Legg, A., Strang, G., and Courtenay-Luck, N. (1981c). *Proc. Natl. Acad. Sci. U.S.A.* **78,** 4392.
Rozengurt, E., Collins, M. K. L., and Keehan, M. (1983a). *J. Cell. Physiol.* **116,** 379.
Rozengurt, E., Stroobant, P., Waterfield, M. D., Deuel, T. F., and Keehan, M. (1983b). *Cell (Cambridge, Mass.)* **34,** 265.
Rozengurt, E., Rodriguez-Pena, M. and Smith, K. A. (1983c). *Proc. Natl. Acad. Sci. U.S.A.* **80,** 7244.
Rozengurt, E., Rodriguez-Pena, A., Coombs, M., and Sinnett-Smith, J. (1984). *Proc. Natl. Acad. Sci. U.S.A.* **81,** 5748.
Rubin, H., and Koide, T. (1976). *Proc. Natl. Acad. Sci. U.S.A.* **73,** 168.
Rubin, H., Vidair, C., and Sanui, H. (1981). *Proc. Natl. Acad. Sci. U.S.A.* **78,** 2350.
Rubin, J. B., Shia, M. A., and Pilch, P. F. (1983). *Nature (London)* **305,** 438.
Russell, D. H. (1981). *Biochem. Biophys. Res. Commun.* **99,** 1167.
Russev, G., and Hancock, R. (1981). *Nucleic Acids Res.* **9,** 4129.
Russev, G., and Hancock, R. (1982). *Proc. Natl. Acad. Sci. U.S.A.* **79,** 3143.
Rydlander, L., Pigon, A., and Edström, J. E. (1980). *Chromosoma* **81,** 101.
Sachs, L. (1980). *Proc. Natl. Acad. Sci. U.S.A.* **77,** 6152.
Sahasrabuddhe, C. G., Adlakha, R. C., and Rao, P. N. (1984). *Exp. Cell Res.* **153,** 439.
Sakai, H. (1978). *Int. Rev. Cytol.* **55,** 23.
Sakai, H., Hiramoto, Y., and Kuriyama, R. (1975). *Dev. Growth Differ.* **17,** 265.
Sakai, H., Mabuchi, I., Shimoda, S., Kuriyama, R., Ogawa, K., and Mohri, H. (1976). *Dev. Growth Differ.* **18,** 211.
Sakai, H., Shimoda, S., and Hiramoto, Y. (1977). *Exp. Cell Res.* **104,** 457.
Salmon, E. D., Leslie, R. J., Saxton, W. M., Karow, M. L., and McIntosh, J. R. (1984). *J. Cell Biol.* **99,** 2165.
Sanger, J. W. (1975). *Proc. Natl. Acad. Sci. U.S.A.* **72,** 2451.
Sanui, H., and Rubin, H. (1982). *Exp. Cell Res.* **139,** 15.
Sasaki, Y., and Hidaka, H. (1982). *Biochem. Biophys. Res. Commun.* **104,** 451.
Sass, H. (1980). *J. Cell Sci.* **45,** 269.
Sass, H. (1982). *Cell (Cambridge, Mass.)* **28,** 269.
Sass, H., and Bautz, E. K. F. (1982). *Chromosoma* **85,** 633.
Satya-Prakash, K. L., Hsu, T. C., and Pathak, S. (1980). *Chromosoma* **81,** 1.
Sawai, T. (1972). *J. Cell Sci.* **11,** 543.
Sawai, T. (1976). *J. Cell Sci.* **21,** 537.
Saxton, W. M., Stemple, D. L., Leslie, R. J., Salmon, E. D., Zavortink, M., and McIntosh, J. R. (1984). *J. Cell Biol.* **99,** 2175.
Schatten, G., and Schatten, H. (1980). *Eur. J. Cell Biol.* **22,** 314.
Scheer, U., and Sommerville, J. (1982). *Exp. Cell Res.* **139,** 410.
Schiff, P. B., Fant, J., and Horwitz, S. B. (1979). *Nature (London)* **277,** 665.
Schlessinger, J., Schreiber, A. B., Levi, A., Lax, I., Libermann, T., and Yarden, Y. (1983). *CRC Crit. Rev. Biochem.* **14,** 93.
Schloss, J. A., Milsted, A., and Goldman, R. D. (1977). *J. Cell Biol.* **74,** 794.
Schnebli, H. P., and Burger, M. M. (1972). *Proc. Natl. Acad. Sci. U.S.A.* **69,** 3825.

Schroeder, T. E. (1972). *J. Cell Biol.* **53,** 419.
Schroeder, T. E. (1981). *Exp. Cell Res.* **134,** 231.
Schuldiner, S., and Rozengurt, E. (1982). *Proc. Natl. Acad. Sci. U.S.A.* **79,** 7778.
Schwartz, D. (1975). *Chromosoma* **52,** 293.
Sealy, L., Hartley, J., Donelson, J., and Chalkley, R. (1981). *J. Mol. Biol.* **145,** 291.
Seifert, R. A., Schwartz, S. M., and Bowen-Pope, D. F. (1984). *Nature* **311,** 669.
Sekas, G., Owen, W. G., and Cook, R. T. (1979). *Exp. Cell Res.* **122,** 47.
Server, A. C., Sutter, A., and Shooter, E. M. (1976). *J. Biol. Chem.* **251,** 1188.
Shen, D. W., Real, F. X., DeLeo, A. B., Old, L. J., Marks, P. A., and Rifkind, R. A. (1983). *Proc. Natl. Acad. Sci. U.S.A.* **80,** 5919.
Sherline, P., and Mascardo, R. N. (1982a). *J. Cell Biol.* **93,** 507.
Sherline, P., and Mascardo, R. (1982b). *J. Cell Biol.* **95,** 316.
Sherline, P., and Mascardo, R. N. (1984). *Exp. Cell Res.* **153,** 109.
Sherline, P., and Schiavone, K. (1978). *J. Cell Biol.* **77,** R9.
Shimada, H., and Terayama, H. (1976). *Dev. Biol.* **54,** 151.
Shioda, M., Nelson, E. M., Bayne, M. L., and Benbow, R. M. (1982). *Proc. Natl. Acad. Sci. U.S.A.* **79,** 7209.
Shoham, J., and Sachs, L. (1974). *Exp. Cell Res.* **85,** 8.
Silver, R. B., Saft, M. S., Taylor, A. L., and Cole, R. D. (1983). *J. Biol. Chem.* **258,** 13287.
Sims, J. L., Sikorski, G. W., Catino, D. M., Berger, S. J., and Berger, N. A. (1982). *Biochemistry* **21,** 1813.
Skog, S., Tribukait, B., and Sundius, G. (1982). *Exp. Cell Res.* **141,** 23.
Skoglund, U., Anderson, K., Björkroth, B., Lamb, M. M., and Daneholt, B. (1983). *Cell (Cambridge, Mass.)* **34,** 847.
Smekens-Etienne, M., Goldstein, J., and Dumont, J. E. (1983). *Eur. J. Biochem.* **130,** 269.
Smith, H. C., and Berezney, R. (1983). *Biochemistry* **22,** 3042.
Smith, J. A., and Martin, L. (1973). *Proc. Natl. Acad. Sci. U.S.A.* **70,** 1263.
Smith, J. C., and Stiles, C. D. (1981). *Proc. Natl. Acad. Sci. U.S.A.* **78,** 4363.
Smith, L. D., and Ecker, R. E. (1971). *Dev. Biol.* **25,** 232.
Snyder, R. D., and Regan, J. D. (1982). *Biochim. Biophys. Acta* **697,** 229.
Solari, A. J. (1972). *Chromosoma* **39,** 237.
Soltys, G. T., and Borisy, G. G. (1985). *J. Cell Biol.* **100,** 1682.
Sommerville, J., Malcolm, D. B., and Callan, H. G. (1978). *Philos. Trans. R. Soc. London, Ser. B* **283,** 359.
Sorsa, V. (1983). *J. Cell Sci.* **64,** 255.
Spierer, A., and Spierer, P. (1984). *Nature (London)* **307,** 176.
Srb, A. M., Owen, R. D., and Edgar, R. S. (1965). "General Genetics," 2nd ed. Freeman, San Francisco.
Stancel, G. M., Prescott, D. M., and Liskay, R. M. (1981). *Proc. Natl. Acad. Sci. U.S.A.* **78,** 6295.
Stauffer, E. (1945). Ph.D. Thesis, University Bern. A. Kundig Genève.
Steck, P. A., Blenis, J., Voss, P. G., and Wang, J. L. (1982). *J. Cell Biol.* **92,** 523.
Stein, G., and Baserga, R. (1970). *Biochem. Biophys. Res. Commun.* **41,** 715.
Stein, G. H., and Yanishevsky, R. M. (1979). *Exp. Cell Res.* **120,** 155.
Stein, G. H., and Yanishevsky, R. M. (1981). *Proc. Natl. Acad. Sci. U.S.A.* **78,** 3025.
Stein, G. H., Park, W., Thrall, C., Mans, R., and Stein, J. (1975). *Nature (London)* **257,** 764.
Stein, R., Gruenbaum, Y., Pollack, Y., Razin, A., and Cedar, H. (1982). *Proc. Natl. Acad. Sci. U.S.A.* **79,** 61.
Steinert, M. (1953). *Biochim. Biophys. Acta* **10,** 427.
Stern, H., and Hotta, Y. (1977). *Philos. Trans. R. Soc. London, Ser. B* **277,** 277.
Stern, H., and Hotta, Y. (1980). *Mol. Cell. Biochem.* **29,** 145.
Summers, K. E., and Kirchner, M. W. (1979). *J. Cell Biol.* **83,** 205.

Sunkara, P. S., Rao, P. N., Nishioka, A., and Brinkley, B. R. (1979a). *Exp. Cell Res.* **119,** 63.
Sunkara, P. S., Wright, D. A., and Rao, P. N. (1979b). *Proc. Natl. Acad. Sci. U.S.A.* **76,** 2799.
Sunkara, P. S., Al-Bader, A. A., Riker, M. A., and Rao, P. N. (1980). *Cell Biol. Int. Rep.* **4,** 1025.
Sunkara, P. S., Chakroborty, B., Wright, D. A., and Rao, P. N. (1981). *Eur. J. Cell Biol.* **23,** 312.
Sunkara, P. S., Chang, C. C., and Prakash, N. J. (1983). *Cell Biol. Int. Rep.* **7,** 455.
Szpirer, C. (1974). *Exp. Cell Res.* **83,** 47.
Tabor, C. W., and Tabor, H. (1976). *Annu. Rev. Biochem.* **45,** 285.
Tarkowski, A. K., and Bałakier, H. (1980). *J. Embryol. Exp. Morphol.* **55,** 324.
Tarnowka, M. A., and Baglioni, C. (1979). *J. Cell. Physiol.* **99,** 359.
Tarnowka, M. A., Baglioni, C., and Basilico, C. (1978). *Cell (Cambridge, Mass.)* **15,** 163.
Tarrant, G. M., and Holliday, R. (1977). *Mol. Gen. Genet.* **156,** 273.
Taylor, J. H. (1974). *Int. Rev. Cytol.* **37,** 1.
Taylor, J. H., and Hozier, J. C. (1976). *Chromosoma* **57,** 341.
Taylor, J. H., Woods, P. S., and Hughes, W. L. (1957). *Proc. Natl. Acad. Sci. U.S.A.* **43,** 122.
Taylor, J. H., Adams, A. G., and Kwek, M. P. (1973). *Chromosoma* **41,** 361.
Taylor, J. H., Wu, M., Erickson, L. C., and Kurek, M. P. (1975). *Chromosoma* **53,** 175.
Telzer, B. R., and Haimo, L. T. (1981). *J. Cell Biol.* **89,** 373.
Telzer, B. R., and Rosenbaum, J. L. (1979). *J. Cell Biol.* **81,** 484.
Telzer, B. R., Moses, M. J., and Rosenbaum, J. L. (1975). *Proc. Natl. Acad. Sci. U.S.A.* **72,** 4023.
Tencer, R. (1978). *Exp. Cell Res.* **116,** 253.
Thomas, G., Siegmann, M., and Gordon, J. (1979). *Proc. Natl. Acad. Sci. U.S.A.* **76,** 3952.
Thompson, L. R., and McCarthy, B. J. (1973). *Biochim. Biophys. Acta* **331,** 202.
Thornley, A. L., and Laurence, E. B. (1976). *Dev. Biol.* **51,** 10.
Thyberg, J. (1984). *Exp. Cell Res.* **155,** 1.
Tissières, A., Mitchell, H. K., and Tracy, U. M. (1974). *J. Mol. Biol.* **84,** 389.
Tomita, Y., Nakamura, T., and Ichihara, A. (1981). *Exp. Cell Res.* **135,** 363.
Toyoshima, S., Iwata, M., and Osawa, T. (1976). *Nature (London)* **264,** 447.
Tseng, B. Y., and Ahlem. C. N. (1982). *J. Biol. Chem.* **257,** 7280.
Tseng, B. Y., and Ahlem, C. N. (1983). *J. Biol. Chem.* **258,** 9845.
Tseng, B. Y., and Goulian, M. (1977). *Cell (Cambridge, Mass.)* **12,** 483.
Tsuneoka, K., Shikita, M., Takatsuki, A., and Tamura, G. (1981). *J. Biochem. (Tokyo)* **90,** 611.
Tsutsui, Y., Chang, S. D., and Baserga, R. (1978). *Exp. Cell Res.* **113,** 359.
Valenzuela, M. S., Mueller, G. C., and DasGupta, S. (1983). *Nucleic Acids Res.* **11,** 2155.
Van Assel, S., and Brachet, J. (1966). *J. Embryol. Exp. Morphol.* **15,** 143.
Van Assel, S., and Brachet, J. (1968). *J. Embryol. Exp. Morphol.* **19,** 261.
Van Assel, S., and Steinert, M. (1971). *Exp. Cell Res.* **65,** 353.
Vandre, D. D., Davis, F. M., Rao, P. N., and Borisy, G. G. (1984). *Proc. Natl. Acad. Sci. USA* **81,** 4439.
Van Ness, J. and Pettijohn, D. E. (1983). *J. Mol. Biol.* **171,** 175.
Van Obberghen, E., Rossi, B., Kowalski, A., and Gazzano, H. (1983). *Proc. Natl. Acad. Sci. U.S.A.* **80,** 945.
van Zoelen, E. J. J., van Der Saag, P. T., and de Laat, S. W. (1981). *Exp. Cell Res.* **131,** 395.
Veigl, M. L., Vanaman, T. C., and Sedwick, W. D. (1984). *Biochim. Biophys. Acta* **738,** 21.
Velazquez, J. M., and Lindquist, S. (1984). *Cell* **36,** 655.
Verly, W. G., Deschamps, Y., Pushpathadam, J., and Desrosiers, M. (1971). *Can. J. Biochem.* **49,** 1376.
Villasante, A., Corces, V. G., Manso-Martinez, R., and Avila, J. (1981). *Nucleic Acids Res.* **9,** 895.
Villereal, M. L. (1981). *J. Cell. Physiol.* **107,** 359.
Vogelstein, B., Pardoll, D. M., and Coffey, D. S. (1980). *Cell (Cambridge, Mass.)* **22,** 79.
Vorobjev, I. A., and Chentsov, Y. S. (1982). *J. Cell Biol.* **98,** 988.
Wagenaar, E. B. (1983). *Exp. Cell Res.* **144,** 393.

Waithe, W. I., Renaud, J., Nadeau, P., and Pallotta, D. (1983). *Biochemistry* **22,** 1778.
Walmsley, M., and Moses, M. J. (1981). *Exp. Cell Res.* **133,** 405.
Wang, C., Gomer, R. H., and Lazarides, E. (1981). *Proc. Natl. Acad. Sci. U.S.A.* **78,** 3531.
Wang, E., Pfeffer, L. M., and Tamm, I. (1981). *Proc. Natl. Acad. Sci. U.S.A.* **78,** 6281.
Wang, T. S. F., Hu, S. Z., and Korn, D. (1984). *J. Biol. Chem.* **259,** 1854.
Wang, Z. W. and Rozengurt, E. (1983). *J. Cell Biol.* **96,** 1743.
Waqar, M. A., and Huberman, J. A. (1973). *Biochem. Biophys. Res. Commun.* **51,** 174.
Warden, C. H., and Friedkin, M. (1984). *Biochim. Biophys. Acta* **792,** 270.
Wasserman, W. J., and Smith, L. D. (1978). *J. Cell Biol.* **78,** R15.
Waterborg, J. H., and Matthews, H. R. (1982). *Exp. Cell Res.* **138,** 462.
Watson, J. D., and Crick, F. H. C. (1953). *Nature (London)* **171,** 737.
Welsh, M. J., Dedman, J. R., Brinkley, B. R., and Means, A. R. (1978). *Proc. Natl. Acad. Sci. U.S.A.* **75,** 1867.
Welsh, M. J., Dedman, J. R., Brinkley, B. R., and Means, A. R. (1979). *J. Cell Biol.* **81,** 624.
White, J. G., and Borisy, G. G. (1983). *J. Theor. Biol.* **101,** 289.
Whitfield, J. F., Boynton, A. L., MacManus, J. P., Sikorska, M., and Tsang, B. K. (1979). *Mol. Cell. Biochem.* **27,** 155.
Wiche, G., Corces, V. G., and Avila, J. (1978). *Nature (London)* **273,** 403.
Wickremasinghe, R. G., Yaxley, J. C., and Hoffbrand, A. V. (1983). *Biochim. Biophys. Acta* **740,** 243.
Wigler, M. H., and Axel, R. (1976). *Nucleic Acids Res.* **3,** 1463.
Willingham, M. C., Wehland, J., Klee, C. B., Richert, N. D., Rutherford, A. V., and Pastan, I. H. (1983). *J. Histochem. Cytochem.* **31,** 445.
Wilson, E. B. (1901). *Arch. Entwicklungsmech. Org.* **12,** 529.
Winclester, G. (1983). *Nature (London)* **303,** 660.
Witkin, E. M. (1976). *Bacteriol. Rev.* **40,** 869.
Wolff, S. H., and Perry, P. (1975). *Exp. Cell Res.* **93,** 23.
Wolstenholme, D. (1973). *Chromosoma* **43,** 1.
Woodcock, D. M., Adams, J. K., and Cooper, I. A. (1982). *Biochim. Biophys. Acta* **696,** 15.
Wright, J. A. (1973). *Exp. Cell Res.* **78,** 456.
Wu, R. S., and Bonner, W. M. (1981). *Cell (Cambridge, Mass.)* **27,** 321.
Yagura, T., Kozu, T., and Seno, T. (1982). *J. Biol. Chem.* **257,** 11121.
Yagura, T., Tanaka, S., Kozu, T., Seno, T., and Korn, D. (1983). *J. Biol. Chem.* **258,** 6698.
Yanishevsky, R. M., and Prescott, D. M. (1978). *Proc. Natl. Acad. Sci. U.S.A.* **75,** 3307.
Yanishevsky, R. M., and Stein, G. H. (1981). *Int. Rev. Cytol.* **69,** 223.
Yarden, Y., Schreiber, A. B., and Schlessinger, J. (1982). *J. Cell Biol.* **92,** 687.
Yoshida, S., Suzuki, R., Masaki, S., and Koiwai, O. (1983). *Biochim. Biophys. Acta* **741,** 348.
Yurov, Y. B., and Liapunova, N. A. (1977). *Chromosoma* **60,** 253.
Zavortink, M., Welsh, M. J., and McIntosh, J. R. (1983). *Exp. Cell Res.* **149,** 375.
Zegarelli-Schmidt, E., and Goodman, R. (1981). *Int. Rev. Cytol.* **71,** 245.
Zeilig, C. E., and Langan, T. A. (1980). *Biochem. Biophys. Res. Commun.* **95,** 1372.
Zeilig, C. E., Johnson, R. A., Sutherland, E. W., and Friedman, D. L. (1976). *J. Cell Biol.* **71,** 515.
Zetter, B. R., Chen, L. B., and Buchanan, J. M. (1976). *Cell* **7,** 407.
Zeuthen, E. (1953). *Arch. Neerl. Zool.* **10,** Suppl. 1, 31.
Zeuthen, E. (1955). *Biol. Bull. (Woods Hole, Mass.)* **108,** 366.
Zeuthen, E., and Hamburger, K. (1972). *Biol. Bull. (Woods Hole, Mass.)* **143,** 699.
Zhimulev, I. F., Belyaeva, E. S., and Semesnin, V. F. (1981). *CRC Crit. Rev. Biochem.* **11,** 303.
Zieve, G. W., and McIntosh, J. R. (1981). *J. Cell Sci.* **48,** 241.
Zieve, G. W., and Solomon, F. (1982). *Cell (Cambridge, Mass.)* **28,** 233.
Zieve, G. W., Heidemann, S. R., and McIntosh, J. R. (1980). *J. Cell Biol.* **87,** 160.
Zylber, E. A., and Penman, S. (1971). *Science* **172,** 947.

# INDEX*



*Boldface numbers indicate volume numbers.

**D**

## H

## I



## J



## K



## L

## M

**Q**



**R**

## T

**Y**



**Z**